Nanotechnology in the Life Sciences

Series Editor
Ram Prasad
Department of Botany
Mahatma Gandhi Central University
Motihari, Bihar, India

Nano and biotechnology are two of the 21st century's most promising technologies. Nanotechnology is demarcated as the design, development, and application of materials and devices whose least functional make up is on a nanometer scale (1 to 100 nm). Meanwhile, biotechnology deals with metabolic and other physiological developments of biological subjects including microorganisms. These microbial processes have opened up new opportunities to explore novel applications, for example, the biosynthesis of metal nanomaterials, with the implication that these two technologies (i.e., thus nanobiotechnology) can play a vital role in developing and executing many valuable tools in the study of life. Nanotechnology is very diverse, ranging from extensions of conventional device physics to completely new approaches based upon molecular self-assembly, from developing new materials with dimensions on the nanoscale, to investigating whether we can directly control matters on/in the atomic scale level. This idea entails its application to diverse fields of science such as plant biology, organic chemistry, agriculture, the food industry, and more.

Nanobiotechnology offers a wide range of uses in medicine, agriculture, and the environment. Many diseases that do not have cures today may be cured by nanotechnology in the future. Use of nanotechnology in medical therapeutics needs adequate evaluation of its risk and safety factors. Scientists who are against the use of nanotechnology also agree that advancement in nanotechnology should continue because this field promises great benefits, but testing should be carried out to ensure its safety in people. It is possible that nanomedicine in the future will play a crucial role in the treatment of human and plant diseases, and also in the enhancement of normal human physiology and plant systems, respectively. If everything proceeds as expected, nanobiotechnology will, one day, become an inevitable part of our everyday life and will help save many lives.

More information about this series at http://www.springer.com/series/15921

Devarajan Thangadurai • Jeyabalan Sangeetha
Ram Prasad
Editors

Nanotechnology for Food, Agriculture, and Environment

Springer

Editors
Devarajan Thangadurai
Department of Botany
Karnatak University
Dharwad, Karnataka, India

Jeyabalan Sangeetha
Department of Environmental Science
Central University of Kerala
Kasaragod, Kerala, India

Ram Prasad
Department of Botany
Mahatma Gandhi Central University
Motihari, Bihar, India

ISSN 2523-8027 ISSN 2523-8035 (electronic)
Nanotechnology in the Life Sciences
ISBN 978-3-030-31937-3 ISBN 978-3-030-31938-0 (eBook)
https://doi.org/10.1007/978-3-030-31938-0

This Springer imprint is published by the registered company Springer Nature Switzerland AG
The registered company address is: Gewerbestrasse 11, 6330 Cham, Switzerland

Preface

Nanotechnology involves the handling of nano-devices and nano-systems by manipulating matter at nanometer scale. Nanoparticles have received much attention because of their unique physiochemical properties. These properties offer new means of miniaturization for existing products, new materials with new properties, production methods, and deep new insights on nature and life. Nanotechnology is one of the rapidly developing frontiers that have greatly impacted life expectancy in almost every arena.

Nanotechnology offers high potential to update and revolutionize conventional agri-food industry and food science. Nanotechnology applications in food engineering promise improved flavor, taste, color, consistency, and texture of food products and increased bioavailability and absorption of nutraceuticals. Novel approaches of food nanotechnology represent the most recent advances in nanostructured materials that have significant impact on food industry. As current food arcade demands novelty in food processing methods, nanotechnology linked with innovative interdisciplinary approaches has empowered essential advancements that have the capacity to revolutionize the food industry. Nanotechnology can assist with settling challenges in food processing industries in implementing and developing systems able to produce quantitative and qualitative foodstuffs that are sustainable, safe, and eco-friendly.

A huge number of innovations in nanotechnology are supplying novel and exclusive applications in biotechnology- and agriculture-related fields. Nanomaterials play a significant part in agriculture through nano-pesticides and compound fertilizers, acting as magic bullets to deliver chemical components that target specific molecules of plant cells. Nanotechnology application in food science is a subject of emerging concern toward the food packaging industry, food quality and safety, and eliminating contaminants in foodborne pathogens. Nanomaterials in agriculture also reduce nutrient losses during enrichment, minimize the expense of chemical objects sprayed by way of smart transmission of active components, and increase harvests through nutrient management and optimized water. There is a high aptitude for the nanotechnology industry in the endowment of advanced solutions for a variety of encounters faced by society and agriculture, both at present and in the future.

Nanotechnology in environmental research is considered to have a crucial role in the advancement of existing science practices and environmental engineering. An overview of the present state of nanotechnology-based devices with applications in environmental science mainly focuses on nanomaterials and polymer nanocomposites. Among these areas, nanotechnology has participated significantly in environmental remediation and conservation by addressing long-term threatening issues for humans. In this context, many nanostructures have been formulated, and several more are in pipeline. No doubt that this field has granted humans with several amenities and has made huge contributions by using fewer chemicals, reduction in energy, less waste, enhancement of atom efficiency, and providing better technology and materials for environmental remedies, promising ecological sustainability; however, certain negative factors still have to be reviewed.

In this book, an interdisciplinary group of researchers summarizes the innovations and applications of nanotechnology and describes challenges and opportunities of nanotechnology in the food, agriculture, and environment sectors, emphasizing the technical, scientific, safety, regulatory, and societal impacts. They also discuss present and future insights of nanotechnology to build consumer confidence in its role in food, agriculture, and the environment.

This book mainly discusses the various applications of nanomolecules in the fields of food, agriculture, and the environment in a comprehensive manner. Chapter 1 describes the production of nanoparticles from fungal resources and their applications in various fields. The importance of nanoparticles in the food industry is discussed in Chaps. 2, 3, and 4, wherein the authors discuss in detail nanoparticle incorporated soy protein, formulation and production of nanoparticles in the food and pharmaceutical sectors, and applications of nanosensors in food safety monitoring. Various applications of nanomaterials in the agricultural sector are elaborately described in Chaps. 5, 6, 7, 8, 9, and 10, whose authors discuss the future perspectives of agronanotechnology and applications of various nanomaterials and nanomolecules to control crop diseases, and thus to improve crop productivity. Chapters 11, 12, 13, 14, and 15 discuss the applications of different nanoparticles in the cleanup of xenobiotics, like heavy metals and dyes, and bioenergy production. In addition, the authors review the ecological risk assessment of nanoparticles and their impact on beneficial insects in Chap. 16.

The contributing authors of this book have been selected for their prominent expertise in the fields of production and application of nanoparticles in the food, agriculture, and environmental sectors. Their acceptance of our invitation and willingness to contribute chapters to this interesting book are gratefully acknowledged. The editors of this book are very much obliged to Mr. Eric Stannard, Senior Editor, Springer Nature, for his industrious, valuable, and prudent support throughout this publication.

This book will serve as a source of scientific information on the formulation, production, and application of nanomolecules in the food, agriculture, and environmental sectors for sustainable development and food security. This book is suitable for students, researchers, academicians, and industrial professionals who are engaged in the field of nanotechnology in food, agriculture, and environmental sectors for the benefit of humankind.

Dharwad, Karnataka, India Devarajan Thangadurai
Kasaragod, Kerala, India Jeyabalan Sangeetha
Motihari, Bihar, India Ram Prasad

Contents

About the Editors

Devarajan Thangadurai is Assistant Professor at Karnatak University, Dharwad, Karnataka, India. He received his PhD in Botany from Sri Krishnadevaraya University in South India as CSIR Senior Research Fellow with funding from the Ministry of Science and Technology, Government of India. He served as a Postdoctoral Fellow at the University of Madeira, Portugal, University of Delhi, India, and ICAR National Research Centre for Banana, India. He is the Recipient of the Best Young Scientist Award with a Gold Medal from Acharya Nagarjuna University, India, and the VGST-SMYSR Young Scientist Award of the Government of Karnataka, Republic of India. He has authored and edited more than 20 books with publishers of national/international reputation. He has also visited 23 countries in Africa, Asia, and Europe for academic visits, scientific meetings, and international collaborations.

Jeyabalan Sangeetha is Assistant Professor in the Central University of Kerala at Kasaragod, Kerala, India. She earned her BSc in Microbiology and PhD in Environmental Science from Bharathidasan University, Tiruchirappalli, Tamil Nadu, India. She holds an MSc in Environmental Science from Bharathiar University, Coimbatore, Tamil Nadu, India. She is the Recipient of Tamil Nadu Government Scholarship and Rajiv Gandhi National Fellowship of the University Grants Commission, Government of India, for her doctoral studies. She served as Dr. D.S. Kothari Postdoctoral Fellow and UGC Postdoctoral Fellow at Karnatak University, Dharwad, South India, in 2012–2016 with funding from the University Grants Commission, Government of India, New Delhi. Her research interests are in the fields of environmental toxicology, environmental microbiology, environmental biotechnology, and environmental nanotechnology.

Ram Prasad, Ph.D. is associated with Department of Botany, Mahatma Gandhi Central University, Motihari, Bihar, India. His research interest includes applied microbiology, plant-microbe-interactions, sustainable agriculture and nanobiotechnology. Dr. Prasad has more than one hundred fifty publications to his credit, including research papers, review articles & book chapters and five patents issued or pending, and edited or authored several books. Dr. Prasad has twelve years of teaching experience and has been awarded the Young Scientist Award & Prof. J.S. Datta Munshi Gold Medal by the International Society for Ecological Communications; FSAB fellowship by the Society for Applied Biotechnology; the American Cancer Society UICC International Fellowship for Beginning Investigators, USA; Outstanding Scientist Award in the field of Microbiology by Venus International Foundation; BRICPL Science Investigator Award and Research Excellence Award etc. He has been serving as editorial board members: Frontiers in Microbiology, Frontiers in Nutrition, Phyton- International Journal of Experimental Botany; Academia Journal of Biotechnology including Series editor of Nanotechnology in the Life Sciences, Springer Nature, USA. Previously, Dr. Prasad served as Assistant Professor Amity University Uttar Pradesh, India; Visiting Assistant Professor, Whiting School of Engineering, Department of Mechanical Engineering at Johns Hopkins University, Baltimore, United States and Research Associate Professor at School of Environmental Science and Engineering, Sun Yat-sen University, Guangzhou, China.

Contributors

Nawfel Abdullah Australian Institute of Innovative Materials (AIIM), University of Wollongong, North Wollongong, NSW, Australia

Rahul Agnihotri Forest Pathology Discipline, Forest Protection Division, Forest Research Institute, Indian Council Forestry Research and Education, Dehradun, Uttarakhand, India

Shubhi Avasthi School of Studies in Botany, Jiwaji University, Gwalior, Madhya Pradesh, India

Ashutosh Bahuguna Department of Food Science and Technology, College of Life and Applied Sciences, Yeungnam University, Gyeongsan-si, Gyeongsangbuk-do, South Korea

Shivanand Bhat Department of Botany, Government Arts and Science College, Karwar, Uttar Kannada, Karnataka, India

Amandeep Brar Department of Microbiology, School of Life Sciences, Central University of Rajasthan, Bandarsindri, Kishangarh, Ajmer, Rajasthan, India

Prashantkumar Chakra Department of Microbiology, Davangere University, Davangere, Karnataka, India

Sunil Kumar Dwivedi Department of Entomology, School of Agriculture, Lovely Professional University, Phagwara, Punjab, India

Kolanthasamy Elango Department of Entomology, Tamil Nadu Agricultural University, Coimbatore, Tamil Nadu, India

Muhammad Farooq Department of Crop Sciences, College of Agricultural and Marine Sciences, Sultan Qaboos University, Muscat, Oman

Department of Agronomy, University of Agriculture, Faisalabad, Pakistan

The UWA Institute of Agriculture and School of Agriculture and Environment, The University of Western Australia, Perth, WA, Australia

Akshita Gaur Forest Pathology Discipline, Forest Protection Division, Forest Research Institute, Indian Council Forestry Research and Education, Dehradun, Uttarakhand, India

Ajay Kumar Gautam School of Agriculture, Abhilashi University, Mandi, Himachal Pradesh, India

Hikmet Geckil Department of Molecular Biology and Genetics, Inonu University, Malatya, Turkey

Pynadathu Jacob George Centre for Human Resource Development, Sam Higginbottom University of Agriculture, Technology and Sciences, Allahabad, Uttar Pradesh, India

Abbas Gungordu Department of Biology, Inonu University, Malatya, Turkey

Ravichandra Hospet Department of Botany, Karnatak University, Dharwad, Karnataka, India

Akbar Hossain Bangladesh Wheat and Maize Research Institute, Dinajpur, Bangladesh

M. Tofazzal Islam Institute of Biotechnology and Genetic Engineering (IBGE), Bangabandhu Sheikh Mujibur Rahman Agricultural University, Gazipur, Bangladesh

Mohd Rafie Bin Johan Nanotechnology and Catalysis Research Centre, Institute for Advanced Studies, University of Malaya, Kuala Lumpur, Malaysia

Malaichamy Kannan Department of Plant Protection, Horticultural College and Research Institute, Periyakulam, Tamil Nadu Agricultural University, Tamil Nadu, India

Govindaraju Kasivelu Centre for Ocean Research, Sathyabama Institute of Science and Technology, Chennai, Tamil Nadu, India

Rout George Kerry PG Department of Biotechnology, Utkal University, Bhubaneswar, Odisha, India

Myunghee Kim Department of Food Science and Technology, College of Life and Applied Sciences, Yeungnam University, Gyeongsan-si, Gyeongsangbuk-do, South Korea

Dileep Kumar Department of Biochemistry, University of Lucknow, Lucknow, India

Manish Kumar Department of Microbiology, School of Life Sciences, Central University of Rajasthan, Bandarsindri, Kishangarh, Ajmer, Rajasthan, India

Rakesh Kumar Department of Biotechnology, Central University of South Bihar, Gaya, India

Chin Wei Lai Nanotechnology and Catalysis Research Centre, Institute for Advanced Studies, University of Malaya, Kuala Lumpur, Malaysia

Shanthala Mallikarjunaiah Centre for Applied Genetics, Department of Zoology, Bangalore University, Jnana Bharathi, Bengaluru, Karnataka, India

Basavaraja Metikurki Department of Pharmaceutical Chemistry, Vivekananda College of Pharmacy, Bengaluru, Karnataka, India

Abhishek Mundaragi Department of Microbiology, Davangere University, Davangere, Karnataka, India

Kalaivani Nadarajah School of Environmental and Natural Resources Sciences, Faculty of Science and Technology, Universiti Kebangsaan Malaysia, Bangi, Selangor, Malaysia

B. Shankar Naik Department of Biology, Government Science College, Basavanahalli Extension, Chikmagalur, Karnataka, India

Murat Ozmen Department of Biology, Inonu University, Malatya, Turkey

Nidhi Pareek Department of Microbiology, School of Life Sciences, Central University of Rajasthan, Bandarsindri, Kishangarh, Ajmer, Rajasthan, India

Vipin Parkash Forest Pathology Discipline, Forest Protection Division, Forest Research Institute, Indian Council Forestry Research and Education, Dehradun, Uttarakhand, India

Om Prakash Department of Fruits and Vegetables Technology, CSIR-Central Food Technological Research Institute, Mysore, Karnataka, India

Mahesh Pattabhiramaiah Centre for Applied Genetics, Department of Zoology, Bangalore University, Jnana Bharathi, Bengaluru, Karnataka, India

Sundharam Preetha Department of Nano Science and Technology, Tamil Nadu Agricultural University, Coimbatore, Tamil Nadu, India

Priyanka Priyanka Department of Biochemistry, University of Lucknow, Lucknow, India

Srinivasan Ramalingam Department of Food Science and Technology, College of Life and Applied Sciences, Yeungnam University, Gyeongsan-si, Gyeongsangbuk-do, South Korea

Shikha Rani Department of Biotechnology, Central University of South Bihar, Gaya, India

Nelson Pynadathu Rumjit Department of Environmental and Water Resources Engineering, School of Civil Engineering, Vellore Institute of Technology, Vellore, Tamil Nadu, India

Jeyabalan Sangeetha Department of Environmental Science, Central University of Kerala, Kasaragod, Kerala, India

Manickam Puratchiveeran Saravanakumar Department of Environmental and Water Resources Engineering, School of Civil Engineering, Vellore Institute of Technology, Vellore, Tamil Nadu, India

Kulvinder Singh Department of Chemistry, School of Basics and Applied Sciences, Maharaja Agrasen University, Baddi, India

Thangavel Tamilnayagan Department of Entomology, Tamil Nadu Agricultural University, Coimbatore, Tamil Nadu, India

Devarajan Thangadurai Department of Botany, Karnatak University, Dharwad, Karnataka, India

Paul Thomas Nanotechnology and Catalysis Research Centre, Institute for Advanced Studies, University of Malaya, Kuala Lumpur, Malaysia

Ajay Tomer Department of Plant Pathology, School of Agriculture, Lovely Professional University, Phagwara, Punjab, India

Preeti Tyagi Nanotechnology and Catalysis Research Centre, Institute for Advanced Studies, University of Malaya, Kuala Lumpur, Malaysia

Vivekanand Vivekanand Centre for Energy and Environment, Malaviya National Institute of Technology, Jaipur, Rajasthan, India

Anurag Yadav College of Basic Sciences and Humanities, Sardarkrushinagar Agricultural University Dantiwada, Banaskantha, India

Kusum Yadav Department of Biochemistry, University of Lucknow, Lucknow, India

Chapter 1
Nanoparticles from Fungal Resources: Importance and Applications

Vipin Parkash, Akshita Gaur, and Rahul Agnihotri

Contents

1.1 Fungi

Fungi are the primitive eukaryotic organisms. Most fungi are microscopic in nature, while some are macroscopic, such as mushrooms, toadstools, puffballs and stinkhorns. Generally, their mode of nutrition is heterotrophic, although some are parasites

V. Parkash (✉) · A. Gaur · R. Agnihotri
Forest Pathology Discipline, Forest Protection Division, Forest Research Institute, Indian Council Forestry Research and Education, Dehradun, Uttarakhand, India

D. Thangadurai et al. (eds.), *Nanotechnology for Food, Agriculture, and Environment*, Nanotechnology in the Life Sciences,
https://doi.org/10.1007/978-3-030-31938-0_1

and saprophytes. Fungi are cosmopolitan in distribution. In ecosystem, fungi play an important and vital role in nutrient cycling, as sources of food, decomposers, symbiosis and protection. From time immemorial, fungi have been utilized as a source of food and harnessed to ferment and preserve foods and beverages. In the twentieth century, human beings have learned to exploit fungi to protect human health through antibiotics, anticholesterol statins and immunosuppressive agents. Industrial utilization of fungi for production of enzymes, acids and biosurfactants with the advent of modern nanotechnology has been started in the 1980s. Fungi can easily be isolated from different sources of environment and also can be cultivated in simple and less nutrient media like Potato Dextrose Agar and Czapek Dox Broth in laboratory, and the maintenance of fungi in laboratory is also very easy. Fungi have more enzyme-secreting activity, and it is easy to isolate and maintain, so they are selected for silver and other nanoparticle production by the researchers (Abdel-Aziz et al. 2018).

1.2 Nanoparticles and Nanotechnology

Nanoparticles (NPs) are very minute particles ranging from 1 to 100 nm and have started to follow the principles of quantum physics rather than classic physics. For example, the same material will acquire different optical/electrical properties at nanoscale when compared to macroscale. Thus, the study of these kinds of materials/particles with significant properties, functions and phenomena due to their small size is known as nanoscience. The use of nanoscience for human welfare in terms of industrial/commercial application through the formation of nanoparticles is termed as 'nanotechnology'.

Nanotechnology finds its way long back in the history. Horikoshi and Serpone (2013) in their revised article listed the chronological sequence of developments in nanotechnology throughout the history. The main historical events along with the origin/country have been shown in Table 1.1.

1.3 Properties of Nanoparticles

Nanoparticles can be considered as a sub-group of colloidal particles. Since, the use of nanotechnology arises from the interchanging properties between classical physics and quantum mechanics, nanotechnology finds its major applications in production of intermediate goods connecting different disciplines, thus, offering the foundation of so-called Nano-Bio-Info-Cogno (NBIC – is an acronym standing for Nanotechnology, Biotechnology, Information technology and Cognitive sciences) convergence according to a report by the Scientific Committee on Emerging and Newly Identified Health Risks (SCENIHR 2007). The properties of nanoparticles are completely different or improved and mainly depend on its size, structure and shape. Alternation in any one of the parameter enables the researchers to form materials of specific use. Nanoparticles have some very significant properties, which are discussed below.

1.3.1 Physical Properties

Nanoparticles are unique due to their physical properties as they possess a large surface area. Since, nanoparticles also have very small size, they absorb greater amount of solar radiation, and the best example where this property is being exploited majorly is the use of zinc oxide (ZnO) NPs in sunscreen lotions with UV-blocking property (Lademann et al. 1999).

1.3.2 Biological Properties

Use of NPs has exponentially increased in food industry as a food packaging material to control the food spoilage from the ambient atmosphere that surrounds the food, keeping it safe from harmful disease-causing microorganisms that may contaminate the food. Claylike NPs slow down the entry of moisture and inhibit the other gas transport through the packaging. NPs exhibit antimicrobial activity which is now being incorporated into paints and wall coating, making products more useful and safer for surfaces of hospitals and medical laboratories (Percival et al. 2007).

1.3.3 Medical Properties

Physical properties are claimed to have a great potential for medical applications such as disease diagnosis, drug delivery system and imaging. Nanoparticles can not only circulate easily throughout the body but also enter the cell. This property enhances the images of organs and tumours (Gao et al. 2004). Anti-inflammatory property of the nanoparticles has been used in medical field as it alters the expression of proteolytic enzymes, suppresses the expression of tumour necrosis factor (TNF) and interleukin-12 (IL-12) and induces apoptosis of inflamed cells (Duan et al. 2015).

1.3.4 Mechanical Properties

Mechanical properties of NPs depend upon the composition of type of bond between the atoms, viz. covalent, ionic and metallic. As a result, NP materials may be strong, tough and ductile, but the presence of impurities will affect the properties. Since NPs are stronger, harder and erosion resistant, they are used as spark plugs in automobile companies. Engine cylinders are coated with nanocrystalline ceramics such as zirconia and alumina that can retain heat much more efficiently and, hence, result in efficient and complete combustion of fuel (Guo et al. 2013).

Table 1.1 Developments in the field of nanotechnology chronologically (Horikoshi and Serpone 2013)

Developments in nanotechnology	Country/people/researcher(s)	Year
Discovery of soluble gold	Egypt and China	1200–1300 BC
Lycurgus cup	Alexandria or Rome	290–325 AD
First book on colloidal gold	F. Antonii	1618
Book published on drinkable gold that contains metallic gold in neutral media	J. von Löwenstern-Kunckel (Germany)	1676
Publication of a complete treatise on colloidal gold	Hans Heinrich Helcher	1718
Synthesis of colloidal gold	M. Faraday (The Royal Institution of Great Britain)	1857
Surface plasmon resonance (SPR)	R. W. Wood (Johns Hopkins University, USA)	1902
Scattering and absorption of electromagnetic fields by a nanosphere	G. Mie (University of Göttingen, Germany)	1908
Transmission electron microscope (TEM)	M. Knoll and E. Ruska (Technical University of Berlin, Germany)	1931
Scanning electron microscope (SEM)	M. von Ardenne (Forschungslaboratorium für Elektronenphysik, Germany)	1937
Microelectromechanical systems (MEMS)	I. Igarashi (Toyota Central R&D Labs, Japan)	1960
The Kubo effect	R. Kubo (University of Tokyo, Japan)	1962
Moore's Law	G. Moore (Fairchild Semiconductor Inc., USA)	1965
The Honda-Fujishima effect	A. Fujishima and K. Honda (University of Tokyo, Japan)	1969
Amorphous heterostructure photodiode created with bottom-up process	E. Maruyama (Hitachi Co. Ltd., Japan)	1972
Concept of nanotechnology proposed	N. Taniguchi (Tokyo University of Science, Japan)	1974
Carbon nanofiber	M. Endo (Shinshu University, Japan)	1976
Amorphous silicon solar cells	D. E. Carlson and C. R. Wronski (RCA, USA)	1976
Quantum hall effect (Nobel Prize)	K. von Klitzing (University of Würzburg, Germany)	1980
Scanning tunnelling microscope (STM) (Nobel Prize)	G. Binnig and H. Rohrer (IBM Zurich Research Lab., Switzerland)	1982
Atomic force microscope (AFM)	G. Binnig (IBM Zurich Research Lab., Switzerland)	1986
Gold nanoparticle catalysis	M. Haruta (Industrial Research Institute of Osaka, Japan)	1987
Atoms controlled with scanning tunnelling microscope (STM)	D. M. Eigler (IBM, USA)	1990

(continued)

Table 1.1 (continued)

Developments in nanotechnology	Country/people/researcher(s)	Year
Carbon nanotubes discovered	S. Iijima (NEC Co., Japan)	1991
Nano-imprinting	S. Y. Chou (University of Minnesota, USA)	1995
Nanosheets	T. Sasaki (National Institute for Research in Inorganic Materials, Japan)	1996
National Nanotechnology Initiative	(NNI), USA	2000
21st Century Nanotechnology Research and Development Act	USA	2003
Nanosciences and Nanotechnologies: An action plan	Europe	2005

1.3.5 Optical Properties

Nanoparticles also often possess unexpected optical properties as they are small enough to confine their electrons and produce quantum effects (Bailey et al. 2004). Best example of this is gold nanoparticles that appear deep red to black in solution.

1.3.6 Electrical Properties

Conductivity is a property of electron in the solids, and resistivity is the inverse of conductivity. Resistivity of metals is very low, while resistivity of nanosized grains is generally high because the electrons get scattered at grain boundaries resulting in high resistance (Camacho and Oliva 2005).

1.4 Classification of Nanoparticles

The classification of nanoparticles is based on different approaches. Hett (2004) had classified the nanoparticles based on different dimensional structures. The one-dimensional nanoparticle includes thin films or manufactured surfaces used in the area of biological sensors, catalysis and solar cells. The two-dimensional nanoparticles include carbon nanotubes (CNTs) which are 1 nm in diameter and 100 nm in length. CNTs can both be single-walled (SWCNTs) and multi-walled (MWCNTs). The use of CNTs lies in their metallic and semiconductor properties with improved capacity in electrical properties, thus, performing as semiconductors. The three-dimensional nanoparticles constitute fullerenes (carbon 60), dendrimers and quantum dots (QDs). Fullerenes are materials in the form of hollow sphere looking like soccer ball made of carbon 60 (C_{60}). These exhibit unique properties in relation to excessive pressure subjection and regaining their original shape on pressure release.

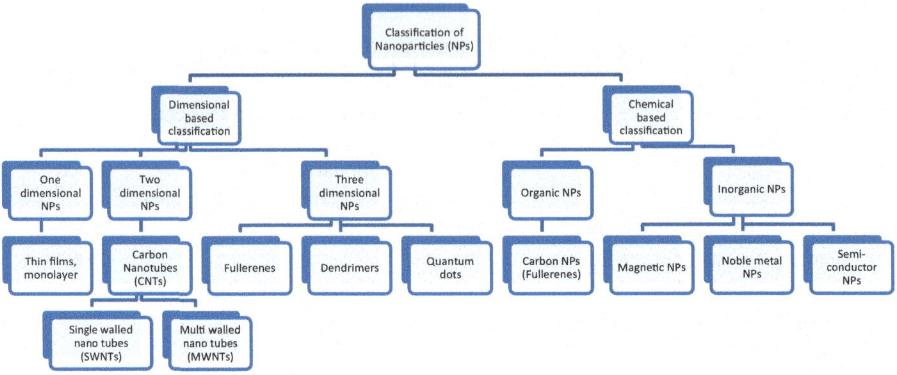

Fig. 1.1 Classification of nanoparticles. (Modified from Pal et al. 2011; Karthika et al. 2015)

Since fullerenes are hollow structures, their application can be related to filling them up with several substances, and thus, they are used in medical applications.

Dendrimers are polymers of 10–100 nm diameter with several functional molecules encapsulated inside, thus indicating their potential application in drug delivery systems. Quantum dots (QDs) are nanocrystals of 2–10 nm diameter and can contain a single electron to thousands. QDs can be used for optical and optoelectronic devices, quantum computing and information storage, whereas colour-coded QDs are used for fast DNA testing (Pal et al. 2011).

In the review given by Karthika et al. (2015), nanoparticles were classified into organic nanoparticles (including fullerenes) and inorganic nanoparticles (magnetic nanoparticles). Recent researches on inorganic nanoparticles, i.e. noble metallic nanoparticles, are gaining prime importance as these find wide application in diverse medical fields. A broad outline of classification of nanoparticles (modified from Pal et al. 2011; Karthika et al. 2015) is given in Fig. 1.1.

1.5 Nanoparticles Synthesis

Nanoparticle synthesis is done through two approaches: top-down and bottom-up approaches (Horikoshi and Serpone 2013). Top-down approach involves breaking up of solid material into nanoparticles by application of external force, whereas in bottom-up approach, nanoparticle formation is done by bonding of atoms of gases and liquids through atomic transformations. Liquid phase methods of bottom-up approach have been the majorly used nanoparticle fabrication methods. Liquid phase methods are subcategorized into liquid/liquid methods and sedimentation methods (Horikoshi and Serpone 2013). Chemical reduction of metal ions is a common method for nanoparticle formation. Now the use of chemicals in these chemical and physical methods may prove toxic as some of these chemicals can remain on synthesized nanoparticles. The application of these synthesized nanoparticles in the field of medical may be risky. Moreover, these methods are not cost-effective and require costly equipment (Agarwal et al. 2017). Thus, a more eco-friendly approach or green synthesis towards synthesis of nanoparticles is required. Therefore, the use of microorganisms and plants for nanoparticle synthesis comes under green synthesis of nanoparticles (Prasad et al. 2016, 2018).

Biogenic synthesis of metal nanoparticles involves bio-reduction of metal salts to elemental metal which are stabilized by organic molecules present in the microbes such as fungi and bacteria. The other way of producing metal nanoparticles is biosorption where metal ions in the aqueous medium are bonded to the surface of the cell wall of the microorganisms. For large-scale production of nanoparticles, fungi and yeasts are preferred over other organisms. When any fungus is exposed to metal salts, i.e. $AgNO_3$ or $AuCl_4$, it produces enzymes and metabolites to protect itself from unwanted foreign matters, and in doing so, the metal ions are reduced to metal nanoparticles (Lloyd 2003). The fungi also produce naphthoquinones and anthraquinones (Baker and Tatum 1998; Medentsev and Alimenko 1998) which act as reducing agents. Thus, a specific enzyme can act on a specific metal, e.g. nitrate reductase is essential for ferric ion reduction to iron nanoparticles. It is well understood that nanomaterials may be beneficial or harmful to living systems (Husen and Siddiqi 2014). For example, Cd (cadmium), Hg (mercury), Pb (lead) and Tl (thallium) nanoparticles are toxic and produce adverse effect in mammals and plants. The toxicity also depends on their shape, size and the nature of the specific metal ion (Husen and Siddiqi 2014). As compared to other microorganisms, use of fungal resources in nanotechnological systems can be of more importance as these can be easily isolated and mass cultured. Metal nanoparticles are synthesized from their salts, and this process can be carried out with the help of fungus due to the presence of proteins and enzymes that act as reducing agents (Siddiqi and Husen 2016). Thus, converting metal ions to less toxic forms leads to their increased tolerance towards high concentration of metal ions (Sastry et al. 2003). Moreover, fungi provide cost-effective approach towards biosynthesis of nanoparticles (Saglam et al. 2016; Prasad 2016, 2017).

1.6 Nanoparticle Synthesis by Fungal Resources

Synthesis of nanoparticles through fungal source involves two mechanisms, i.e. extracellular and intracellular. Extracellular synthesis happens with the help of reduction of metal ions done in the presence of enzymes secreted by fungal cells. Intracellular enzymes also help in reduction of metal ions to nanoparticles, and this can be achieved by involving trapping of metal ions on the surface of cells and reducing them in the presence of enzymes. The particle size is specific to the fungus from which it is synthesized and to the metal which is being reduced. Incubation time varies from fungus to fungus (Siddiqi and Husen 2016). The detailed outline of nanoparticle biosynthesis from fungal resource is shown in Fig. 1.2.

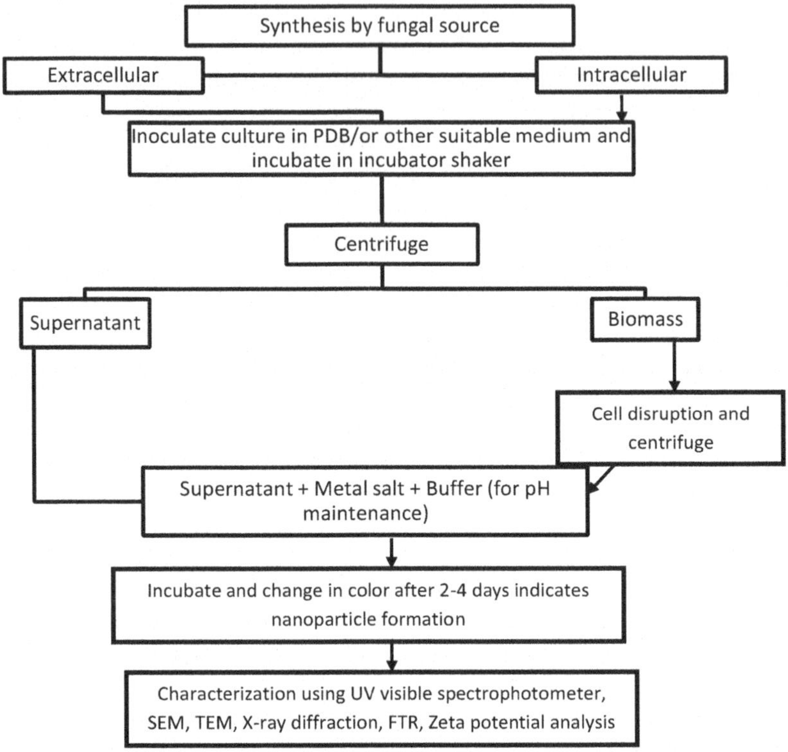

Fig. 1.2 Schematic representation of nanoparticle biosynthesis from fungal resource. (Modified from Prathna et al. 2010; Tidke et al. 2014; Molnár et al. 2018)

Nachiyar et al. (2015) had synthesized gold nanoparticles from three different endophytic fungal isolates within the range of 15–35 nm indicating the fact that endophytic fungi can be used as prospective synthesizers for nanoparticle synthesis. Mukherjee et al. (2002) worked on the extracellular fabrication of nanoparticles from the fungus, *Fusarium oxysporum*, and had synthesized gold nanoparticles. The synthesis of different nanoparticles through fungal source is discussed below.

1.6.1 Silver Nanoparticles (Ag-NPs)

When silver metal is reduced to particle size of 1–100 nm scale, the properties of the noble metal change drastically and, thus, Ag-NPs are formed. Silver nanoparticles possess distinctive physico-chemical and biological properties as compared to macro-sized silver metal. The electrical and thermal conductivity gets increased with more improved chemical stability. These enhanced properties increase their significance in application areas of microelectronics and medical imaging and as antimicrobial agents (Monteiro et al. 2009). Tran and Le (2013) analysed the trend in terms of growing research on silver nanoparticles and noticed a significant upward drift in the publication per year indicating the growing interest of researchers towards silver nanoparticles all over the world. They also analysed that chemistry and materials science were the prime research areas in which the studies related to silver nanoparticles (Ag-NPs) have been focused. In terms of research done by researchers in different countries on silver nanoparticles, maximum records have been noted from the USA followed by China and then India. Around 9.8% of research articles on Ag-NPs all over the world have been published from India (Tran and Le 2013).

Ag-NPs can be synthesized by chemical synthesis, physical synthesis, photochemical synthesis and biological synthesis. Chemical synthesis is the mostly used method for synthesis of Ag-NPs. Physical methods involve evaporation-condensation or arc-discharge method. These methods are good for large-scale production of Ag-NPs; however, the initial instruments' cost is high and may be not feasible. Photochemical synthesis involves the use of different light sources during reduction process. Photochemical synthesis of Ag-NPs is beneficial in terms of clean course of action and versatility (NPs can be fabricated in different media like emulsion, glasses, surfactant micelles and polymer films). Biological synthesis involves reduction of silver salt (usually $AgNO_3$) with the help of biotic agents like fungi, bacteria, yeasts, algae and plants. Biosynthesis is an environmentally friendly approach and also cost-effective. Also, the rate of biosynthesis of NPs is much faster as compared to other techniques. Table 1.2 provides a comprehensive list of fungi from which Ag-NPs have already been synthesized.

Ag-NPs are known for their usage as antimicrobial agents. Their effect against different pathogenic microbes including bacteria, fungi and even viruses has been evaluated previously. Morones et al. (2005) tested Ag-NPs against different Gram-negative bacteria like *Escherichia coli* and *Vibrio cholera*, and a maximum concentration of 75 $\mu g\ ml^{-1}$ was enough for inhibiting the growth of all tested bacteria significantly. Lara et al. (2010) tested the antibacterial effect of Ag-NPs against drug-resistant bacterial pathogens (multidrug-resistant *Pseudomonas aeruginosa*, ampicillin resistant *E. coli* and erythromycin resistant *Streptococcus pyogenes*) and found that Ag-NPs were effective against these bacteria. Monteiro et al. (2011) worked on the antifungal effect of Ag-NPs against disease-causing pathogens, i.e. *Candida albicans* and *C. glabrata*, and found that Ag-NPs were able to control these pathogens at a very low concentration of 0.4–3.3 $\mu g\ ml^{-1}$. Ag-NPs inhibit HIV-1 virus from binding to host cells (Elechiguerra et al. 2005).

Also, Ag-NPs exhibit toxicity in vitro at certain concentrations probably by interference with mitochondrial respiratory chain (Tran and Le 2013), and in vivo toxicity is observed, may be due to oxidative stress, DNA damage and apoptosis. Thus, Ag-NPs offer a great deal of opportunities in areas where, up till now, no significant way has been developed as they have been screened to be more effective when compared to conventional methods and also are cost-effective.

Table 1.2 List of fungi from which Ag-NPs have been already synthesized

Fungus used for synthesis	Size (nm)	Shape	References
Rhizopus stolonifer	5–50	Spherical	Afreen and Ranganath (2011)
Phoma glomerata	60–80	Spherical	Birla et al. (2009)
Fusarium oxysporum	50	Spherical	Karbasian et al. (2008)
F. solani	16.23	Spherical	Ingle et al. (2009)
Pestalotia sp.	10–40	Spherical	Raheman et al. (2011)
Pleurotus sajor-caju	5–50	Spherical	Nithya and Ragunathan (2009)
Alternaria alternata	20–60	Spherical	Gajbhiye et al. (2009)
Trichoderma sp.	8–60	Spherical	Kaler et al. (2013)
Penicillium fellutanum	5–25	Spherical	Kathiresan et al. (2009)
P. brevicompactum	58.35–76.80	Spherical	Shaligram et al. (2009)
Aspergillus clavatus	10–25	Spherical, hexagonal	Verma et al. (2010)
A. flavus	12–22	Spherical	Jain et al. (2010)
Cylindrocladium Floridanum	25	Spherical	Banu and Balasubramanian (2014)
Volvariella volvacea	20–150	Spherical, hexagonal	Philip (2009)
Verticillium lecanii	20–50	Spherical	Castro-Longoria et al. (2012)

1.6.2 Gold Nanoparticles (Au-NPs)

When reduction of $HAuCl_4$ or some other Aurum salt (like $AuCl_3$) in the presence of enzymes or other reducing agents to form particles in the scale of 1–100 nm leading to change in basic properties of the noble metal, the gold nanoparticles are formed. Au-NPs appear wine red to purplish in solution. They exhibit different shapes like spherical ring, sub-octahedral spheres, icosahedral, tetrahedral, hexagonal, decahedral, octahedral and nanorods (Alaqad and Saleh 2016).

Au-NP properties vary according to their size and shape. They have valuable optical properties, large surface to volume ratio and electronic features, thus enabling them to act as an important means in bio-nanotechnology. Surface plasmon resonance (SPR) and the ability of Au-NPs to quench fluorescence are their most important physical properties. Table 1.3 illustrates a comprehensive list of fungi from which Au-NPs have already been synthesized.

El-Sayed et al. (2005) worked on the use of gold nanoparticles in detection of cancerous cells. Gold nanoparticles when in conjugation with anti-EGFR (anti-epidermal growth factor receptor) antibody were able to bind homogeneously and specifically with cancer-type cells with 600 times greater affinity. Thus, the detection of cancerous cells can be more effective with the help of gold nanoparticles. Gold nanoparticles find wide applications in the field of biology. Sperling et al. (2008) classified applications of Au-NPs into four classes, namely, delivering,

Table 1.3 List of fungi from which Au-NPs have been already synthesized

Fungus used for synthesis	Size (nm)	Shape	References
Helminthosporium solani	2–70	Rods, triangles, pentagons, stars, pyramids	Kumar et al. (2008)
Rhizopus oryzae	10	Multishaped	Das et al. (2009)
Fusarium oxysporum	20–40	Multishaped	Anitha and Palanivelu (2011)
F. semitectum	18–80	Multishaped	Sawle et al. (2008)
Neurospora crassa	10–200, 6–23, 3–12	Triangles, hexagons, pentagons, quasi-spheres	Quester et al. (2013)
Aspergillus fumigatus, *A. flavus*	17.76–26.00	Triangles, hexagons, spherical	Gupta and Bector (2013)
Penicillium sp.	30–50	Spherical	Du et al. (2011)
Sclerotium rolfsii	25	Triangle, decahedral, spherical	Narayanan and Sakthivel (2011)

labelling, heating and sensing. For labelling, they are used in immunostaining, in single particle tracking and as contrast agents for X-rays. In the case of delivering, Au-NPs can be used in gene guns and for uptake by cells. The property of Au-NPs acting as heat source can be utilized for anticancer therapy through the concept of hyperthermia and for opening of chemical bonds triggered optically (photo-induced heating). Sensing can be used through the properties of Au-NPs like surface-enhanced Raman scattering, electron transfer, surface plasmons and fluorescence quenching. Moreover, Au-NPs are non-toxic nanoparticles which give them a large benefit over silver nanoparticles (Guo et al. 2014).

1.6.3 Other Nanoparticles

The most commonly studied nanoparticles are of silver and gold. Other than these, NPs are synthesized from metals like platinum and copper, non-metals like selenium and transition metal oxides like ZnO and SnO_2. The list of these types of nanoparticles which have been biosynthesized from fungal resources is tabulated in Table 1.4.

Platinum nanoparticles are utilized in fields such as nanocatalysts, electrical conductivity, optics and nonlinear optics (Stepanov et al. 2014). Rajan et al. (2016) also tested the antimicrobial efficiency of zinc oxide nanoparticles and found them to be strong antimicrobial agents against pathogenic microorganisms. Magnetite nanoparticles offer various applications like high-gradient magnetic separation (HGMS), magnetic resonance tomography (MRT), magnetically guided drug delivery and many mechano-electrical applications (Blaney 2007). Cadmium-containing quantum dots contain excess potential in the field of medical treatment, but on the

Table 1.4 List of biosynthesized nanoparticles (other than Ag-NPs and Au-NPs) from fungal resources

Nanoparticles	Fungus used for synthesis	Size (nm)	Shape	References
CdS (Cadmium sulphide)	*Fusarium oxysporum*	5–20	Spherical	Ahmad et al. (2002)
Magnetite	*Coriolus versicolor*	100	Spherical	Chen et al. (2011)
	Schizosaccharomyces pombe	1–1.5	Hexagonal lattice	Kowshik et al. (2002)
Se	*Alternaria alternata*	30–35	Spherical	Sarkar et al. (2011)
$SrCO_3$	*F. oxysporum*	Varying size	Needle shaped	Rautaray et al. (2004)
CdSe	*F. oxysporum*	9–15	Spherical	Kumar et al. (2007)
Ti	*F. oxysporum*	2–6	Quasi-spherical	Bansal et al. (2006)
	F. oxysporum	5–15	Quasi-spherical	Bansal et al. (2005)
$BaTiO_3$	*F. oxysporum*	4–5	Quasi-spherical	Bansal et al. (2006)
Pt	*F. oxysporum*	5–30	–	Syed and Ahmad (2012)
	Neurospora crassa	20–110	Spherical	Castro-Longoria et al. (2012)
ZnO	*Aspergillus fumigatus*	60–80	Spherical	Rajan et al. (2016)

other hand, toxic cadmium in nano-form can be a limitation. Thus, proper evaluation for the use of nanoparticles of these elements is required before their use (Rzigalinski and Strobl 2009).

1.7 Applications of Nanoparticles

Nanoparticles (NPs) find a wide range of applications in different fields. NPs are used in the formation of nanomaterials which find application in areas of electronics, optics and devices (both mechanical and medicinal). These devices work on the properties like inter-atomic interactions and quantum mechanics. Tunnelling and atomistic disorder effect controls the properties of these devices. Nanophotonics is a field which utilizes the properties of light and light-matter interaction.

Nanoparticles exhibit antimicrobial properties and, thus, find wide applications in the field of disease control. There are so many reports of different metal nanoparticles being used in plant disease control (Abd-Elsalam and Prasad 2018, 2019). There are different mechanisms involved in their action against pathogens. Either they directly act as antimicrobial agents or alter the nutritional status of host, thus, activating their defence mechanisms. Nanoparticles of copper and silver can be directly toxic to microorganisms, whereas nanoparticles of manganese, zinc and silicon act in altering host defence (Elmer et al. 2018).

Previous studies revealed that silver nanoparticles cause the disruption of transport system which mainly includes ion influx system. Fungal hyphae treated with silver nanoparticles lead to its damage causing plasmolysis. Dysfunction of ion influx system causes accumulation of silver ions which affects various metabolism and cellular process like respiration. Silver ions are also known to produce reactive oxygen species (ROS) that damage protein, lipids and nucleic acids. Treatment with silver nanoparticles causes the DNA to lose its ability to replicate, resulting in inactivated ribosomal expression and other cellular process like ATP production (Abdal Dayem et al. 2017; Aziz et al. 2016, 2019). Some of other broader areas in which nanotechnology applications are used are listed below.

1.7.1 Catalysis

Nanoparticles that are synthesized from fungi possess unique biocatalytic properties. Mishra et al. (2014) stated that biosynthesized gold nanoparticles from *Trichoderma viride* have remarkable catalytic properties. These gold nanoparticles show antimicrobial properties and reduce nitrophenol to 4-aminophenol in the presence of $NaBH_4$. In order to improve the enzyme activity, fungal nanoparticles are also used in enzyme immobilization.

1.7.2 Wound Healing

Ag-NPs synthesized from the fungus, *Aspergillus niger*, possess wound healing activity. Sundaramoorthi et al. (2009) synthesized Ag-NPs in an experimental rat model. The wound healing activity of Ag-NPs was checked by measuring the contraction of wound and duration of epithelialisation in dose in accordance with time.

1.7.3 Textile Fabrics

Staphylococcus aureus and *Escherichia coli* are responsible for nosocomial infections by the means of cotton fabrics. Ag-NPs, which are synthesized by *Lecanicillium lecanii*, when integrated with cotton fabrics inhibit the growth of *S. aureus* and *E. coli*. This cotton fabric cloth can be used to prevent nosocomial infections (Namasivayam and Avimanyu 2011). Duran et al. (2007) have demonstrated the antibacterial activity of *F. oxysporum* against *S. aureus* when incorporated with cotton fibres.

1.7.4 Vegetables and Food Preservation

Fayaz et al. (2009) showed that when thin film of sodium alginate was integrated with Ag-NPs synthesized from *Trichoderma viride*, it increases the life of carrots and peas and shows antibacterial activity. The increased shelf life was measured by comparing with control in terms of soluble protein content and weight loss.

1.7.5 Molecular Detection

Bansod et al. (2013) have developed a new PCR assay by conjugating Ag and Au nanoparticles of *Fusarium oxysporum* with master mix and DNA sample of *Candida sp.* from low concentrated DNA. This bioconjugate nano-PCR is highly specific and sensitive as compared to conventional one.

1.8 Conclusion

Nanoparticles can be used in biocontrol of pathogens and for forest disease management in the field of forest pathology. Nanoparticle synthesis is mostly achieved through the use of chemicals by reduction which can be a source of toxic materials entering into the environment, thereby, causing a threat to environment and human life. Therefore, mycosynthesis can be a great option for green synthesis of nanoparticles as a fabrication through fungal resources which is fast, cheap and more sustainable in nature.

References

Abdel-Aziz SM, Prasad R, Hamed AA, Abdelraof M (2018) Fungal nanoparticles: A novel tool for a green biotechnology? In: Fungal Nanobionics: Principles and Applications (eds. Prasad R, Kumar V, Kumar M, Wang S), Springer Singapore Pte Ltd. 61–87

Abdal Dayem A, Hossain MK, Lee SB, Kim K, Saha SK, Yang GM, Choi HY, Cho SG (2017) The role of reactive oxygen species (ROS) in the biological activities of metallic nanoparticles. Int J Mol Sci 18(1):120. https://doi.org/10.3390/ijms18010120

Abd-Elsalam KA, Prasad R (2018) Nanobiotechnology Applications in Plant Protection. Springer International Publishing (ISBN 978-3-319-91161-8) https://www.springer.com/us/book/9783319911601

Abd-Elsalam K, Prasad R (2019) Nanobiotechnology Applications in Plant Protection. Volume 2. Springer International Publishing (ISBN 978-3-030-13295-8) https://www.springer.com/gp/book/9783030132958

Afreen RV, Ranganath E (2011) Synthesis of monodispersed silver nanoparticles by *Rhizopus stolonifer* and its antibacterial activity against MDR strains of *Pseudomonas aeruginosa* from burnt patients. Int J Environ Sci 1(7):1582–1592

Agarwal H, Kumar SV, Rajeshkumar S (2017) A review on green synthesis of zinc oxide nanoparticles-an eco-friendly approach. Resour Eff Technol 3(4):406–413
Ahmad A, Mukherjee P, Mandal D, Senapati S, Khan MI, Kumar R, Sastry M (2002) Enzyme mediated extracellular synthesis of CdS nanoparticles by the fungus, *Fusarium oxysporum*. J Am Chem Soc 124(41):12108–12109
Alaqad K, Saleh TA (2016) Gold and silver nanoparticles: synthesis methods, characterization routes and applications towards drugs. J Environ Anal Toxicol 6(384):2161–0525
Anitha TS, Palanivelu P (2011) Synthesis and structural characterization of polydisperse silver and multishaped gold nanoparticles using *F. oxysporum* MTCC 284. Dig J Nanomater Biostruc 6(4):1587–1595
Aziz N, Pandey R, Barman I, Prasad R (2016) Leveraging the attributes of *Mucor hiemalis*-derived silver nanoparticles for a synergistic broad-spectrum antimicrobial platform. Front Microbiol 7:1984. https://doi.org/10.3389/fmicb.2016.01984
Aziz N, Faraz M, Sherwani MA, Fatma T, Prasad R (2019) Illuminating the anticancerous efficacy of a new fungal chassis for silver nanoparticle synthesis. Front Chem 7:65. https://doi.org/10.3389/fchem.2019.00065
Bailey RE, Smith AM, Nie S (2004) Quantum dots in biology and medicine. Physica E 25(1):1–2
Baker RA, Tatum JH (1998) Novel anthraquinones from stationary cultures of *Fusarium oxysporum*. J Ferment Bioeng 85:359–361
Bansal V, Rautaray D, Bharde A, Ahire K, Sanyal A, Ahmad A, Sastry M (2005) Fungus-mediated biosynthesis of silica and titania particles. J Mater Chem 15(1):2583–2589
Bansal V, Poddar P, Ahmad A, Sastry M (2006) Room-temperature biosynthesis of ferroelectric barium titanate nanoparticles. J Am Chem Soc 128(36):11958–11963
Bansod S, Bonde S, Tiwari V, Bawaskar M, Deshmukh S, Gaikwad S, Gade A, Rai M (2013) Bioconjugation of gold and silver nanoparticles synthesized by *F. oxysporum* and their use in rapid identification of *Candida* species by using bioconjugate-nano-polymerase chain reaction. J Biomed Nanotechnol 9(12):1962–1971
Banu NA, Balasubramanian C (2014) Myco-synthesis of silver nanoparticles using *Beauveria bassiana* against dengue vector, *Aedes aegypti* (Diptera: Culicidae). Parasitol Res 113(8):2869–2877
Birla SS, Tiwari VV, Gade AK, Ingle AP, Yadav AP, Rai MK (2009) Fabrication of silver nanoparticles by *Phoma glomerata* and its combined effect against *Escherichia coli*, *Pseudomonas aeruginosa* and *Staphylococcus aureus*. Lett Appl Microbiol 48(2):173–179
Blaney L (2007) Magnetite (Fe_3O_4): properties, synthesis, and applications. Lehigh Rev 15:33–81
Camacho JM, Oliva AI (2005) Morphology and electrical resistivity of metallic nanostructures. Microelectron J 36(3–6):555–558
Castro-Longoria E, Moreno-Velázquez SD, Vilchis-Nestor AR, Arenas-Berumen E, Avalos-Borja M (2012) Production of platinum nanoparticles and nano-aggregates using *Neurospora crassa*. J Microbiol Biotechnol 22(7):1000–1004
Chen GQ, Zou ZJ, Zeng GM, Yan M, Fan JQ, Chen AW, Yang F, Zhang WJ, Wang L (2011) Coarsening of extracellularly biosynthesized cadmium crystal particles induced by thioacetamide in solution. Chemosphere 83(9):1201–1207
Das SK, Das AR, Guha AK (2009) Gold nanoparticles: microbial synthesis and application in water hygiene management. Langmuir 25(14):8192–8199
Du L, Xian L, Feng JX (2011) Rapid extra−/intracellular biosynthesis of gold nanoparticles by the fungus *Penicillium* sp. J Nanopart Res 13:921–930
Duan WX, He MD, Mao L, Qian FH, Li YM, Pi HF, Liu C, Chen CH, Lu YH, Cao ZW, Zhang L (2015) NiO nanoparticles induce apoptosis through repressing SIRT1 in human bronchial epithelial cells. Toxicol Appl Pharmacol 286(2):80–91
Duran N, Marcato PD, De Souza GIH, Alves OL, Esposito E (2007) Antibacterial effect of silver nanoparticles produced by fungal process on textile fabrics and their effluent treatment. J Biomed Nanotechnol 3(2):203–208

Elechiguerra JL, Burt JL, Morones JR, Camacho-Bragado A, Gao X, Lara HH, Yacaman MJ (2005) Interaction of silver nanoparticles with HIV-1. J Nanobiotechnol 3(6). https://doi.org/10.1186/1477-3155-3-6

Elmer WH, Ma C, White JC (2018) Nanoparticles for plant disease management. Curr Opin Environ Sci Health 6:66–70

El-Sayed IH, Huang X, El-Sayed MA (2005) Surface plasmon resonance scattering and absorption of anti-EGFR antibody conjugated gold nanoparticles in cancer diagnostics: applications in oral cancer. Nano Lett 5(5):829–834

Fayaz MA, Balaji K, Girilal M, Kalaichelvan PT, Venkatesan R (2009) Mycobased synthesis of silver nanoparticles and their incorporation into sodium alginate films for vegetable and fruit preservation. J Agric Food Chem 57(14):6246–6252

Gajbhiye M, Kesharwani J, Ingle A, Gade A, Rai M (2009) Fungus mediated synthesis of silver nanoparticles and their activity against pathogenic fungi in combination with fluconazole. Nanomed Nanotechnol Biol Med 5(4):382–386

Gao X, Cui Y, Levenson RM, Chung LW, Nie S (2004) *In vivo* cancer targeting and imaging with semiconductor quantum dots. Nat Biotechnol 22(8):969–976

Guo D, Xie G, Luo J (2013) Mechanical properties of nanoparticles: basics and applications. J Phys D Appl Phys 47:013001. https://doi.org/10.1088/0022-3727/47/1/013001

Guo Q, Guo Q, Yuan J, Zeng J (2014) Biosynthesis of gold nanoparticles using a kind of flavonol: Dihydromyricetin. Colloids Surf A Physicochem Eng Asp 441:127–132

Gupta S, Bector S (2013) Biosynthesis of extracellular and intracellular gold nanoparticles by *Aspergillus fumigatus* and *A. flavus*. Antonie Van Leeuwenhoek 103(5):1113–1123

Hett A (2004) Nanotechnology: small matter, many unknowns. Swiss Reinsurance Company, Zurich

Horikoshi S, Serpone N (2013) Introduction to nanoparticles. In: Microwaves in nanoparticle synthesis: fundamentals and applications, 1st edn. Wiley, New York, pp 1–24

Husen A, Siddiqi KS (2014) Carbon and fullerene nanomaterials in plant system. J Nanobiotechnol 12:16. https://doi.org/10.1186/1477-3155-12-16

Ingle A, Gade A, Bawaskar M, Rai M (2009) *Fusarium solani*: a novel biological agent for the extracellular synthesis of silver nanoparticles. J Nanopart Res 11(8):2079–2085

Jain N, Bhargava A, Majumdar S, Tarafdar JC, Panwar J (2010) Extracellular biosynthesis and characterization of silver nanoparticles using *Aspergillus flavus* NJP08: a mechanism perspective. Nanoscale 3(2):635–641

Kaler A, Jain S, Banerjee UC (2013) Green and rapid synthesis of anticancerous silver nanoparticles by *Saccharomyces boulardii* and insight into mechanism of nanoparticle synthesis. Biomed Res Int. https://doi.org/10.1155/2013/872940

Karbasian M, Atyabi SM, Siadat SD, Momem SB, Norouzian D (2008) Optimizing nano-silver formation by *F. oxysporum* (PTCC 5115) employing response surface methodology. Am J Agric Biol Sci 3(1):433–437

Karthika D, Vadakkan K, Ashwini R, Shyamala A, Hemapriya J, Vijayanand S (2015) Prodigiosin mediated biosynthesis of silver nanoparticles (AgNPs) and evaluation of its antibacterial efficacy. Int J Curr Microbiol App Sci 4(11):868–874

Kathiresan K, Manivannan S, Nabeel AM, Dhivya B (2009) Studies on silver nanoparticles synthesized by a marine fungus *Penicillium fellutanum* isolated from coastal mangrove sediment. Colloids Surf B Biointerfaces 71(1):133–137

Kowshik M, Deshmukh N, Vogel W, Urban J, Kulkarni SK, Paknikar KM (2002) Microbial synthesis of semiconductor CdS nanoparticles, their characterization, and their use in fabrication of an ideal diode. Biotechnol Bioeng 78(5):583–588

Kumar SA, Ansary AA, Ahmad A, Khan MI (2007) Extracellular biosynthesis of CdSe quantum dots by the fungus *F. oxysporum*. J Biomed Nanotechnol 3(2):190–194

Kumar SA, Peter YA, Nadeau JL (2008) Facile biosynthesis, separation, conjugation of gold nanoparticles to doxorubicin. Nanotechnology 19(49):495101. https://doi.org/10.1088/0957-4484/19/49/495101

Lademann J, Weigmann HJ, Rickmeyer C, Barthelmes H, Schaefer H, Mueller G, Sterry W (1999) Penetration of titanium dioxide microparticles in a sunscreen formulation into the horny layer and the follicular orifice. Skin Pharmacol Physiol 12(5):247–256

Lara HH, Ayala-Núñez NV, Turrent LDCI, Padilla CR (2010) Bactericidal effect of silver nanoparticles against multidrug-resistant bacteria. World J Microbiol Biotechnol 26(4):615–621

Lloyd JR (2003) Microbial reduction of metals and radionuclides. FEMS Microbiol Rev 27:411–425

Medentsev AG, Alimenko VK (1998) Naphthoquinone metabolites of the fungi. Photochemistry 47:935–959

Mishra A, Kumari M, Pandey S, Chaudhry V, Gupta KC, Nautiyal CS (2014) Biocatalytic and antimicrobial activities of gold nanoparticles synthesized by *Trichoderma* sp. Bioresour Technol 166:235–242

Molnár Z, Bódai V, Szakacs G, Erdélyi B, Fogarassy Z, Sáfrán G, Varga T, Konya Z, Toth-Szeles E, Szucs R, Lagzi I (2018) Green synthesis of gold nanoparticles by thermophilic filamentous fungi. Sci Rep 8(1):3943. https://doi.org/10.1038/s41598-018-22112-3

Monteiro DR, Gorup LF, Takamiya AS, Ruvollo-Filho AC, de Camargo ER, Barbosa DB (2009) The growing importance of materials that prevent microbial adhesion: antimicrobial effect of medical devices containing silver. Int J Antimicrob Agents 34(2):103–110

Monteiro DR, Gorup LF, Silva S, Negri M, de Camargo ER, Oliveira R, Barbosa DB, Henriques M (2011) Silver colloidal nanoparticles: antifungal effect against adhered cells and biofilms of *Candida albicans* and *Candida glabrata*. Biofouling 27(7):711–719

Morones JR, Elechiguerra JL, Camacho A, Holt K, Kouri JB, Ramírez JT, Yacaman MJ (2005) The bactericidal effect of silver nanoparticles. Nanotechnology 16(10):2346–2353. https://doi.org/10.1088/0957-4484/16/10/059

Mukherjee P, Senapati S, Mandal D, Ahmad A, Khan MI, Kumar R, Sastry M (2002) Extracellular synthesis of gold nanoparticles by the fungus *Fusarium oxysporum*. Chembiochem 3(5):461–463

Nachiyar V, Sunkar S, Prakash P (2015) Biological synthesis of gold nanoparticles using endophytic fungi. Der Pharma Chem 7(11):31–38

Namasivayam SKR, Avimanyu (2011) Silver nanoparticle synthesis from *Lecanicillium lecanii* and evolutionary treatment on cotton fabrics by measuring their improved antibacterial activity with antibiotics against *Staphylococcus aureus* (ATCC 29213) and *E. coli* (ATCC 25922) strains. Int J Pharm Pharm Sci 3(4):190–195

Narayanan KB, Sakthivel N (2011) Facile green synthesis of gold nanostructures by NADPH-dependent enzyme from the extract of *Sclerotium rolfsii*. Colloids Surf A Physicochem Eng Asp 380(1–3):156–161

Nithya R, Ragunathan R (2009) Synthesis of silver nanoparticle using *Pleurotus sajor caju* and its antimicrobial study. Dig J Nanomater Bios 4(4):623–629

Pal SL, Jana U, Manna PK, Mohanta GP, Manavalan R (2011) Nanoparticle: an overview of preparation and characterization. J Appl Pharm Sci 1(6):228–234

Percival SL, Bowler PG, Dolman J (2007) Antimicrobial activity of silver-containing dressings on wound microorganisms using an *in vitro* biofilm model. Int Wound J 4(2):186–191

Philip D (2009) Biosynthesis of Au, Ag and Au-Ag nanoparticles using edible mushroom extract. Spectrochim Acta Part A 73(2):374–381

Prasad R (2016) Advances and Applications through Fungal Nanobiotechnology. Springer, International Publishing Switzerland (ISBN: 978-3-319-42989-2)

Prasad R (2017) Fungal Nanotechnology: Applications in Agriculture, Industry, and Medicine. Springer Nature Singapore Pte Ltd. (ISBN 978-3-319-68423-9)

Prasad R, Pandey R, Barman I (2016) Engineering tailored nanoparticles with microbes: quo vadis. WIREs Nanomed Nanobiotechnol 8:316–330. https://doi.org/10.1002/wnan.1363

Prasad R, Jha A, Prasad K (2018) Exploring the Realms of Nature for Nanosynthesis. Springer International Publishing (ISBN 978-3-319-99570-0 https://www.springer.com/978-3-319-99570-0

Prathna TC, Mathew L, Chandrasekaran N, Raichur AM, Mukherjee A (2010) Biomimetic synthesis of nanoparticles: science, technology and applicability. In: Mukherjee A (ed) Biomimetics learning from nature. InTech, China, pp 1–20

Quester K, Avalos-Borja M, Vilchis-Nestor AR, Camacho-López MA, Castro-Longoria E (2013) SERS properties of different sized and shaped gold nanoparticles biosynthesized under different environmental conditions by *Neurospora crassa* extract. PLoS One 8(10):77486. https://doi.org/10.1371/journal.pone.0077486

Raheman F, Deshmukh S, Ingle A, Gade A, Rai M (2011) Silver nanoparticles: novel antimicrobial agent synthesized from an endophytic fungus *Pestalotia* sp. isolated from leaves of *Syzygium cumini* (L). Nano Biomed Eng 3(3):174–178

Rajan A, Cherian E, Baskar G (2016) Biosynthesis of zinc oxide nanoparticles using *Aspergillus fumigatus* JCF and its antibacterial activity. Int J Mod Sci Technol 1:52–57

Rautaray D, Sanyal A, Adyanthaya SD, Ahmad A, Sastry M (2004) Biological synthesis of strontium carbonates crystals using the fungus *F. oxysporum*. Langmuir 20(16):6827–6833

Rzigalinski BA, Strobl JS (2009) Cadmium-containing nanoparticles: perspectives on pharmacology and toxicology of quantum dots. Toxicol Appl Pharmacol 238(3):280–288

Saglam N, Yesilada O, Cabuk A, Sam M, Saglam S, Ilk S, Emul E, Celik PA, Gurel E (2016) Innovation of strategies and challenges for fungal nanobiotechnology. In: Prasad R (ed) Advances and applications through fungal Nanobiotechnology. Springer, Cham, pp 25–46

Sarkar J, Dey P, Saha S, Acharya K (2011) Mycosynthesis of selenium nanoparticles. IET Micro Nano Lett 6(8):599–602

Sastry M, Ahmad A, Khan MI, Kumar R (2003) Biosynthesis of metal nanoparticles using fungi and actinomycete. Curr Sci 85(2):162–170

Sawle BD, Salimath B, Deshpande R, Bedre MD, Prabhakar KB, Venkataraman A (2008) Biosynthesis and stabilization of Au and Au–Ag alloy nanoparticles by fungus, *F. semitectum*. Sci Technol Adv Mater 9(3):035012. https://doi.org/10.1088/1468-6996/9/3/035012

SCENIHR (Scientific committee on emerging and newly identified health risks) (2007) Modified opinion (after public consultation) on the appropriateness of the risk assessment methodology in accordance with the technical guidance documents for new and existing substances for assessing the risks of nanomaterials. European Commission Health and Consumer Protection Directorate-General. Synthesis report: http://ec.europa.eu/health/ph_risk/documents/synth_report.pdf

Shaligram NS, Bule M, Bhambure R, Singhal RS, Singh SK, Szakac SG, Pandey A (2009) Biosynthesis of silver nanoparticles using aqueous extract from the compactin producing fungal strain. Process Biochem 44(8):939–943

Siddiqi KS, Husen A (2016) Fabrication of metal nanoparticles from fungi and metal salts: scope and application. Nanoscale Res Lett 11(1):98. https://doi.org/10.1186/s11671-016-1311-2

Sperling RA, Gil PR, Zhang F, Zanella M, Parak WJ (2008) Biological applications of gold nanoparticles. Chem Soc Rev 37(9):1896–1908

Stepanov AL, Golubev AN, Nikitin SI, Osin YN (2014) A review on the fabrication and properties of platinum nanoparticles. Rev Adv Mater Sci 38(2):160–175

Sundaramoorthi C, Kalaivani M, Mathews DM, Palanisamy S, Kalaiselvan V, Rajasekaran A (2009) Biosynthesis of silver nanoparticles from *Aspergillus niger* and evaluation of its wound healing activity in experimental rat model. Int J Pharm Tech Res 1(4):1523–1529

Syed A, Ahmad A (2012) Extracellular biosynthesis of platinum nanoparticles using the fungus *F. oxysporum*. Colloids Surf B Biointerfaces 97:27–31

Tidke PR, Gupta I, Gade AK, Rai M (2014) Fungus-mediated synthesis of gold nanoparticles and standardization of parameters for its biosynthesis. IEEE Trans Nanobioscience 13(4):397–402

Tran QH, Le AT (2013) Silver nanoparticles: synthesis, properties, toxicology, applications and perspectives. Adv Nat Sci Nanosci Nanotechnol 4(3):033001. https://doi.org/10.1088/2043-6254/aad12b

Verma VC, Kharwar RN, Gange AC (2010) Biosynthesis of antimicrobial silver nanoparticles by the endophytic fungus *Aspergillus clavatus*. Nanomedicine 5(1):33–40

Chapter 2
Nanoparticle-Incorporated Soy Protein Isolate Films

Shikha Rani and Rakesh Kumar

Contents

2.1 Introduction

Nanoparticles are particulate dispersions or solid particles with sizes that range between 10 and 1000 nm. The name nanoparticle itself dictates its meaning, as the prefix "nano" is adapted from the Latin *nanus,* which means literally *dwarf* or *very small.* If we talk of SI unit, it denotes the reduction of size by 10^9 times. Hence, nanometer is used to measure the nanosized world (1 nm is equal to 10^9 m), and it

S. Rani · R. Kumar (✉)
Department of Biotechnology, Central University of South Bihar, Gaya, India
e-mail: rakeshkr@cusb.ac.in

D. Thangadurai et al. (eds.), *Nanotechnology for Food, Agriculture, and Environment*, Nanotechnology in the Life Sciences,
https://doi.org/10.1007/978-3-030-31938-0_2

comprehends structures whose size is somewhere in between the molecular and microscopic dimensions (commonly >1 nm and < 100 nm) (Pal et al. 2011).

Nanoparticles can be prepared by emulsion-solvent evaporation method, in which the polymer solution of interest is emulsified in an aqueous solution. Then that polymer solvent is evaporated and evaporation process induces the polymer precipitation which further forms the nanosphere. Other methods of nanoparticle preparation are salting-out method, emulsion-diffusion method, and solvent displacement/precipitation method (Mohanraj and Chen 2006). Depending on the preparation method, they can be termed as nanoparticles, nanocapsules, nanodot, or nanospheres. Both the nanospheres and nanocapsules are widely used in medicine as drug delivery systems; the former is used as a matrix for uniform dispersion of drug, while the latter surrounds the drug by unique polymeric membrane (Rajput 2015).

From the above explanations, it is clear that the preparation of nanoparticles is based on reduction of the size of particle from a micrometer to a nanometer, and this reduction results in dramatic changes of the properties of nanoparticles. The changes in properties influence antimicrobial activity, electrical conductivity, mechanical properties, physical properties, and chemical reactivity.

Inorganic and metallic nanoparticles exhibit considerably enhanced chemical, physical, and biological properties due to their nanoscale size (Brigger et al. 2012). Size, shape, and integrity of nanoparticles play very important roles in their functionality. Reduction of the particle size from micrometer to nanometer significantly changes its physical and chemical properties, and this reduced size of nanoparticles sometimes results in completely different and novel phenomenon. This can be evident from the fact that antimicrobial properties of nano-Ag is higher than micro-Ag particles. Some nanosized particles inhibit some microbes significantly, as they can enter in the microbial cell due to their small size and thus can be used as effective bactericidal formulation (Sondi and Salopek-Sondi 2004). Zinc oxide nanoparticles exhibit outstanding biocidal action against a large spectrum of Gram-positive and Gram-negative bacteria (Stoimenov et al. 2002; Padmavathy and Vijayaraghavan 2008; Espitia et al. 2012; Sirelkhatim et al. 2015).

Characterization of nanoparticle is a very important step as it determines the nature of nanoparticle. It also states whether the particle is toxic or not. Nanoparticles are basically characterized by their size, morphology, and surface charge, with the help of scanning electron microscopy (SEM), transmission electron microscopy (TEM), and atomic force microscopy (AFM). Stability and distribution of nanoparticle within a system are based on several factors such as their average particle diameter, size distribution, and charge. In order to find out the potential toxicity of nanoparticle, there is a need to find out their overall shape, and electron microscopy technique is a suitable tool for this task. Apart from shape and size, there is one more major factor that determines the physical stability and polymer dispersion—surface charge of nanoparticles. Surface charges also determine the in vivo performance of nanoparticle.

2.2 Soy Protein

Soybean (*Glycine max*) belongs to leguminous plant family and is closely connected to clover, peas, and alfalfa. The origin of this plant is basically from Asia, and it has two Asian wild relatives. The color of soybean flowers is either violet or red. Soybeans are grown in clusters of pods, each pod having 3 to 5 beans. An average of 60–80 pods is cultivated from one soybean plant, which means one soybean plant produces around 300–400 beans. Shapes of beans are either oval or round. Soybeans are found in a variety of colors like black, violet, green, and yellow brown (Gianessi and Carpenter 2000). Soybeans are less expensive, are abundantly available, and are a good source of nutrition. Due to its very good qualities, it is used for a variety of purposes, for example, it is used as oil seed, one of the major foods for pets, feed for aquarium, and a biofuel, and most importantly as a human food source.

There are several advantages of soy proteins such as it contains all essential amino acids with an impressive balance of them (Table 2.1). It contains some components which are physiologically very beneficial as these compounds are known to reduce the level of bad cholesterol, lowering the danger of hyperlipidemia. Due to its capacity to reduce the level of cholesterol, it reduces the chances of cardiovascular diseases. Soybeans are known for their gelling and emulsifying abilities; they also retain good water-holding and oil-holding capacities.

Soybeans also contain some harmful compounds like trypsin inhibitors and hemagglutinin. In order to deactivate them, it is crucial to heat the soybean before use. There are other benefits of heating soybeans as the heating process induces the denaturation of soy protein, so the amino acids are easily available. Heating eventually softens the soybean tissue and also reduces that odor. Other than these benefits, soybean gets sterilized by heating and thus becomes safer for use.

On an average, the soybean contains up to 40% protein and 20% oil. In food industry SPI (Soy Protein Isolate) are used on a large scale; thus, there is a need to isolate it. For isolation of soy protein, the oils are needed to be removed at low temperature. Protein content of SPI which are extractable comprises the storage

Table 2.1 Types and concentrations of amino acids present in soy protein

Aromatic (% concentration)	Aliphatic (% concentration)	Positively charged (% concentration)	Negatively charged (% concentration)	Uncharged (% concentration)
Phenylalanine (4.5)	Glycine (4.7)	Lysine (8.7)	Aspartate (13.1)	Serine (6.8)
Tryptophan (0.2)	Proline (4.2)	Arginine (6.1)	Glutamate (18.6)	Threonine (2.3)
Tyrosine (4.7)	Alanine (4.5)	Histidine (4.7)		Cysteine (1.9)
	Valine (3)			
	Leucine (6.8)			
	Methionine (2.6)			
	Isoleucine (2.8)			

Data in bracket represent the percentage amount of amino acid globulin and whey protein fractions. To separate both the fractions, the whole fraction needed to be acidified at pH 4.5–4.8 which is the isoelectric point of soy proteins. Thus by acidifications, one can precipitate major soybean storage protein (Nishinari et al. 2014).

SPI is a combination of numerous proteins which can be separated by gravitational force under centrifugal field. Based on their sedimentation coefficients, these proteins can be classified into four major categories: 2S, 7S, 11S, and 15S. 7S (β-conglycinin) and 11S (glycinin) represent more than 80% among these categories (Saio et al. 1969).

Soy protein structures diversify its physicochemical properties which ultimately results in the variety of functionalities in food systems. The functionalities of protein are generally determined by some of the major key factors such as water uptake properties, protein conformation, and size of protein; other than these factors, some of the extrinsic factors also play a major role in protein functionality such as the interaction behavior of protein with protein and other biomolecules, temperature, pH, and ionic strength. Soy protein has some major functions which are listed below (Hettiarachchy and Kalapathy 1998):

(a) *Water interaction properties*: Interaction of soy protein with water depends on the hydrogen bonds or dipole-dipole interaction and nature of the side chain of amino acid. Basically the protein structure of soy protein determines the water hydration properties as well as water uptake property of soy protein.
(b) *Water uptake capacity*: The side chain of soy protein, the conformation of protein, and the surface charge largely determine the water uptake property of soy protein.
(c) *Water solubility*: The solubility of protein is very important to obtain the desired functions of protein. The higher the solubility, the better the functionality. As solubility is an important factor for functionality of soy protein, it can be enhanced by introducing protein in basic condition, while acidic condition precipitates it out.
(d) *Matrix forming properties*: Soy protein can form matrix when exposed to suitable environment under the influence of several extrinsic as well as intrinsic factors. Matrix can be formed when protein interacts with protein itself in covalent form. Matrix formation property of soy protein leads to the gelation and film formation.
(e) *Gelation*: Soy proteins can be used for the formation of gels which acts as a structural matrix and can hold water, sugars, and different flavors.
(f) *Film formation*: Soy protein can form film under influence of pH and temperature. For this soy protein is firstly concentrated and then casted on a flat surface under constant temperature and ultimately dried out to form film.

Synthetic plastics are posing a great threat to the environment as they are nonrenewable and nonbiodegradable (Tian and Xu 2011). To overcome these problems, there is a great need of suitable alternative, which can be used as packaging material. Packaging materials should exhibit less water uptake, high mechanical strength, longer shelf-life, and renewability. Natural products like polysaccharides, proteins, and lipids qualify in some of the major criteria of a good packaging material as they are biodegradable and renewable and thus are considered as the attractive alternative of synthetic plastics (Coa et al. 2007; Shi and Dumont 2014; Garrido et al. 2016). Among all natural products, SPI attracts more attention as it contains 90–95% of protein along with almost all essential amino acids as discussed above in Table 2.1. SPI are obtained from natural resources and thus are renewable; they have a good polymerization property; thus, the film-forming capacity is also good; and SPI also attain high mechanical strength with some sort of antimicrobial properties (Kumar et al. 2002; Song et al. 2011).

2.3 Nanoparticles

As the basics are already discussed in the introduction part, we can go through some of the major nanoparticles used so far with soy protein as a reinforcing agent. Based on the presence and nature of side groups, nanoparticles can be categorized as either non-functional or functional nanoparticles. Some of the most used non-functionalized nanoparticle that we will deal in this chapter are carbon nanoparticles and silver nanoparticles. Non-functionalized nanoparticles meant the nanoparticles without any reactive moiety. The functionalized/reactive moiety ones are nano-silica (SiO_2), TiO_2, montmorillonite (MMT), and carbon nanotubes.

2.3.1 Non-functionalized/Absence of Reactive Moiety Nanoparticles

Carbon nanoparticles (CNPs) exhibit great significance in optical and biochemical fields (Zhu et al. 2009; Guo et al. 2013). CNPs are nearly spherical in shape with a good distribution. Average sizes of CNPs prepared by subjecting to heating at 180 °C for 5 h and 150 °C for 7 h were 82.71 and 24.94 nm, respectively. CNPs as a kind of nanoparticles have advantages over other nanoparticles due to the water solubility, active functional groups on the surface, interfacial properties, good compatibility with SPI, and nontoxicity (Li et al. 2016). Ag nanoparticles of different shapes and dimensions can be prepared by physical, chemical, or biological synthesis. It has been reported by Tran and Le (2013) that one can prepare Ag nanoparticle of dimension ranging from 2 nm to 150 nm with spherical, rod, octagonal, hexagonal, triangle, flower-like shapes by the abovementioned methods. Ag nanoparticles show excellent antimicrobial properties against bacteria, fungi, and virus (Zhang et al. 2016).

2.3.2 *Functionalized/Reactive Moiety Nanoparticles*

Nano-SiO_2 particles are amorphous white powder (in its agglomerate state) having a diameters of 20–60 nm. The incorporation of SiO_2 in the matrix either synthetic or nonsynthetic significantly improves the mechanical properties of nanoparticles-incorporated film (Tian 2012; Liu et al. 2017a). It has been reported by Ai et al. (2007) that nano-SiO_2 upon incorporation in SPI film improves the morphology and properties of SPI films.

Aluminum-silicate mineral montmorillonite (MMT) has a great reinforcing effect, and it can form nanocomposites with various polymeric matrices. The reinforcing effects are generally improved with uniform distribution of MMT nanoclay within the polymeric matrix. This is due to the fact that uniform distribution caused even dispersion of particle and this kind of dispersion gives rise to intercalated or exfoliated structure (Echeverría et al. 2014).

Carbon nanotubes (CNTs) have received considerable attention as reinforcing agent. There are two main types of CNTs: single-walled nanotubes (SWNTs), which represent the type where single graphene sheet is rolled into a cylinder, and multiwalled nanotubes (MWNTs), which are being considered as stacking of concentric layers of several graphene layers in the form of cylinders with an interspacing of 0.34 nm. CNTs exhibit high flexibility, high aspect ratio (typically ca. 300–1000), extraordinary mechanical strength, and high electrical conductivity at their nanometric scale. Due to these properties, it is considered as ideal reinforcing fillers for polymer matrices so as to achieve high performance and special functions. The homogeneous dispersion of CNTs and strong interfacial interaction and adhesion between the matrix and fillers are the critical factors influencing the reinforcing effects. However, strong van der Waals interactions among pristine CNTs result in their bundling and aggregation, preventing their facile individualization (Xiang et al. 2017).

TiO_2 is an oxide form of Ti and is found naturally. It is also known as titanium (IV) oxide, titanic acid anhydride, titania, titanic anhydride, or Ti white. TiO_2 is bright white pigment with high refractive index, and it is used worldwide on a large scale. Approximately, four million tons of this pigment in both nano- and micro-form are consumed annually worldwide (Malathi et al. 2017).

2.4 SPI as a Film

The SPI can molded to form a film, and for this particular purpose of film formation, the main necessity is having proteins with high molecular weight, and this can be accomplished by the process of polymerization. As discussed above, soy protein is composed of four types of protein moieties: 2S, 7S, 11S, and 15S; although all these four types have polymerizing capacity, the best out of them are

7S and 11S as they form disulfide linkages, and thus the films formed with these moiety are smooth and firm. There are plenty of work published on the film-forming capacity of 7S-RG and 11S-RG (Briggs and Wolf 1957; Sun et al. 1999). The literature reports published on film-forming ability of 7S-RG and 11S-RG also indicate that the films formed by 11S-RG are stronger with lower water uptake property than that of 7S-RG as their tensile strength is higher. This kind of work somehow suggests that the two proteins (11S and 7S) are composed of different sets of amino acids. Soy film formation is solely dependent on some major factors; these are pH, temperature, and the plasticizer. The particular pH of film-forming solution is necessary to maintain during film formation as it increases the solubility of soy protein, but the lower or higher pH tends to influence the deformity of protein as the higher pH influences denaturation, while lower pH possibly increases the hydrophilicity. Another major factor to be considered is the temperature; as proteins need to be linearized for polymerization, the temperature should be optimal (65 °C). Higher temperature leads to denaturation and lower temperature hinders polymerization hence proteins may not get linearized at lower temperature.

Another major necessity of soy film formation is the use of plasticizers, as it makes the film smooth, flexible, and uniform. Film can be formed without plasticizers too, but those films are quite rough and brittle, and they eventually lose their properties (Kumar et al. 2009). Generally, the plasticizers are of low molecular weight compound with higher or lower hydrophilicity. Some popular plasticizers are a glycerol, ethylene glycol, triethylene glycol, poly(ethylene glycol), butane diols (Wang et al. 1995; Swain et al. 2004), urea (Mo and Sun 2001), acetamide (Liu and Zhang 2006), and thiodiglycol (Kumar and Zhang 2009). The plasticizers which are polyols based have hydroxyl groups. These hydroxyl groups form hydrogen bonds with polymer; thus, they form polymer-plasticizer bonds instead of polymer-polymer bond (Eswaranandam et al. 2004). This kind of interaction improves the quality of film. One of the most commonly used plasticizers among polyols is glycerol. Glycerol has three hydroxyl groups; thus, glycerol-plasticized soy films show higher affinity toward water; in addition, there are reports that indicate the increased water sensitivity and decreased tensile strength of film upon incorporation of other plasticizers (Chen and Zhang 2005; Chen et al. 2005). Water sensitivity and affinity toward water are major limitations of soy film as they increase the moisture content, so the film becomes vulnerable to microbial attack and eventually the shelf-life of film decreases dramatically. In order to minimize it, several types of additives either chemical based, nanomaterial based, or natural based are incorporated into the films. Groups of research team are constantly working to find out the most suitable additives to incorporate into films. Water uptake can be reduced by two methods, either by some chemical modifications or with the help of cross-linking reactions. Hence, incorporation of cross-linkers such as nanoparticles can significantly improve the water uptake property of the films.

The nature of amino acids present in soy film matrix can significantly monitor the release or retention of incorporated nanomaterials (Ou and Kwok 2004; Kumar et al. 2016). The mechanical properties of soy film are largely dependent on the method of film preparation and the nature of added additive. The performance of soy film is considerably dependent on the intermolecular interactions which take place by hydrogen bonding and hydrophobic interactions of side chain amino acid in the film network (Sun et al. 1999; Rhim et al. 2000; Dawson et al. 2003).

There are two methods of soy protein film preparation. They can be prepared by either compression molding or solution casting. Compression molding is mostly acquired for the preparation of soy film. In this method, different concentrations of plasticizers such as glycerol (~30% w/w with respect to SPI) are mixed with SPI powder separately in an electronic mixer for about 15 min (Kumar and Zhang 2009). For the preparation of nanoparticle-based additive-incorporated SPI film, nanoparticles are also added to the mixer, during the mixing process. In this way, nanoparticles get incorporated with soy protein. The resulting mixtures are then subjected to the same conditions as discussed above. This is an efficient way to prepare plasticized SPI film. There is one more method to incorporate nanoparticles in SPI film by soaking the prepared neat SPI film in a solution of nanoparticle containing glycerol for some hours followed by washing of the film by deionized water to remove excess nanoparticles. In order to make the film hydrophobic, films are again soaked in any hydrophobic solution of interest. This way, a strong and hydrophobic nanoparticle-incorporated SPI film can be obtained (Xie et al. 2016).

Solution casting is also a reliable method to prepare SPI film. In this method, about 5–8 g of SPI is mixed with water along with glycerol 30% (w/w) of SPI under constant stirring to form film-forming solution (Eswaranandam et al. 2004). In this method, pH and temperature during film formation play a major part, so the pH of the solution is adjusted to 10 with the help of suitable base so as to achieve high level of unfolding. And the solution should be stirred on a particular temperature ranging between 60 °C and 70 °C. Nanoparticles can directly be added to the solution in stirring condition. Stirring is necessary to mix additive well in solution so they can form the desired interaction with soy protein. The prepared solutions are then casted on leveled Teflon-coated glass plates or simple glass plate coated with silicone oil. After drying at ambient conditions (50–60 °C) for about 24 h, the films are peeled off from the plates.

Both the methods of film preparation are efficient in their own ways. Comparison of the results and qualities of the prepared films showed that the method of film preparation can significantly affect the material properties of the film. Both the mechanical and physical properties of the films prepared by these two methods differed. The films prepared by solution casting showed less water resistance and low hydrophobicity, whereas the one prepared by compression molding showed a relatively high mechanical strength, firm and smooth surface, and high transmittance (Garrido et al. 2016).

2.5 Structural and Reinforcement Effect of Nanoparticle in SPI Film

Upon incorporation in SPI film, nanoparticle shows significant effect on the structure and reinforcement of film based on their particulate nature. To find out the actual effects of nanoparticles, the prepared films are characterized by several methods or assays; some of them are listed below.

2.5.1 FTIR

FTIR analysis is based on the vibration which occurs due to different molecular interactions. Vibrations can be of several types, say, for example, stretching vibration, bending vibration, and asymmetric stretching vibration. For the characterization of film, both the neat and nanoparticle-incorporated SPI films are subjected to FTIR spectroscopy in the form of film. Data of this technique gives information about interaction between SPI and nanoparticle-incorporated film. Data shows that the neat SPI film shows a broad N-H stretching and O-H stretching band between 3200 and 3400 cm^{-1} which are actually assigned to amide A or amide I of soy protein films (Xu et al. 2015). Similarly, the peaks at 1538 and 1260 cm^{-1} are assigned to the amide II and amide III bands of protein, respectively (Li et al. 2016). The carbonyl (–C=O) band present in all the proteins is represented at 1631 cm^{-1} in neat SPI film.

By knowing this much detail of the banding pattern of neat film, one can easily find out any discrepancy in the banding pattern of nanoparticle-incorporated SPI film by comparing it with the data of a neat one. Presence of new kind of vibration clearly indicates the different patterns of interaction, and these kinds of interactions are most possibly induced by nanoparticles incorporated, it can be said theoretically.

2.5.2 Transmittance

Generally, SPI film is yellow in color, and transmittance of SPI in visible range increases with increasing wavelength. Incorporation of nanoparticle may increase or decrease the transmittance. Decrease or increase in transmittance is also dependent on the time and speed of homogenization as higher level of speed and time can allow the even distribution of particle in the film-forming solution so the thickness may increases and the film eventually became opaque (Malathi et al. 2017). With increase in opacity, the color of film becomes darker.

2.5.3 Water Uptake

Water uptake test is necessary to find out the water resistance of the prepared film. The lesser the water uptake, the better the film in terms of mechanical properties and vice versa. Water uptake of film is measured by testing method such as ASTM D570–81. For this purpose, the films are preconditioned at 50 °C for 24 h and weighed. After that, the films are dipped in water for 24 h, dried with paper towels to remove the excess water from the surface, and weighed. The total difference in mass before and after dipping in water is the actual water uptake test. Although it is a simple test, its significance is high. It is observed that the nanoparticle as in case of MMT significantly lowers the water uptake of film and it is a beneficial sign in order to prepare water-resistant film (Echeverría et al. 2014).

By incorporating certain types of nanoparticles in situ, it is also possible to get a superhydrophobic film. $Cu_3(PO_4)_2{\cdot}3H_2O$ nanoflowers are formed by controlling growth time and content of copper sulfate. Optimum pH value of the solution gives superhydrophobic self-cleaning property to the SPI film (Fig. 2.1) (Xie et al. 2016).

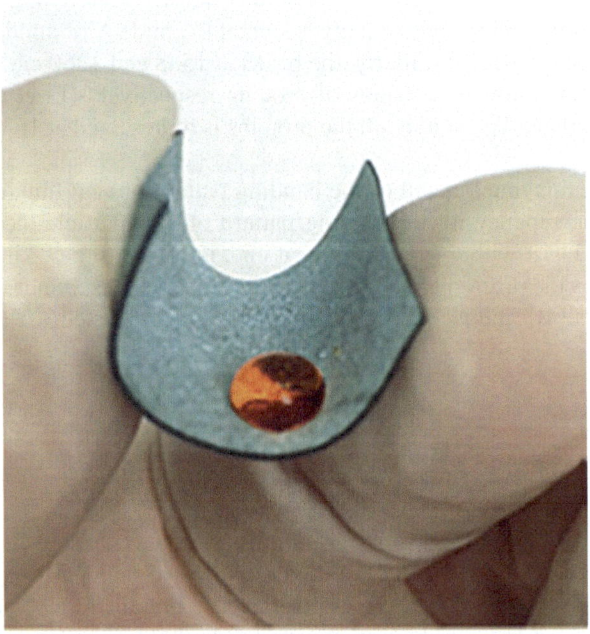

Fig. 2.1 Superhydrophobic property of soy film surface fabricated with $Cu_3(PO_4)_2{\cdot}3H_2O$ nanoflowers. (Reprinted with permission from Xie et al. (2016), Copyright {2018} American Chemical Society)

2.5.4 Mechanical Properties

Tensile strength determines the actual mechanical strength of the film, which means the higher the tensile strength, the stronger the film. It is greatly dependent on Young's modulus and elongation at break. We can say that the tensile property of the film can give an idea about the quality of film. A good film is highly flexible, so by testing the elongation at break, we can validate the overall flexibility of film.

Figure 2.2 depicts about the mechanical properties of SPI/MMT sheets (Chen and Zhang 2006). We can see in the graph that there is a decrease in elongation at break with the increasing concentration of MMT. Young's modulus also increased from 180.2 to 587.6 MPa with an increase of the MMT content from 0 to 20 wt % (Chen and Zhang 2006). It is also clear from Fig. 2.2 that the tensile property increases with increasing concentration of MMT up to 16% and shows a decreasing behavior after 16% MMT concentration. The reason behind good mechanical property of MMT-incorporated SPI film is probably the strong electrostatic and hydrogen bonding interactions between the soy protein and the highly dispersed MMT

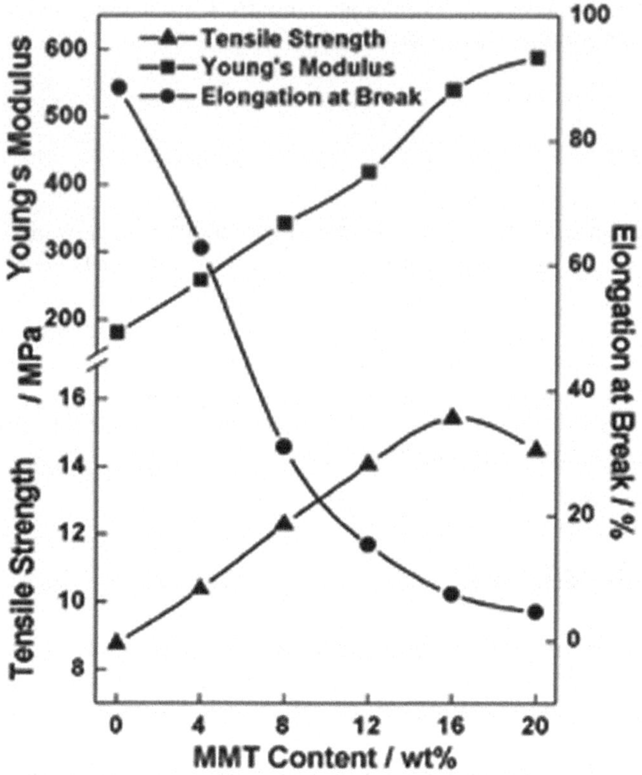

Fig. 2.2 Mechanical properties of MWCNT-incorporated SPI. (Reprinted with permission from Chen and Zhang (2006), Copyright {2018} American Chemical Society)

layers which is sufficient to restrict the segmental motion of the soy protein, and this restriction led to the enhancement of the modulus and tensile strength of the soy protein plastics.

Size of nanoparticle also exerts significant effect on the mechanical properties of SPI film. Zheng et al. (2007) showed the effect of MWNT with different diameters (<10, 10–15, 20–40, 40–60, and > 60 nm) on the mechanical properties of the nano-composite of MWNT and SPI. Interestingly the MWNT with diameter up to 10 to 15 nm showed positive effect on mechanical strength. On the contrary, as the diameter of MWNT increased up to 40–60 nm, the mechanical properties dramatically decreased and again started increasing with a diameter > 60 nm of MWNT.

There are also reports which show the effect of nano-SiO_2 contents on the mechanical properties of SPI-SiO_2 nano-composite sheets (Ai et al. 2007). Interestingly, this report gives a clear idea that the concentration of nanomaterial exerts dramatic effects on film. As the 4 wt% of nano-SiO_2 significantly increased the tensile strength and modulus, 8% of nano-SiO_2 sharply decreased the tensile strength, but the elongation and reinforcement effect was optimum at 8 wt% of nano-SiO_2. Similarly, citric acid-starch nanoparticles (CSN) upturn the tensile properties of CSN/SPI nanocomposites. Both tensile strength and Young's modulus got increased from 3.7 MPa and 60 MPa at 3 and 4% CSN loading and reached optimum value of 5.6 MPa and 130 MPa, respectively.

2.5.5 XRD

X-ray diffraction study is an efficient way of structural characterization of any biomolecule. Amorphous nature of SPI gives broad peaks on XRD analysis. But few nanoparticles are crystalline in nature. Hence, it is interesting to characterize samples by XRD after the incorporation of nanoparticles in SPI. Addition of nanoparticles like MMT decreased the amorphous nature of SPI. XRD patterns of MMT-incorporated SPI film showed two basal reflection diffraction peaks at $2\theta = 8.96°$ and 19.97° corresponding to the α-helix and β-sheet structures of the SPI secondary conformation, respectively (Xu et al. 2015).

2.5.6 SEM and TEM

Scanning electron microscopy (SEM) and transmission electron microscopy (TEM) are techniques to observe the finest details of structure of the given molecule. It is a best tool for physical observations like morphology analysis. Soy films are also characterized with the help of SEM and TEM. The nanoflowers of $Cu_3(PO_4)_2 \cdot 3H_2O$ are generated on SPI film from SEM; it is easy to observe the changes created by these nanoflowers on SPI film (Fig. 2.3).

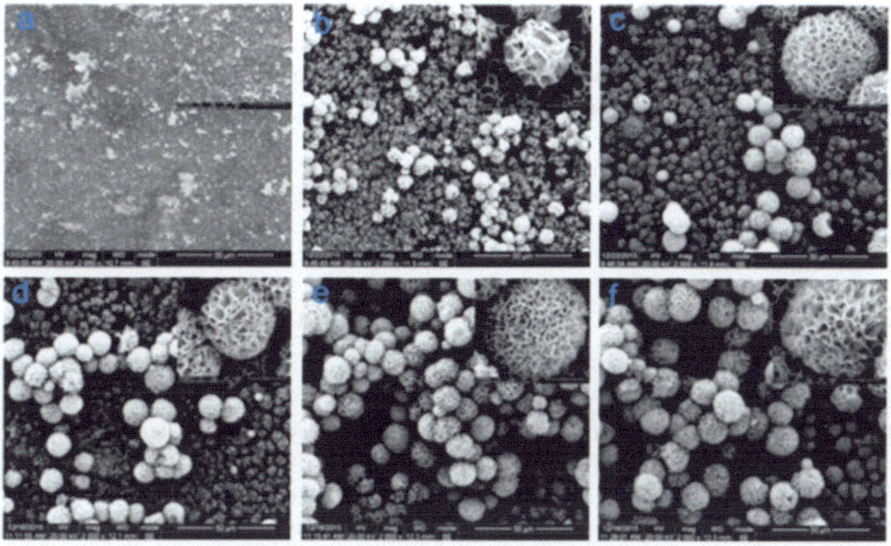

Fig. 2.3 SEM image of the nanoflower-incorporated SPI film (**b-f**) and control (**e**). (Reprinted with permission from Xie et al. (2016), Copyright {2018} American Chemical Society)

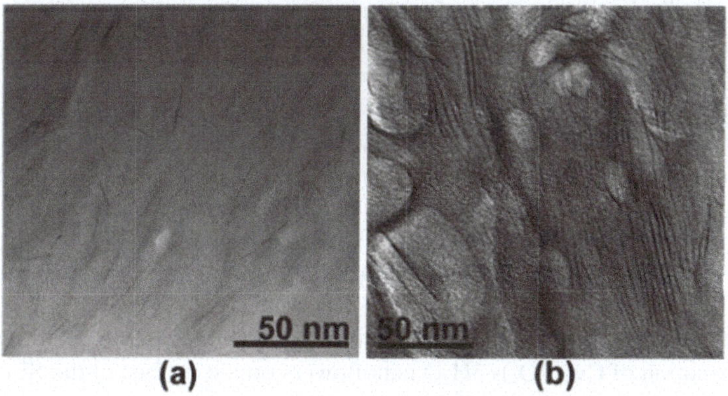

Fig. 2.4 TEM images of SPI/MMT plastics, (**a**) MS-8 and (**b**) MS-16. (Reprinted with permission from Chen and Zhang (2006), Copyright {2018} American Chemical Society)

The cross section of MWNT-incorporated nanocomposite sheets showed a surface with donut-like and raised blob-like objects, while the native SPI shows a fractured lamellae-like structure under scanning electron microscope (Zheng et al. 2007). Cross section of nano-SiO_2 sheet showed a heterogeneous morphology as compared to neat film (Ai et al. 2007). Interestingly, the SPI film containing 8 wt % nano-silica showed a heterogeneous morphology unlike 4% nano-silica-containing SPI film. The results from SEM confirmed that the 8 wt% nano-silica-containing SPI film have better reinforcement effect as compared to 4 wt%.

TEM image of the microstructure of SPI/MMT plastics is also reported and is shown in Fig. 2.4 (Chen and Zhang 2006). The image showed that the lower content of MMT diminished the dimension of silicate layer about 30 nm in length and 1 nm in thickness. This result indicated the highly *exfoliated* layered structure of MMT the soy protein molecules (Fig. 2.4a). On the other hand, *intercalated* tactoids with a *d*-spacing of about 6 nm were observed at higher contents of MMT in SPI (Fig. 2.4b).

2.6 Mode of Interactions Between Nanoparticles and Soy Protein

The interaction behavior of nanoparticle and SPI is a key factor for determining the film properties. Soy protein is a biomacromolecule that is made up of amino acids with different side groups; thus, its interaction with the added nanoparticles is quite obvious. For example, the hydroxyl groups of nano-SiOx provide the interacting sites for the formation of hydrogen bonds with each other, leading to the agglomeration of nano-SiOx and suggesting their strong hydrophilicity. This example indicates that the interaction behavior of nanoparticles and SPI strongly determines the overall properties of resulting film. MMT and carbon nanotubes (CNTs) are widely accepted reinforcing fillers for polymer matrices to attain high performance and exceptional functions (Moniruzzaman and Winey 2006). Some nanofillers such as chitin (Zheng et al. 2003), industrial lignin (Huang et al. 2003), layered silicate (Chen and Zhang 2006; Yu et al. 2007), carbon nanoparticles (Li et al. 2016), and carbon nanotube (Zhang et al. 2006) showed noticeable reinforcing effects in soy protein polymers upon blending. MMT-incorporated nanocomposites of SPI film can be intercalated and exfoliated. Exfoliated nanocomposites are widely used in fundamental research these days. These exfoliated nanocomposites can be formed by solution intercalation, melt intercalation, and in situ intercalative polymerization (Ray and Okamoto 2003).

One of the best examples of the interaction between nanoparticle and soy film is for the formation of $Cu_3(PO_4)_2{\cdot}3H_2O$ nanoflowers on the surface of the SPI film. As this work leads to a superhydrophobic and self-cleaning film, the interaction pattern of soy and nanoparticle must be interesting. For this interaction pattern, a four step mechanisms (Fig. 2.5) have been proposed by the researcher (Xie et al. 2016), which are as follows:

1. Absorption of Cu^{2+} by SPI with the help of coordination interaction
2. Formation of primary $Cu_3(PO_4)_2{\cdot}3H_2O$ nanocrystals in the presence of phosphate at the nucleation location where SPI interacted with Cu^{2+}
3. Enlargement of crystal size which appeared on surface as lamellar networks
4. Crimpling of lamella tip induced by repulsion between lamellae which eventually results in the formation of nanoflowers

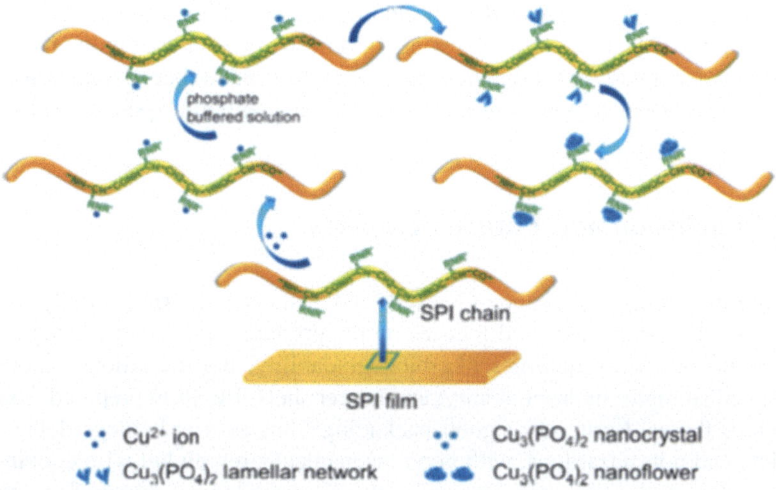

Fig. 2.5 Mechanism of soy film surface fabrication with $Cu_3(PO_4)_2{\cdot}3H_2O$ nanoflowers. (Reprinted with permission from Xie et al. (2016), Copyright {2018} American Chemical Society)

One of the major mechanisms of the interaction between soy and nanoparticle is the cross-linking reaction which takes place by hydrogen boding. One of the examples of this kind of interaction is soy protein incorporated with citric acid-modified starch nanoparticles (CSN). Tian and Xu prepared the nanoparticles with an average size of 82 nm and incorporated in SPI (Tian and Xu 2011). The cross-linking of CSN with soy protein took place by hydrogen bonding.

Carbon nanoparticles (CNPs) have benefits over other nanoparticles because of their water solubility, active functional groups on the surface, interfacial properties, good compatibility with SPI, and nontoxicity. As the CNPs are water soluble and also possess good interfacial effect, they showed better compatibility with soy protein. The nanoparticles got absorbed strongly in the film due to interfacial absorption. A large amount of -OH and -COOH is also present on the surface of CNPs. This enables CNPs to bind with the active side groups such as NH_2, OH, and COOH in SPI. This kind of interaction improved the overall performance of SPI film incorporated with CNPs (Li et al. 2016).

2.7 Antimicrobial Effects of Nanoparticles in Soy Protein Film

SPI films also contain some sort of antimicrobial properties, but they are vulnerable to microbial attack as their self-antimicrobial effects are not that much efficient. Silver nanoparticles are well known for their antimicrobial effects because they inhibit microbes by several mechanisms. The antimicrobial efficacy of soy protein incorporated with silver nanoparticles has been suggested. SPI film with silver nanoparticles is effective against both Gram-positive and Gram-negative bacteria. Zhang et al. (2013) used solution blowing to form soy protein-containing nanofibers which were decorated with silver nanoparticles. These nanofibers demonstrated significant antibacterial activity against *E. coli* colonies without exposure to UV light.

2.8 Conclusion and Future Prospects

In the present scenario, demand of packed food is increasing exponentially, and this hike in demand of packed food focuses on the production of good packaging film with certain necessary qualities like biodegradability; non-hazardous, impressive mechanical strength; hydrophobicity; and longer shelf-life. Film prepared from SPI suits in all the said criteria of good packaging film, as it is biodegradable, cost-effective, and non-hazardous, with good mechanical strength but with certain limitations such as hydrophilicity, and there is also room for improvement of mechanical strength.

From this chapter, we can conclude that the properties of film incorporated with functionalized and non-functionalized nanoparticles showed a bit different characteristics, as the functionalized nanoparticles like nano-SiO_2 and MMT showed better performance in terms of mechanical strength. The $Cu_3(PO_4)_2{\cdot}3H_2O$ nanoflower-incorporated SPI film showed superhydrophobic properties. Functionalized nanoparticles also increased the opacity of film.

On the contrary, non-functionalized nanoparticles, that is, carbon nanoparticle-incorporated SPI film, had less effect on the mechanical properties of film as well as the water resistance of the film (Moniruzzaman and Winey 2006). Recently, researchers have tried to incorporate other additives in presence of nanoparticles and SPI. In one of the study, eco-friendly and high-potential hybrid SPI-based nanocomposites were fabricated by using the hybrid of carboxymethylated chitosan (CMCS) and halloysite nanotubes (HNTs) in SPI and 1,2,3-propanetriol-diglycidyl-ether (PTGE) solution. In another study, the authors have used PVA instead of CMC to prepare hybrid nanocomposites (Liu et al. 2017b, c). The mechanical properties of as fabricated hybrid nanocomposites increased having potential to be used in the field of sustainable and environmentally friendly packaging.

References

Ai F, Zheng H, Wei M, Huang J (2007) Soy protein plastics reinforced and toughened by SiO_2 nanoparticles. J Appl Polym Sci 105(3):1597–1604

Brigger I, Dubernet C, Couvreur P (2012) Nanoparticles in cancer therapy and diagnosis. Adv Drug Deliv Rev 64:24–36

Briggs DR, Wolf WJ (1957) Studies on the cold-insoluble fraction of the water-extractable soybean proteins. I. Polymerization of the 11 S component through reactions of sulfhydryl groups to form disulfide bonds. Arch Biochem Biophys 72(1):127–144

Cao N, Fu Y, He J (2007) Mechanical properties of gelatin films cross-linked, respectively, by ferulic acid and tannin acid. Food Hydrocoll 21(4):575–584

Chen P, Zhang L (2005) New evidences of glass transitions and microstructures of soy protein plasticized with glycerol. Macromol Biosci 5(3):237–245

Chen P, Zhang L (2006) Interaction and properties of highly exfoliated soy protein/montmorillonite nanocomposites. Biomacromolecules 7(6):1700–1706

Chen P, Zhang L, Cao F (2005) Effects of moisture on glass transition and microstructure of glycerol plasticized soy protein. Macromol Biosci 5(9):872–880

Dawson PL, Hirt DE, Rieck JR, Acton JC, Sotthibandhu A (2003) Nisin release from films is affected by both protein type and film-forming method. Food Res Int 36(9–10):959–968

Echeverría I, Eisenberg P, Mauri AN (2014) Nanocomposites films based on soy proteins and montmorillonite processed by casting. J Membr Sci 449:15–26

Espitia PJP, Soares NDFF, dos Reis Coimbra JS, de Andrade NJ, Cruz RS, Medeiros EAA (2012) Zinc oxide nanoparticles: synthesis, antimicrobial activity and food packaging applications. Food Bioprocess Tech 5(5):1447–1464

Eswaranandam S, Hettiarachchy NS, Johnson MG (2004) Antimicrobial activity of citric, lactic, malic, or tartaric acids and nisin incorporated soy protein film against *Listeria monocytogenes*, *Escherichia coli* O157: H7, and *Salmonella gaminara*. J Food Sci 69(3):79–84

Garrido T, Leceta I, Cabezudo S, Guerrero P, de la Caba K (2016) Tailoring soy protein film properties by selecting casting or compression as processing methods. Eur Polym J 85:499–507

Gianessi LP, Carpenter JE (2000) Report on agricultural biotechnology: benefits of transgenic soybeans. National Center for Food and Agricultural Policy, Washington, DC. https://www.iatp.org/sites/default/files/Agricultural_Biotechnology_Benefits_of_Transge.pdf

Guo Y, Wang Z, Shao H, Jiang X (2013) Hydrothermal synthesis of highly fluorescent carbon nanoparticles from sodium citrate and their use for the detection of mercury ions. Carbon 52:583–589

Hettiarachchy NS, Kalapathy U (1998) Functional properties of soy proteins. ACS Symp Ser 708:80–95. https://doi.org/10.1021/bk-1998-0708.ch006

Huang J, Zhang L, Chen P (2003) Effects of lignin as a filler on properties of soy protein plastics. II. Alkaline lignin. J Appl Polym Sci 88(14):3291–3297

Kumar R, Zhang L (2009) Soy protein films with the hydrophobic surface created through non-covalent interactions. Ind Crop Prod 29(2–3):485–494

Kumar R, Choudhary V, Mishra S, Varma IK, Mattiason B (2002) Adhesives and plastics based on soy protein products. Ind Crop Prod 16(3):155–172

Kumar R, Wang L, Zhang L (2009) Structure and mechanical properties of soy protein materials plasticized by thiodiglycol. J Appl Polym Sci 111:970–977

Kumar R, Anandjiwala RD, Kumar A (2016) Thermal and mechanical properties of mandelic acid-incorporated soy protein films. J Therm Anal Calorim 123(2):1273–1279

Li Y, Chen H, Dong Y, Li K, Li L, Li J (2016) Carbon nanoparticles/soy protein isolate bio-films with excellent mechanical and water barrier properties. Ind Crop Prod 82:133–140

Liu D, Zhang L (2006) Structure and properties of soy protein plastics plasticized with acetamide. Macromol Mater Eng 291(7):820–828

Liu R, Liu D, Liu Y, Song Y, Wu T, Zhang M (2017a) Using soy protein SiOx nanocomposite film coating to extend the shelf life of apple fruit. Int J Food Sci Technol 52(9):2018–2030

Liu X, Kang H, Wang Z, Zhang W, Li J, Zhang S (2017b) Simultaneously toughening and strengthening soy protein isolate-based composites via carboxymethylated chitosan and halloysite nanotube hybridization. Materials 10(6):653. https://doi.org/10.3390/ma10060653

Liu X, Song R, Zhang W, Qi C, Zhang S, Li J (2017c) Development of eco-friendly soy protein isolate films with high mechanical properties through HNTS, PVA, and PTGE synergism effect. Sci Rep 7:44289. https://doi.org/10.1038/srep44289

Malathi AN, Kumar N, Nidoni U, Hiregoudar S (2017) Development of soy protein isolate films reinforced with titanium dioxide nanoparticles. Int J Agric Environ Biotechnol 10(1):141–148

Mo X, Sun X (2001) Thermal and mechanical properties of plastics molded from urea-modified soy protein isolates. J Am Oil Chem Soc 78(8):867–872

Mohanraj VJ, Chen Y (2006) Nanoparticles – a review. Trop J Pharm Res 5(1):561–573

Moniruzzaman M, Winey KI (2006) Polymer nanocomposites containing carbon nanotubes. Macromolecules 39(16):5194–5205

Nishinari K, Fang Y, Guo S, Phillips GO (2014) Soy proteins: a review on composition, aggregation and emulsification. Food Hydrocoll 39:301–318

Ou S, Kwok KC (2004) Ferulic acid: pharmaceutical functions, preparation and applications in foods. J Sci Food Agric 84(11):1261–1269

Padmavathy N, Vijayaraghavan R (2008) Enhanced bioactivity of ZnO nanoparticles - an antimicrobial study. Sci Technol Adv Mater 9(3):035004. https://doi.org/10.1088/1468-6996/9/3/035004

Pal SL, Jana U, Manna PK, Mohanta GP, Manavalan R (2011) Nanoparticle: an overview of preparation and characterization. J Appl Pharm Sci 1(6):228–234

Rajput N (2015) Methods of preparation of nanoparticles - a review. Int J Adv Eng Technol 7(6):1806–1811

Ray SS, Okamoto M (2003) Polymer/layered silicate nanocomposites: a review from preparation to processing. Prog Polym Sci 28(11):1539–1641

Rhim JW, Gennadios A, Handa A, Weller CL, Hanna MA (2000) Solubility, tensile, and color properties of modified soy protein isolate films. J Agric Food Chem 48(10):4937–4941

Saio K, Kamiya M, Watanabe T (1969) Food processing characteristics of soybean 11S and 7S proteins: part 1. Effect of difference of protein components among soybean varieties on formation of tofu-gel. Agric Bio Chem 33(9):1301–1308

Shi W, Dumont MJ (2014) Bio-based films from zein, keratin, pea, and rapeseed protein feedstocks. J Mater Sci 49(5):1915–1930

Sirelkhatim A, Mahmud S, Seeni A, Kaus NHM, Ann LC, Bakhori SKM, Mohamad D (2015) Review on zinc oxide nanoparticles: antibacterial activity and toxicity mechanism. Nano-Micro Lett 7(3):219–242

Sondi I, Salopek-Sondi B (2004) Silver nanoparticles as antimicrobial agent: a case study on *E. coli* as a model for gram-negative bacteria. J Colloid Interface Sci 275(1):177–182

Song F, Tang DL, Wang XL, Wang YZ (2011) Biodegradable soy protein isolate-based materials: a review. Biomacromolecules 12(10):3369–3380

Stoimenov PK, Klinger RL, Marchin GL, Klabunde KJ (2002) Metal oxide nanoparticles as bactericidal agents. Langmuir 18(17):6679–6686

Sun XS, Kim HR, Mo X (1999) Plastic performance of soybean protein components. J Am Oil Chem Soc 76(1):117–123

Swain SN, Rao KK, Nayak PL (2004) Biodegradable polymers. III. Spectral, thermal, mechanical, and morphological properties of crosslinked furfural–soy protein concentrate. J Appl Polym Sci 93(6):2590–2596

Tian H (2012) Processing and properties of soy protein/silica nanocomposites fabricated *in situ* synthesis. J Compos Mater 46:427–435

Tian H, Xu G (2011) Processing and characterization of glycerol-plasticized soy protein plastics reinforced with citric acid-modified starch nanoparticles. J Polym Environ 19(3):582–588

Tran QH, Le AT (2013) Silver nanoparticles: synthesis, properties, toxicology, applications and perspectives. Adv Nat Sci Nanosci Nanotechnol 4(3):033001. https://doi.org/10.1088/2043-6262/4/3/033001

Wang S, Zhang S, Jane J, Sue HJ (1995) Effects of polyols on mechanical-properties of soy-protein plastics. Polym Mater Sci Eng 209:57–58

Xiang A, Guo G, Tian H (2017) Fabrication and properties of acid treated carbon nanotubes reinforced soy protein nanocomposites. J Polym Environ 25(3):519–525

Xie WY, Song F, Wang XL, Wang YZ (2016) Development of copper phosphate nanoflowers on soy protein toward a superhydrophobic and self-cleaning film. ACS Sustain Chem Eng 5(1):869–875

Xu F, Dong Y, Zhang W, Zhang S, Li L, Li J (2015) Preparation of cross-linked soy protein isolate-based environmentally-friendly films enhanced by PTGE and PAM. Ind Crop Prod 67:373–380
Yu J, Cui G, Wei M, Huang J (2007) Facile exfoliation of rectorite nanoplatelets in soy protein matrix and reinforced bionanocomposites thereof. J Appl Polym Sci 104(5):3367–3377
Zhang J, Jiang L, Zhu L, Jane JL, Mungara P (2006) Morphology and properties of soy protein and polylactide blends. Biomacromolecules 7(5):1551–1561
Zhang Y, Lee MW, An S, Sinha-Ray S, Khansari S, Joshi B, Yoon SS (2013) Antibacterial activity of photocatalytic electrospun titania nanofiber mats and solution-blown soy protein nanofiber mats decorated with silver nanoparticles. Catal Commun 34:35–40
Zhang XF, Liu ZG, Shen W, Gurunathan S (2016) Silver nanoparticles: synthesis, characterization, properties, applications, and therapeutic approaches. Int J Mol Sci 17(9):1534–1538
Zheng H, Tan ZA, Ran Zhan Y, Huang J (2003) Morphology and properties of soy protein plastics modified with chitin. J Appl Polym Sci 90(13):3676–3682
Zheng H, Ai F, Wei M, Huang J, Chang PR (2007) Thermoplastic soy protein nanocomposites reinforced by carbon nanotubes. Macromol Mater Eng 292(6):780–788
Zhu H, Wang X, Li Y, Wang Z, Yang F, Yang X (2009) Microwave synthesis of fluorescent carbon nanoparticles with electrochemiluminescence properties. Chem Commun 34:5118–5120

Chapter 3
Formulation, Characterization, and Potential Application of Nanoemulsions in Food and Medicine

Ashutosh Bahuguna, Srinivasan Ramalingam, and Myunghee Kim

Contents

A. Bahuguna · S. Ramalingam · M. Kim (✉)
Department of Food Science and Technology, College of Life and Applied Sciences, Yeungnam University, Gyeongsan-si, Gyeongsangbuk-do, South Korea
e-mail: foodtech@ynu.ac.kr

D. Thangadurai et al. (eds.), *Nanotechnology for Food, Agriculture, and Environment*, Nanotechnology in the Life Sciences,
https://doi.org/10.1007/978-3-030-31938-0_3

3.1 Introduction

An emulsion is defined as the dispersion of two immiscible phases (Becher 2001). Based on their droplet size, stability, and appearance, emulsions can be classified into three main groups, namely, macroemulsions, nanoemulsions, and microemulsions. Conventional macroemulsions are thermodynamically unstable turbid suspension of droplet with sizes between 0.1 and 100 μm and with polydispersity index (PDI) >0.4. On the other hand, nanoemulsions have a droplet size between 5 and 200 nm and are kinetically stable and usually transparent in nature with PDI <0.1–0.2. In contrast, microemulsions are thermodynamically stable and transparent suspensions with a droplet size 1–100 nm and PDI <0.1 (Fernandez et al. 2004; Thakur et al. 2013; Gupta et al. 2016; Kadir 2017). Although nano- and microemulsions possess a similar droplet size, they differ in their basic compositions and preparation methods. One of the main differences is the amount of surfactant used; mostly nanoemulsions are prepared using a lower amount of surfactant (5–10% w/w) compared to that employed for microemulsions (>20% w/w). Due to low amount of surfactant, nanoemulsions are more attractive and demanding for industrial applications.

Nanoemulsions principally consist of three main components, i.e., water, oil, and a surfactant/emulsifying agent. The emulsifier is an important constituent since it not only allows to attain a small droplet size by decreasing the interfacial tension (surface energy) per unit area between the oil and water phase but also stabilizes the nanoemulsion by repulsive electrostatic interactions and steric hindrance. In recent years, nanoemulsions have gained enormous attention due to their wide applicability in various industries. Nanosized emulsions offer considerable advantages associated with their large surface area, which has an obvious effect on their bioavailability. In addition, they may serve as protective shields for the bioactive components encapsulated within and also improve the solubility of lipophilic compounds in aqueous media.

3.2 Types of Nanoemulsions

Nanoemulsions can be broadly classified into three major groups (Kriwet and Müller-Goymann 1995):

(i) Oil-in-water emulsions (O/W): the oil droplets are surrounded by the surfactant and ramify into the water phase (Fig. 3.1).
(ii) Water-in-oil emulsions (W/O): the water droplets are distributed throughout the continuous oil phase. These are commonly known as reverse micelles.
(iii) Bi-continuous nanoemulsions: microdomains of either water or oil are interspersed within the system.

3.3 Constituents of Nanoemulsions

As mentioned above, there are three main components of nanoemulsions, i.e., water, oil, and a surfactant/emulsifier (Liu et al. 1991; Kriwet and Müller-Goymann 1995; Trotta 1999). The different types of oils used for the preparation of nanoemulsions can be classified into two groups based on their viscosity, i.e., high or low, and include castor oil, corn oil, coconut oil, evening primrose oil, mineral oil, olive oil, peanut oil, triglycerides, and hexadecane. Mixing of oil in water yields a temporary emulsion, which separates into two distinct phases by coalescence. The addition of a surfactant to an oil-water suspension increases the stability of the biphasic system. Common emulsifiers used for the formation of nanoemulsions are spans, tweens, hydrophilic colloids like acacia, bentonite, and bee gum, as well as polysorbate 80

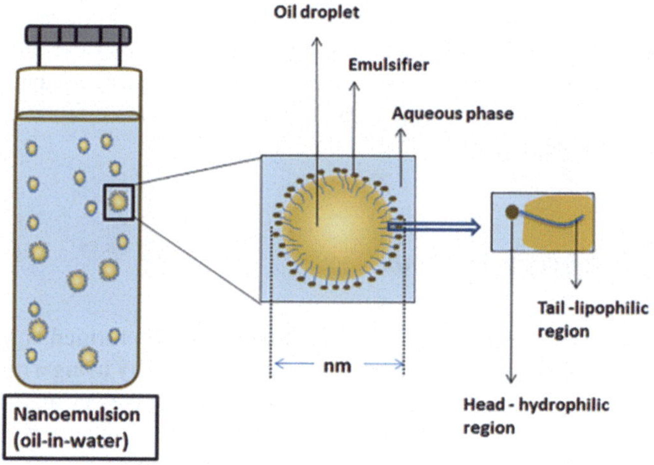

Fig. 3.1 Representation of an oil-in-water emulsion (O/W). (Figure reprinted with permission (Prakash et al. 2018) of the Elsevier Copyright Clearance Center (2018))

and sorbitan oleate. An ideal surfactant must possess the following properties (Jaiswal et al. 2015): (i) nontoxic in nature and chemically stable, (ii) able to reduce the surface tension below 10 dynes/cm, (iii) easily adsorbed in the dispersed phase to prevent coalescence, (iv) aiding to build a sufficient zeta potential (surface charge), and (v) effective at low concentrations.

The solubility of a particular type of surfactant in a given nanoemulsion can be evaluated in terms of hydrophilic-lipophilic balance (HLB), which highly depends on the concentration and phase volume of both water and oil (Myers 1999). Utilizing the solubility parameter (SP) value, HLB is used to characterize most oil phases. The value of HLB can be derived mathematically by using the equation:

$$\mathrm{HLB} = 4 \times \left[\frac{\mathrm{SP} + 7}{8} \right]$$

This equation gives an idea about selecting an optimal emulsifying agent as well as the stability of the emulsion. In general, a low HLB value is indicative of O/W emulsions, while a high HLB value is indicative of W/O emulsions. In other words, the HLB indicates the solubility of the surfactant in the oil and water phase. A low HLB suggests that the surfactant is more lipophilic, while a high HLB value suggests that it has a hydrophilic nature. In general, HLB values higher than 7 are typically suitable for W/O emulsions, while HLB values lower than 7 are suitable for O/W emulsions (Sajjadi et al. 2003; Tadros et al. 2004).

3.4 Methodology for the Preparation of Nanoemulsions

Nanoemulsions are non-equilibrated systems; therefore, they need energy and a surfactant to form. They can be formed either by high- or low-energy methods (Anton and Vandamme 2009) (Fig. 3.2). Traditionally, high-energy methods are employed; however, low-energy methods are also attractive due to several advantages such as the mild conditions employed, which lead to reduced adverse effects on the encapsulated bioactive components.

3.4.1 High-Energy Methods

Nanoemulsions are nonequilibrium systems that cannot be formed spontaneously, but need an external energy. Such energy is usually applied by using a high-pressure homogenizer, high shear stirring, and ultrasonication (Graves et al. 2005; Mason et al. 2006; Jafari et al. 2007). A device able to produce an input energy in the range from 10^8 to 10^{10} W/kg is required as external energy source. For the large-scale production of nanoemulsions, a high-pressure homogenizer is the most frequently

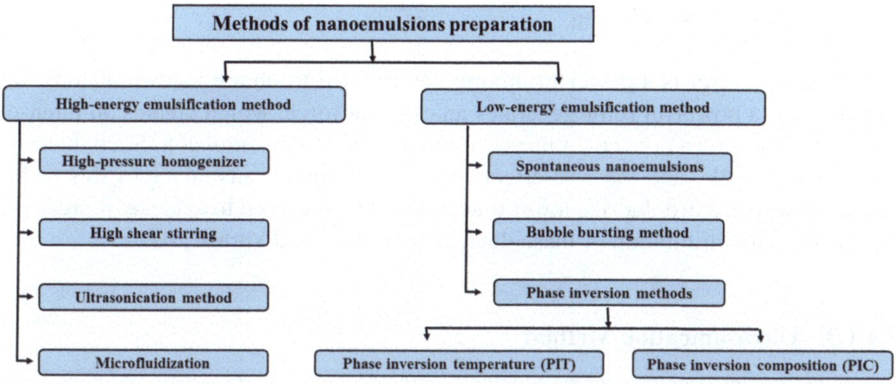

Fig. 3.2 Different high- and low-energy methods for the preparation of nanoemulsions

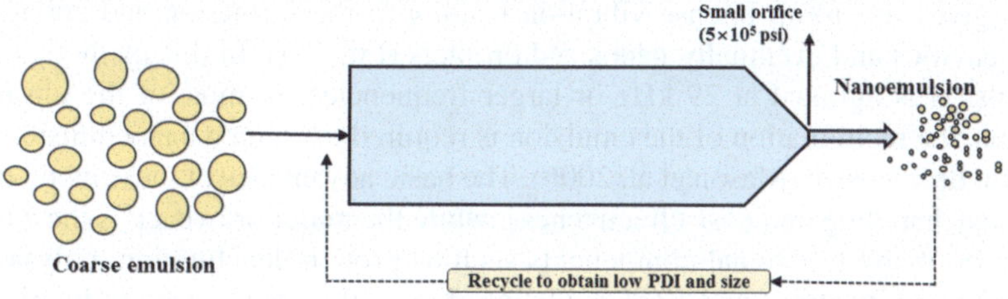

Fig. 3.3 Ray diagram representing a high-pressure homogenizer/piston homogenizer for the preparation of nanoemulsions

used device (Solans et al. 2005). However, in recent times, ultrasounds generated by a sonicator have emerged as a cost-effective substitute (Kaltsa et al. 2013).

3.4.1.1 High-Pressure Homogenizer

A high-pressure homogenizer or piston homogenizer is the most common device to produce nanosized emulsions with a droplet size up to 1.0 nm (Fig. 3.3). In a high-pressure homogenizer, a macroemulsion is forced to pass through a small orifice under a pressure in the range from 5×10^2 to 5×10^3 psi (Chime et al. 2014). During the process, forces like deep turbulence, hydraulic shear, and cavitation work collectively to form nanoemulsion. The desired size and homogeneity of a nanoemulsion are achieved by passing it many times through a high-pressure homogenizer. An ideal nanoemulsion consists of a monodispersed sample with a PDI lower than 0.08. Undoubtedly, a high-pressure homogenizer is the most effective device for nanoemulsion preparation. However, the basic disadvantage of this method is the generation of heat during the preparation (Setya et al. 2014), which may have an adverse effect on heat-labile components such as nucleic acids, proteins, and enzymes (Floury et al. 2000; Chime et al. 2014).

3.4.1.2 High Shear Stirring

A high shear stirrer is a powerful mixing device used to obtain nanoemulsions. In such device, a powerful rotor generates an external force, which creates an intense agitation that in turn converts a macroemulsion into a nanoemulsion. Such devices are easy to handle, and the cost-effectiveness is the major advantage of this technique. However, a droplet size lower than 200–300 nm is hard to achieve, representing the principal limitation of these devices (Koroleva and Yurtov 2012).

3.4.1.3 Ultrasonication Method

In the ultrasonication method, the external energy is generated by applying sound waves that create an intense agitation in the solution. The ultrasonication energy is provided by a sonicator probe/sonotrode at the contact point of the macroemulsion, which gives rise to an intense vibration leading to the formation and collapse of vapor cavities and eventually nanosized droplets (Fig. 3.4). In this method, a macroemulsion is agitated at 29 kHz or larger frequencies. In most of the ultrasonic systems, the recirculation of the emulsion is required to attain a nanoemulsion with uniform droplet size (Mason et al. 2006). The basic advantages of these methods are their easy handling and cost-effectiveness, while the major drawback is the adverse effects on many functional components such as protein denaturation, polysaccharide depolymerization, and lipid oxidation due to the sound waves (Jafari et al. 2007; McClements and Rao 2011).

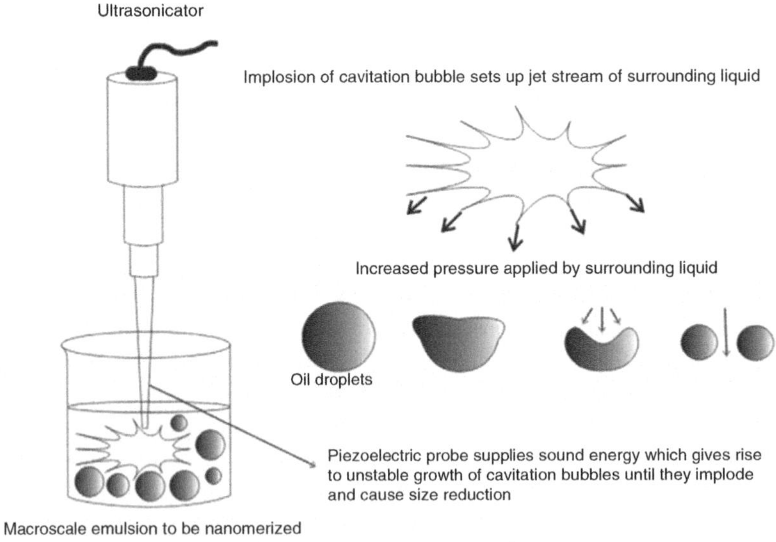

Fig. 3.4 Ultrasonication method for the preparation of nanoemulsions. (Figure reprinted with permission (Singh et al. 2017) of the Elsevier Copyright Clearance Center (2018))

3.4.1.4 Microfluidization

Microfluidization is the most widely accepted technique for the preparation of nanoemulsions in the pharmaceutical industry. In this technique, a macroemulsion is suspended and passed through a narrow bifurcated pipe at high pressure (150–650 MPa); it then collides in the interaction chamber, thus leading to the formation of a nanoemulsion (Fig. 3.5). A homogenous nanoemulsion of the desired size can be produced by repeating the process several times by applying different pressures (Chime et al. 2014; Jaiswal et al. 2015).

3.4.2 Low-Energy Methods

Homogenous small sized nanoemulsions can be prepared by using low-energy emulsification methods such as phase inversion component and phase inversion temperature (Fig. 3.2). These methods exploit the physiochemical properties of the system to generate the emulsion (Caldero et al. 2011). Although low-energy methods

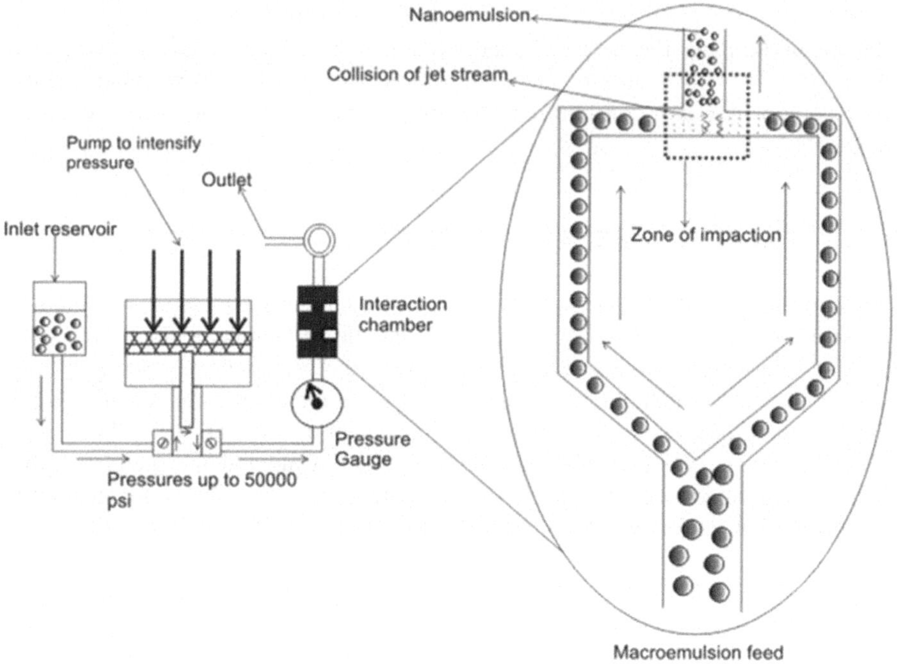

Fig. 3.5 Diagram representing micro-desulfurizer device for the preparation of nanoemulsions. (Figure reprinted with permission (Singh et al. 2017) of the Elsevier Copyright Clearance Center (2018))

are more effective in producing nanoemulsions as compared to high-energy ones, nonetheless, they have certain limitations as only definite types of components such as proteins and polysaccharides can be used. In order to overcome such issue, many synthetic surfactants are used.

3.4.2.1 Spontaneous Nanoemulsions

Spontaneous nanoemulsions are prepared without any external pressure at a constant temperature without phase transition of the system during emulsification (Solans and Sole 2012). This method depends on the interfacial tension, viscosity of the interfacial phase transition region, type of surfactant, and surfactant concentration. Such emulsions are commonly known as self-emulsifying drug delivery systems in the pharmaceutical industry. During the spontaneous nanoemulsion formation, the oil phase containing an aqueous soluble surfactant is mixed with water; upon movement of the water-dispersible substance from the oil to the water phase, an interfacial turbulence is generated, thereby resulting in the spontaneous formation of small oil droplets (McClements and Rao 2011).

3.4.2.2 Phase Inversion Method

This method utilizes the chemical energy that is released during the emulsification process. The varied degree of phase transition can be achieved by changing the basic composition at a constant temperature or by constant composition at varied temperature.

3.4.2.3 Phase Inversion Temperature (PIT)

In this method, emulsions are formed by varying the temperature while maintaining constant amount of the basic constituents. Nonionic surfactants with altering solubility in relation to temperature are required for the preparation of such types of emulsions. Thus, emulsions are formed by changing the temperature of the system, which in turn changes the solubility of the surfactant in water and oil (Chime et al. 2014). In order to prepare nanoemulsions using the PIT method, it is mandatory that the sample temperature reaches the PIT or HLB level, where the emulsification is promoted by an extremely low interfacial tension.

3.4.2.4 Phase Inversion Composition (PIC)

According to this methodology, emulsions are prepared at a constant temperature with a varying amount of either water, oil, water plus surfactant, or oil plus surfactant. The PIC method is more suitable for industrial applications as compared to PIT.

3.4.3 *Bubble Bursting Method*

In the bubble bursting method, gas is bubbled from the bottom of the water plus surfactant phase. As the bubbles reach the interphase of oil and water, they burst resulting in the spray of oil droplets in the water phase. A constant repetition of this process results in the formation of homogenous nanoemulsions (Gupta et al. 2016).

3.5 Firmness of Nanoemulsions

The stability of nanoemulsions is a key factor for their long-term storage and application, which highly depends on phenomena like coalescence, flocculation, sedimentation, creaming, and Ostwald ripening. The principal limiting factor to the nanoemulsion stability is the choice of surfactant, which makes the emulsion stable against heating, cooling, as well as changes of pH and ionic strength (McClements and Rao 2011). A surfactant can provide stability in various ways. For instance, an ionic surfactant imparts an electric charge to the nanodroplets present in the emulsion creating a sufficient repulsive force, which hinders the droplet aggregation. In contrast, a nonionic surfactant provides a steric barrier, thus preserving the structure of the nanoemulsion (Silva et al. 2012). Owing to their unique size, nanoemulsions display a great stability against coalescence, flocculation, sedimentation, or creaming. A small droplet within a nanoemulsion exhibits a Brownian movement rather than being subjected to the gravitational force, thus averting sedimentation or creaming (Klang et al. 2012). In addition, small droplets resist to adhesion due to the electronic repulsion and steric hindrance avoiding flocculation. The Ostwald ripening in which small droplets aggregate to form a larger one is the only serious concern regarding the stability of nanoemulsions. In the process of Ostwald ripening due to the variation of the droplet radius, a change in the chemical potential of the material inside the droplet occurs; consequently, the free energy of the system is reduced, leading to a decrease of the interfacial area; as a result, small droplets start to aggregate to form bigger ones, destabilizing the nanoemulsion. The rate of Ostwald ripening can be reduced by providing a kinetic barrier using less polar chain lipids. In addition, the use of high molecular weight oils, which are associated with the diffusion coefficient, can also be used to prevent the Ostwald ripening (Kadir 2017).

3.6 Characterization of Nanoemulsions

The characterization of nanoemulsions involves the assessments of their morphology, droplet size, charge (zeta potential), PDI, viscosity, refractive index, and thermodynamic stability.

3.6.1 Morphology

The morphology of nanoemulsions is often determined by scanning electron microscopy (SEM) and transmission electron microscopy (TEM) (Klang et al. 2012). SEM provides 3D images of the droplets, while TEM allows to obtain higher resolution images of the dispersed phase (Fig. 3.6).

3.6.2 Size

Dynamic light scattering (DLS) analysis is the most commonly employed technique for the examination of the droplet size of nanoemulsions, and this method relies on the fluctuations of the scattering intensity owing to the Brownian motion of particles (Jaiswal et al. 2015). In addition, electron microscopy can also be employed to analyze the droplet size (Alka et al. 2007; Farhan et al. 2008).

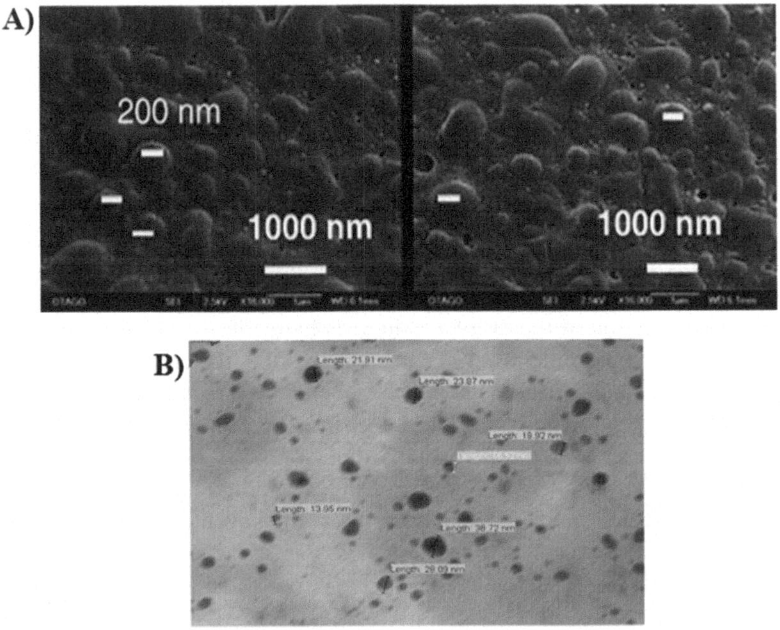

Fig. 3.6 Electron microscopic images of a nanoemulsion: (**a**) Cryo SEM image (Figure reprinted with permission (Saupe et al. 2006) of the Elsevier Copyright Clearance Center (2018)) and (**b**) TEM image. (Figure reprinted with permission (Thomas et al. 2017) of the Elsevier Copyright Clearance Center (2018))

3.6.3 Zeta Potential (Surface Charge)

The zeta potential denotes the charge on the droplets present within a nanoemulsion and can be determined by using a zeta particle size analyzer (Erol and Hans-Hubert 2005). Zeta potential is directly associated with the stability of nanoemulsion, indicating a similar charge repulsion between adjacent particles within the suspension. It can fall in both negative and positive ranges, which highly depends on the nature of the surfactant, electrolyte concentration, size, morphology of the particles, and pH of the solution. When the zeta potential value is low (either with negative or positive charges), the attraction between particles increases resulting in sedimentation or flocculation. A zeta potential value between 0 and ± 30 mV indicates instability of the nanoemulsion. In order to achieve the formation of a stable nanoemulsion, the zeta potential value must be higher than ±30 mV (Silva et al. 2012).

3.6.4 Polydispersity Index (PDI)

The PDI, which represents the uniformity of the droplet size within an emulsion, can be evaluated by DLS analysis. A higher value of PDI suggests heterogeneity of the system. In contrast, a lower PDI value indicates a uniformity of the droplets within the emulsion. The numerical value of PDI is calculated by the ratio of the mean standard deviation of the droplet size to the average droplet diameter (Jaiswal et al. 2015). The PDI also gives an insight on the stability of the formed nanoemulsion. The numeric value of PDI falls in the range from 0 (perfectly uniform preparation) to 1 (highly heterogeneous preparation). PDI values below 0.3 are acceptable for free drug delivery applications (Danaei et al. 2018).

3.6.5 Viscosity

Viscosity is one of the important parameters to evaluate the nanoemulsion stability and release of the active components. It can be measured by a Brookfield-type rotary viscometer at different temperatures and shear rates.

3.6.6 Refractive Index

The chemical stability of nanoemulsions at different storage time points is determined by measuring the refractive index using an Abbe refractometer.

3.6.7 *Dye Test*

A dye test is used to assess the successful preparation of O/W or W/O nanoemulsions. When a water soluble dye is added to an O/W nanoemulsion, it uniformly colors such emulsion. In contrast, in W/O emulsions, the dye imparts the color only to the dispersed phase, which can be easily detected by microscopic analysis (Jaiswal et al. 2015).

3.6.8 *Thermodynamic Stability*

The thermodynamic stability of nanoemulsions is characterized through heating and cooling cycles, centrifugation, and repeated freeze and thaw cycles (Kazimiera et al. 2009).

3.6.8.1 Heating and Cooling Cycles

Six alternating cycles of heating (45 °C) and cooling (4 °C) are performed, and finally the nanoemulsion is examined for any change in size and viscosity. An ideal nanoemulsion resists against any changes during this process.

3.6.8.2 Centrifugation

A nanoemulsion is centrifuged at 3500 rpm for 20 min and further examined for any sedimentation and phase separation. An ideal nanoemulsion does not show any changes in its structure after centrifugation.

3.6.8.3 Freeze and Thaw Cycles

Finally, to confirm the stability of nanoemulsions, freeze and thaw cycles are carried out. The nanoemulsion is stored at a freezing temperature (−21 °C) for 24 h followed by thawing at room temperature. Several cycles of freezing and thawing are repeated, and the emulsion is examined for any changes. An ideal nanoemulsion is stable during this process and does not undergo any changes related to size, shape, viscosity, and phase separation.

3.7 Application of Nanoemulsions

Nanotechnology is emerging as a large-scale business due to its wide application in various fields. Its overall market reached 1 trillion USD in 2015, and it is constantly growing at a rapid pace (Neethirajan and Jayas 2011). The application of nanoemulsions is also gaining popularity in various areas such as the cosmetic, medical, drug delivery, and food sectors.

3.7.1 *Antimicrobial Nanoemulsions*

Antimicrobial nanoemulsions are constituted of O/W droplets that exhibit broad-spectrum activities against various Gram-positive and Gram-negative bacteria, fungi like *Candida*, dermatophytes, viruses like herpes simplex, and spores such as anthrax. Such nanoemulsions are thermodynamically driven to fuse with a lipid-containing organism. Owing to its unique size, surface charge, and formulation, nanoemulsions facilitate passive transport through cell membrane, fusion with phospholipid bilayer, and partition in aqueous phase. Additionally, an electrostatic interaction between the cationic charges of the nanoemulsion and the anionic charges of a pathogen facilitates a direct association of emulsions with pathogen (Donsì and Ferrari 2016) (Fig. 3.7). When a nanoemulsion is adequately fused with the pathogen, a part of the energy is released, which together with the entrapped active ingredient damages the pathogen lipid membrane leading to cell death (Patil and Bhutkar 2016).

Nanoemulsions prepared by using citral as functional ingredient displayed a wide spectrum of antimicrobial potency against a group of Gram-positive and Gram-negative bacteria (Lu et al. 2018).

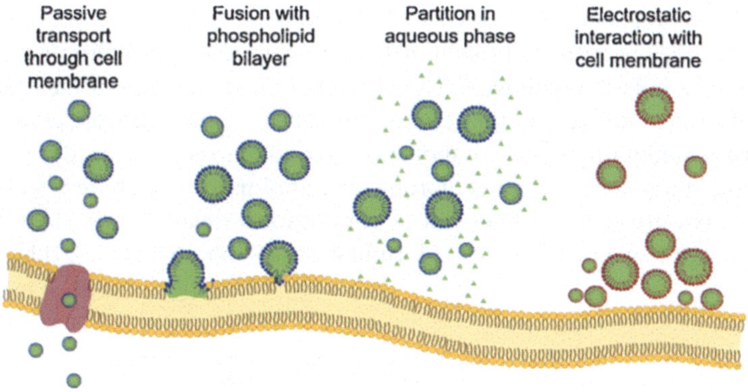

Fig. 3.7 Various routes of nanoemulsions interaction with microbial cell membrane. (Figure reprinted with permission (Donsì and Ferrari 2016) of the Elsevier Copyright Clearance Center (2018))

In order to achieve spore eradication, the addition of a spore germinator or direct sporicidal substance is added to the nanoemulsions. Once the spores are germinated by the spore germinator, the active ingredient present in the nanoemulsion easily attacks the pathogen and kills it. A direct sporicidal effect of nanoemulsions BCTP and BCTP401 against the spores of *Bacillus* species has been successfully documented (Hamouda et al. 1999).

3.7.2 Preventive Agents Against Bioterrorism

Many reports describe nanoemulsions as prophylactic agents to prevent biological attacks against anthrax and Ebola. A broad-spectrum nanoemulsion has been tested against the spores of anthrax by the US Army in 1999. It was further scrutinized by RestOps in 2001 as a chemical decontamination agent. A similar technology was tested against *Clostridium botulinum* spores. The company NanoStat™ (Nanobio Corp.) marketed a nanoemulsion in the form of liquid, spray, and foam to decontaminate a large number of materials (Subhashis et al. 2011).

3.7.3 Disinfectant Cleaners

Nanoemulsions have been employed to develop nontoxic disinfectant cleaners. Nanoemulsion-based disinfectants have been found to be effective against a large spectrum of viruses, bacteria, and fungi without posing any hazard to the environment. A nanoemulsion is made of nanospheres; therefore, only a small amount of active ingredient such as parachlorometaxylenol is required for the preparation of an effective disinfectant.

The mode of action of such disinfectants is based on the electrostatic attraction between the microspheres (cations) and the anionic microbial surface, followed by the release of the encapsulated functional component (e.g., parachlorometaxylenol) that effectively kills the pathogen. The efficiency of such types of disinfectants is based on the direct delivery of the active components to the pathogen; consequently, a reduced amount of functional components is required to achieve the desired results. For example, the amount of parachlorometaxylenol used in such disinfectants was found to be two times lower than that of conventional disinfectants, suggesting the high efficacy of nanoemulsion-based formulations (Subhashis et al. 2011).

3.7.4 Nanoemulsions in Delivery of Vaccine

Nanoemulsions have proved effective to deliver inactivated viruses to a target and serve as suitable vectors. Many recent studies showed that HIV infects the mucosal immune system. Hence, developing mucosal immunity might provide protection against the HIV infection. Oil-based nanoemulsions are excellent vehicles for the interaction with the mucosal layer due to their unique composition. Many nanoemulsion-based vaccines against influenza virus and HIV are under clinical trials. The nanoemulsion transfers the antigen (inactivated pathogen) to the mucosal surface, where it is taken by the nearby antigen-presenting cells. This leads to the stimulation of the humoral immune system upon activation of specific B lymphocytes that further develop a memory and thus boost the immune system for future challenges. In many studies, the successful use of nanoemulsions for the prevention of influenza infections has been shown by a single mucosal exposure of the nanoemulsions mixed with the virus. In some recent studies, the nasal mucosa exposure of nanoemulsions comprising recombinant gp120 antigen showed a great promise as anti-HIV vaccine. Many clinical trials are currently underway aiming to the development of vaccines against anthrax and hepatitis B infections (Patil and Bhutkar 2016).

3.7.5 Nanoemulsions for Oral Delivery

Nanoemulsions provide several advantages over traditional formulations by enhancing the absorption potency and decreasing drug toxicity. In particular, oil and water emulsion is suitable for enhancing the bioavailability of the hydrophobic drug, thus having a tremendous effect on the potency. Nanoemulsions have been proved to be effective for the oral delivery of steroids, hormones, diuretics, and antibiotics. In the case of the drug primaquine, a 25% lower dose was required against *Plasmodium berghei* infection when it was delivered as nanoemulsion compared to the conventional type. Furthermore, a higher content (45%) of the drug was detected in the liver of tested mice treated with nanoemulsion as compared to the plain drug, suggesting an enhanced bioavailability and accumulation imposed in the form of the nanoemulsion (Mishra et al. 2014).

3.7.6 Nanoemulsions in Cell Culture

Cell culture technology is used for many in vitro assays to produce various important biological substances including antibodies, hormones, and recombinant proteins. To optimize the cell growth, many components are needed in the culture media. However, it is nearly impossible to provide any oil-soluble substance to the media due to its low solubility in water. Nanoemulsions represent a way to supplement lipophilic substances to the media, which can be further absorbed by the cells. The advantages of using nanoemulsions in cell culture technology include an enhanced uptake of oil-soluble components, which leads to an improved cellular growth, that eventually affects the production of the desired products such as antibodies or enzymes and proteins (Patil and Bhutkar 2016).

3.7.7 Nanoemulsions as Ocular Drug Delivery

Conventional eye drops suffer from poor bioavailability and thus reduced therapeutic response, owing to the lacrimal secretion and nasolacrimal drainage occurring in the eyes (Sieg and Robinson 1977). Most of the drug is drained away without absorption in the eyes; as a result, a high amount or repeated applications of the medication are required to achieve the desired effect. In order to improve the absorption efficiency of drug, different methodologies can be employed. For instance, nanoemulsions provide an advantage of stimulating the penetration of the drug into the deeper layer of the ocular structure. An available nanoemulsion for the antiglaucoma, containing dorzolamide hydrochloride, has been shown to enhance the bioavailability by slowing down the release of the drug, thus affording a prolonged effect compared to conventional eye drops (Hussein et al. 2009).

3.7.8 Nanoemulsions as Transdermal Drug Delivery System

The delivery of a drug through the skin to the circulation represents a convenient and effective method, which offers several advantages such as a steady release of the drug for a prolonged time, self-administration in contrast to the parental route, and elimination of the drug input at any time by simple removal of the transdermal patch. Furthermore, it allows to avoid the gastrointestinal route associated with oral delivery. Transdermal drugs have been developed against many diseases such as anxiety, depression, and Alzheimer's and Parkinson's diseases. A drug can penetrate through the skin either via the hair follicles, via sweat ducts, or directly across the stratum corneum. For an effective drug absorption, these barriers must be crossed. Nanosized emulsions are capable of crossing the pores of the skin and thus transferring their active components into the circulation (Ravi and Padma 2011). Caffeine is a common oral drug for the treatment of different types of cancer. A nanoemulsion containing caffeine showed a greater promise as a transdermal delivery system compared to the conventional ones in many in vitro studies (Shakee and Ramadan 2010). Similar to caffeine, an indomethacin-based nanoemulsion showed a great promise as a transdermal delivery system against inflammatory-related diseases (Barakat et al. 2011).

3.7.9 Nanoemulsions in Cancer Therapy and Targeted Drug Delivery

Another important application of nanoemulsions is in the controlled drug delivery and targeting. To cure cancer, magnetic nanoemulsion poses a new possibility which can deliver Foscan (photosensitizer) inside the tissue layer. A hyperthermic condition subsequently generated the huge amount of reactive oxygen species form Foscan that effectively kills the tumor cells. This approach of nanoemulsion can be utilized for cancer treatment in conjunction with photodynamic therapy (Primo et al. 2007).

3.7.10 Nanoemulsions in Parenteral Drug Delivery

Parenteral delivery is the most common route for the administration of drugs with low bioavailability and narrow therapeutic index. The administration of a parenteral drug in the form of a nanoemulsion may protect the active component from the enzymatic degradation and permit the release of the drug in a sustainable manner for a longer time. The parental drug chlorambucil, which is used against breast and ovarian cancer, was delivered more effectively within a nanoemulsion compared to its free form (Ganta et al. 2010). The effectiveness of parenteral nanoemulsions for the delivery of drugs such as dexamethasone, diazepam, propofol, flurbiprofen, etomidate, and prostaglandin E1 has been well documented in the literature (Chime et al. 2014).

3.7.11 Nanoemulsions in Cosmetics

In case of cosmetics, the main issues are the uneven distribution and absorption of the bioactive component through the skin layer. These drawbacks can be overcome by using a nanoemulsion due to its small droplet size and high surface area, which eventually improve the absorption. The role of O/W nanoemulsions in the improvement of the permeability and skin whitening properties of ellagic acid has been demonstrated (Zhang et al. 2018).

3.7.12 Nanoemulsions for Nutraceuticals

Most of the natural compounds particularly belonging to essential oil displayed a great promise as a therapeutic agent. However, their physicochemical instability is a fundamental concern for their industrial application. Nanoemulsions can effectively overcome with this issue through the encapsulation of the active components. Physicochemical instability and functional behavior of carvacrol, an essential oil of thyme, have been improved by utilizing it in the form of oil-in-water nanoemulsion (Khan et al. 2018). Similarly, the efficacy of a nanoemulsion to increase the anti-inflammatory activity of curcumin compared to the free form has been documented (Patil and Bhutkar 2016).

3.7.13 Nanoemulsions in Food

Food quality, preservation, and safety are major concerns of the food industry. Approximately, more than 400 nano-based companies have been dealing in the food sector up to 2012 (Silva et al. 2012). In the food industry, different types of chemical substances are used to inactivate some enzymes and the microbial population that decreases the shelf life of food materials. However, occasionally such chemical compounds resulted to be corrosive and ineffective, particularly for the foods with high organic loads. Furthermore, their long-term consumption can also be associated with toxicological implications (Amaral and Bhargava 2015). Presently, most of the food industries focused on the use of natural compounds, which are generally regarded as safe for human consumption. In particular, essential oils have gained great attention as food preservatives owing to their broad antifungal, antibacterial, and antiviral activities (Burt 2004). Moreover, many components of essential oils possess antioxidant (Misharina and Samusenko 2008), antidiabetic (Tahir et al. 2016), and anti-inflammatory (Hajhashemi et al. 2003) properties, thus providing additional beneficial functions to the supplemented food. Although essential oil emerges as an alternative to the chemical method, but still there are many limitations of using it in the food, including their high volatility, strong odor, and low water solubility (Liang et al. 2012). However, some of these issues can be overcome by using essential oils in the form of nanoemulsions. Although in the preparation of nanoemulsions for food, a great precaution has to be taken by utilizing the only food, grade surfactant and oil. There are many successful examples showing the utility of nanoemulsion in foods. A study demonstrated the effective role of an oregano oil-based nanoemulsion to protect lettuce against *Listeria monocytogenes*, *Salmonella typhimurium*, and *Escherichia coli* (Bhargava et al. 2015). The protection of broccoli and radish seed by employing a carvacrol nanoemulsion has also been demonstrated by Landry et al. (2015). In addition, the same emulsion protected mung bean and alfalfa seeds against food pathogenic microbes *S. typhimurium* and *E. coli* (Landry et al. 2014).

Another study carried out in plumps highlighted the role of a lemongrass oil-based nanoemulsion to provide protection against *S. typhimurium* and *E. coli* (Kim et al. 2013). No adverse effects were detected with regard to the flavor and glossiness of these essential oil-supplemented nanoemulsions. However, a decreased ethylene production was observed, which prolonged the ripening process.

Demand of minimally processed foods is growing nowadays owing to its high nutritive value and absence of any harsh physical and chemical treatment. Essential oil-based nanoemulsions emerged as effective natural preservative agents for minimally processed fruits and vegetables. There are many successful reports elucidating the potential of essential oil nanoemulsions in fresh and minimally processed foods. A preservative role of oregano oil (Bhargava et al. 2015) and carvacrol/eugenol (Ruengvisesh et al. 2015) nanoemulsions for fresh lettuce and spinach leaves, respectively, has been well documented. Besides, few studies demonstrated a successful application of edible film coating of nanoemulsions in minimally processed food system (Prakash et al. 2018). In one such study, sodium alginate coating supplemented with lemongrass nanoemulsion demonstrated inhibition of *E. coli* and extension of the self-life of apple (Prakash et al. 2018). Many essential oil-based nanoemulsions have been proved effective against the formation of biofilm. Inhibitory role of *Eucalyptus globulus* essential oil nanoemulsion against the biofilm formation of *Candida* species has been well documented (Quatrin et al. 2017). The role of *Citrus medica* L. var. *sarcodactylis* essential oil-supplemented nanoemulsion against biofilm of *Staphylococcus aureus* was noticed. Similar inhibitory effect of cumin oil nanoemulsion was observed against the biofilm production of *E. coli* and *S. aureus* (Amrutha et al. 2017). Interestingly, an enhanced anti-biofilm activity of essential oil was noticed in the form of nanoemulsion compared to their free form (Lou et al. 2017).

Encapsulation and delivery of a bioactive functional component are two of the major objectives of using nanoemulsions in the food sector. Nanoemulsions can be used in the food industry to supplement the ingredients that possess low solubility in water, e.g., β-carotene (Gupta et al. 2016). In many studies, it has been shown that nanoemulsions allow easy digestion of many encapsulated food supplements. In particular, the use of curcumin in the form of a nanoemulsion led to easy and high digestion compared to its free form (Gupta et al. 2016).

Many functional components such as lycopene, lutein, β-carotene, coenzyme Q10, omega-3, vitamins A, D3, and E, isoflavones, and phytosterols have been encapsulated in nanoemulsions and successfully used in beverages by Nutralean. The scientist of Nutralean claimed an improved bioavailability of the encapsulated compounds in the form of the nanoemulsion (Silva et al. 2012). Furthermore, it was also observed that nanoemulsions preserve and protect the flavor of compounds from enzymatic hydrolysis as well as temperature and pH variations.

An improved stability of functional components (e.g., DL-α-tocopheryl acetate, omega-3 fatty acids, coenzyme Q10, and vitamins A, D, D3, E, and K) and coloring agents (e.g., chlorophyll, β-carotene, curcumin, lutein, apocarotenal, and sweet pepper extract) in beverages was demonstrated by Aquanova (Silva et al. 2012).

The company Unilever produced a healthier ice cream by using a nanoemulsion with the aim of reducing the fat content from 16% to 1%. It was reported that the reduction of the fat content did not change the taste of the ice cream (Silva et al. 2012).

3.8 Summary and Conclusion

Nanoemulsions offer several advantages due to their nanosize, enhanced surface area, and the possibility to be formulated as foams, liquids, and sprays. Additionally, they provide an ideal way for the dissolution and administration of hydrophobic substances. Currently, nanoemulsions are used in various areas of medicine and food sector and have been shown to have a tremendous potential by enhancing the bioavailability and shelf life of the encapsulated components. Continuous research is ongoing to improve the stability and functional properties of nanoemulsions in order to enhance their applicability. In food system, applicability of nanoemulsions still poses challenges both in terms of production cost and safety. This issue should be addressed in the future to accept it the terms of food safety.

References

Alka AJA, Baboota S, Shakeel F, Shafiq S (2007) Design development and evaluation of novel nanoemulsion formulations for transdermal potential of Celecoxib. Acta Pharma 57:315–332

Amaral DMF, Bhargava K (2015) Essential oil nanoemulsions and food applications. Adv Food Technol Nutr Sci 4:84–87

Amrutha B, Sundar K, Shetty PH (2017) Spice oil nanoemulsions: potential natural inhibitors against pathogenic *E. coli* and *Salmonella* spp. from fresh fruits and vegetables. LWT-Food Sci Technol 79:52–159

Anton N, Vandamme T (2009) The universality of low-energy nano-emulsification. Int J Pharm 377(1–2):42–147

Barakat N, Fouad E, Elmedany A (2011) Formulation design of indomethacin-loaded nanoemulsion for transdermal delivery. Pharmaceutica Analytica Acta S2:002. https://doi.org/10.4172/2153-2435.S2-002

Becher P (2001) Emulsions: theory and practice. Oxford University Press, New York

Bhargava K, Conti DS, da Rocha SR, Zhang Y (2015) Application of an oregano oil nanoemulsion to the control of foodborne bacteria on fresh lettuce. Food Microbiol 47:69–73

Burt S (2004) Essential oils: their antibacterial properties and potential applications in foods-a review. Int J Food Microbiol 94:223–253

Caldero G, Maria JGC, Solans C (2011) Formation of polymeric nano-emulsions by a low energy method and their use for nanoparticle preparation. J Colloid Interface Sci 353:406–411

Chime SA, Kenechukwu FC, Attama AA (2014) Nanoemulsions-advances in formulation, characterization and applications in drug delivery. In: Sezer D (ed) Application of nanotechnology in drug delivery. InTech, Croatia, pp 77–126

Danaei M, Dehghankhold M, Ataei S, Hasanzadeh Davarani F, Javanmard R, Dokhani A, Khorasani S, Mozafari MR (2018) Impact of particle size and polydispersity index on the clinical applications of lipidic nanocarrier systems. Pharmaceutics 10(2):57. https://doi.org/10.3390/pharmaceutics10020057

Donsì F, Ferrari G (2016) Essential oil nanoemulsions as antimicrobial agents in food. J Biotechnol 233:106–120

Erol Y, Hans-Hubert B (2005) Design of a phytosphingosine containing, positively-charged nanoemulsion as a colloidal carrier system for dermal application of ceramides. Eur J Pharm Biopharm 60(1):91–98

Farhan AJ, Mushir A, Faiyaz S, Cushman T, Roop KK, Sheikh S (2008) Investigation of nanoemulsion system for transdermal delivery of domperidone: *ex vivo* and *in vivo* studies. Curr Nanosci 4(4):381–390

Fernandez P, Andre V, Rieger J, Kühnle A (2004) Nano-emulsion formation by emulsion phase inversion. Colloids Surf A Physicochem Eng Asp 251:53–58

Floury J, Desrumaux A, Lardieres J (2000) Effect of high-pressure homogenization on droplet size distributions and rheological properties of model oil-in-water emulsions. Innov Food Sci Emerg Technol 1(2):127–134

Ganta S, Deshpande D, Korde A, Amiji M (2010) A review of multifunctional nanoemulsion systems to overcome oral and CNS drug delivery barriers. Mol Membr Biol 27(7):260–273

Graves S, Meleson K, Wilking J (2005) Structure of concentrated nanoemulsions. J Chem Phys 122(13):134703. https://doi.org/10.1063/1.1874952

Gupta A, Eral HB, Hatton TA, Doyle PS (2016) Nanoemulsions: formation, properties and applications. Soft Matter 12:2826–2841

Hajhashemi V, Ghannadi A, Sharifa B (2003) Anti-inflammatory and analgesic properties of the leaf extracts and essential oil of *Lavandula angustifolia* Mill. J Ethnopharmacol 89(1):67–71

Hamouda T, Hayes MM, Cao Z, Tonda R, Johnson K, Wright DC, Brisker J, Baker JR (1999) A novel surfactant nanoemulsion with broad-spectrum sporicidal activity against *Bacillus* species. J Infect Dis 180:1939–1949

Hussein OA, Salama HA, Ghorab M, Mahmoud AA (2009) Nanoemulsion as a potential ophthalmic delivery system for dorzolamide hydrochloride. AAPS Pharm Sci Tech 10(3):808–819

Jafari SM, He Y, Bhandari B (2007) Optimization of nanoemulsion production by microfluidization. Eur Food Res Technol 225(5–6):733–741

Jaiswal M, Rupesh Dudhe R, Sharma PK (2015) Nanoemulsion: an advanced mode of drug delivery system. 3Biotech 5:123–127

Kadir C (2017) Review on nanoemulsions: preparation methods and stability. Trakya University J Eng Sci 18(1):73–83

Kaltsa O, Michon C, Yanniotis S, Mandala I (2013) Ultrasonic energy input influence on the production of sub-micron o/w emulsions containing whey protein and common stabilizers. Ultrason Sonochem 20(3):881–891

Kazimiera A, Katarzyna Z, Agnieszka H, Adam J (2009) Biocompatible nanoemulsions of dicephalicaldonamide-type surfactants: formulation, structure and temperature influence. J Colloid Interface Sci 334:87–95

Khan I, Bahuguna A, Kumar P, Bajpai VK, Kang SC (2018) *In vitro* and *in vivo* antitumor potential of carvacrol nanoemulsion against human lung adenocarcinoma A549 cells via mitochondrial mediated apoptosis. Sci Rep 8:144. https://doi.org/10.1038/s41598-017-18644-9

Kim IH, Lee H, Kim JE, Song KB, Lee YS, Chung DS, Min SC (2013) Plum coatings of lemongrass oil-incorporating carnauba wax-based nanoemulsion. J Food Sci 78:E1551–E1559

Klang V, Matsko N, Valenta C, Hofer F (2012) Electron microscopy of nanoemulsions: an essential tool for characterization and stability assessment. Micron 43:85–103

Koroleva MY, Yurtov EV (2012) Nanoemulsions: the properties, methods of preparation and promising applications. Russ Chem Rev 81(1):21–43

Kriwet K, Müller-Goymann CC (1995) Diclofenac release from phospho-lipid drug systems and permeation through excised human stratum corneum. Int J Pharm 125(2):231–242

Landry KS, Chang Y, McClements DJ, McLandsborough L (2014) Effectiveness of a novel spontaneous carvacrol nanoemulsion against *Salmonella enterica* Enteritidis and *Escherichia coli* O157: H7 on contaminated mung bean and alfalfa seeds. Int J Food Microbiol 187:15–21

Landry KS, Micheli S, McClements DJ, McLandsborough L (2015) Effectiveness of a spontaneous carvacrol nanoemulsion against *Salmonella enterica* Enteritidis and *Escherichia coli* O157: H7 on contaminated broccoli and radish seeds. Food Microbiol 51:10–17

Liang R, Xu S, Shoemaker CF, Li Y, Zhong F, Huang Q (2012) Physical and antimicrobial properties of peppermint oil nanoemulsions. J Agric Food Chem 60:7548–7555

Liu P, Kurihara-Bergstrom T, Good WR (1991) Cotransport of estradiol and ethanol through human skin *in vitro*: understanding the permeant/enhancer flux relationship. Pharm Res 8:938–944

Lou Z, Chen J, Yu F, Wang H, Kou X, Ma C, Zhu S (2017) The antioxidant, antibacterial, antibiofilm activity of essential oil from *Citrus medica* L. var. *sarcodactylis* and its nanoemulsion. LWT – Food Sci Technol 80:371–377

Lu WC, Huang DW, Wang C, Yeh CH, Tsai JC, Huang YT, Li PH (2018) Preparation, characterization, and antimicrobial activity of nanoemulsions incorporating citral essential oil. J Food Drug Anal 26:82–89

Mason TG, Graves SM, Wilking JN, Lin MY (2006) Extreme emulsification: formation and structure of nanoemulsions. Condens Matter Phys 9(1):193–199

McClements DJ, Rao J (2011) Food-grade nanoemulsions: formulation, fabrication, properties, performance, biological fate, and potential toxicity. Crit Rev Food Sci Nutr 51:285–330

Misharina TA, Samusenko AL (2008) Antioxidant properties of essential oils from lemon, grapefruit, coriander, clove, and their mixtures. Appl Biochem Microbiol 45(4):438–442

Mishra RK, Soni GC, Mishra RP (2014) A review article: on nanoemulsion. World J Pharm Pharm Sci 3(9):258–274

Myers D (1999) Surfaces, interfaces, and colloids: principles and applications. Wiley-VCH, New York

Neethirajan S, Jayas D (2011) Nanotechnology for the food and bioprocessing industries. Food Bioprocess Tech 4(1):39–47

Patil PA, Bhutkar BR (2016) Biomedical application of nanoemulsion-A features review. Int J Res Methods 3:37–58

Prakash A, Baskaran R, Paramasivam N, Vadivel V (2018) Essential oil based nanoemulsions to improve the microbial quality of minimally processed fruits and vegetables: a review. Food Res Int 111:509–523

Primo FL, Michieloto L, Rodrigues MAM, Macaroff PP, Morais PC, Lacava ZGM, Bently MVLB, Tedesco AC (2007) Magnetic nanoemulsions as drug delivery system for Foscan: skin permeation and retention *in vitro* assays for topical application in photodynamic therapy (PDT) of skin cancer. J Magn Magn Mater 31:354–357

Quatrin PM, Verdi CM, de Souza ME, de Godoi SN, Klein B, Gundel A, Santos RCV (2017) Antimicrobial and antibiofilm activities of nanoemulsions containing *Eucalyptus globulus* oil against *Pseudomonas aeruginosa* and *Candida* spp. Microb Pathogenesis 112:230–242

Ravi TPU, Padma T (2011) Nanoemulsions for drug delivery through different routes. Res Biotechnol 2(3):1–13

Ruengvisesh S, Loquercio A, Castell-Perez E, Taylor TM (2015) Inhibition of bacterial pathogens in medium and on spinach leaf surfaces using plant-derived antimicrobials loaded in surfactant micelles. J Food Sci 80:M2522–M2529

Sajjadi S, Zerfa M, Brooks BW (2003) Phase inversion in p-xylene/water emulsions with the non-ionic surfactant pair sorbitan monolaurate/polyoxyethylene sorbitan monolaurate (Span 20/ Tween 20). Colloids Surf A Physicochem Eng Asp 218:241–254

Saupe A, Gordon KC, Rades T (2006) Structural investigations on nanoemulsions, solid lipid nanoparticles and nanostructured lipid carriers by cryo-field emission scanning electron microscopy and Raman spectroscopy. Int J Pharm 314:56–62

Setya S, Talegaonkar S, Razdan BK (2014) Nanoemulsions: formulation methods and stability aspects. World J Pharm Pharm Sci 3(2):2214–2228

Shakee F, Ramadan W (2010) Transdermal delivery of anticancer drug caffeine from water-in-oil nanoemulsions. Colloids Surf B Biointerfaces 75(1):356–362

Sieg JW, Robinson JR (1977) Vehicle effects on ocular drug bioavailability II: evaluation of pilocarpine. J Pharm Sci 66:1222–1228

Silva HD, Cerqueira MA, Vicente AA (2012) Nanoemulsions for food applications: development and characterization. Food Bioprocess Technol 5:854–867

Singh Y, Meher JG, Raval K, Khan FA, Chaurasia M, Jain NK, Chourasia MK (2017) Nanoemulsion: concepts, development and applications in drug delivery. J Control Release 252:28–49

Solans C, Izquerdo P, Nolla J, Azemar N, Garcia-Celma MJ (2005) Nano-emulsions. Curr Opin Colloid Interface Sci 10:102–110

Solans C, Sole I (2012) Nano-emulsions: formation by low-energy methods. Curr Opin Colloid Interface Sci 17:246–254

Subhashis D, Satayanarayana J, Gampa VK (2011) Nanoemulsion – a method to improve the solubility of lipophilic drugs. Pharmanest – Int J Adv Pharm Med 2:72–83

Tadros T, Izquierdo P, Esquena J, Solans C (2004) Formation and stability of nano-emulsions. Adv Colloid Interf Sci 108-109:303–318

Tahir HU, Sarfraz RA, Ashraf A, Adil S (2016) Chemical composition and antidiabetic activity of essential oils obtained from two spices (*Syzygium aromaticum* and *Cuminum cyminum*). Int J Food Prop 19(10):2156–2164

Thakur N, Walia MK, Kumar SLH (2013) Nanoemulsion in enhancement of bioavailability of poorly soluble drugs: a review. Pharmacophore 4(1):15–25

Thomas L, Zakir F, MirzaMA AMK, Ahmad FJ, Iqbal Z (2017) Development of Curcumin loaded chitosan polymer based nanoemulsion gel: *In vitro, ex vivo* evaluation and *in vivo* wound healing studies. Int J Biol Macromol 101:569–579

Trotta M (1999) Influence of phase transformation on indomethacin release from microemulsions. J Control Release 60:399–405

Zhang H, Zhao Y, Ying X, Peng Z, Guo Y, Yao X, Chen W (2018) Ellagic acid nanoemulsion in cosmetics: the preparation and evaluation of a new nanoemulsion method as a whitening and anti-aging agent. IEEE Nanotechnol Mag 12:14–20. https://doi.org/10.1109/mnano.2017.2780859

Chapter 4
Nanosensors for Food Safety and Environmental Monitoring

Kulvinder Singh

Contents

4.1 Introduction

Food and environment are two highly intertwined factors sitting at the centre of human survival and well-being. The effects of these resources on humans have been well established, and the complications arising from the fluctuations are also sufficiently documented. Environmental factors like air and water are thus among the top carriers of majority of communicable diseases. These factors also govern the food quality due to their inseparable nature and thus penetrate the human lives again indirectly. Thus, it is of utmost importance to maintain the quality of food and environmental factors. There are several factors, biological and chemical, which are continuously infiltrating these systems and prove harmful if they enter the bodies. Industrial revolution has played a central role in the degradation of environment and entry of such androgenic harmful chemicals in natural resources. It is estimated that tonnes of industrial and municipal wastewater is dumped in water bodies with-

K. Singh (✉)
Department of Chemistry, School of Basics and Applied Sciences, Maharaja Agrasen University, Baddi, India
e-mail: kulvinderchem@mau.edu.in

D. Thangadurai et al. (eds.), *Nanotechnology for Food, Agriculture, and Environment*, Nanotechnology in the Life Sciences,
https://doi.org/10.1007/978-3-030-31938-0_4

out any treatment, and agricultural discharges are often fortified with drugs and pesticides which tend to deteriorate its quality (Mateo-Sagasta et al. 2017). According to the World Health Organization (WHO) reports, around 600 million cases of food- and waterborne disease were reported in 2010 alone from just 31 causative agents causing 420,000 deaths (WHO 2015). In addition, an annual number of over 300,000 cases were reported resulting from only a few chemical mediators found in food and water resources. Further, certain chemicals have accumulating effects, magnifying over time to lethal amounts even when their basal concentrations are within permitted levels. Such chemicals are further maintained under strict guidelines to minimize the exposure and associated bioaccumulation.

With the emergence of nanotechnology, multitude of affordable rapid and specific sensing techniques has been developed. To maintain the quality of these parameters, we require robust tools to reveal the presence of such contaminants and at times quantify them at their minute amounts. One of such tools, "nanosensors", is of paramount importance for the quality assessment and assurance of food, water and other environmental factors. Enhancement of these nanosensors with good affinity chemo−/bioreceptors to deliver reproducible and accurate results along with multiplexing is another niche being explored by various researchers worldwide.

4.2 Sensors and Nanosensors

A sensor is a compound system with the capability to respond to a target in qualitative or quantitative manner. The target can be anything from a physical parameter, small inorganic molecules, gases or biomolecules like proteins, DNA or even whole cells. Sensors measuring physical parameters like temperature, pressure, potential, mass and light intensity are simpler and operate by measuring the changes in the physical properties of a standard substance like expansion of mercury in response to temperature changes in a thermometer or rise in mercury level in response to change in atmospheric pressure in classic barometer. In contrast, the sensors quantifying analytes require more sophisticated components to pertain the sensor with highest levels of fidelity. The fidelity is provided by using a biological moiety, and such sensors are called *biosensors*. By definition, a biosensor is a two-component system comprising of a biological molecule called a *receptor* and a second *transducer* element. A receptor in a biosensor can be any biological molecule with an intrinsic affinity towards a certain entity. This affinity provides the biosensor the ability to distinguish and specifically interact with the targeted analytes from a mixture of compounds. Various receptors have been applied for the development of biosensors, beginning with the application of enzymes in the invention of glucose sensors by Leland L. Clark in the 1960s. Depending on the type of molecules used, biosensors can be classified as (a) enzyme-based biosensors, (b) antibody-based biosensors (immunosensors), (c) aptamer-based biosensors (aptasensors), (d) peptide-based biosensors, (e) natural receptor-based biosensors and (f) gene-based biosensors (genosensors).

Enzymes have a limited repertoire of interacting molecules and can be used for only those analytes against which a recognizing enzyme is available. Antibodies, as bioreceptors, had a wider target range than enzymes due to the flexibility of antibody generation against molecules which are antigenic, although generation of large amounts of antibodies required heavy animal rearing and were slower for antibody-based biosensors to be commercially successful. Hybridoma technology developed by Milstein and Köhler in 1975 allowed the production of identical antibodies from a single parent culture at large scales creating a forward thrust in the antibody-based *immunosensors*. *Aptamers*, single-stranded nucleic acid-based receptors, are screened from a large combinatorial library by an iterative method called systematic evolution of ligands by exponential enrichment (SELEX), which was independently developed by two research groups in 1990 (Ellington and Szostak 1990; Tuerk and Gold 1990). Since then, several aptamers have been developed for a variety of target analytes, and subsequent application in their sensing has been shown. Such sensors which use aptamers as their sensing element are termed *aptasensors*. Further, *peptide-based receptors* can also be screened from a combinatorial library of different sequences by a method called phage display. In the method, the peptides library is hosted in bacteriophages and is made to be expressed in the coat protein of the phage. The phage particles are exposed to the target analytes bound to a surface. The binding phages are then eluted and amplified by infecting in bacteria. The whole process is repeated multiple times to separate the peptides with highest affinities from rest of the library and finally sequenced to find the peptide sequence. The technique is relatively recent and however has the promising advantages of SELEX methodology. Further, proteins that occur naturally in the cells can be also be used as receptors, and the respective biosensors are called natural receptor-based biosensors. Besides these, a final category of nucleic acid-based sensors exists which rely on the complementary Watson–Crick pairing and thus can strictly be used to sense other nucleic acids only. The target DNA are usually a particular sequence of a genome, and these types of *genosensors* are used as a rapid and efficient alternative to PCR-based DNA analysis techniques.

The function of transducer is to convert the interaction between receptor and its target analyte into a measurable output signal. Change in colour and electronic properties are the simpler and usually applied categories of transducers. However, since the dawn of nanotechnology, several nanomaterials are exploited for their role as transducers to provide superior properties to the sensors. Nanomaterials are substances having at least one spatial dimension in the *nano* regime, i.e. measuring between 1 and 100 nanometres. Due to their small sizes, nanomaterials display the phenomenon of quantum confinement and have very high surface areas per unit volume. Quantum confinement relays several novel properties to the materials, which are starkly different from their bulk counterparts, and the high surface areas provide more interaction sites and reaction points. On the basis of transducing element, a sensor can be broadly categorized in three categories, viz. optical sensors, electrochemical sensors and mechanical sensors, each of which relies on changes in the particular property of the transducing material. These will be elaborated in further sections of the chapter.

4.3 Major Types of Nanosensors

Lately, with the emergence of rapid industrialization, a growing demand for monitoring and adopting a sustainable approach in real-time has garnered the attention of both governments and citizens alike. Hence, to mitigate any unforeseen pollutant epidemics due to natural or androgenic causes, there is an urgent need to develop cost-effective, multiplexed, stable and portable sensing techniques for a very early recognition of detrimental pollutants. Currently used approaches undertaken by technicians in the field of food, environment or medicine rely on analytical tools such as high-performance liquid chromatography (HPLC), gas chromatography (GC), mass spectrometry (MS), nucleic acid polymerase chain reaction (PCR) and enzyme-linked immunosorbent assays (ELISA) which are heavily dependent on sophisticated instrumentation and costly reagents. As aforementioned, nanomaterial-based sensing platforms and assays are being explored massively by researchers worldwide as an improved alternative to conventional tools and methodologies. These nanosensors are comparatively facile, stable and faster and require way less reagents in detection than conventionally available techniques. Broadly on the basis of physical phenomena via which the output of interaction between the target and analyte is interpreted, the nanosensors are classified into optical, electrochemical and mechanical (Fig. 4.1).

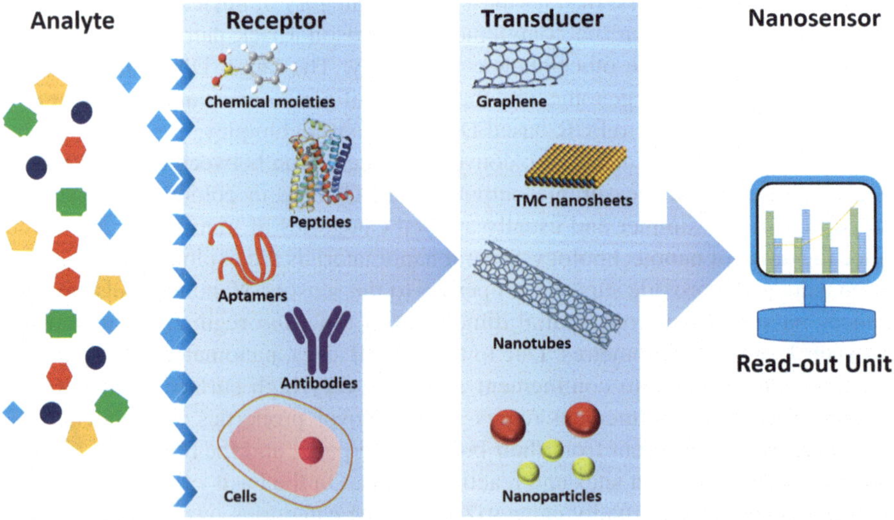

Fig. 4.1 Schematic showing essential components of a nanosensor and their widely used examples

4.3.1 Optical Nanosensors

One of the most prominently used sensory parameters is that of light. Perception of different radiations in the visible range of electromagnetic spectrum, i.e. colours, is one of the primary indications of presence or absence of an analyte. This sensation can be amplified many-folds to increase the sensitivity of detection with the help of nanoparticles. Metal nanoparticles, viz. gold, silver, copper, nickel, palladium and iron, have been shown to have enhanced electronic and plasmonic properties. These nanosized metal particles show size-dependent properties like high surface-to-area ratio, localized surface plasmon resonance, super-paramagnetism and quantum confinement that are in contrast significantly from that of micron-sized or larger particles. Optical nanosensors harvest this very size-dependent property and are significantly visible in variation of colourations of metallic nanoparticles with increase in size from 5 nm and above gradually. The most exploited nanomaterials are gold and silver, prominently gold nanoparticles (GNPs), in all forms, viz. spheres, hollow spheres, nanorods, nanocubes, prisms and other exotic shapes like thorns, corals, etc. Especially, spherical GNPs can be easily synthesized via chemical reduction of Au^{3+} to Au^{0} ions with the help of sodium citrate, which plays a dual role of reducing agent as well as capping agent during formation. This method synthesizes stable GNPs owing to capping of negatively charged citrate over the nanoparticles preventing their aggregation and further aiding in surface functionalization with biomolecules (Turkevich et al. 1951). These coloured nanoparticles of colloidal gold (<20 nm) have the ability to colour switch from red state absorbing around 520 nm to aggregated blue state occurring due to the plasmon coupling between nanoparticles, which is the basis of many colorimetric nanosensors. As depicted in Fig. 4.2, pathogen in food and water can be easily detected by functionalizing the GNPs with bioreceptors (aptamer/antibody) and subsequent agglomeration of nanoparticles upon bacteria-GNP interaction. This red shift of colouration can be analysed both with the naked eye for higher concentrations or can be quantified using a spectrophotometer. Numerous researchers have used this approach for detection of food adulterants and pathogens in complex matrices. One of the recent novel studies fabricated a paper-based immunochromatographic assay using monoclonal antibodies, generated by using 2-(methoxycarbonylamino)-3H-benzimidazole-5-carboxylic acid as the hapten, for simultaneous detection of benzimidazoles and metabolite residues in milk samples (Guo et al. 2018a). They developed a strip that can detect 11 benzimidazoles including albendazole, albendazole s-oxide, albendazole sulfone, fenbendazole, fenbendazole sulfone, flubendazole, mebendazole, parbendazole, oxfendazole, oxibendazole and carbendazim in a short test time of under 15 min in milk samples with limit of detection (LoD) as low as 0.7 ng/mL also. For estimation of pathogens likely to be found in fresh and rich food sources like milk and meats, screening and monitoring in short time span by both local vendors and food processing companies have come to be as an urgent need. To mitigate epidemics from bacteria and their toxins in contaminated packaged foods, visual aids in form of small paper-based colorimetric sensors are a great

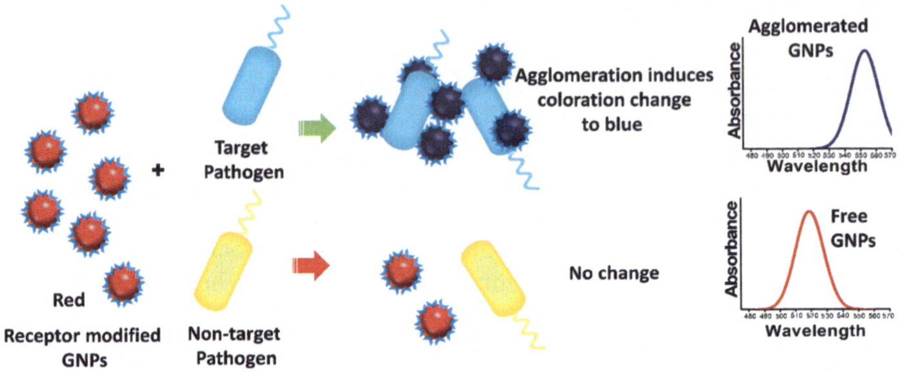

Fig. 4.2 Schematic representation of gold nanoparticle-based optical nanosensor for detection of bacterial species

advancement. On similar colorimetric sensing lines but a with a twist, for estimation of most prevalent enteropathogen *Salmonella typhimurium*, a visual-specific GNP-based method was developed using GNPs conjugated bifunctional oligonucleotide probe and an aptamer. The anti-salmonella aptamers with higher affinity to the cells bind and release the oligo-GNPs turning test solution from blue to red. They showed a good dynamic detection range from 10^1 to 10^6 cfu/mL in milk and shrimp samples (Xu et al. 2018b).

The numerous physicochemical properties of metal nanoparticles, viz. size-based colouration, localized surface plasmon resonance (LSPR), surface-enhanced Raman scattering (SERS), fluorescence quenching and even catalytic ability, have catapulted these high-performance nanoparticles into highlight since the beginning of nanotechnology. Exploiting the phenomenon of surface plasmon resonance, a multiplexed biosensor assay based on a competitive inhibition format with secondary antibodies (Ab2) conjugated to gold nanoparticles (GNPs) used as a signal amplification tag for three *Fusarium* toxins, viz. deoxynivalenol (DON), zearalenone (ZEA) and T-2 toxin, has been developed and successfully validated (Hossain and Maragos 2018). They showed that the signal was amplified up to 25 times for DON, 90 times for ZEA and 12 times for T-2 toxin assay using the conjugated Ab2-GNPs and having LoD of 15 μg/kg, 24 μg/kg and 12 μg/kg for DON, ZEA and T-2 toxin in wheat samples, respectively.

In fluorescent platforms, either nanomaterials can be used as quenchers or those with good photoluminescent properties can be utilized as fluorescent labels. This brings to mention quantum dots (QDs), which are inorganic fluorescent nanoparticles having sizes in the range of 1–10 nm and are the next biggest transducer utilized in nanosensing assays. These are synthesized from both top-down and bottom-up approaches like pyrolysis, hydrothermal, acid treatment, exfoliation and electrochemical synthesis. These can be synthesized from semiconductor materials like cadmium selenide/telluride, molybdenum sulphide/selenide, tungsten sulphide/selenide or the all-time favourite of nano-scientists carbon dots. These tiny nanoparticles reach the quantum level and showcase quantum confinement of electrons and hole carriers which confers to ability to have excellent and very stable photoluminescent properties. In addition to easy generation of functional groups on surface, these tend to have low toxicity and have been suggestively used as alternatives for chemical fluorophores for both sensing and bioimaging. Currently, QDs have been applied as fluorescent markers in numerous nanosensor assays in order to diagnose pathogens and their related toxins.

Adding to the various nanosensors being developed using nanomaterials, another major category is of quenching of fluorescent dyes by either nanoparticles or 2D nanomaterials like graphene and its forms graphene oxide (GO) and reduced graphene oxide (rGO). These nanomaterials have good radiation absorption properties tuneable as per the need of the nanosensing assay. The biggest achievement of a fluorescence-based sensor is nullification of background non-specific signal, which in turn increases the sensitivity of the system and helps in detection of small molecules like pesticides with ease. In a typical aptasensing assay employing this strategy (Fig. 4.3), aptameric strands are coupled with short complimentary strands which are usually labelled with a fluorophore. Upon interaction of the target with the aptamer, the target with high affinity to the aptamer displaces the complimentary strand, which is then free to associate itself with planar GO nanosheets in the system. These GO nanosheets then in turn absorb the resonant energy transferred from the fluorophore and quench its emission. The drop-in fluorescence is proportional to the target antigen bound to the aptamer, which in turn is used to quantify the analyte concentration. Various nanosensing assays for fungal toxins have also been studied using quencher approach. Using a more facile reverse approach, nano-scientists were able to detect as less as 1 nM of aflatoxin B1, a secondary fungal metabolite of *Aspergillus flavus*, using GO in an aptamer-based nanoassay featuring the quench-

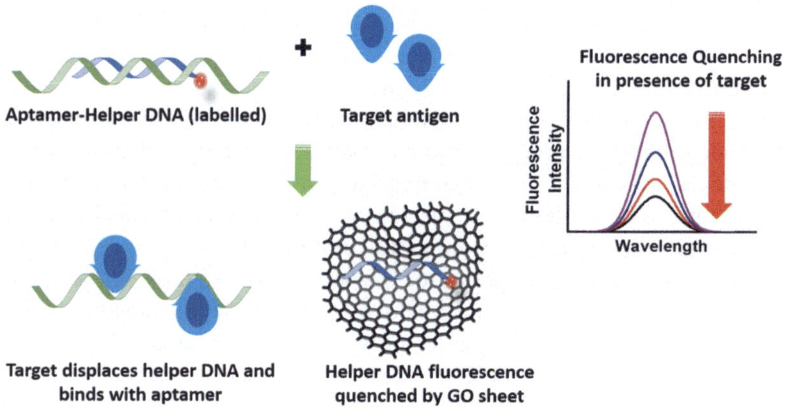

Fig. 4.3 Schematic representation of aptamer displacement assay using fluorophore labelled helper DNA and graphene oxide nanosheets as quencher

ing aptamer-ligated cadmium telluride quantum dots. While the fluorescence of the aptamer-modified QDs was strongly quenched by GO, the addition of aflatoxin B1 to the system restored its fluorescence as the aptamer separates from the sheet to bind to the target toxin and validated this in both water and oil (Lu et al. 2014). Researchers have utilized these approaches to detect a multitude of contaminants ranging from bacteria like *Pseudomonas* to even small antibiotic molecules present in food samples. Further studies for multiplexed detection can be explored for an overall sensory profile of any food material, which will negate the need of multiple sampling in a single batch.

4.3.2 Electrochemical Nanosensors

Food materials are an easy source for contaminants; especially bacterial and fungal growth in such rich mediums is uninhibited unless specific preservation steps are undertaken. Still many a times even washed fruits and vegetables, cured meats and pasteurized milk may be spoilt either at the packager's end, during shipment or at the consumer's end in a short time period. This becomes a challenge for the consumer to identify whether the commodity is in safe edible condition just from the looks or smell of it. This becomes the responsibility of both the provider and the consumer to able to easily identify the situation before it becomes a reason for any gastrointestinal disease or discomfort. As mentioned before, the complex matrices of foods with a lot of interfering carbohydrate and protein molecules present in the immediate environment of the target molecule render it difficult to detect the target without a prior pre-processing in labs. Thus, a need for easy to use point-of-care (PoC) devices is essential to lessen the burden. For this specific purpose, in addition to colorimetric and luminescence-based nanosensors, electrochemical sensors have been extensively used for food safety and environmental monitoring based on their portability and ease of operation. Although electrochemical sensors were the first kind of sensors commercially used by masses (glucose sensors), these sensors before the implementation of nanomaterials inclined to suffer from certain issues like electrode fouling, by-product formation, surface leeching and transducer corrosion. To alleviate erroneous and cross-reactive sensing, electroactive nanomaterials with good functional moieties on surface aid the ligation of specific receptors, which efficiently capture targets from a milieu of molecules.

The exceptional surface properties of nanomaterials like high surface area, tensile strength and superior electronic conductivity have together aided in fabrication of electrochemical and electrical platforms like FETs (Chen et al. 2019), eNose and eTongue (Haddi et al. 2014) for detection of bio-interactions via electrochemical reduction-oxidation (redox) or electron-hole pair signals. These nanosensor sensing mechanisms are dependent on the changes in the current/potential alteration arising from faradaic processes due to the redox reaction at the electrode surface or ohmic response of the transistor between contacts. The interaction between receptor and target analyte is analysed with the help of an oscilloscope or potentiometer. These minute electrical differences generated by the bio-interactions are measured and correlated with standardized calibration curves for the target analyte in question. Depending on the parameter measured, electrochemical nanosensors are classified as *amperometric* or *impedimetric* sensors, and field-effect transistors (FET) are often considered as their third category (Kaur and Shorie 2019). Amperometric sensors measure the changes in current values under the influence of a second variable parameter. The variable is generally time or voltage, and the respective amperometric sensors are called *chronoamperometric* or *voltammetric sensors*. Chronoamperometric sensors work by measuring the current changes at a fixed potential over time and are often used in systems which result in consumption/production of certain electrochemical species. Chronoamperometric sensors are frequent with application of enzymes as the bioreceptor as the biointeraction is followed by a chemical reaction, the by-products of which are often electrochemically active, giving rise to increase in the current values. For other types of bioreceptors, such labels are used which mimic the function of enzymes to replicate such reactions. In contrast, voltammetric sensors function at a variable potential and measure the current changes resulting from biointeraction on the electrode. The second category, impedimetric sensors, relies on the changes in the charge-transfer abilities of the system and thus is a label-free technique. The label in impedimetric sensors is the analyte molecules itself, which, depending on their own conductance, facilitate or hinder the charge transfer between the electrolyte and the electrode.

In a typical electrochemical nanosensor, the surface of the working electrode is modified with an electroconductive nanomaterial, which helps in faster and more generous transfer of electrons from the surface to the detector and a receptor for analyte interaction. The electrochemical cell functions by movement of electrons from the electrolyte to the surface of the electrode. On the basis of the approach taken, three common techniques are used: (1) direct electron transfer from the captured analyte or analyte-receptor interaction is measured, and a current to potential plot is made showing the oxidation/reduction current change peaks over a range of potential applied; (2) sandwich approach using a secondary receptor for the target which is either electroactive in nature or labelled with an electroactive compound like ferrocene; or (3) impedance or resistance of the electronic charge transferred from the electrolyte or electroactive nanomaterial present in the system. This methodology is usually utilized for detection of bulky biological materials like pathogens and proteins, which are electrochemically inert in nature and hamper the flow of current in an electrochemical cell. The receptor-analyte interactions are usually likely to decrease in capacitance and increase in resistance (Fig. 4.4). The most prevalent presentation for estimating the analyte concentration in an impedimetric setup is Nyquist and Bode plots. In the former, the imaginary part of the impedance component (Z'') and the real part of impedance component (Z') are plotted to generate a semi-circle curve. In the latter, the log of impedance ($|Z|$) and phase shift (φ) on the y-axis are plotted against the log of excitation frequency.

A variety of novel sensors using electrochemical platform for pathogens, toxins and even virus have been fabricated. In view of recent epidemics, these have come up as an affordable approach in sensing by practitioners even in remote areas of the

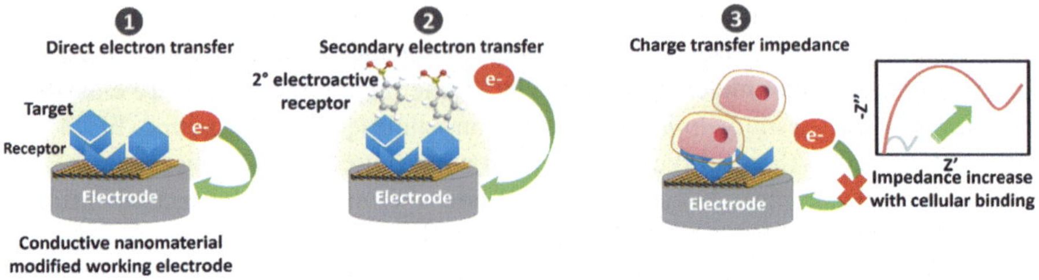

Fig. 4.4 Representation of amperometric detection of analyte species based on direct electron transfer of electroactive targets (1) and through electroactive mediator labels (2) on a nanomaterial modified electrode; general depiction of impedance based label-free detection of cellular species (3)

world. These nanosensor fabrications involve modification of the working electrode with various conductive nanoparticles, 1D nanotubes and 2D nanosheets like graphene and transition metal dichalcogenides (TMDs) and functionalization of these transducers with specific receptors. A modification in either part of the sensor can render in more specific and sensitive to contaminant analyte. A simple paper-based platform was fabricated for multiplex detection of paraoxon, 2,4-dichlorophenoxyacetic acid and atrazine by exploring their property to inhibit butyrylcholinesterase, alkaline phosphatase and tyrosinase, respectively (Arduini et al. 2019). Their device was developed by integrating two different paper-based screen-printed electrodes and multiple filter paper-based pads to load enzymes and enzymatic substrates. The versatile analysis of different pesticides was carried by folding and unfolding the filter paper-based structure, without any addition of reagents and any sample treatment (i.e. dilution, filtration, pH adjustment). They correlated this degree of inhibition to pesticide concentration and measured the enzymatic activity chronoamperometrically using a portable potentiostat. For the detection of 2,4-dichlorophenoxyacetic acid and atrazine, these paper-based electrodes were modified with carbon black and for the detection of paraoxon decorated with Prussian blue nanoparticles. They showed that this simple platform was able to detect these pesticides in ppb levels in both test and field conditions. Likewise, dual detection of two extensively used pesticides acetamiprid and atrazine was done using an impedimetric biosensing platform wherein they sputtered platinum nanoparticles using e-beam lithography in a bridge-like arrangement between interdigitated electrodes (Madianos et al. 2018). These bridged platinum nanowires were then immobilized with aptamers, and the hindrance upon analyte binding was recorded and showed a linear range of response in the range of 10 pM to 100 nM for acetamiprid with LoD of 1 pM and of atrazine with a from 100 pM to 1 μM with LoD of 10 pM, respectively.

Another type of specific receptor compatible with electrochemical technique has been fabricated with the help of molecular imprinting called *molecular imprinted polymer* (MIP), also known as plastic antibodies (Fig. 4.5). This is a facile method for large-scale synthesis of polymeric-specific receptors that are generated via in situ reaction of the target analyte with the monomers with various

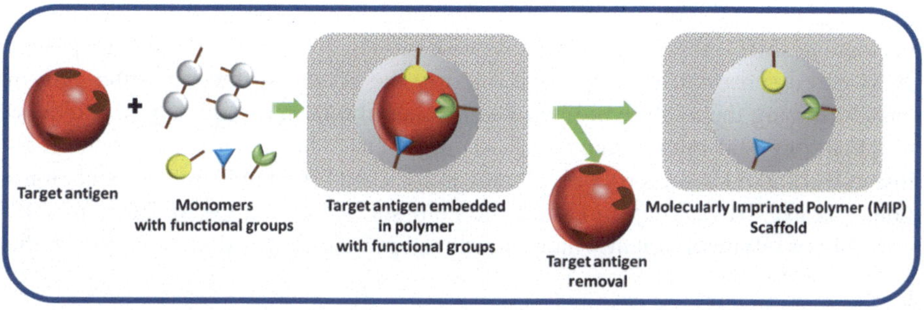

Fig. 4.5 Scheme showing the generalized route followed for the production of molecular imprinted polymers (MIP)

functional moieties. These functional groups can non-covalently bind with the target's exposed groups lending a one-on-one direct interaction at molecular level. This template-driven methodology has helped synthesize polymer scaffolds that can detect specific recognition sites on the target surface and has arisen as a possible replacement for large and bulky antibodies, given the fact that these are easy to synthesize for a large variety of targets and have a long shelf life. The only lacuna lies in cross-reactivity with closely placed species and molecules with single amino acid or nucleobase change. As in case of hybrid nanomaterials, MIP in conjunction with graphene oxide was used to modify a commercially available disposable electrode surface, and these were able to detect zika virus in a low-cost electrochemical format (Tancharoen et al. 2019). Their biosensor was functional in both buffer and blood environments and was validated to detect as low as 2×10^{-4} pfu/mL with a little higher specific LoD of 2×10^{-2} pfu/mL in the presence of dengue virus.

4.3.3 *Mechanical Nanosensors*

Mechanical sensors are the most sophisticated category of label-free diagnostics which functions on the changes in mass. Compared to optical and electrochemical nanosensors, mechanical nanosensing platforms confer higher sensitivity at even attomolar detection levels. In classical approach, the binding of analyte creates an increase in mass of the system which in turn affects the properties of the system. However, in recent times several approaches have been reported which showcase intriguing application of nanomaterials causing a decrease in mass. The change in mass can affect several properties, depending on which mechanical sensors are divided in *cantilever*-based, *quartz crystal microbalance* (QCM)-based and *piezoelectric* mechanical sensors. In cantilever-based sensors, the receptor is used to modify the tip of cantilever, and the binding of analyte creates a bend in the cantilever, and the mass gain is indirectly measured by monitoring the shift in the laser. Similarly, in piezoelectric sensors, the bending caused by analyte binding changes the conductivity of the sensors, which causes a shift in the flowing current (Fig. 4.6). As shown in Fig. 4.6, nanomaterials can be used to enhance the mechanical shift, enhancing the signal/noise ratio to give a more reliable response. QCM-based sensors use receptors to modify the crystal surface, and the gain in mass following the analyte biding changes the vibrational frequency of the crystal. Quartz crystal microbalance (QCM) nanosensor has been used for the real-time detection of kaempferol polyphenol antioxidant found in fruits and vegetables (Gupta et al. 2014). They modified the gold surface of the QCM chip via self-assembly of allyl mercaptan which introduced polymerizable double bonds on the

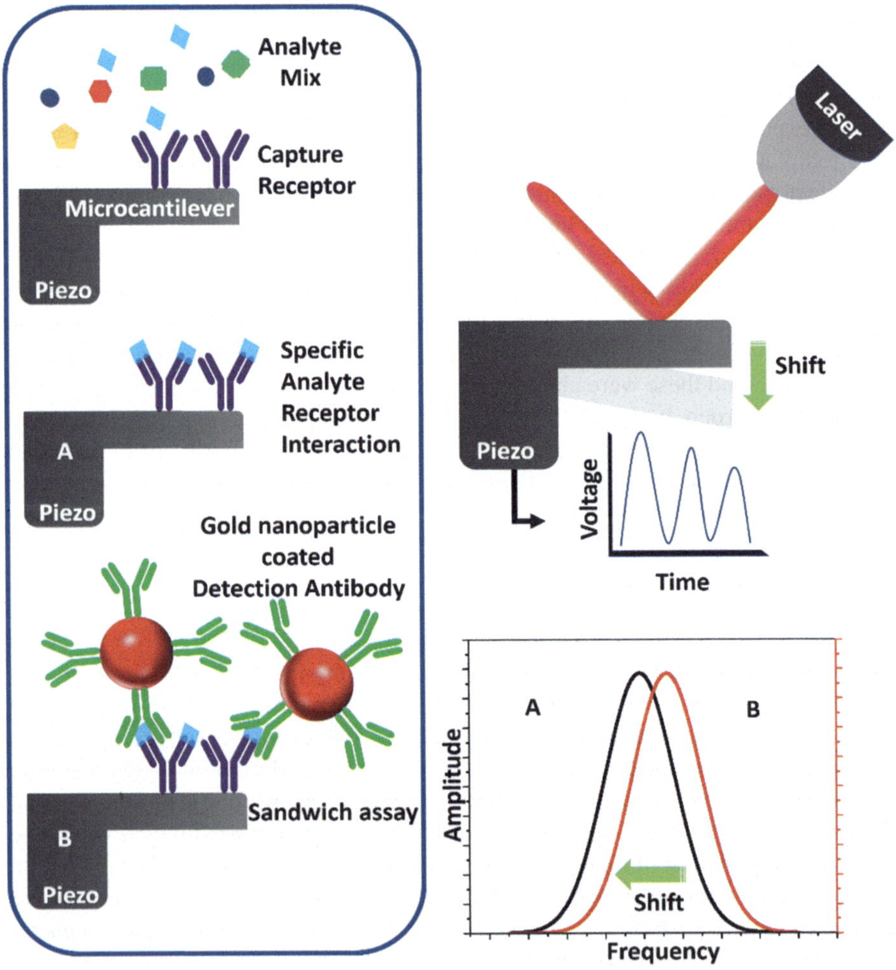

Fig. 4.6 Schematic representation of cantilever-based mechanical sensor; left panel shows the stepwise progression of assay in a dual antibody-based sandwich assay; right panel shows the detection of cantilever deflection via piezoelectric effects or frequency shifts

chip surface for imprinting of KAE-imprinted poly(2-hydroxyethyl methacrylate-methacryloylamidoaspartic acid) film on the gold surface. They showed a linear range from 2.0×10^{-10} to 1.5×10^{-9} M with LoD as low as 6.0×10^{-11} M even in real samples such as orange and apple juices in the presence of cross-reactive species of quercetin, myricetin and apigenin. Dual detection mode has been exploited in one of the studies, where they fabricated a double template imprinted biomimetic dendritic nanofiber-based piezoelectric sensor for the analysis of organopesticides, dichlorodiphenyltrichloroethane and hexachlorobenzene (Prasad and Jauhari 2015). For this they immobilized 2,5-thiophenecarbonyldichloride molecules on surface of gold quartz crystal which were then attached with dendron monomer molecules followed by free radical polymerization of the same in presence of the target analyte. These self-assembled imprinted nanofibers on the surface helped generate cavities for specific analyte binding with LoD of 0.75 and 0.69 ng/mL for the two pesticides, respectively.

4.4 Sensing of Major Food and Environmental Contaminants

Food and water sensing required for monitoring contaminants has become a necessity in preview of rising health concerns especially in third-world countries, where outflow of effluents into environment is unfortunately currently unchecked. Nanosensors have shown a new horizon in detection and sensing of contaminants ranging from small molecules like pesticides, pharmaceuticals like drugs and antibiotics, heavy metals, toxic organic and volatile compounds to biologically active viruses, toxins and bacteria. These have aided heavily in diagnostics, food screening and environmental scrutiny of pollutants with the immense research and input of scientists worldwide. Various biosensing strategies have been developed exploring the unique physicochemical properties of nanomaterials, and the addition of excellent high affinity stable receptors has increased their potential in both lab and field. The following section recaps the different types of contaminants detected via nanosensors using various nanotechnologies available.

4.4.1 Applications of Nanosensors for Chemical Contaminants

Contamination by synthetic chemicals and compounds in the nature is one of the foremost worldwide food and environment threats. These belong to a variety of groups ranging from heavy metal waste generated from industrial androgenic processes, pesticides, herbicides and insecticides as a result of excessive agricultural application and pharmaceutical over-usage of antibiotics and drugs (FAO 2016). These contaminants usually enter into the food chain leading to magnification of toxins in the system. These contaminate both our food sources, viz. crops and livestock and their products, and freshwater sources, namely, rivers. Most of the agricultural wastes generated even after biodegradation form harmful toxic compounds which accumulate in our plants and soil. Several nanosensors have been developed for detection of toxic organic compounds, viz. aldehydes (Duan et al. 2019), phenols and their derivatives (Sadeghi et al. 2013; Hu et al. 2014; Qi et al. 2018; Ren et al. 2018), biphenyls, hydrazine (Teymoori et al. 2018), hydroquinone (Wang et al. 2018b), melamine (Chang et al. 2017; Jigyasa and Rajput 2018; Li et al. 2018a), chloropropane (Fang et al. 2019), trihalomethanes (Wujcik et al. 2016) and 2,4,6-trinitrotoluene (Priyanka et al. 2014; Malik et al. 2019).

A multitude of synthetic fertilizers and pesticides are leeched into soil reaching the water reserves below ground level. Nanosensors for these phosphates or chloro derivatives of organopesticides have been extensively worked upon, especially for 2,4-dichlorophenoxyacetic acid (Wagner et al. 2018; Zhang et al. 2019), acetamiprid (Bahreyni et al. 2018; Guo et al. 2018c; Saberi and Rezaei 2019), atrazine (Sang et al. 2018), bitertanol (Cao et al. 2018), carbofuran (Campos et al. 2017), chlorpyrifos (Dou et al. 2018; Islam et al. 2019), clopyralid (Wang et al. 2018a), cyfluthrin (Li et al. 2018b), diuron (Sun et al. 2018b), malathion (Bala et al. 2017; Li et al. 2019), paraoxon (Aghaie et al. 2019; Peng et al. 2019), paraquat (Ghalkhani et al. 2019) and thiram (Sun et al. 2019a) to name a few. These poisonings associated with inappropriate usage have resulted in a high mortality with up to million deaths annually (WHO 2011). In view of excessive industrial waste generation, heavy metal pollution has arisen as a worldwide concern as it infiltrates both the land and water and accumulates in flora and fauna have lethal toxicological effects (Köhler and Triebskorn 2013). Major heavy metal ions like bismuth (Mortazavi and Farmany 2017), cadmium (Qian et al. 2017), chromium (Hu et al. 2018b; Rahbar et al. 2018), copper (Elmizadeh et al. 2017; Deymehkar et al. 2018), lead (Ikram et al. 2019; Memon et al. 2019), mercury (Zhang et al. 2018; Niu et al. 2019; Shorie et al. 2019) and radioactive waste (Dewangan et al. 2019; Saha et al. 2019) have been successfully captured and detected using nanofabricated sensing platforms.

Many broad-spectrum antibiotics like chloramphenicol (Miao et al. 2015; Javidi et al. 2018; Muhammad et al. 2018), kanamycin (Liu et al. 2018a), neomycin (Shi et al. 2018), cefixime (Akhgari et al. 2017), sparfloxacin (Geng et al. 2019), metronidazole (Mehrzad-Samarin et al. 2017) and amoxicillin (Chullasat et al. 2018; ul Ain et al. 2018) have serious potential side effects. These are usually supplemented in poultry and livestock feed to increase their life, but their penetration in our food has resulted in evolution of resistant strains and superbugs. Several biosensors based on nanomaterial-functionalized platforms have been fabricated other compounds like cocaine (Beigloo et al. 2016), morphine (Masteri-Farahani and Mosleh 2019) and volatile gases (Hebbar et al. 2017; Tomer et al. 2017; Bao et al. 2018; Dong et al. 2019) also.

4.4.2 Applications of Nanosensors for Biological Contaminants

Biological contaminants in raw fruits, vegetables and meat products are pathogenic microbes such as bacteria, fungi and viruses. These pathogens may be introduced into food and drinking water via faecal contamination, soil or pests. Even in canned or processed foods, these may enter through poor handling practices. Several receptors for targeting of bacterial pathogens and their enterotoxins have been generated and used for optimal detection even up to a single cell. Major pathogens, viz. *Salmonella* sp. (Xu et al. 2018a; Hu et al. 2019; Zou et al. 2019), shiga-toxin producing *Escherichia* sp. (Kaur et al. 2017; Liu et al. 2018b; Shorie and Kaur 2018; Zhu et al. 2018), *Listeria* sp. (Qiu et al. 2019), *Campylobacter* sp. (Shams et al. 2019), *Shigella* sp. (Feng et al. 2019), *Vibrio* sp. (Guo et al. 2018b; Sun et al. 2019b) and *Pseudomonas* sp. (Hu et al. 2018a), have been detected via nanosensors. Bacteriotoxins have been diagnosed in food, water and patient samples following this similar trend by employing superior nanomaterials (Zhang et al. 2016; Wang et al. 2017; Mousavi Nodoushan et al. 2019; Tan et al. 2019; Xie et al. 2019). Not only bacterial but viral infections responsible for hepatitis and gastro-related diseases have also been sensed with the aid of nanomaterial-enhanced sensors (Han et al. 2016; Ilkhani and Farhad 2018; Manzano et al. 2018; Xiang et al. 2018a; Khater et al. 2019; Tancharoen et al. 2019).

Algal and fungal toxins are naturally produced and grow on numerous foodstuffs such as cereals and nuts. These chemically stable compounds include aflatoxin (Guo et al. 2019), anatoxin (Lin et al. 2017), atrazine (Tortolini et al. 2016), microcystin (Bilibana et al. 2016), saxitoxin (Bratakou et al. 2017; Sun et al. 2018a), ochratoxin A (Song et al. 2018; Xiang et al. 2018b) and patulin (Fang et al. 2016), which are produced during storage in humid conditions and are linked with long-term diseases like cancer and liver and kidney damage that have been sensed successfully with the help of various nanosensors (Fig. 4.7).

Fig. 4.7 Common categories of contaminants in food, water and environmental resources

4.5 Conclusion

The chapter brings into focus sensing platforms fabricated from nanomaterials with excellent optical, magnetic, electrical or electrochemical properties showing immense enhancement over conventional systems employing bulk materials. These nanosensors with the added advantage of chemical and biological receptors aid in efficient and specific binding with target molecules in presence of even complex food matrices for rapid detection of contaminants in food, water and other environmental sources. The rapidity in seconds, sensitivity in pico-molar ranges, specificity for single molecules and easy of usage by even untrained personal have helped in bringing the nanotechnology-based sensors a notch up from conventional microbiological, immunological or spectroscopic techniques. As mentioned above, nanosensing platforms can assist food industries, environmental agencies and even untrained personnel in detection of food contaminants ranging from toxic chemicals to pathogens. The challenge lies in implementation of these nanosensors to further the routine monitoring of food and environment. Further exploration of hybrid nanomaterials with high responsiveness and room temperature-stable bioreceptors is needed to advance food and environmental nanosensors.

References

Aghaie A, Khanmohammadi A, Hajian A, Schmid U, Bagheri H (2019) Nonenzymatic electrochemical determination of paraoxon ethyl in water and fruits by graphene-based NiFe bimetallic phosphosulfide nanocomposite as a superior sensing layer. Food Anal Methods 12:1545. https://doi.org/10.1007/s12161-019-01486-8

Akhgari F, Samadi N, Farhadi K (2017) Fluorescent carbon dot as nanosensor for sensitive and selective detection of cefixime based on inner filter effect. J Fluoresc 27:921–927

Arduini F, Cinti S, Caratelli V, Amendola L, Palleschi G, Moscone D (2019) Origami multiple paper-based electrochemical biosensors for pesticide detection. Biosens Bioelectron 126:346–354

Bahreyni A, Yazdian-Robati R, Ramezani M, Abnous K, Taghdisi SM (2018) Fluorometric aptasensing of the neonicotinoid insecticide acetamiprid by using multiple complementary strands and gold nanoparticles. Microchim Acta 185:272

Bala R, Dhingra S, Kumar M, Bansal K, Mittal S, Sharma RK, Wangoo N (2017) Detection of organophosphorus pesticide – malathion in environmental samples using peptide and aptamer based nanoprobes. Chem Eng J 311:111–116

Bao Y, Xu P, Cai S, Yu H, Li X (2018) Detection of volatile-organic-compounds (VOCs) in solution using cantilever-based gas sensors. Talanta 182:148–155

Beigloo F, Noori A, Mehrgardi MA, Mousavi MF (2016) Label-free and sensitive impedimetric nanosensor for the detection of cocaine based on a supramolecular complexation with β-cyclodextrin, immobilized on a nanostructured polymer film. J Iran Chem Soc 13:659–669

Bilibana MP, Williams AR, Rassie C, Sunday CE, Makelane H, Wilson L, Ntshongontshi N, Jijana AN, Masikini M, Baker PGL, Iwuoha EI (2016) Electrochemical aptatoxisensor responses on nanocomposites containing electro-deposited silver nanoparticles on poly(Propyleneimine) dendrimer for the detection of microcystin-LR in freshwater. Sensors (Switzerland) 16:1–13

Bratakou S, Nikoleli GP, Siontorou CG, Nikolelis DP, Karapetis S, Tzamtzis N (2017) Development of an electrochemical biosensor for the rapid detection of saxitoxin based on air stable lipid films with incorporated anti-STX using graphene electrodes. Electroanalysis 29:990–997

Campos BB, Contreras-Cáceres R, Bandosz TJ, Jiménez-Jiménez J, Rodríguez-Castellón E, da Silva JCGE, Algarra M (2017) Carbon dots coated with vitamin B12 as selective ratiometric nanosensor for phenolic carbofuran. Sens Actuators B Chem 239:553–561

Cao X, Zhao F, Jiang Z, Hong S, Zhang C, She Y, Jin F, Jin M, Wang J (2018) Rapid analysis of bitertanol in agro-products using molecularly imprinted polymers-surface-enhanced Raman spectroscopy. Food Anal Methods 11:1435–1443

Chang K, Wang S, Zhang H, Guo Q, Hu X, Lin Z, Sun H, Jiang M, Hu J (2017) Colorimetric detection of melamine in milk by using gold nanoparticles-based LSPR via optical fibers. PLoS One 12:1–12

Chen X, Liu Y, Fang X, Li Z, Pu H, Chang J, Chen J, Mao S (2019) Ultratrace antibiotic sensing using aptamer/graphene-based field-effect transistors. Biosens Bioelectron 126:664–671

Chullasat K, Nurerk P, Kanatharana P, Davis F, Bunkoed O (2018) A facile optosensing protocol based on molecularly imprinted polymer coated on CdTe quantum dots for highly sensitive and selective amoxicillin detection. Sens Actuators B Chem 254:255–263

Dewangan PK, Khan F, Shrivas K, Sahu V (2019) Determination of uranium in environmental sample by nanosensor graphene quantum dots. J Radioanal Nucl Chem 320:757. https://doi.org/10.1007/s10967-019-06512-x

Deymehkar E, Taher MA, Karami C, Arman A (2018) Synthesis of SPR nanosensor using gold nanoparticles and its application to copper (II) determination. SILICON 10:1329–1336

Dong X, Wu K, Zhu W, Wu P, Hou J, Wang Z, Li R, Wu J, Liu Z, Guo X (2019) TiO_2 nanotubes/g-C_3N_4 quantum dots/rGO Schottky heterojunction nanocomposites as sensors for ppb-level detection of NO_2. J Mater Sci 54:7834–7849

Dou X, Zhang L, Liu C, Li Q, Luo J, Yang M (2018) Fluorometric competitive immunoassay for chlorpyrifos using rhodamine-modified gold nanoparticles as a label. Microchim Acta 185:41

Duan H, Deng W, Gan Z, Li D, Li D (2019) SERS-based chip for discrimination of formaldehyde and acetaldehyde in aqueous solution using silver reduction. Microchim Acta 186:175

Ellington AD, Szostak JW (1990) *In vitro* selection of RNA molecules that bind specific ligands. Nature 346:818–822

Elmizadeh H, Soleimani M, Faridbod F, Bardajee GR (2017) Ligand-capped CdTe quantum dots as a fluorescent nanosensor for detection of copper ions in environmental water sample. J Fluoresc 27:2323–2333

Fang G, Wang H, Yang Y, Liu G, Wang S (2016) Development and application of a quartz crystal microbalance sensor based on molecularly imprinted sol-gel polymer for rapid detection of patulin in foods. Sens Actuators B Chem 237:239–246

Fang M, Zhou L, Zhang H, Liu L, Gong ZY (2019) A molecularly imprinted polymers/carbon dots-grafted paper sensor for 3-monochloropropane-1,2-diol determination. Food Chem 274:156–161

Feng J, Shen Q, Wu J, Dai Z, Wang Y (2019) Naked-eyes detection of *Shigella flexneri* in food samples based on a novel gold nanoparticle-based colorimetric aptasensor. Food Control 98:333–341

Food and Agriculture Organisation of the United Nations (2016) Guidelines on highly hazardous pesticides. World Health Organisation, Rome

Geng Y, Guo M, Tan J, Huang S, Tang Y, Tan L, Liang Y (2019) The fabrication of highly ordered fluorescent molecularly imprinted mesoporous microspheres for the selective sensing of sparfloxacin in biological samples. Sens Actuators B Chem 281:821–829

Ghalkhani M, Maghsoudi S, Saeedi R, Khaloo SS (2019) Ultrasensitive quantification of paraquat using a newly developed sensor based on silver nanoparticle-decorated carbon nanotubes. J Iran Chem Soc 16:1301. https://doi.org/10.1007/s13738-019-01605-6

Guo L, Wu X, Liu L, Kuang H, Xu C (2018a) Gold nanoparticle-based paper sensor for simultaneous detection of 11 benzimidazoles by one monoclonal antibody. Small 14:1701782

Guo Z, Jia Y, Song X, Lu J, Lu X, Liu B, Han J, Huang Y, Zhang J, Chen T (2018b) Giant gold nanowire vesicle-based colorimetric and SERS dual-mode immunosensor for ultrasensitive detection of *Vibrio parahemolyticus*. Anal Chem 90:6124–6130

Guo Z, Zhao X, Jiao S, Jiao Z, Zhang H, Zhu D (2018c) A turn-on biosensor-based aptamer-mediated carbon quantum dots nanoaggregate for acetamiprid detection in complex samples. Food Anal Methods 12:668–676

Guo P, Yang W, Hu H, Wang Y, Li P (2019) Rapid detection of aflatoxin B1 by dummy template molecularly imprinted polymer capped CdTe quantum dots. Anal Bioanal Chem 411:2607. https://doi.org/10.1007/s00216-019-01708-2

Gupta VK, Yola ML, Atar N (2014) A novel molecular imprinted nanosensor based quartz crystal microbalance for determination of kaempferol. Sens Actuators B Chem 194:79–85

Haddi Z, Mabrouk S, Bougrini M, Tahri K, Sghaier K, Barhoumi H, El Bari N, Maaref A, Jaffrezic-Renault N, Bouchikhi B (2014) E-nose and e-tongue combination for improved recognition of fruit juice samples. Food Chem 150:246–253

Han KN, Choi JS, Kwon J (2016) Three-dimensional paper-based slip device for one-step point-of-care testing. Sci Rep 6:1–7

Hebbar VS, Javey A, Chen Y-Z, Kiriya D, Hettick M, Ota H, Fahad HM, Gao W, Shiraki H, Amani M, Chueh Y-L, Zhang C (2017) Room temperature multiplexed gas sensing using chemical-sensitive 3.5-nm-thin silicon transistors. Sci Adv 3:e1602557

Hossain MZ, Maragos CM (2018) Gold nanoparticle-enhanced multiplexed imaging surface plasmon resonance (iSPR) detection of *Fusarium* mycotoxins in wheat. Biosens Bioelectron 101:245–252

Hu P, Zhou X, Wu Q (2014) A new nanosensor composed of laminated samarium borate and immobilized laccase for phenol determination. Nanoscale Res Lett 9:1–7

Hu J, Fu K, Bohn PW (2018a) Whole-cell *Pseudomonas aeruginosa* localized surface plasmon resonance aptasensor. Anal Chem 90:2326–2332

Hu L, Zhang Q, Gan X, Yin W, Fu W (2018b) Switchable fluorescence of MoS_2 quantum dots: a multifunctional probe for sensing of chromium(VI), ascorbic acid, and alkaline phosphatase activity. Anal Bioanal Chem 410:7551–7557

Hu J, Jiang YZ, Tang M, Wu LL, Xie HY, Zhang ZL, Pang DW (2019) Colorimetric-fluorescent-magnetic nanosphere-based multimodal assay platform for *Salmonella* detection. Anal Chem 91:1178–1184

Ikram F, Qayoom A, Aslam Z, Shah MR (2019) Epicatechin coated silver nanoparticles as highly selective nanosensor for the detection of Pb^{2+} in environmental samples. J Mol Liq 277:649–655

Ilkhani H, Farhad S (2018) A novel electrochemical DNA biosensor for Ebola virus detection. Anal Biochem 557:151–155

Islam S, Shukla S, Bajpai VK, Han YK, Huh YS, Ghosh A, Gandhi S (2019) Microfluidic-based graphene field effect transistor for femtomolar detection of chlorpyrifos. Sci Rep 9:1–7

Javidi M, Housaindokht MR, Verdian A, Razavizadeh BM (2018) Detection of chloramphenicol using a novel apta-sensing platform based on aptamer terminal-lock in milk samples. Anal Chim Acta 1039:116–123

Jigyasa A, Rajput JK (2018) Bio-polyphenols promoted green synthesis of silver nanoparticles for facile and ultra-sensitive colorimetric detection of melamine in milk. Biosens Bioelectron 120:153–159

Kaur H, Shorie M (2019) Nanomaterial based aptasensors for clinical and environmental diagnostic applications. Nanoscale Adv 1:2123. https://doi.org/10.1039/C9NA00153K

Kaur H, Shorie M, Sharma M, Ganguli AK, Sabherwal P (2017) Bridged rebar graphene functionalized aptasensor for pathogenic *E. coli* O78:K80:H11 detection. Biosens Bioelectron 98:486–493

Khater M, de la Escosura-Muñiz A, Quesada-González D, Merkoçi A (2019) Electrochemical detection of plant virus using gold nanoparticle-modified electrodes. Anal Chim Acta 1046:123–131

Köhler H-R, Triebskorn R (2013) Wildlife ecotoxicology of pesticides: can we track effects to the population level and beyond? Science 341:759–765

Li H, Somerson J, Xia F, Plaxco KW (2018a) Electrochemical DNA-based sensors for molecular quality control: continuous, real-time melamine detection in flowing whole milk. Anal Chem 90:10641–10645

Li X, Jiao HF, Shi XZ, Sun A, Wang X, Chai J, Li DX, Chen J (2018b) Development and application of a novel fluorescent nanosensor based on FeSe quantum dots embedded silica molecularly imprinted polymer for the rapid optosensing of cyfluthrin. Biosens Bioelectron 99:268–273

Li D, Wang S, Wang L, Zhang H, Hu J (2019) A simple colorimetric probe based on anti-aggregation of AuNPs for rapid and sensitive detection of malathion in environmental samples. Anal Bioanal Chem 411:2645. https://doi.org/10.1007/s00216-019-01703-7

Lin ZT, Gu J, Li CH, Lee TR, Xie L, Chen S, Cao PY, Jiang S, Yuan Y, Hong X, Wang H, Wang D, Wang X, Jiang GB, Heon M, Wu T (2017) A nanoparticle-decorated biomolecule-responsive polymer enables robust signaling cascade for biosensing. Adv Mater 29:1–8

Liu J, Zeng J, Tian Y, Zhou N (2018a) An aptamer and functionalized nanoparticle-based strip biosensor for on-site detection of kanamycin in food samples. Analyst 143:182–189

Liu Y, Chen X, Wang M, Ma Q (2018b) A visual electrochemiluminescence resonance energy transfer/surface plasmon coupled electrochemiluminescence nanosensor for shiga toxin-producing: *Escherichia coli* detection. Green Chem 20:5520–5527

Lu Z, Chen X, Wang Y, Zheng X, Li CM (2014) Aptamer based fluorescence recovery assay for aflatoxin B1 using a quencher system composed of quantum dots and graphene oxide. Microchim Acta 182:571–578

Madianos L, Tsekenis G, Skotadis E, Patsiouras L, Tsoukalas D (2018) A highly sensitive impedimetric aptasensor for the selective detection of acetamiprid and atrazine based on microwires formed by platinum nanoparticles. Biosens Bioelectron 101:268–274

Malik M, Padhye P, Poddar P (2019) Down conversion luminescence-based nanosensor for label-free detection of explosives. ACS Omega 4:4259–4268

Manzano M, Viezzi S, Mazerat S, Marks RS, Vidic J (2018) Rapid and label-free electrochemical DNA biosensor for detecting hepatitis a virus. Biosens Bioelectron 100:89–95

Masteri-Farahani M, Mosleh N (2019) Functionalization of graphene quantum dots with anti-morphine: design of selective nanosensor for detection of morphine. Mater Lett 241:206–209

Mateo-Sagasta J, Marjani S, Turral H, Burke J (2017) Water pollution from agriculture: a global review. Food and Agriculture Organisation of the United Nations, Rome

Mehrzad-Samarin M, Faridbod F, Dezfuli AS, Ganjali MR (2017) A novel metronidazole fluorescent nanosensor based on graphene quantum dots embedded silica molecularly imprinted polymer. Biosens Bioelectron 92:618–623

Memon AG, Zhou X, Xing Y, Wang R, Liu L, Khan M, He M (2019) Label-free colorimetric nanosensor with improved sensitivity for Pb^{2+} in water by using a truncated 8–17 DNAzyme. Front Environ Sci Eng 13:1–6

Miao Y, Gan N, Li T, Zhang H, Cao Y, Jiang Q (2015) A colorimetric aptasensor for chloramphenicol in fish based on double-stranded DNA antibody labeled enzyme-linked polymer nanotracers for signal amplification. Sens Actuators B Chem 220:679–687

Mortazavi SS, Farmany A (2017) MWCNTs@rubeanic acid nanosensor for ultra sensitive quantification of bismuth in groundwater and soil. Russ J Electrochem 53:223–226

Mousavi Nodoushan S, Nasirizadeh N, Amani J, Halabian R, Imani Fooladi AA (2019) An electrochemical aptasensor for staphylococcal enterotoxin B detection based on reduced graphene oxide and gold nano-urchins. Biosens Bioelectron 127:221–228

Muhammad N, Rahman A, Younis MA, Subhani Q, Shehzad K, Cui H, Zhu Y (2018) Porous SnO_2 nanoparticles based ion chromatographic determination of non-fluorescent antibiotic (chloramphenicol) in complex samples. Sci Rep 8:1–11

Niu X, He Y, Li X, Zhao H, Pan J, Qiu F, Lan M (2019) A peroxidase-mimicking nanosensor with Hg^{2+}-triggered enzymatic activity of cysteine-decorated ferromagnetic particles for ultrasensitive Hg^{2+} detection in environmental and biological fluids. Sens Actuators B Chem 281:445–452

Peng J, Yin W, Shi J, Jin X, Ni G (2019) Magnesium and nitrogen co-doped carbon dots as fluorescent probes for quenchometric determination of paraoxon using pralidoxime as a linker. Microchim Acta 186:24

Prasad BB, Jauhari D (2015) A dual-template biomimetic molecularly imprinted dendrimer-based piezoelectric sensor for ultratrace analysis of organochlorine pesticides. Sens Actuators B Chem 207:542–551

Priyanka, Shorie M, Bhalla V, Pathania P, Raman SC (2014) Nanobioprobe mediated DNA aptamers for explosive detection. Chem Commun 50:1080–1082

Qi J, Li B, Wang X, Fu L, Luo L, Chen L (2018) Rotational paper-based microfluidic-chip device for multiplexed and simultaneous fluorescence detection of phenolic pollutants based on a molecular-imprinting technique. Anal Chem 90:11827–11834

Qian J, Wang K, Wang C, Ren C, Liu Q, Hao N, Wang K (2017) Ratiometric fluorescence nanosensor for selective and visual detection of cadmium ions using quencher displacement-induced fluorescence recovery of CdTe quantum dots-based hybrid probe. Sens Actuators B Chem 241:1153–1160

Qiu Q, Chen H, Ying S, Sharif S, You Z, Wang Y, Ying Y (2019) Simultaneous fluorometric determination of the DNAs of *Salmonella enterica, Listeria monocytogenes* and *Vibrio parahemolyticus* by using an ultrathin metal-organic framework (type Cu-TCPP). Microchim Acta 186:93. https://doi.org/10.1007/s00604-019-3226-y

Rahbar N, Salehnezhad Z, Hatamie A, Babapour A (2018) Graphitic carbon nitride nanosheets as a fluorescent probe for chromium speciation. Microchim Acta 185:101

Ren X, Cheshari EC, Qi J, Li X (2018) Silver microspheres coated with a molecularly imprinted polymer as a SERS substrate for sensitive detection of bisphenol A. Microchim Acta 185:242

Saberi Z, Rezaei B (2019) Fluorometric label-free aptasensor for detection of the pesticide acetamiprid by using cationic carbon dots prepared with cetrimonium bromide. Microchim Acta 186:273. https://doi.org/10.1007/s00604-019-3378-9

Sadeghi R, Karimi-Maleh H, Khalilzadeh MA, Beitollahi H, Ranjbarha Z, Zanousi MBP (2013) A new strategy for determination of hydroxylamine and phenol in water and waste water samples using modified nanosensor. Environ Sci Pollut Res 20:6584–6593

Saha A, Neogy S, Rao DRM, Deb SB, Saxena MK, Tomar BS (2019) Colorimetric and visual determination of ultratrace uranium concentrations based on the aggregation of amidoxime functionalized gold nanoparticles. Microchim Acta 186:183

Sang S, Guo X, Liu R, Wang J, Guo J, Zhang Y, Yuan Z, Zhang W (2018) A novel magnetoelastic nanobiosensor for highly sensitive detection of atrazine. Nanoscale Res Lett 13:414

Shams S, Bakhshi B, Tohidi Moghadam T, Behmanesh M (2019) A sensitive gold-nanorods-based nanobiosensor for specific detection of *Campylobacter jejuni* and *Campylobacter coli*. J Nanobiotechnol 17:43

Shi Q, Huang J, Sun Y, Deng R, Teng M, Li Q, Yang Y, Hu X, Zhang Z, Zhang G (2018) A SERS-based multiple immuno-nanoprobe for ultrasensitive detection of neomycin and quinolone antibiotics via a lateral flow assay. Microchim Acta 185:84

Shorie M, Kaur H (2018) Microtitre plate based cell-SELEX method. Bio-Protocol 8:1–9

Shorie M, Kaur H, Chadha G, Singh K, Sabherwal P (2019) Graphitic carbon nitride QDs impregnated biocompatible agarose cartridge for removal of heavy metals from contaminated water samples. J Hazard Mater 367:629–638

Song D, Yang R, Fang S, Liu Y, Long F, Zhu A (2018) SERS based aptasensor for ochratoxin A by combining Fe_3O_4@Au magnetic nanoparticles and Au-DTNB@Ag nanoprobes with multiple signal enhancement. Microchim Acta 185:491

Sun A, Chai J, Xiao T, Shi X, Li X, Zhao Q, Li D, Chen J (2018a) Development of a selective fluorescence nanosensor based on molecularly imprinted-quantum dot optosensing materials for saxitoxin detection in shellfish samples. Sens Actuators B Chem 258:408–414

Sun J, Gan T, Zhai R, Fu W, Zhang M (2018b) Sensitive and selective electrochemical sensor of diuron against indole-3-acetic acid based on core-shell structured SiO_2@Au particles. Ionics (Kiel) 24:2465–2472

Sun H, Mei Q, Shikha S, Liu J, Zhang J, Zhang Y (2019a) White-light emissive upconversion nanoparticles for visual and colorimetric determination of the pesticide thiram. Microchim Acta 186:106

Sun Y, Duan N, Ma P, Liang Y, Zhu X, Wang Z (2019b) Colorimetric aptasensor based on truncated aptamer and trivalent dnazyme for *Vibrio parahemolyticus* determination. J Agric Food Chem 67:2313–2320

Tan F, Xie X, Xu A, Deng K, Zeng Y, Yang X, Huang H (2019) Fabricating and regulating peroxidase-like activity of eggshell membrane-templated gold nanoclusters for colorimetric detection of staphylococcal enterotoxin B. Talanta 194:634–642

Tancharoen C, Sukjee W, Thepparit C, Jaimipuk T, Auewarakul P, Thitithanyanont A, Sangma C (2019) Electrochemical biosensor based on surface imprinting for zika virus detection in serum. ACS Sensors 4:69–75

Teymoori N, Raoof JB, Khalilzadeh MA, Ojani R (2018) An electrochemical sensor based on CuO nanoparticle for simultaneous determination of hydrazine and bisphenol a. J Iran Chem Soc 15:2271–2279

Tomer VK, Singh K, Kaur H, Shorie M, Sabherwal P (2017) Rapid acetone detection using indium loaded WO_3/SnO_2 nanohybrid sensor. Sens Actuators B Chem 253:703–713

Tortolini C, Bollella P, Antiochia R, Favero G, Mazzei F (2016) Inhibition-based biosensor for atrazine detection. Sens Actuators B Chem 224:552–558

Tuerk C, Gold L (1990) Systematic evolution of ligands by exponential enrichment: RNA ligands to bacteriophage T4 DNA polymerase. Science 249:505–510

Turkevich J, Stevenson PC, Hillier J (1951) A study of the nucleation and growth processes in the synthesis of colloidal gold. Discuss Faraday Soc 11:55

Ul Ain N, Anis I, Ahmed F, Shah MR, Parveen S, Faizi S, Ahmed S (2018) Colorimetric detection of amoxicillin based on querecetagetin coated silver nanoparticles. Sens Actuators B Chem 265:617–624

Wagner S, Bell J, Biyikal M, Gawlitza K, Rurack K (2018) Integrating fluorescent molecularly imprinted polymer (MIP) sensor particles with a modular microfluidic platform for nanomolar small-molecule detection directly in aqueous samples. Biosens Bioelectron 99:244–250

Wang Y, Fry HC, Skinner GE, Schill KM, Duncan TV (2017) Detection and quantification of biologically active botulinum neurotoxin serotypes A and B using a förster resonance energy transfer-based quantum dot nanobiosensor. ACS Appl Mater Interfaces 9:31446–31457

Wang Q, Li S, Li J (2018a) A molecularly imprinted sensor with enzymatic enhancement of electrochemiluminescence of quantum dots for ultratrace clopyralid determination. Anal Bioanal Chem 410:5165–5172

Wang Y, Yue Q, Tao L, Zhang C, Li CZ (2018b) Fluorometric determination of hydroquinone by using blue emitting N/S/P-codoped carbon dots. Microchim Acta 185:550

World Health Organisation (2011) World health report 2010. World Health Organisation, Geneva

World Health Organisation (2015) Estimates of the global burden of foodborne diseases. World Health Organisation, Geneva

Wujcik EK, Duirk SE, Chase GG, Monty CN (2016) A visible colorimetric sensor based on nanoporous polypropylene fiber membranes for the determination of trihalomethanes in treated drinking water. Sens Actuators B Chem 223:1–8

Xiang Q, Huang J, Huang H, Mao W, Ye Z (2018a) A label-free electrochemical platform for the highly sensitive detection of hepatitis B virus DNA using graphene quantum dots. RSC Adv 8:1820–1825

Xiang Y, Camarada MB, Wen Y, Wu H, Chen J, Li M, Liao X (2018b) Simple voltammetric analyses of ochratoxin a in food samples using highly-stable and anti-fouling black phosphorene nanosensor. Electrochim Acta 282:490–498

Xie X, Tan F, Xu A, Deng K, Zeng Y, Huang H (2019) UV-induced peroxidase-like activity of gold nanoclusters for differentiating pathogenic bacteria and detection of enterotoxin with colorimetric readout. Sens Actuators B Chem 279:289–297

Xu X, Ma X, Wang H, Wang Z (2018a) Aptamer based SERS detection of *Salmonella typhimurium* using DNA-assembled gold nanodimers. Microchim Acta 185:325

Xu Z, Bi X, Huang Y, Che Z, Chen X, Fu M, Tian H, Yang S (2018b) Sensitive colorimetric detection of *Salmonella enteric* serovar *typhimurium* based on a gold nanoparticle conjugated bifunctional oligonucleotide probe and aptamer. J Food Saf 38:e12482

Zhang CH, Liu LW, Liang P, Tang LJ, Yu RQ, Jiang JH (2016) Plasmon coupling enhanced raman scattering nanobeacon for single-step, ultrasensitive detection of cholera toxin. Anal Chem 88:7447–7452

Zhang P, Lyu Z, Viktorova J, Offenhäusser A, Feng L, Mayer D (2018) Nanoparticle stripe sensor for highly sensitive and selective detection of mercury ions. Biosens Bioelectron 117:450–456

Zhang Z, Ma X, Jia M, Li B, Rong J, Yang X (2019) Deposition of CdTe quantum dots on microfluidic paper chips for rapid fluorescence detection of pesticide 2,4-D. Analyst 144:1282–1291

Zhu T, Hu Y, Yang K, Dong N, Yu M, Jiang N (2018) A novel SERS nanoprobe based on the use of core-shell nanoparticles with embedded reporter molecule to detect *E. coli* O157:H7 with high sensitivity. Microchim Acta 185:30

Zou D, Jin L, Wu B, Hu L, Chen X, Huang G, Zhang J (2019) Rapid detection of *Salmonella* in milk by biofunctionalised magnetic nanoparticle cluster sensor based on nuclear magnetic resonance. Int Dairy J 91:82–88

Chapter 5
Advances in Agronanotechnology and Future Prospects

Kalaivani Nadarajah

Contents

5.1 Introduction

The agricultural industry is facing several challenges such as climate change, urbanization, industrialization, and nonsustainable agricultural practices. To further escalate the problem, an increase in global population is likely to result in an exponentially increase in demand over the next 30 years (Gruère et al. 2011). With the depleting natural energy resources, agricultural produce is no longer just viewed as a food source but as material for other commerce and manufacturing industries (Gruère et al. 2011). In developing countries where agriculture is a main contributor toward their GDP, critical issues faced include lack of arable land, cost of fertilizers and pesticides, lack of agricultural land, poverty, and mal-

K. Nadarajah (✉)
School of Environmental and Natural Resources Sciences, Faculty of Science and Technology, Universiti Kebangsaan Malaysia, Bangi, Selangor, Malaysia
e-mail: vani@ukm.edu.my

D. Thangadurai et al. (eds.), *Nanotechnology for Food, Agriculture, and Environment*, Nanotechnology in the Life Sciences,
https://doi.org/10.1007/978-3-030-31938-0_5

nutrition. Therefore, besides changes in policies and governance covering these issues, there is also a need to develop technologies to address these challenges to food security (Biswal et al. 2012).

Modern agriculture is dependent on advances made in agricultural technologies. One of the latest technologies to prominently transform the agro-based industry is the use of nanotechnology. Nanotechnology allows for the use of materials to alter the physical and chemical properties of any substance. Through the development of nanomaterials as well as nano-based devices, various novel applications in plant biotechnology and agriculture may be established (Scrinis and Lyons 2007). Advances in the field of agricultural nanotechnology has contributed towards reducing the hazards produced by chemicals such as fertilizers and pesticides, reducing water utilization, enhancing pests and disease management, and nutrient and soil management (Prasad et al. 2017; Sangeetha et al. 2017a, b). Through nanotechnology, it is hoped that agricultural inputs (water, fertilizers, pesticides) may be reduced while still increasing yield and reducing costs (Chen and Yada 2011; Gruère et al. 2011; Fakruddin et al. 2012; Tarafdar et al. 2013). Research and development in nanotechnology is set to take the agricultural industry to the next level through the utilization of this technology in generating animal production inputs, genetically modified crops, chemical pesticides, and precision farming techniques (Prasad et al. 2014, 2017). Nanoparticles are part and parcel of the abovementioned application through encapsulation, entrapment, polymers, and others (Scott and Chen 2003).

As mentioned above, nanotech addresses the issue of hazards to ensure human well-being by providing an alternative ecofriendly, controlled, and reduced risk agricultural technology (Kah 2015; Prasad et al. 2017). Smart nanotools revolutionized agricultural practices to improve both quantity and quality of produce (Liu and Lal 2015; Servin et al. 2015). To date, some of the potential applications of these nanomaterials have been reviewed (Nair et al. 2010; Jampílek and Kráľová 2015). For this technology to take off in the agricultural industry, more research on the application, feasibility, and efficiency of this technology needs to be conducted to convince the critics of this technology in agriculture (Laborie 2009). The future utilization of nanomaterials has raised concerns with regard to their disposal, and numerous universal projects these days are contemplating the hazard identified with their utilization. Nanotechnologies are promising advances with the capacity to make gigantic changes in nourishment and agrarian frameworks. Despite the fact that the hazard appraisal of the utilization of nanoparticles in agribusiness is not well-characterized, nanotechnology-supported products and applications are better equipped to manage and conserve the inputs in agriculture. With optimization and better understanding, this technology is predicted to resolve most of the challenges faced by the agricultural industry. In the following sections, we will look into the various applications of nanomaterials in agriculture and the challenges and future prospects of this technology.

5.2 What Is Nanotechnology?

Nanotechnology is a technology that has made a significant impact on many industries and daily life (Scott and Chen 2003; Gruère et al. 2011). Nanoparticles, which has biological, chemical, and physical properties, are slowly but surely making an impact on human well-being. According to EPA, nanoparticles are between 1 and 100 nm. Due to its minute size and properties, this material is suitable for use in many innovative and novel applications that address societal needs and problems (Nakache et al. 1999). There are many agriculturally dependent countries that have started looking into the use of nanotechnology like China. Products developed for use in agriculture in China are likely to hit the fields in the next decade. While there are products being developed with nanomaterials, their use in the field is largely dependent on factors such as acceptance by the farming community, risk assessment, effectiveness in the environment, cost and profit margins, and market demand. In the following sections, we address the various potential applications of nanotechnology and nanomaterial in agriculture dealing with both plant and animal agricultural industry. Some of these applications have been applied in the field, and some are still in the experimental stages needing more tests in field to determine their efficacy and also their cost-effectiveness in the environment and in the market. However, in the years to come, it is almost impossible not to look into the use of nanotechnology in the agricultural arena due to the many benefits nanomaterials possess (Ditta 2012).

5.3 Overview of Nanotechnology Applications in Agriculture

In the years to come, with the shrinking land and the increasing population, the agricultural industry is under pressure from lack of arable land and the increasing cost of inputs such as water fertilizer and pesticides. In order to address this issue, there is a need to introduce precision farming to overcome the constraints and to maximize agricultural produce. One way of introducing precision farming is through the use of advanced technologies such as nanotechnology. The current use of nanotechnology has been in fix situations such as in factory settings. However, in the agricultural system, the setting is fluid. Environment changes and the materials used need to have flexibility to adapt and change with the environment to be responsive and provide positive change. This will require the development of specific nanomaterials suitable for agriculture.

Nanoporous materials have been developed with the capability to store and slowly release water when needed in agricultural regions that face water stress. This soil enhancer is able to evenly distribute water in the field (Prasanna 2007; Mousavi and Rezaei 2011). In addition to water storage and release, nanomaterials have also been used in the delivery of chemicals such as pesticides and fertilizers. Nanomaterials are able to detect and deliver with high precision nutrients and pesticides. Through this precision delivery mechanism, the carriers are able to reduce the incidence of excessive chemical usage that not only increases costs but also has a detrimental effect on the environment through the agricultural waste runoffs (DeRosa et al. 2010; Chen and Yada 2011). Through proper use of chemicals in the agricultural industry, we can reduce damage to the plants and agricultural ecology where the effect to environment is reduced, providing an eco-friendly solution to the agricultural industry (El Beyrouthya and El Azzi 2014).

While there has been research in the use of nanotechnology in agriculture, the application of this technology has received some negative response due to concerns on environmental security, food safety, and their disposal (Takeuchi et al. 2014). Therefore, the risk assessment on the use of this material needs to be sufficient to convince farmers and the public of the general good. However, based on the knowledge behind the technology, it is clear that its application is wide, providing a way forward in addressing increase in production, management, and conservation of inputs. This will provide a twenty-first-century solution to the twenty-first-century problems (Mukhopadhyay 2014). Figure 5.1 provides the various sources of nanotechnology and how these materials may be utilized in agriculture.

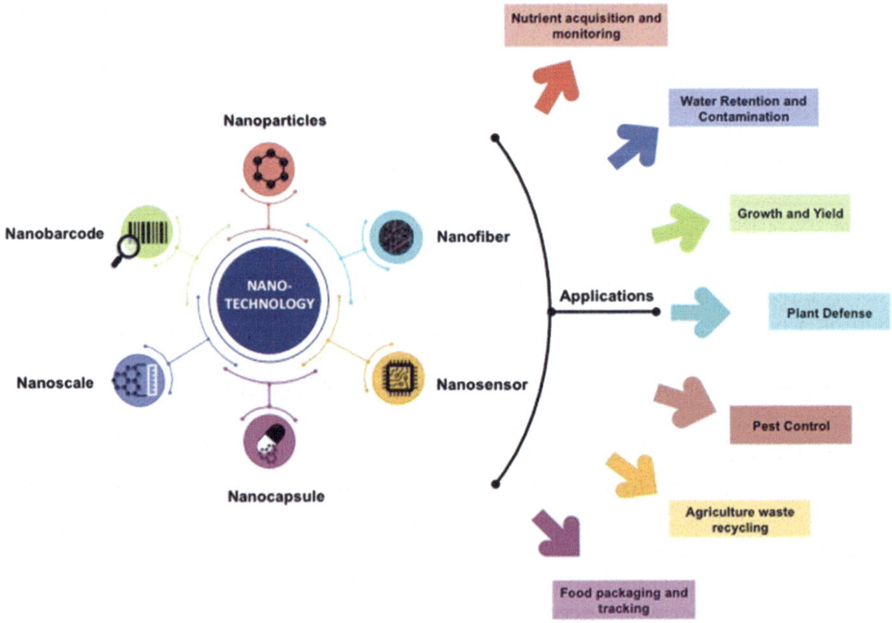

Fig. 5.1 Nanotechnology products and their application in agriculture

5.4 Application of Nanotechnology in Agriculture

In the following subsections, we will look into the main uses of nanotechnology in agriculture with examples.

5.4.1 Nanomaterials in Nutrient Acquisition

Nanofertilizers are able to enhance nutrient acquisition and assist with intensified sustainable agriculture. This application as nanofertilizer is achieved through encapsulation of product or the utilization of biosensors which are able to determine amounts of fertilizer needed in any given soil condition. Fertilizer can be encapsulated within nanoparticles to increase duration of effectiveness through (i) the use of nanoporous materials, (ii) polymer coating, and (iii) nanoscale emulsions (Rai et al. 2012). The ability of these nanoparticles to increase surface area and thus improve absorbance of nutrients has reduced wastage of fertilizers in farms. These compounds are also efficient in reducing loss of nutrients through emissions, leaching, and assimilation via microorganisms (DeRosa et al. 2010). Nutrient delivery systems via nanoparticles have been shown to be effective in tomatoes (Khodakovskaya et al. 2009) and ryegrass (Lin et al. 2010). These encapsulated compounds exhibited sustained release while preventing premature conversion into substances that cannot be utilized by the plant. This can be easily achieved through the use of biosensors. In addition, the slow controlled release of nutrients is also bound to reduce the toxicity effect of fertilizers in the soil (Rai et al. 2012).

One commonly used material for agricultural delivery systems is zeolite. This material has been utilized to improve nutrient delivery, slow release of chemicals, and water management/retention in soil, thus resulting in better soil quality. Due to the ability of zeolite to hold nutrients in the root zone, this results in better and more effective fertilization over a longer period of time resulting in improved plant yield. In addition, this material is long lasting and is able to provide a permanent solution to nutrient and water management issues. Further, when zeolites are linked to nanobiosensors, this will enable the system to respond toward deficiencies in plant and soil. In dealing with farm animals too, nanoparticles have been utilized in assisting with nutrient acquisition and digestion. Casein micelles have been used to provide nutrients for weanlings. These casein particles are able to undergo proteolytic cleavage in the stomach and therefore pose no threat to the farm animals (Semo et al. 2007; Haham et al. 2012; Day et al. 2015). However, in designing nanoparticles for nutrient supplementation, we must address the coherent challenges of oral administration which includes the different pH and unique environment of the gastrointestinal tract (Ban et al. 2015). Akbari and Wu (2016) also put forward the use of canola protein cruciferin as nanoparticles to encapsulate and protect hydrophobic and hydrophilic bioactive compounds from the harsh surrounding in the stomach. Nanotechnocrates are able to study the regulation of hormones in plant physiology. McLamore et al. (2010) investigated the regulation of auxin in plant development and growth. Nanosensors are able to monitor auxin levels and ensure the adaptation of the plant to the root environment (McLamore et al. 2010).

5.4.2 Nanoparticle in Growth Promotion

Nanoparticles also contribute to growth enhancement in plants (Table 5.1). One particle that has been reported to induce growth is carbon. Khodakovskaya et al. (2009) reported improved seed germination and increased biomass in tomato plants. The induced germination rate may be due to increased absorption of water from the surrounding by tissue-bound nanoparticles. Similarly, Tiwari et al. (2014) reported increased seed germination and growth from application of carbon nanoparticles. Carbon nanoparticles increased cell elongation in root and shoot of various plant systems in both light and dark conditions (Lin and Xing 2007; Canas et al. 2008; Wang et al. 2012). In the past years, the use of metal oxide nanoparticles has been studied in their efficiency in eliciting improved germination and growth in plants. Silver nanoparticles have been shown to increase root elongation in maize, cabbage, and barley (Pokhrel and Dubey 2013). Gold nanoparticles increased crop yield through improved chlorophyll content (Gopinath et al. 2014), improved antioxidant properties, and increased seed germination (Kumar et al. 2013). Iron oxides enhanced growth and biomass in crops like spinach. The content of iron in spinach was reported to increase significantly (more than 11 times) when treated with iron nanoparticles (Almeelbi and Bezbaruah 2014). Iron oxide nanoparticles resulted in elongated root cells in both pumpkin and soybean (López-Moreno et al. 2010) which consequently brought about growth and biomass increase (Sheykhbaglou et al. 2010).

Novel metal oxides such as titanium oxide can contribute to plant growth through enhanced seed germination (Suman et al. 2010; Feizi et al. 2013). In spinach, seed treatment with titanium nanoparticles increased fresh and dry weight (Yang et al. 2007). Similarly, seed germination and growth were stimulated in canola, by the application of TiO_2 nanoparticles (Mahmoodzadeh et al. 2013). It is believed that

Table 5.1 Some examples of nanoparticles with application in yield and growth enhancement of plants

Nanoparticle	Application	References
Zinc oxide	Increased shoot and root growth; increased yield	Barrena et al. (2009), Burman et al. (2013)
Alumina	Improved seed germination, root and shoot length; increased biomass	Barrena et al. (2009)
Iron oxide	Improved biomass and yield	Dhoke et al. (2013)
Titanium oxide	Increased chlorophyll and net photosynthesis	Mahmoodzadeh et al. (2013)
Carbon	Increase in growth	Khodakovskaya et al. (2013)

TiO_2 nanoparticles improved plant growth by enhancing Rubisco activities (Yang et al. 2007). Together with SiO_2, TiO_2 improves nitrate reductase activity leading to stimulated antioxidant activity in soybean (Lu et al. 2002). SiO_2 has the potential to improve photosynthesis, transpiration, and other physiological functions (Xie et al. 2012). Most of the abovementioned metal ions have been studied in different plant systems and show potential in growth enhancement. However, further study is required to determine efficiency and safety of these metals and metal ions in their application. Figure 5.1 provides some examples of these applications.

5.4.3 Nanomaterials in Plant Disease Management

In management of plant diseases, the most crucial part is understanding the process of infection and proliferation and thereafter looking into modes by which this may be mitigated. However, this has been a problem until recent years where through the use of nanofabrication we were able to study the interaction between plant and pathogen and therefore better understand how the microbes move. Through nanofabrication of the xylem system, we were able to better understand the mechanism involved in the microbes inhabiting the xylem. Previously, any such study would involve the destruction of tissue before any clear picture may be garnered on the process. The nanofabrication method took us into the entirety of the process in ways which were virtually impossible through the conventional method of analysis (Cursino et al. 2009; Zaini et al. 2009; Ditta 2012). Further, the application of nanosensors was also useful in the diagnosis of soil-borne diseases via the detection of the differential oxygen depletion/utilization status in the soil between the good and bad microbes. Similarly, the presence of plant viruses can also be detected through nanosensors in the soil (Ingale and Chaudhari 2013). By conducting this quantitative analysis, we are able to decide which microbes are suitable for the soil and also to predict an outbreak of soil-borne diseases ahead of time through a semiquantitative method of analysis. Therefore, biosensors are useful means to diagnose soil conditions based on semiquantitative approach (Rai et al. 2012).

In addressing fungal pathogens, it is advisable to use combinatorial nanoparticles in combating the fungal pathogens (Mishra et al. 2014). One commonly used nanoparticle in treating fungal infections is silver where it has been efficiently used in control of *Magnaporthe grisea* in the form of isoprothiolane or some other derivative (Rabab and El-Shafey 2013; Marziye Aboli et al. 2014). *Alternaria alternata* and *Botrytis cinerea* have been treated with copper nanoparticles, which have suppressed their growth efficiently (Ouda 2014). Some other nanoparticles such as magnesium oxide and zinc oxide have been used against *Fusarium oxysporum*, *Penicillium expansum*, *A. alternata*, and *Rhizopus stolonifer* (Wani and Shah 2012; Ramy and Ahmed 2013). These nanoparticles were effective at a concentration-dependent manner. The inhibitory concentration differed between organisms. Nanoparticles when used on fungi showed structural damages to the cell wall when observed under scanning electron microscopy. In a study conducted by Medda et al. (2015), silver nanoparticles were able to result in inhibitory effect against germination and budding of fungi. In addition to structural damage, there was also alteration in the lipid, protein, and sugar content. Similarly, silver nanoparticles have bactericidal activities against a wide range of organisms. Some of the organisms affected by silver are *Agrobacterium tumefaciens* (Chowdhury et al. 2014), *Erwinia cacticida*, *Citrobacter freundii* (Paulkumar et al. 2014), *Xanthomonas axonopodis*, *Ralstonia solanacearum* (Aravinthan et al. 2015), and *Xanthomonas axonopodis* pv. *punicae* (Bhor et al. 2014). Some of these silver nano particles were synthesized from extracts of plant materials (Prasad 2014, Prasad et al. 2018).

In the case of animal farming however, nanoparticles have been seen as a feasible way of administering antibiotics to prevent infections and infiltration of pathogens into hosts. Antibiotics have become a real problem in animal agriculture where the unregulated use of antibiotics has resulted in resistant bugs. Globally, legislation is in place restricting the use of prophylactic antibiotic (Yang et al. 2016a, b). However, if the use of antibiotics is to be restricted and regulated, we need to develop methods that impede disease transmission effectively, and this is where the nanoparticles are a good solution (Huijskens et al. 2016). Research has shown that metal nanoparticles due to their charge are able to lyase negatively charged membranes of bacteria (Gahlawat et al. 2016). Silver nanoparticles are potent against hemorrhagic enteritis-inciting *Escherichia coli* O157:H7 and yeast (Kim et al. 2007). Further modified bismuth nanofilms are utilized in the detection of *E. coli* (Zhang et al. 2007). Besides silver, copper has also been used as an oral delivery system with antimicrobial properties with efficient absorption at the intestinal mucosa (Solaiman et al. 2001; Zhao et al. 2007). In addition, zinc oxide (ZnO) and magnesium oxide (MgO) nanoparticles have been used in many antimicrobial products due to their ability in inhibiting microorganisms (Shah and Towkeer 2010). Shi et al. (2006) have also shown that nanoclay can be used to reduce the effect of microbial toxins on poultry. Therefore, these results show that there is some potential in the use of nano-antimicrobials in reducing disease incidence in farm animals and requires further study to determine the level of effectiveness compared to feed antibiotics. These nanoparticles are able to enhance the potential of antibiotics at lower doses. In administration of antibiotics, there has always been the ability of resistant microbes to cleave antibiotics into inactive forms. One such example is the cleavage of β-lactam ring by certain bacteria. However, by coating β-lactam antibiotic with nanoparticles, the cleavage ability of microbes is inhibited (Kalita et al. 2016). Therefore, this method is suitable in dealing with antibiotic-resistant organisms such as MRSA. Further, in studying the effectiveness and the safety of this material, experimentation on murine model showed that there was no toxicity when applied systemically of topically (Greenhalgh and Turos 2009). Hydrogels developed with nanoparticles such as qPDMAEMA have been reported to exhibit biocidal activity lyasing or distorting the membrane cell wall of bacteria (Xu et al. 2015). qPDMAEMA prevents bacterial adhesion, colonization, and biofilm formation. However, in surface application and oral administration, toxicity and long-term effects will have to be determined (Baker et al. 2013a, b).

5.4.4 Nanoparticles in Management of Pests

Other than managing the airborne and soil-borne pathogens in agricultural systems, nanoparticles are also introduced in the control of insect pests that are a major concern in the agricultural sector (Khot et al. 2012; Bhattacharyya et al. 2016). There have been serious concerns over the past decades on the use of pesticides to control insect pests in the field. The runoffs from the field post-application of these chemicals have resulted in chemical-contaminated water source that is harmful to the environment not forgetting the effects it has on the farm hands (Krupke et al. 2012). Recent reports have highlighted the toxicity of these chemicals where there have been implications to cause health issues such as neurological effects, respiratory diseases, cancer, Parkinson disorder, fetal diseases, infertility, diabetes, and genetic disorders (Hu et al. 2015). As an ecofriendly solution, nanotechnology has been introduced into crop protection against pest including weeds. As in the case with fertilizers, the nanoparticles may be utilized to stably deliver pesticides and other agrochemicals in any agricultural system. Their slow release mechanism also proves useful especially with regard to slow release of pesticide into the environment continuously and at a controlled and safer concentration. This results in less frequent application which is good for costs, human safety, and environment. Through green synthesis approach, both plant and microbes have become good candidates for the development of nanoparticles (Prasad 2014; Prasad et al. 2016, 2018). These materials are a more ecofriendly source for nanoparticles synthesis as it addresses the issue of chemical toxicity to environment and all living matter. This is the advantageous over ecotoxic conventional chemical methods (Khodakovskaya et al. 2012).

The encapsulated, slow-release pesticides have enhanced specificity, stability, and permeability that provide pest control efficacy for longer duration at less concentration levels then manual treatment (Bhattacharyya et al. 2016; Nuruzzaman et al. 2016). This provides a human, economic, and environmentally safe solution for pest control in agriculture (de Oliveira et al. 2014; Kah and Hofmann 2014; Bhattacharyya et al. 2016; Grillo et al. 2016). One example of the use of nanoparticles in pest control is through the use of porous silica loaded with water-soluble pesticide with the ability to slow release (Liu et al. 2006). The nanosilica component is able to interfere with the cuticular lipid barrier found in insect pests (Ulrichs et al. 2005). Other than silica, silver, aluminum, zinc, and titanium have been used in the management of rice weevil and *Bombyx mori* (Goswami et al. 2010). Nanoscale alumina has been used effectively against two insect pests: *Rhyzopertha dominica* (F.) and *S. oryzae* L. (Stadler et al. 2010). The nanoaluminum particles have shown high mortality of insect pests. Other than the individual application of these nanoparticles, combinatorial applications of compounds such as nanoparticles of silver-zinc combined were tested against *Aphis nerii* Boyer resulting in high level of mortality (Rouhani et al. 2012). In the recent years, a large number of chemical companies have started marketing nanoparticle-based pesticides such a Karate ZEON, Subdue MAXX, Primo MAXX, Banner MAXX, Ospray's Chyella, and Penncap-M (Gouin 2004; Otsuka et al. 2012). BASF and Sygenta are two major companies that have been and continue to be involved in the generation of nanoencapsulated pesticide and herbicides.

Other than insect pests, weeds are another group of pests that pose a problem in the agricultural industry. The growth of these weeds has either been controlled by herbicides in modern farming systems or through weeding in the conventional farms. The chemical constituents of these herbicides have been reported to have residual toxicity. Nanoherbicides are chemically stable and soluble and due to their slow release have less of a toxic effect on the environment. Recent studies have looked into nanoparticles as carriers of herbicides and have shown promise in reduced mobility in soil, thus causing genotoxicity (Grillo et al. 2014; Pereira et al. 2014). The use of chitosan cross-linked with herbicide has less toxic effect to the environment. Further, Grillo et al. (2015) studied the effect of hybrid herbicides (made from a concoction of three herbicides) in nanopackaging, which brought about systemic slow release of the herbicide to the environment (Huang et al. 2018).

5.4.5 Nanotechnology and Recycling of Water in Agriculture

One of the most crucial and critical ingredients in agriculture is water. It is an important and costly commodity in agriculture. However, in most farming communities, the water that is used for agriculture comes from contaminated water source. This contamination may be from any form of solid or liquid waste coming from domestic and industrial waste disposal. These contaminants are deleterious to the crop as well as to the soil fertility (Chong et al. 2010; Fatta-Kassinos et al. 2011). Then, there is the fact that the waters that result from agricultural practices are also contaminated with various chemical compounds contributing to groundwater pollution through fertilizers, pesticides, and other agricultural chemicals. Wastewater treatment requires consideration on water quality and quantity, where treatment, reuse, and safety needs to be coupled with proper monitoring. There are various ways of treating wastewater for use. However, some of these techniques may not be feasible for certain countries and locations (Oller et al. 2011). The recent implementation of nanotechnology in treatment of wastewater includes the use of certain membranes, nanoporous materials, and other nanomaterials (Prasad and Thirugnanasanbandham 2019). These materials are efficient in removal of chemical as well as biological contaminants. Two commonly used nanoparticle in water treatment are the nanoscale zero-valent iron which is cost-effective with strong absorption, and the other is the magnetic nanoparticle which is charge with paramagnetic properties and therefore able to absorb contaminants easily (Girginova et al. 2010; Fan et al. 2012; Qu et al. 2013). As in previous applications, nanoparticles like zeolite, metal oxides, alumina, silver, zinc, carbon nanotubes and fibers, enzymes, and titanium are also good for water treatments. Some of the ways that may be utilized to improve water quality and safety are as follows:

(i) *Purification from biological hazards*: In most developed countries, methods such as ozone, irradiation, and chlorine dioxide have been used to reduce or kill microorganisms. As some of these infrastructures are expensive, cheaper yet affective methods such as nanoparticles have been utilized in certain countries. Particles such as silver are bactericidal and viricidal (Nangmenyi and Economy 2009). Photocatalysis is also another method in the nanoscale that has been utilized in purification, decontamination, and deodorization of water and removal of organic contaminants (Li et al. 2009).

(ii) *Removal and detection of pollutants*: Nanoparticles or combinatorial nanoparticles have been used to remove chemical contaminants from agriculture-derived wastewater (Diallo et al. 2009; Farmen 2009). Biosensors have also been used to detect nanoscale contaminants in the water supply (Chen and Yada 2011).

(iii) *Wireless nanosensors*: These sensors are useful means of monitoring issues such as water storage, water distribution, water holding capacity, water utilization efficiency of plants, and water release, on demand fertilizers and pesticides. Biosensors that utilize the photosystem II can bind to herbicides and are potentially useful in monitoring pollutants in the environment (Giardi et al. 2005).

5.4.6 Recycling Agricultural Waste for Nanotechnology Application

There are a large group of researchers who are focused on looking into the use of green technology in addressing current and future problems. These products need to be safe for market and should preferably be biodegradable. One good source to develop green technology-based product is agricultural waste (Sangeetha et al. 2017). The amount of agricultural waste has been increasing over the years due to the increased output from the agricultural industry. Most agricultural waste comes from the stem, leaves, empty fruit bunch, and other plant material. All of these are rich in lignocellulosic material (Abdul Khalil et al. 2012). The cellulosic material obtained from agricultural waste has been used to produce nanocellulose. In addition to nanocellulose, there have been initiatives to extract noble metals from plant waste for use in drug delivery systems (Mahal et al. 2013), cancer therapy (Jain et al. 2006), and tissue/tumor imaging (Peng et al. 2009). Gold nanoparticles isolated from grape waste have affinity and in vitro stability in human breast cancer cells making it suitable for use in cancer studies (Krishnaswamy et al. 2014). The ability of these grape byproduct-derived gold nanoparticles to disperse water can revolutionize the application of nanotechnology in drug and nutraceutical delivery (Chen et al. 2013). The rice industry through rice milling has also provided cellulose that is processed into nanofibers that can serve as absorbents for fertilizers and pesticides. These highly absorbent nanofibers allow for targeted application when and where needed. In addition, through the burning of rice husks to produce thermal or biofuel, a large amount of nanosilica is produced which can be utilized into mak-

Table 5.2 Some applications of nanoparticles derived from plant waste

Nanoparticle	Plants	Application	References
CuO	*Gum carava*, *Malva* sp., and *Phyllanthus* sp.	Antimicrobial application against Gram-positive and Gram-negative bacteria	Padil and Černík (2013), Acharyulu et al. (2014), Awwad et al. (2015)
ZnO	*Camellia sinensis*	Antimicrobial	Senthilkumar and Sivakumar (2014)
$AgNO_3$	*Croton sparsiflorus*	Antimicrobials against Gram-positive and Gram-negative bacteria	Kathiravan et al. (2015)
TiO_2	*Oryza sativa*	Potential photocatalyst	Ramimoghadam et al. (2014)

ing many different types of materials such as glass and concrete (Hiregoudar 2014). Like the examples given above, there is much that can be done with agricultural waste other than biofuel. The waste needs to be examined for substances like cellulose or metals that may be a source for application in nanotechnology (Ditta 2012). Table 5.2 has some applications of waste derived from agriculture.

5.5 Constraints

While nanotechnology may provide a solution to many of the issues faced by the agricultural industry to date, still further research is needed to assuage the questions that the public and policy makers have with regard to the effect of these materials on human, ecosystem, and environment. Some of the key concerns that plague the application of nanotechnology in the agricultural industry are as follows:

(i) *Nontarget interactions*: There is the possibility that these nanoparticles may also interact with nontarget surfaces or cells. For instance, if a nanoparticle is being used as an antimicrobial agent, these particles may also act on nontarget organisms or even on other compounds creating nondesired outcomes (Chaudhary 2017).
(ii) *Effect on human and environment*: Although much research is being directed toward the development of new nanoparticles in various industries, there is insufficient research on determining the effect of these nanoparticles on human and environment. However, Mukhopadhyay (2014) has reported that nanotechnology is the way forward as it has enormous potential that remains to be harnessed for the agricultural industry.
(iii) *Cost effect*: This technology is a state of the art and therefore not a cheap alternative to be adopted by all agricultural nations. Investment by the government and the industry players is limited, and this can limit the adoption of this technology. Research in this area also requires funding which is also limited.
(iv) *Public awareness*: Making the general public and policy makers aware of the application and providing evidence on the safety and positive outcomes of utilizing this platform is lacking.

(v) *Regulations and ethical concerns*: This is a new platform for the agroindustry, and therefore, regulations needs to be in place to ensure all safety procedures have been adhered to and the products using nanomaterials are labeled accordingly.

Each one of the abovementioned constraints can deter or slower the process of acceptance of this technology in the agroindustry.

5.6 Future Perspectives

Through this chapter, we have seen the various areas in which nanotechnology may be applied. Nanotechnology has the ability to be applied in precision farming; in reducing waste and contamination; in improving utilization of resources such as water, fertilizer, and pesticides; and in much more. It would appear that this technology may provide solutions for sustainable agriculture in the years to come. As the years progress and new knowledge is added into the field of nanotechnology, there will be further expansion of the areas in which nanomaterials may be utilized in agriculture. The expanse by which the existing materials may be utilized may be enhanced over the years. Therefore, it is important to address the constraints mentioned above. As long as these matters are left unresolved, the advancement of this technology in agroindustry will be hampered. The following are the areas in which nanotechnology may be utilized in agroindustry:

1. Advance delivery systems that may be utilized in the application of fertilizers, pesticides, and herbicides.
2. Nanosensors that may be utilized in smart agriculture through the effective usage of water, fertilizers, nutrients, herbicides, and others.
3. Nanoscale biopolymers that may be utilized to disinfect and neutralize heavy metals and toxins.
4. Synthesis of enhanced nanomaterials taking into account changes to structure, surface chemistry, size, dosage, human, and environment (these should be carefully evaluated to ensure that it meets requirement for safe and enhanced performance).
5. Smart packaging and preservation of food produce to enhance shelf life and quality of food products.

5.7 Conclusion

In conclusion, nanotechnology definitely has a role to play in agriculture. Its potential is vast and is bound to have a broader base of application with more research and funds applied to determining its application in the agroindustry. However, while many research groups are looking into new nanoparticles and new applications for these compounds, there is still a bit of resistance to the adoption of these products in the agricultural industry due to the safety and environmental impact concerns mentioned under constraints. Research has to be focused on addressing these concerns, and by so doing, we hope that more and more nanomaterials will be commercialized for use in the agroindustry. The responsibility lies with the researchers and the scientific community to do the necessary research to convince the public of its potential and also to alleviate their concerns on the negative impacts of these products. Therefore, proper risk assessment analysis should accompany any development of nanomaterial for use in the agricultural industry to ensure that stakeholders are well informed on any products surfacing from any company or university. In addition to ensuring that stakeholders are kept well informed on the products and their impact thereafter, the success of nanotechnology is also dependent on market demand, profit margins, and the readiness of the agricultural industry to adopt the new and advanced nanotechnology. We believe that this is the way of the future and is surely a way of addressing the green revolution. The adoption of this material in green technology will provide sustainable precision farming for enhanced production to meet the increasing need for agricultural produce worldwide.

Acknowledgments Thanks to Ilakiya Sharanee Kumar for technical and graphical assistance.

References

Abdul Khalil HPS, Bhat AH, Ireana Yusra AF (2012) Green composites from sustainable cellulose nanofibrils: a review. Carbohydr Polym 87:963–979

Acharyulu N, Dubey R, Swaminadham V, Kollu P, Kalyani R, Pammi S (2014) Green synthesis of CuO nanoparticles using *Phyllanthus amarus* leaf extract and their antibacterial activity against multidrug resistance bacteria. Intl J Eng 3:639–641

Akbari A, Wu J (2016) Cruciferin coating improves the stability of chitosan nanoparticles at low pH. J Mater Chem 4:4988–5001

Almeelbi T, Bezbaruah A (2014) Nanoparticle-sorbed phosphate: iron and phosphate bioavailability studies with *Spinacia oleracea* and *Selenastrum capricornutum*. ACS Sustain Chem Eng 2:1625–1632

Aravinthan A, Govarthanan M, Selvam K, Praburaman L, Selvankumar T, Balamurugan R, Kim JH (2015) Sunroot mediated synthesis and characterization of silver nanoparticles and evaluation of its antibacterial and rat splenocyte cytotoxic effects. Int J Nanomedicine 10:1977–1983

Awwad A, Albiss B, Salem N (2015) Antibacterial activity of synthesized copper oxide nanoparticles using *Malva sylvestris* leaf extract. SMU Med J 2:91–101

Baker S, Harini BP, Rakshith D, Satish S (2013a) Marine microbes: invisible nanofactories. J Pharm Res 6:383–388

Baker S, Rakshith DK, Kavitha S, Santosh P, Kavitha HU, Rao Y, Satish S (2013b) Plants: emerging as nanofactories towards facile route in synthesis of nanoparticles. Bioimpacts 3:111–117

Ban JJ, Lee M, Im W, Kim M (2015) Low pH increases the yield of exosome isolation. Biochem Biophys Res Commun 461:76–79

Barrena R, Casals E, Colón J, Font X, Sánchez A, Puntes V (2009) Evaluation of the ecotoxicity of model nanoparticles. Chemosphere 75:850–857

Bhattacharyya A, Duraisamy P, Govindarajan M, Buhroo AA, Prasad R (2016) Nanobiofungicides: Emerging trend in insect pest control. In: Advances and Applications through Fungal Nanobiotechnology (ed. Prasad R), Springer International Publishing Switzerland 307–319

Bhor G, Deokar G, Hinge S, Singh L, Nalwade A (2014) Green synthesis of silver nanoparticles using fruit extract of bitter gourd and evaluation of their antibacterial activity against human and plant pathogenic bacteria. Int J Adv Scientific Technical Res 4:506–518

Biswal SK, Nayak AK, Parida UK, Nayak P (2012) Applications of nanotechnology in agriculture and food sciences. Intl Sci Inno Disc 2:21–36

Burman U, Saini M, Kumar P (2013) Effect of zinc oxide nanoparticles on growth and antioxidant system of chickpea seedlings. Toxicol Environ Chem 95:605–612

Cañas JE, Long M, Nations S, Vadan R, Dai L, Luo M (2008) Effects of functionalized and nonfunctionalized single-walled carbon nanotubes on root elongation of select crop species. Environ Toxil Chem 27:1922–1931

Chaudhary M (2017) Nanotechnology: resource management for sustainable agriculture. Ind Res J Genet Biotech 9:310–313

Chen H, Yada R (2011) Nanotechnologies in agriculture: new tools for sustainable development. Trends Food Sci Technol 22:585–594

Chen J, Liu S, Ye R, Cai Y et al (2013) Angiotensin-I converting enzyme (ACE) 642 inhibitory tripeptides from rice protein hydrolysate: purification and 643 characterization. J Funct Foods 5:1684–1692

Chong MN, Jin B, Chow CW, Saint C (2010) Recent developments in photocatalytic water treatment technology: a review. Water Res 44:2997–3027

Chowdhury S, Basu A, Kundu S (2014) Green synthesis of protein capped silver nanoparticles from phytopathogenic fungus *Macrophomina phaseolina* (Tassi) gold with antimicrobial properties against multidrug-resistant bacteria. Nanoscale Res Lett 9:365. https://doi.org/10.1186/1556-276X-9-365

Cursino L, Li Y, Zaini PA, Fuente DL et al (2009) Twitching motility and biofilm formation are associated with tonB1 in *Xylella fastidiosa*. FEMS Microbiol Lett 299:193–199

Day L, Williams R, Otter D, Augustin M (2015) Casein polymorphism heterogeneity influences casein micelle size in milk of individual cows. J Dairy Sci 98:3633–3644

DeRosa MC, Monreal C, Schnitzer M, Walsh R, Sultan Y (2010) Nanotechnology in fertilizers. Nat Nanotech 5:91. https://doi.org/10.1038/nnano20102

Dhoke SK, Mahajan P, Kamble R, Khanna A (2013) Effect of nanoparticles suspension on the growth of mung (*Vigna radiata*) seedlings by foliar spray method. Nanotechnol Dev 3:1. https://doi.org/10.1038/nnano20102

Diallo M, Street A, Sustich R, Duncan J, Savage N (2009) Nanotechnology applications for clean water: solutions for improving water quality. Nanotech Appl Clean Water 585–587

Ditta A (2012) How helpful is nanotechnology in agriculture? Adv Nat Sci: J Nanosci Nanotechnol 3:33002

El Beyrouthya M, El Azzi D (2014) Nanotechnologies: novel solutions for sustainable agriculture. Adv Crop Sci Technol 2:e118

Fakruddin M, Hossain Z, Afroz H (2012) Prospects and applications of nanobiotechnology: a medical perspective. J Nanobiotechnology 10:31

Fan W, Gao W, Zhang C, Tjiu WW, Pan J (2012) Hybridization of graphene sheets and carbon-coated Fe_3O_4 nanoparticles as a synergistic adsorbent of organic dyes. J Mater Chem 22(48):25108

Farmen L (2009) Commercialization of nanotechnology for removal of heavy metals in drinking water. In: Nanotechnology applications for clean water. Elsevier, London, pp 115–130

Fatta-Kassinos D, Kalavrouziotis I, Koukoulakis P, Vasquez M (2011) The risks associated with wastewater reuse and xenobiotics in the agroecological environment. Sci Total Environ 409:3555–3563

Feizi H, Kamali M, Jafari L, Moghaddam PR (2013) Phytotoxicity and stimulatory impacts of nanosized and bulk titanium dioxide on fennel (*Foeniculum vulgare* mill). Chemosphere 91:506–511

Gahlawat G, Shikha S, Chaddha BS, Chaudhuri SR, Mayilraj S, Choudhury AR (2016) Microbial glycolipoprotein-capped silver nanoparticles as emerging antibacterial agents against cholera. Microb Cell Factories 15:25. https://doi.org/10.1186/s12934-016-0422-x

Gopinath K, Gowri S, Karthika V, Arumugam A (2014) Green synthesis of gold nanoparticles from fruit extract of *Terminalia arjuna*, for the enhanced seed germination activity of *Gloriosa superba*. J Nanostructure Chem 4:115. https://doi.org/10.1007/s40097-014-0115-0

Giardi MT, Guzzella L, Euzet P, Rouillon R, Esposito D (2005) Detection of herbicide subclasses by an optical multibiosensor based on an array of photosystem II mutants. Environ Sci Technol 39:5378–5384

Girginova PI, Daniel-da-Silva AL, Lopes CB, Figueira P, Otero M, Amaral VS, Pereira E, Trindade T (2010) Silica coated magnetite particles for magnetic removal of Hg^{2+} from water. J Colloid Interface Sci 345:234–240

Goswami A, Roy I, Sengupta S, Debnath N (2010) Novel applications of solid and liquid formulations of nanoparticles against insect pests and pathogens. Thin Solid Films 519:1252–1257

Gouin S (2004) Microencapsulation: industrial appraisal of existing technologies and trends. Trends Food Sci Technol 15:330–347

Greenhalgh K, Turos E (2009) *In vivo* studies of polyacrylate nanoparticle emulsions for topical and systemic applications. Nanomedicine 5:46–54

Grillo R, Pereira AE, Nishisaka CS, de Lima R, Oehlke K, Greiner R, Fraceto LF (2014) Chitosan/tripolyphosphate nanoparticles loaded with paraquat herbicide: an environmentally safer alternative for weed control. J Hazard Mater 278:163–171

Grillo R, Clemente Z, de Oliveira JL, Campos EV, Chalupe VC, Jonsson CM, de Lima R, Sanches G, Nishisaka CS, Rosa AH, Oehlke K, Greiner R, Fraceto LF (2015) Chitosan nanoparticles loaded the herbicide paraquat: the influence of the aquatic humic substances on the colloidal stability and toxicity. J Hazard Mater 286:562–572

Grillo R, Abhilash PC, Fraceto LF (2016) Nanotechnology applied to bio-encapsulation of pesticides. Nanosci Nanotechnol 16:1231–1234

Gruère G, Narrod C, Abbott L (2011) Agricultural, food, and water nanotechnologies for the poor. International Food Policy Research Institute, Washington, DC

Haham M, Ish-Shalom S, Nodelman M, Duek I, Segal E, Kustanovich M, Livney YD (2012) Stability and bioavailability of vitamin D nanoencapsulated in casein micelles. Food Funct 3:737–744

Hiregoudar S (2014) Application of nanotechnology in enhancing quality of agricultural produce. http://www.isssonline.in/isss2014/agri.html

Hu R, Huang X, Huang J, Li Y, Zhang C, Yin Y, Chen Z, Jin Y, Cai J, Cui F (2015) Long and short term health effect of pesticides exposure: A cohort Study from China. PLoS One 10(6):e0128766. https://doi.org/10.1371/journalpone0128766

Huang Y, Reddy KN, Fletcher RS, Pennington D (2018) UAV low-altitude remote sensing for precision weed management. Weed Technol 32:2–6

Huijskens E, Smit L, Rossen J, Heederik D, Koopmans M (2016) Evaluation of patients with community-acquired pneumonia caused by zoonotic pathogens in an area with a high density of animal farms. Zoonoses Public Health 63:160–166

Ingale AG, Chaudhari A (2013) Biogenic synthesis of nanoparticles and potential applications: an eco-friendly approach. J Nanomed Nanotechol 4:1–7

Jain PK, Lee KS, El-Sayed IH, El-Sayed MA (2006) Calculated absorption and scattering properties of gold nanoparticles of different size, shape, and composition: applications in biological imaging and biomedicine. J Phys Chem Biophys B 110:7238–7248

Jampílek J, Králová K (2015) Application of nanotechnology in agriculture and food industry, its prospects and risks. Ecol Chem Eng S22:321–361

Kah M (2015) Nanopesticides and nanofertilizers: emerging contaminants or opportunities for risk mitigation? Front Chem 3:64. https://doi.org/10.3389/fchem201500064

Kah M, Hofmann T (2014) Nanopesticide research: current trends and future priorities. Environ Int 63:224–235

Kathiravan V, Ravi S, Ashokkumar S, Velmurugan S, Elumalai K, Khatiwada CP (2015) Green synthesis of silver nanoparticles using *Croton sparsiflorus* Morong leaf extract and their antibacterial and antifungal activities. Spectrochim Acta A Mol Biomol Spectrosc 139:200–205

Khodakovskaya M, Dervishi E, Mahmood M, Xu Y, Li Z, Watanabe F, Biris AS (2009) Carbon nanotubes are able to penetrate plant seed coat and dramatically affect seed germination and plant growth. ACS Nano 3:3221–3227

Khodakovskaya MV, De Silva K, Biris AS, Dervishi E, Villagarcia H (2012) Carbon nanotubes induce growth enhancement of tobacco cells. ACS Nano 6:2128–2135

Khodakovskaya MV, Kim BS, Kim JN, Alimohammadi M, Dervishi E, Mustafa T, Cernigla CE (2013) Carbon nanotubes as plant growth regulators: effects on tomato growth, reproductive system, and soil microbial community. Small 9:115–123

Khot LR, Sankaran S, Maja JM, Ehsani R, Schuster EW (2012) Applications of nanomaterials in agricultural production and crop protection: a review. Crop Prot 35:64–70

Kim JS, Kuk E, Yu KN, Kim JH, Park SJ, Lee HJ, Kim SH, Park YK, Park YH, Hwang CY, Kim YK, Lee YS, Jeong DH, Cho MH (2007) Antimicrobial effects of silver nanoparticles. Nanomedicine 3:95–101

Kalita S, Kandimalla R, Sharma KK, Kataki AC, Deka M, Kotoky J (2016) Amoxicillin functionalized gold nanoparticles reverts MRSA resistance. Mat Sci EngC 61:720–727

Krishnaswamy K, Vali H, Orsat V (2014) Value-adding to grape waste: green synthesis of gold nanoparticles. J Food Eng 142:210–220

Krupke CH, Hunt GJ, Eitzer BD, Andino G, Given K (2012) Multiple routes of pesticide exposure for honey bees living near agricultural fields. PLoS One 7:e29268. https://doi.org/10.1371/journalpone0029268

Kumar A, Zhang X, Liang X-J (2013) Gold nanoparticles: emerging paradigm for targeted drug delivery system. Biotechnol Adv 31:593–606

Lin D, Xing B (2007) Phytotoxicity of nanoparticles: inhibition of seed germination and root growth. Environ Pollut 150:243–250

Lin D, Tian X, Wu F, Xing B (2010) Fate and transport of engineered nanomaterials in the environment. J Environ Qual 39:1896–1908

Laborie MPG (2009) Bacterial cellulose and its polymeric nanocomposites. In: Lucia LA, Rojas OJ (eds) The nanoscience and Technology of Renewable Biomaterials. Wiley, Chichester

López-Moreno ML, de la Rosa G, Hernández-Viezcas JA, Castillo-Michel H, Botez CE, Peralta-Videa JR, Gardea-Torresdey JL (2010) Evidence of the differential biotransformation and genotoxicity of ZnO and CeO_2 nanoparticles on soybean (*Glycine max*) plants. Environ Sci Technol 44:7315–7320

Li W, Bai Y, Liu C, Yang Z, Feng X, Lu X, Laak NK, Chan KY (2009) Highly thermal stable and highly crystalline anatase TiO_2 for photocatalysis. Environ Sci Technol 43:5423–5428

Liu R, Lal R (2015) Potentials of engineered nanoparticles as fertilizers for increasing agronomic productions. Sci Total Environ 514:131–139

Liu F, Wen LX, Li ZZ, Yu W, Sun HY, Chen JF (2006) Porous hollow silica nanoparticles as controlled delivery system for water-soluble pesticide. Mater Res Bull 41:2268–2275

Lu C, Zhang C, Wen J, Wu G, Tao M (2002) Research of the effect of nanometer materials on germination and growth enhancement of *Glycine max* and its mechanism. Soybean Sci 21:168–171

Mahal A, Khullar P, Kumar H, Kaur G, Singh N, Jelokhani-Niaraki M, Bakshi MS (2013) Green chemistry of zein protein toward the synthesis of bioconjugated nanoparticles: understanding unfolding, fusogenic behavior, and hemolysis. ACS Sustain Chem Eng 1:627–639

Mahmoodzadeh H, Nabavi M, Kashefi H (2013) Effect of nanoscale titanium dioxide particles on the germination and growth of canola (*Brassica napus*). J Ornament Horticult Plants 3(1):25–32

Marziye Aboli P, Yazdan M, Roostaei A, Khani M, Negahdari M, Rahimi G (2014) Evaluation of the antifungal effect of magnesium oxide nanoparticles on *Fusarium oxysporum* f. Sp. *lycopersici*, pathogenic agent of tomato. Eur J Exp Biol 4(3):151–156

McLamore ES, Diggs A, Calvo Marzal P, Shi J, Blakeslee JJ, Peer WA, Murphy AS, Porterfield DM (2010) Non-invasive quantification of endogenous root auxin transport using an integrated flux microsensor technique. Plant J 63:1004–1016

Medda S, Hajra A, Dey U, Bose P, Mondal NK (2015) Biosynthesis of silver nanoparticles from *Aloe vera* leaf extract and antifungal activity against *Rhizopus* sp and *Aspergillus* sp. Appl Nanoscience 5:875–880

Mishra A, Kumari M, Pandey S, Chaudhry V, Gupta K, Nautiyal C (2014) Biocatalytic and antimicrobial activities of gold nanoparticles synthesized by *Trichoderma* sp. Bioresour Technol 166:235–242
Mousavi SR, Rezaei M (2011) Nanotechnology in agriculture and food production. J Appl Environ Biol Sci 1:414–419
Mukhopadhyay SS (2014) Nanotechnology in agriculture: prospects and constraints. Nanotechnol Sci Appl 7:63. https://doi.org/10.2147/NSAS39409
Nair R, Varghese SH, Nair BG, Maekawa T, Yoshida Y, Kumar DS (2010) Nanoparticulate material delivery to plants. Plant Sci 179:154–163
Nakache E, Poulain N, Candau F, Orecchioni AM, Irache JM (1999) Biopolymer and polymer nanoparticles and their biomedical applications. In: Nalwa HS (ed) Handbook of nanostructured materials and nanotechnology. Academic Press, New York, pp 577–635
Nangmenyi G, Economy J (2009) Nanometallic particles for oligodynamic microbial disinfection. In: Nanotechnology applications for clean water solutions for improving water quality water. Elsevier, London, pp 3–15
Nuruzzaman M, Rahman MM, Liu Y, Naidu R (2016) Nanoencapsulation, nano-guard for pesticides: a new window for safe application. J Agric Food Chem 64:1447–1483
de Oliveira JL, Campos EVR, Bakshi M et al (2014) Application of nanotechnology for the encapsulation of botanical insecticides for sustainable agriculture: prospects and promises. Biotechnol Adv 32:1550–1561
Oller I, Malato S, Sánchez-Pérez J (2011) Combination of advanced oxidation processes and biological treatments for wastewater decontamination - a review. Sci Total Environ 409:4141–4166
Otsuka H, Nagasaki Y, Kataoka K (2012) PEGylated nanoparticles for biological and pharmaceutical applications. Adv Drug Deliv Rev 64:246–255
Ouda SM (2014) Antifungal activity of silver and copper nanoparticles on two plant pathogens, *Alternaria alternata* and *Botrytis cinerea*. Res J Microbiol 9:34–42
Padil VVT, Černík M (2013) Green synthesis of copper oxide nanoparticles using gum karaya as a biotemplate and their antibacterial application. Int J Nanomedicine 8:889. https://doi.org/10.2147/IJNS40599
Paulkumar K, Gnanajobitha G, Vanaja M, Rajeshkumar S, Malarkodi C, Pandian K, Annadurai G (2014) *Piper nigrum* leaf and stem assisted green synthesis of silver nanoparticles and evaluation of its antibacterial activity against agricultural plant pathogens. Sci World J. https://doi.org/10.1155/2014/829894
Peng G, Tisch U, Adams O, Hakim M, Shehada N, Broza YY, Billan S, Abdah-Bortnyak R, Kuten A, Haick H (2009) Diagnosing lung cancer in exhaled breath using gold nanoparticles. Nat Nanotech 4:669. https://doi.org/10.1038/nnano2009235
Pereira AE, Grillo R, Mello NF, Rosa AH, Fraceto LF (2014) Application of poly (epsilon-caprolactone) nanoparticles containing atrazine herbicide as an alternative technique to control weeds and reduce damage to the environment. J Hazard Mater 268:207–215
Pokhrel LR, Dubey B (2013) Evaluation of developmental responses of two crop plants exposed to silver and zinc oxide nanoparticles. Sci Total Environ 452:321–332
Prasad R (2014) Synthesis of silver nanoparticles in photosynthetic plants. Journal of Nanoparticles, Article ID 963961, https://doi.org/10.1155/2014/963961
Prasad R, Bhattacharyya A, Nguyen QD (2017) Nanotechnology in sustainable agriculture: recent developments, challenges, and perspectives. Front Microbiol 8:1014. https://doi.org/10.3389/fmicb201701014
Prasad R, Jha A, Prasad K (2018) Exploring the Realms of Nature for Nanosynthesis. Springer International Publishing (ISBN 978-3-319-99570-0) https://www.springer.com/978-3-319-99570-0
Prasad R, Kumar V, Prasad KS (2014) Nanotechnology in sustainable agriculture: present concerns and future aspects. Afr J Biotechnol 13(6):705–713
Prasad R, Pandey R, Barman I (2016) Engineering tailored nanoparticles with microbes: quo vadis. WIREs Nanomed Nanobiotechnol 8:316–330. https://doi.org/10.1002/wnan.1363

Prasad R, Thirugnanasanbandham K (2019) Advances Research on Nanotechnology for Water Technology. Springer International Publishing. https://www.springer.com/us/book/9783030023805

Prasanna BM (2007) Nanotechnology in agriculture. Indian Agricultural Research Institute, New Delhi

Qu X, Alvarez PJ, Li Q (2013) Applications of nanotechnology in water and wastewater treatment. Water Res 47:3931–3946

Rabab MAE, El-Shafey RAS (2013) Inhibition effects of silver nanoparticles against rice blast disease caused by *Magnaporthe grisea*. Egypt J Agric Res 91:1271–1283

Rai V, Acharya S, Dey N (2012) Implications of nanobiosensors in agriculture. J Biomater Nanobiotechnol 3:315. https://doi.org/10.4236/jbnb2012322039

Ramimoghadam D, Bagheri S, Abd Hamid SB (2014) Biotemplated synthesis of anatase titanium dioxide nanoparticles via lignocellulosic waste material. Biomed Res Int. https://doi.org/10.1155/2014/205636

Ramy SY, Ahmed OF (2013) *In vitro* study of the antifungal efficacy of zinc oxide nanoparticles against *Fusarium oxysporum* and *Penicillium expansum*. Afr J Microbiol Res 7:1917–1923

Rouhani M, Samih MA, Kalantari S (2012) Insecticide effect of silver and zinc nanoparticles against *Aphis nerii* Boyer De Fonscolombe (Hemiptera: Aphididae). Chil J Agr Res 72:590–594

Sangeetha J, Thangadurai D, Hospet R, Purushotham P, Manowade KR, Mujeeb MA, Mundaragi AC, Jogaiah S, David M, Thimmappa SC, Prasad R, Harish ER (2017) Production of bionanomaterials from agricultural wastes. In: Nanotechnology (eds. Prasad R, Kumar M, Kumar V), Springer Nature Singapore Pte Ltd. 33–58

Sangeetha J, Thangadurai D, Hospet R, Harish ER, Purushotham P, Mujeeb MA, Shrinivas J, David M, Mundaragi AC, Thimmappa AC, Arakera SB, Prasad R (2017a) Nanoagrotechnology for soil quality, crop performance and environmental management. In: Nanotechnology (eds. Prasad R, Kumar M, Kumar V), Springer Nature Singapore Pte Ltd. 73–97

Sangeetha J, Thangadurai D, Hospet R, Purushotham P, Karekalammanavar G, Mundaragi AC, David M, Shinge MR, Thimmappa SC, Prasad R, Harish ER (2017b) Agricultural nanotechnology: Concepts, benefits, and risks. In: Nanotechnology (eds. Prasad R, Kumar M, Kumar V), Springer Nature Singapore Pte Ltd. 1–17

Scott NR and Chen H (2003) Nanoscale science and engineering or agriculture and food systems. Roadmap Report of National Planning Workshop 2002, 18–19 November, 2002, Washington, DC. http://www.nseafscornell.edu/webroadmap.pdf

Scrinis G, Lyons K (2007) The emerging nano-corporate paradigm: nanotechnology and the transformation of nature, food and Agri-food systems. Intl J SociolAgricul Food 15:22–44

Semo E, Kesselman E, Danino D, Livney YD (2007) Casein micelle as a natural nano-capsular vehicle for nutraceuticals. Food Hydrocoll 21:936–942

Senthilkumar S, Sivakumar T (2014) Green tea (*Camellia sinensis*) mediated synthesis of zinc oxide (ZnO) nanoparticles and studies on their antimicrobial activities. Int J Pharm Pharm Sci 6:461–465

Servin A, Elmer W, Mukherjee A, Torre-Roche R, Hamdi H, White JC, Bindraban P, Dimkpa C (2015) A review of the use of engineered nanomaterials to suppress plant disease and enhance crop yield. J Nanopart Res 17:92. https://doi.org/10.1007/s11051-015-2907-7

Shah MA, Towkeer A (2010) Principles of Nanosciences and nanotechnology. Narosa Publishing House, New Delhi

Sheykhbaglou R, Sedghi M, Shishevan MT, Sharifi RS (2010) Effects of nano-iron oxide particles on agronomic traits of soybean. Notulae Scientia Biologicae 2:112–113

Shi Y, Xu Z, Feng J, Wang C (2006) Efficacy of modified montmorillonite nanocomposite to reduce the toxicity of aflatoxin in broiler chicks. Anim Feed Sci Technol 129:138–148

Solaiman S, Maloney M, Qureshi M, Davis G, D'Andrea G (2001) Effects of high copper supplements on performance, health, plasma copper and enzymes in goats. Small Ruminant Res 41:127–139

Stadler T, Buteler M, Weaver DK (2010) Novel use of nanostructured alumina as an insecticide. Pest Manag Sci 66:577–579

Suman, Prasad R, Jain VK, Varma A (2010) Role of nanomaterials in symbiotic fungus growth enhancement. Curr Sci 99:1189–1191

Takeuchi MT, Kojima M, Luetzow M (2014) State of the art on the initiatives and activities relevant to risk assessment and risk management of nanotechnologies in the food and agriculture sectors. Food Res Int 64:976–981

Tarafdar J, Sharma S, Raliya R (2013) Nanotechnology: interdisciplinary science of applications. Afr J Biotechnol 12:219–226

Tiwari D, Dasgupta-Schubert N, Cendejas LV, Villegas J, Montoya LC, García SB (2014) Interfacing carbon nanotubes (CNT) with plants: enhancement of growth, water and ionic nutrient uptake in maize (*Zea mays*) and implications for nanoagriculture. Appl Nanosci 4:577–591

Ulrichs C, Mewis I, Goswami A (2005) Crop diversification aiming nutritional security in West Bengal - Biotechnology of stinging capsules in nature's waterblooms. Ann Tech Issue of State Agri Technologists Service Assoc., pp 1–18

Wang X, Han H, Liu X, Gu X, Chen K, Lu D (2012) Multi-walled carbon nanotubes can enhance root elongation of wheat (*Triticum aestivum*) plants. J Nanopart Res 14(6):1–10

Wani A, Shah M (2012) A unique and profound effect of MgO and ZnO nanoparticles on some plant pathogenic fungi. Int J Appl Pharm Sci Res 2:4. https://doi.org/10.7324/JAPS20122307

Xie Y, Li B, Zhang Q, Zhang C (2012) Effects of nano-silicon dioxide on photosynthetic fluorescence characteristics of *Indocalamus barbatus* McClure. J Nanjing Forest Univ (Natural Science Edition) 2:59–63

Xu LQ, Li NN, Chen JC, Fu GD, Kang E-T (2015) Quaternized poly (2-(dimethylamino) ethyl methacrylate)-grafted agarose copolymers for multipurpose antibacterial applications. RSC Adv 5:61742–61751

Yang F, Liu C, Gao F, Su M, Wu X, Zheng L, Hong F, Yang P (2007) The improvement of spinach growth by nano-anatase TiO_2 treatment is related to nitrogen photoreduction. Biol Trace Elem Res 119:77–88

Yang F, Hong F, You W, Liu C, Gao F, Wu C, Yang P (2016a) Influence of nano-anatase TiO_2 on the nitrogen metabolism of growing spinach. Biol Trace Elem Res 110:179–190

Yang L, Tang Z, Gao L, Li T, Chen Y, Liu L, Han L, Li X, Dong Q, Wei Q (2016b) The augmented prophylactic antibiotic could be more efficacious in patients undergoing transrectal prostate biopsy: a systematic review and meta-analysis. Int Urol Nephrol 48:1197–1207

Zaini PA, De La Fuente L, Hoch HC, Burr TJ (2009) Grapevine xylem sap enhances biofilm development by *Xylella fastidiosa*. FEMS Microbiol Lett 295:129–134

Zhang W, Tang H, Geng P, Wang Q, Jin L, Wu Z (2007) Amperometric method for rapid detection of *Escherichia coli* by flow injection analysis using a bismuth nano-film modified glassy carbon electrode. Electrochem Commun 9:833–838

Zhao J, Harper A, Estienne M, Webb K Jr, McElroy A, Denbow D (2007) Growth performance and intestinal morphology responses in early weaned pigs to supplementation of antibiotic-free diets with an organic copper complex and spray-dried plasma protein in sanitary and nonsanitary environments. J Animal Sci 85:1302–1310

Chapter 6
Nanobiotechnology and its Application in Agriculture and Food Production

Priyanka Priyanka, Dileep Kumar, Anurag Yadav, and Kusum Yadav

Contents

P. Priyanka · D. Kumar · K. Yadav (✉)
Department of Biochemistry, University of Lucknow, Lucknow, India

A. Yadav
College of Basic Sciences and Humanities, Sardarkrushinagar Agricultural University
Dantiwada, Banaskantha, India

D. Thangadurai et al. (eds.), *Nanotechnology for Food, Agriculture, and Environment*, Nanotechnology in the Life Sciences,
https://doi.org/10.1007/978-3-030-31938-0_6

6.1 Introduction

Nanobiotechnology integrates nanotechnology with biology. Nanobiotechnology is a new, emergent, and interdisciplinary field of research that refers to the use of nanotechnology to modify living organisms to enable the combinations of biological and nonbiological materials. The word "nano" refers to scale between 1 and 100 nm, and the term nanobiotechnology was coined by Lynn W. Jelinski, a biophysicist from Cornell University, USA. Nanobiotechnology finds application in genetic engineering and breeding programs (Scrinis and Lyons 2007). Variety of materials are used to synthesize nanoparticles (NPs) such as metal oxides, silicates, magnetic materials, semiconductor quantum dots (QDs), lipids, polymers, and emulsions (Niemeyer and Doz 2001; Oskam 2006; Puoci et al. 2008; Prasad et al. 2017a).

The use of nanobiotechnology is relevant in the current perspective because the widespread use of chemical pesticides and fertilizers has contaminated the environment, is hazardous for human health, and affects the biodiversity. To combat the problem, several nanotechnology-based products such as nanofertilizers, nanoherbicides, nanopesticides, nanosensors, nanofungicides and nanoinsecticides enhanced seed germination, and genetically modified crops are being developed (Jampilek and Kralova 2015; Bhattacharyya et al. 2016; Prasad et al. 2014, 2017a).

Nanotechnology holds great potential to increase global food production and food quality. Important areas of nanotechnology application in the food sector are food processing, food packaging (active, antimicrobial and smart packaging), nanoadditives, nanoencapsulation, nanosensors for contamination detection in food, and controlled release of nutraceuticals (Chellaram et al. 2014; Prasad et al. 2017b).

6.2 Application of Nanobiotechnology in Agriculture

Conventional agriculture uses agrochemicals, such as fertilizers, pesticides, fungicides, insecticides, and herbicides for crop growth and protection. However, their overuse is damaging the ecosystem by adding toxins in ground and surface water (Mukhopadhyay 2014). However, nanobiotechnology holds to promise to solve the agricultural problems and to restrict chemical overuse (Fig. 6.1). The applications of nanobiotechnology are described in this chapter (Table 6.1).

6.2.1 Nanofertilizers

Fertilizers are chemical compounds that provide nutrients to plant (Bottoms and Emerson 2013). Chemical fertilizers are in use since the early 1950s and have become the synonym of fertilzers. However, excessive use of chemical fertilizers has rattled the problem of water contamination, decreased soil fertility and

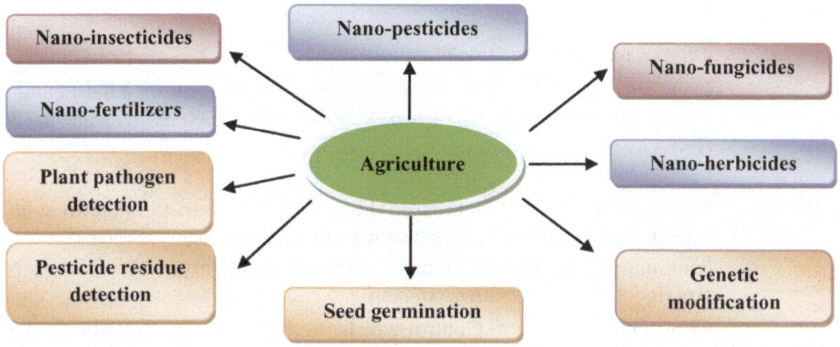

Fig. 6.1 Application of nanobiotechnology in agriculture. (Adapted from Mehrazar et al. 2015)

Table 6.1 Nanoparticle applications in agriculture

Nanomaterial	Type of nanoparticle	Application	References
Nanofertilizer	Synthetic apatite NP	Increase the plant growth rate and seed yield	Liu and Lal (2014)
	Iron chelate	Increase in wet weight and maximum leaf surface index and aerial organs dry weight	Moghadam et al. (2012)
	FeS_2	Breakdown starch in seeds resulting in increased in growth	Srivastava et al. (2014)
	EDTA-coated Fe_3O_4	Increased iron content in sunflower	Shahrekizad et al. (2015)
	Fe_3O_4	Increased the availability of iron to the plants	Rui et al. (2016)
	Fe_3O_4	Increase the iron and protein content and also used for the treatment of chlorosis	Siva and Benita (2016)
	Cu	Enhanced leaf area, chlorophyll content, fresh and dry weight and root dry weight of wheat	Hafeez et al. (2015)
	Nanochitosan-NPK fertilizer	Increased the growth and production of wheat growing in sandy soil	Abdel-Aziz et al. (2016)
	Nanozeourea fertilizer	Increased N content and enhanced the crude protein	Manikandan and Subramanian (2016)
	SiO_2 nanofertilizer	Enhanced the nitrogen and phosphorus content	Yassen et al. (2017)
	Zinc/boron nanofertilizer, prepared by loading $ZnSO_4$ and H_3BO_3 on a chitosan NPs emulsion	Increased the uptake of zinc, chlorophyll content and photosynthesis of the coffee was increased	Wang and Nguyen (2018)

(continued)

Table 6.1 (continued)

Nanomaterial	Type of nanoparticle	Application	References
	Chitosan–polymethacrylic acid (PMAA) NP	Starch accumulation at the root tip of pea. Major proteins such as convicilin, vicilin, and legumin β upregulated	Khalifa and Hasaneen (2018)
Nanoherbicides	Poly(epsilon-caprolactone) NP containing herbicide atrazine	Increased the mobility of atrazine in the soil resulting in increase herbicide effectiveness against *Brassica* sp.	Pereira et al. (2014)
	Chitosan/tripolyphosphate NPs loaded with paraquat herbicide	Control weeds	Grillo et al. (2014)
	Atrazine loaded nanocapsule	Decreased the root and shoot growth in *B. pilosa* and decrease in the photosystem II activity	Sousa et al. (2018)
Nanopesticides	Nanosilica	Effective against nuclear polyhedrosis virus (BmNPV) in silkworm industry	Barik et al. (2008)
	SiO_2	Effective against *Sitophilus oryzae*	Debnath et al. (2011)
	SiO_2	Effective against the stored grain pest *Corcyra cephalonica*	Vani and Brindhaa (2013)
	Ag	Effective against *Sitophilus oryzae* L. (pest of rice)	Abduz Zahir et al. (2012)
	Ag	Effective against the first to fourth instar larvae and pupae of the cotton bollworm (*Helicoverpa armigera*)	Durga Devi et al. (2014)
	CuO	Control cotton leafworm (*Spodoptera littorals*)	Shaker et al. (2016)
	ZnO	Delay in the larval and pupal development period of *Callosobruchus maculates*	Malaikozhundan et al. (2017)
Nanoinsecticides	ZnO	Effective against *Trialeurodes vaporariorum* (greenhouse whitefly)	Khooshe-Bast et al. (2016)
	Bio-silver and Au	Decreased the body weight of *Pericallia ricini* larvae	Sahayaraj et al. (2016)
	Ag and Zn	Effective against *Aphis nerii*	Rouhani et al. (2012)
	Ag	Control cotton leafworm (*Spodoptera littoralis*)	El-bendary and El-Helaly (2013)
	Nanoalumina dust	More effective against *S. oryzae* than *R. dominica*	Buteler et al. (2015)
	SiO_2	Effective against *Rhyzopertha dominica* F., and *Tribolium confusum* Jacquelin du Val.	Ziaee and Ganji (2016)

(continued)

Table 6.1 (continued)

Nanomaterial	Type of nanoparticle	Application	References
Nanofungicide	Ag	Effective against 18 different fungi (*Alternaria alternate*, *A. brassicicola*, *A. solani*, *Botrytis cinerea*, *Cladosporium cucumerinum*, *Corynespora cassiicola*, *Cylindrocarpon destructans*, *Didymella bryoniae*, *Fusarium oxysporum* f. Sp. *cucumerinum*, *F. oxysporum* f. Sp. *lycopersici*, *F. oxysporum*, *F. solani*, *Fusarium* sp., *Glomerella cingulata*, *Monosporascus cannonballus*, *Pythium aphanidermatum*, *P. ythium spinosum*, and *Stemphylium lycopersici)*	Kim et al. (2012)
	Ag	Effective against *Bipolaris sorokiniana* in wheat	Mishra et al. (2014)
	Ag	Effective against *Candida* spp., *Bipolaris sorokiniana*, *Magnaporthe grisea* and powdery mildews	Jo et al. (2009); Kim et al. (2009); Panacek et al. (2009); Lamsal et al. (2011)
	Chitosan	Effective against *F. solani*, and *A. niger*	Ing et al. (2012)
	Cu	Inhibited the activity of plant pathogenic fungi: *Phoma destructiva* (DBT-66), *Curvularia lunata* (MTCC 2030), *A. alternata* (MTCC 6572) and *F. oxysporum* (MTCC 1755)	Kanhed et al. (2014)
	Ag and Cu	Effective against two plant pathogenic fungi *Alternaria alternata,* and *Botrytis cinerea*	Ouda (2014)
	ZnO	Effective against two pathogenic fungi (*Botrytis cinerea*, and *Penicillium expansum*)	He et al. (2011)
	ZnO and MgO	Effective against *A. alternate*, *F. oxysporum*, *Rhizopus stolonifer*, and *Mucor plumbeus*	Wani and Shah (2012)
	Cu–chitosan	Effective against *A. solani* and *F. oxysporum* in tomato	Saharan et al. (2015)
	Nano trifloxystrobin 25% + tebuconazole 50%	Effective against *Macrophomina phaseolina*	Kumar et al. (2016)
	Combination between chitosan (CS) and silver (Ag), Ag@CS NPs	Effective against *Pyricularia oryzae* caused rice blast	Pham et al. (2018)

(continued)

Table 6.1 (continued)

Nanomaterial	Type of nanoparticle	Application	References
	ZnO	Effective against fungal phytopathogens, namely *A. alternata*, *A. niger*, *B. cinerea*, *F. oxysporum*, and *P. expansum*	Jamdagni et al. (2018)
Seed germination	Carbon nanotubes	Increased the germination of tomato seeds	Khodakovskaya et al. (2009)
	ZnO	Enhanced the germination rate	Singh et al. (2013)
	ZnO	Enhanced seed germination in rice and increased radical and plumule length	Upadhyaya et al. (2017)
	Au	Increased the seed germination rate of *Gloriosa superba*	Gopinath et al. (2014)
	Nanosilicon dioxide	Enhanced seed germination (22.16%), germination mean time (3.98%), and seed germination index in tomato	Siddiqui and Al-Whaibi (2014)
	Ag	Increased seed germination rate percent by 95% in *Boswellia ovalifoliolata*	Savithramma et al. (2012)
	Ag	Showed the highest percentage of seed germination in *Pennisetum glaucum*	Parveen and Rao (2014)
	TiO_2	Increased water absorption by the seeds resulting in accelerated seed germination	Feizi et al. (2012)
	Anatase	Enhance the germination of parsley seeds	Dehkourdi and Mosavi (2013)
	Binary mixtures of six metal oxide NPs (TiO_2, Fe_2O_3, CuO, NiO, Co_3O_4 and ZnO)	Enhanced seed germination in *Brassica*	Ko et al. (2017)
	Ag	Highest seed germination (98.6%) followed by copper (69.6%), and gold (56.5%), respectively	Hussain et al. (2017)
	SiO_2	Enhanced seed germination	Alsaeedi et al. (2018)

initiated eutrophication. Nanofertilizers hold the promise to alleviate the environmental problems caused by the use of chemical fertilizers. Nanofertilizers can improve plant growth as they contain nutrients and growth-promoting substances encapsulated in nanopolymers, chelates or emulsions. Nanofertilizers enjoy environmental and other benefits like less contamination of water bodies, reduced soil toxicity, delayed and steady nutrient release, increased production yield, improved photosynthesis, enhanced fertilizer effect period, and increased soil nutrients (Naderi and Danesh-Shahkari 2013; Mehrazar et al. 2015).

Several examples of successful application of metal-based nanofertilizers for improved plant growth are available in the scientific literature. For example, iron-chelated nanofertilizer application on two varieties of spinach, Varamin 88 and Viroflayand, respectively at 4 kg ha^{-1} showed 58% and 47% improvement in wet weight along with enhanced leaf surface index and aerial organs (Moghadam et al. 2012). In addition, the EDTA-coated Fe_3O_4 NPs increased iron content in sunflower (Shahrekizad et al. 2015). Similarly, iron pyrite NP treatment on spinach seeds improved plant growth and enhanced breakdown of starch in seeds (Srivastava et al. 2014). Iron oxide NPs improve peanut growth by augmenting the availability of iron (Rui et al. 2016). The Fe_2O_3 NP application increases the plant iron and protein content and reduces the frequency of chlorosis (Siva and Benita 2016). Copper (Cu) NP application increased the growth and yield of wheat cultivar Millat-2011 by enhancing leaf area, chlorophyll content, fresh and dry weight, and root dry weight (Hafeez et al. 2015).

Synthetic apatite NP application increases the soybean plant growth rate (32.6%) and seed yield (20.4%), respectively (Liu and Lal 2014). The nanozeourea fertilizer enhanced crude protein of maize plant growing in black soil by 26.1% and in red soils by 36.1%, which was higher than conventional urea fertilizer application (Manikandan and Subramanian 2016). The SiO_2 nanofertilizer application increased growth and yield in cucumber through improved nitrogen and phosphorus content in the plant (Yassen et al. 2017).

Recently, the demand for chitosan NP has increased based on their role as nanofertilizer. Chitosan is a biodegradable and biocompatible polymer of randomly distributed β-(1 → 4)-linked D-glucosamine and N-acetyl-D-glucosamine units. Several examples show an impressive plant growth result due to chitosan NP application. Nanochitosan-NPK fertilizer increases the growth and production of wheat growing in sandy soil (Abdel-Aziz et al. 2016). Application of zinc-boron nanofertilizer, prepared by loading $ZnSO_4$ on chitosan NP emulsion, on coffee leaves increased the level of Zn, N, and P uptake and improved chlorophyll content (Wang and Nguyen 2018). Chitosan–polymethacrylic acid (PMAA) NP application on pea plant increased starch accumulation in root tips. In addition, synthesis of proteins such as convicilin, vicilin, and legumin β was upregulated (Khalifa and Hasaneen 2018).

6.2.2 *Nanoherbicides*

Weeds are unwanted plants that reduce crop yield thus need timely removal from agricultural lands. Conventional herbicides kill weeds from the aboveground plant parts and do not remove tubers or roots that sprout another plant under favorable conditions (Ali et al. 2014). In addition, the absence of moisture in herbicides render them less effective in rainfed agriculture systems (Subramanian and Tarafdar 2011).

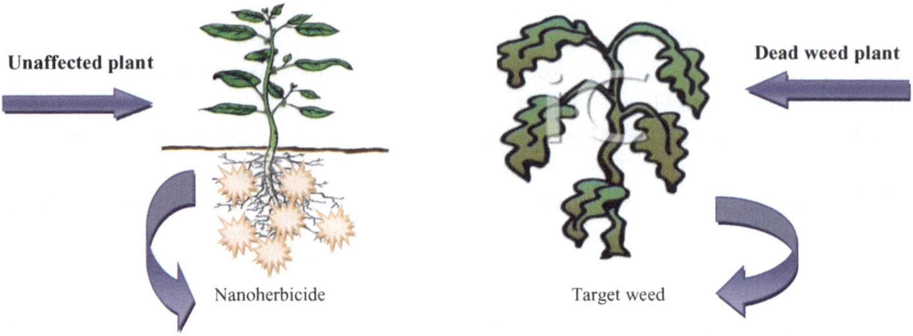

Fig. 6.2 Pathway showing the use of nanoherbicide to target weed plant. (Adapted from Elrahman and Mostafa 2015)

Nanoherbicides application could be an effective method to remove weeds (Fig. 6.2). The NP-based herbicide has several advantages over conventional herbicides such as increased water solubility, lower weed resistance, and decreased toxicity. NPs of poly(epsilon-caprolactone) containing the herbicide atrazine were tested on target (*Brassica* sp.) and nontarget (*Zea mays*) plants. Encapsulated herbicide did not harm a nontarget plant (*Zea mays*). The NPs increase the mobility of atrazine in the soil, resulting in increased herbicide effectiveness against *Brassica* sp. (Pereira et al. 2014). Chitosan/tripolyphosphate NP-based nanoherbicides are also under development. Chitosan/tripolyphosphate NPs loaded with paraquat herbicide are less toxic to crops and are safe in controlling weeds (Grillo et al. 2014). The 2000 g ha^{-1} application of atrazine-loaded nanocapsule effectively decreased the root and shoot growth in *B. pilosa* and reduced photosystem II activity (Sousa et al. 2018).

6.2.3 Nanopesticides

Plant pests are a major threat to crop production and require stringent control measures. The conventional methods for pest control employ several types of chemical pesticides, most of which are xenobiotic in nature. In addition, the excessive use of pesticides is causing adverse effects on organisms, environment, and nitrogen-fixing ability of plants (Ghormade et al. 2011; Elrahman and Mostafa 2015; Bhattacharyya et al. 2016).

NP-based pesticides enjoy several advantages over conventional pesticides. Nanosilica derived from the plant is effective in controlling the infection of nuclear polyhedrosis virus (BmNPV) of silkworm (Barik et al. 2008). Silica NPs used against *Sitophilus oryzae* provide 90% pest mortality (Debnath et al. 2011). In another study, 70–80-nm sized silica NP provided 100% mortality against *Corcyra cephalonica* (Vani and Brindhaa 2013). The Ag NPs, synthesized from aqueous leaf extracts of *Euphorbia prostrate*, demonstrated 100% mortality rate against *Sitophilus oryzae* L., a rice pest (Abduz Zahir et al. 2012). The Ag NPs synthesized from leaf extract of *Euphorbia hirta* provide protection against larvae and pupae of the cotton bollworm (*Helicoverpa armigera*). In addition, the longevity of male and female pest decreases after Ag NP treatment on larvae and pupae of the cotton bollworm (Durga Devi et al. 2014). The use of CuO NPs controls cotton leafworm larvae (*Spodoptera littorals*) with mortality of 100% (Shaker et al. 2016). The treatment of *Bacillus thuringiensis*-coated ZnO NPs delay the larval and pupal development period of *Callosobruchus maculatus*. The results show that Bt ZnO NPs are effective nanopesticide against *C. maculatus* (Malaikozhundan et al. 2017).

6.2.4 Nanoinsecticides

Insects are vectors of various plant diseases that damage crops and therefore require active control. Several chemical-based insecticides have been used to inhibit their reproduction or kill them (Ragaei and Sabry 2014). However, chemical-based insecticides suffer from several drawbacks as they are degraded by light, temperature, microorganism, and hydrolysis. Therefore, only a small quantity of these insecticides reaches the target site. As a result, repeated application of insecticides is necessary to control the insects, which increase the cost of crop production. In addition, the use of insecticides is known to damage ecosystems and human health (Perlatti et al. 2013).

To combat the practical problems associated with field application of chemical insecticide, several NP-based insecticides are being formulated and tested. For example, the ZnO NPs provide mortality rate of 91.6% against *Trialeurodes vaporariorum* (Greenhouse whitefly) (Khooshe-Bast et al. 2016). The use of biosilver and gold NPs affects the growth and decreases the body weight of *Pericallia ricini* larvae. Therefore, Ag and Au NPs could be used to control insects (Sahayaraj et al. 2016). The Ag and Zn NP application against *Aphis nerii* at 700 mg mL^{-1} provides the highest insect mortality rate (Rouhani et al. 2012). The Ag NP application on cotton leafworm (*Spodoptera littoralis*) provides more than 50% mortality of leafworms at 250 ppm and higher concentrations (El-Bendary and El-Helaly 2013). The nanoalumina dust, synthesized using a modified glycine-nitrate combustion process, displayed fair mortality rate against *Sitophilus oryzae* and *Rhyzopertha dominica* (pests of stored grain) (Buteler et al. 2015). The two SiO_2 NPs, namely Aerosil® and Nanosav, provided high-mortality rate of *Rhyzopertha dominica* and *Tribolium confusum* (Ziaee and Ganji 2016).

6.2.5 Nanofungicides

Phytopathogens attack plant tissues at different stages of growth. Some fungal pathogens like *Fusarium* spp., *Botrytis cinerea*, and *Phythophora* spp. infect aerial and few infect ground plant parts, causing a huge loss in crop yield. Conventional methods for phytopathogen control involve chemical fungicide application, which is highly toxic and nonbiodegradable in nature, contaminates the environment, and affects human health. To counter the problems associated with chemical fertilizer application in the field, the development of nanomaterial-based eco-friendly fungicide is underway for agriculture sustainability and greener environment (Abd-Elsalam and Alghuthaymi 2015). For achieving the goal, several nanofungicides have been tested against many plant pests.

Inhibition of several fungal pathogens by Ag NPs indicates their antifungal properties (Kim et al. 2012). For example, silver-based NPs were found to be effective against 18 fungal species (*Alternaria alternata*, *A. brassicicola*, *A. solani*, *Botrytis cinerea*, *Cladosporium cucumerinum*, *Corynespora cassiicola*, *Cylindrocarpon destructans*, *Didymella bryoniae*, *Fusarium oxysporum* f. sp. *Cucumerinum*, *F. oxysporum* f. sp. *lycopersici*, *F. solani*, *Fusarium* sp., *Glomerella cingulata*, *Monosporascus cannonballus*, *Pythium aphanidermatum*, *P. spinosum*, and *Stemphylium lycopersici*). The Ag NPs were also found effective against *Bipolaris sorokiniana* infection in wheat (Mishra et al. 2014). The antifungal activity of Ag NPs was also found against *Candida* spp. (Kim et al. 2009; Panacek et al. 2009), *Bipolaris sorokiniana*, *Magnaporthe grisea* (Jo et al. 2009), and powdery mildews (Lamsal et al. 2011). In addition, Ag and Cu NPs, used against *A. alternata* and *Botrytis cinerea*, showed maximum inhibition of the fungal hyphae growth at the concentration of 15 mg L^{-1} of Ag and Cu NPs (Ouda 2014).

High-molecular-weight chitosan NPs show antifungal activity against *F. solani* and *Aspergillus niger* (Ing et al. 2012). Cu–chitosan NP application at 0.12% concentration against *A. solani* and *F. oxysporum* in tomato caused 70.5% and 73.5% inhibition of fungal mycelia and spore germination by 61.5% and 83.0%, respectively (Saharan et al. 2015). Cu NPs inhibit the activity of plant pathogenic fungi like *Phoma destructive* (DBT-66), *Curvularia lunata* (MTCC 2030), *A. alternata* (MTCC 6572), and *F. oxysporum* (MTCC 1755). Since Cu NP can be quickly and conveniently synthesized using C-TAB-IPA method, it holds potential to be used as a commercial fungicide (Kanhed et al. 2014).

The ZnO NP application against *Botrytis cinerea* and *Penicillium expansum* causes deformation in fungal hyphae and inhibition of *B. cinerea* growth. In addition, ZnO NPs inhibit the growth of conidiophores and conidia of *P. expansum*, leading to the death of fungal hyphae (He et al. 2011). The ZnO and magnesium oxide (MgO)-based NP application has been reported to reduce spore germination of *Alternaria alternata*, *Fusarium oxysporum*, *Rhizopus stolonifer*, and *Mucor plumbeus* (Wani and Shah 2012).

More recently, the NP combinations are under use for effective nanofungicide preparation. For example, the combination of 25% nano trifloxystrobin with 50% tebuconazole (75 WG) fungicide made through ball milling produced better antifungal activity in comparison to conventional fungicide against *Macrophomina phaseolina* (Kumar et al. 2016). The combination of chitosan (CS) and silver (Ag) (also abbreviated as Ag@CS NPs) displayed significant antifungal activity against *Pyricularia oryzae*, responsible for rice blast (Pham et al. 2018). The ZnO NP combination with ciprofloxacin and ceftazidime demonstrated antifungal property against *A. alternata*, *A. niger*, *B. cinerea*, *F. oxysporum*, and *P. expansum*. The maximum antifungal activity was achieved when 0.25 mg mL^{-1} ZnO NPs was combined with 8 μg mL^{-1} ciprofloxacin and 32 μg mL^{-1} ceftazidime (Jamdagni et al. 2018).

6.2.6 Seed Science: Enhancing Seed Germination

Seed yield is the most important factor in determining crop productivity. Usually, the laboratory-tested germination seeds are distributed to farmers for cropping. However, most of the times the provided seeds show viability rate below the claim. The NP-based methods are being developed to improve the seed germination of stored seeds (Manimaran 2015).

The NPs of C, Zn, Au, Ag, and Si have been claimed to improve seed germination of several plants. For example, the carbon nanotube application at 10–40 μg mL^{-1} solvent increases tomato seed germination (Khodakovskaya et al. 2009). The ZnO NP application enhances the germination rate in many plants (Singh et al. 2013). The ZnO NP application at 15 mg L^{-1} enhances rice seed germination (Upadhyaya et al. 2017).

The Au and Ag are the nanoparticles of choice for many reasons. Au NP application at 1000 μM increased the *Gloriosa superba* germination rate (Gopinath et al. 2014). The Ag NP treatment increased seed germination percentage rate of *Boswellia ovalifoliolata* by 95% (Savithramma et al. 2012). The Ag NP treatment at 50 mg L^{-1} on *Pennisetum glaucum* seeds shows the highest germination percentage (93.33%) (Parveen and Rao 2014).

The TiO_2 NPs are preferred in plant sciences due to photocatalytic activity, stability, and lower costs. The 10 ppm nanosize TiO_2 treatment decreases the mean germination time (34%) in spinach seeds by facilitating water absorption (Feizi et al. 2012). Anatase NP (TiO_2) application at 40 mg mL^{-1} enhances the germination of parsley seeds (92.46%) (Dehkourdi and Mosavi 2013). TiO_2 NPs applied at 2000 mg L^{-1} in canola plants have been reported to enhance seed germination (Mahmoodzadeh et al. 2013).

Nanosilicon dioxide ($nSiO_2$) application at 8 g L^{-1} enhanced tomato seed germination percent (22.16%), germination meantime (3.98%), and seed germination index (22.15%) (Siddiqui and Al-Whaibi 2014). The SiO_2 (10–20 nm) and Mo NP (>100 nm) application in rice seeds also shows better germination (Adhikari et al. 2013). The 200 ppm silica NPs were reported to enhance cucumber seed germination (Alsaeedi et al. 2018).

In a recent study, several metallic NPs were reported to enhance the germination percentage of a medicinal plant *Artemisia absinthium*. In the study, the germination percentage of 98.6% was achieved with Ag NPs followed by Cu NPs (69.6%) and Au NPs (56.5%), respectively (Hussain et al. 2017). In another study, binary mixtures of six metal oxide NPs (TiO_2, Fe_2O_3, CuO, NiO, Co_3O_4, and ZnO) were used for evaluating seed germination in brassica. The 2–6% seed germination enhancement was achieved at 1000 and 2000 mg L^{-1} concentration of Ti, Co, and Fe metal oxide NPs. The CuO NP was found to be most toxic for brassica seed germination (Ko et al. 2017).

6.2.7 Nanobiosensors for Plant Pathogen Detection

Traditional microscopy-based and culture-dependent methods are available for detecting plant pathogens, which are time-consuming and often inaccurate (Fletcher et al. 2006). In addition, the immunological and molecular methods available for pathogen detection suffer from longer test time, weak signal strength, and bulky instrumentation (Kashyap et al. 2016). However, nanobiosensor application in plant pathology could provide a novel approach for detecting plant pathogens. The methods of pathogen detection are simple, rapid, and accurate (Sharon et al. 2010; Ismail et al. 2017). For example, fluorescent silica NPs, combined with antibody molecules were used to detect *Xanthomonas axonopodis* pv. *vesicatoria* that causes bacterial spot disease in Solanaceae family plants (Yao et al. 2009). Copper NP-modified gold electrode was used to detect oilseed rape infected with fungal pathogen *Sclerotinia sclerotiorum* (Wang et al. 2010). The TiO_2 or SnO NPs were used for detecting p-ethylguaiacol present in fruits and plants infected with fungi *Phytophthora cactorum* (Fang et al. 2014). Quantum dots fret-based biosensor was used to detect *Candidatus* phytoplasma aurantifolia that causes witche's broom disease of lime (Rad et al. 2012).

Nanobiosensors are also effective in detecting plant viral disease. The immunosensor shows high specificity and sensitivity for virus detection. For example, a chemiresistive sensor based on polypyrrole (PPy) nanoribbon was used to detect plant viruses (Chartuprayoon et al. 2013). Nanorod-based fiber-optic particle plasmon resonance immunosensor was utilized to detect orchid viruses (*Cymbidium mosaic virus* and *Odontoglossum ringspot virus*), which has several advantages such as faster analysis, good reproducibility, and lower detection limit than ELISA (Lin et al. 2014). A nanowire-based biosensor was developed for detecting plant diseases (Ariffin et al. 2014). Carbon nanotube-based Cu NPs are used to develop biosensor for detecting *Begomovirus* (CLCuKoV-Bur). The developed sensor detects viruses up to 0.01 ng μ L^{-1} DNA concentration (Tahira et al. 2018).

6.2.8 Nanobiosensors for Pesticide Residue Detection

Nowadays pesticides are in heavy use for crop improvement. When pesticides degrade in environment, some toxic residues are released in the soil which may enter into the food chain through contaminated water and soil. Henceforth, accurate pesticide detection methods are required to measure the soil and plant contamination status. Conventional methods of pesticide residue detection include mass spectrometry, GC, HPLC, and UV-Vis spectrometry. However, these methods require skilled technicians and are time-consuming, therefore, not suitable for field analysis. As a result, newer methods are required for accurate pesticide residue detection in crop fields. Several nanobiosensor-based methods are under development that are cheaper, fast and do not require sample pretreatment (Vimala et al. 2016). For example, an electrochemical magnetoimmunosensing method was developed for detecting atrazine residue in samples (Zacco et al. 2006). Nanocomposite ZrO_2/Au film electrode is used for detecting parathion (Wang and Li 2008). Amperometric biosensor based on assembling acetylcholinesterase on poly(dimethylsiloxane)-poly(diallydimethylammonium)/gold NP composite film is used for detecting organophosphates (Zhao et al. 2009). The MnO_2 nanosheet-carbon dots are used for detecting organophosphorus pesticides (Yan et al. 2018). An optical biosensor was developed for detecting organophosphate pesticide using CdTe as fluorescent probe (Sun et al. 2011). Core-shell nanosensors based on localized surface plasmon resonance (LSPR) were developed for detecting atrazine (Yang et al. 2014). Recently, a fluorescence sensor was synthesized by combining copper (II) oxide and multiwall carbon nanotubes for detecting glyphosate. The detection was based on catalytic inhibition activity of the copper (II) oxide and multiwall carbon nanotubes (Chang et al. 2016). Sahoo et al. (2018) used zinc oxide quantum dots for detecting aldrin, tetradifon, glyphosate, and atrazine in water.

6.2.9 *Production of Genetically Modified Crops*

Nanobiotechnology also finds application in gene modification. For example, mesoporous silica nanoparticle (MSN) was used to transport DNA and chemicals into plant cells and leaves (Torney et al. 2007). The MSNs were used as carriers to deliver Cre recombinase protein into *Zea mays* cells for genome editing (Martin-Ortigosa et al. 2014). A fluorescent-conjugated polymer NP (CPN) was used for delivering siRNAs to knockdown a target gene in the cellulose biosynthesis pathway (NtCesA-1a and NtCesA1b) in plant protoplasts (Silva et al. 2010). Calcium phosphate NPs were used to deliver pBI121 harboring GFP driven by 35S promoter-encoding plasmid DNA into tobacco (Ardekani et al. 2014).

6.3 Application of Nanobiotechnology in Food Industry

The application of nanobiotechnology in the food sector is a newer concept with immense potential. Nanobiotechnology use in food has improved its taste, color, and flavor. Its application has also enhanced absorption and bioavailability of nutrients as well as health supplements. New food packaging materials are being developed using nanobiotechnology with improved mechanical strength, oxygen barrier, and antimicrobial properties. Nanosensors are used for detecting microbial contamination and toxic substances in food. The major areas of nanobiotechnology application in food sectors include food processing, food packaging, nanoemulsions, nanoencapsulations, nanoadditives, and nutraceutical delivery (Singh et al. 2017) (Fig. 6.3).

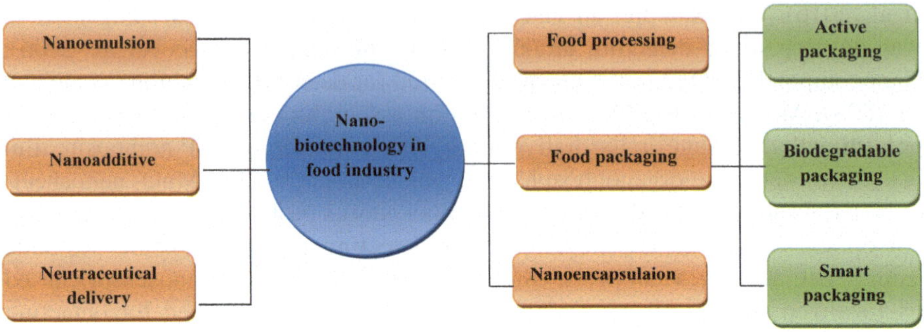

Fig. 6.3 Application of nanobiotechnology in food industry. (Adapted from Ravichandran 2010)

6.3.1 Food Processing

Food processing is done to maintain the nutritional quality of the food and enhance its shelf life. As a result, processed foods are less likely to get spoiled than fresh foods. In addition, such foods can be easily transported to long distances. Nowadays, commercially marketed food is processed using enzyme immobilization nanofiltration and other methods (Dasgupta et al. 2015) to improve the shelf life. Some of the common nanobiotechnology methods used in food processing are described further.

6.3.1.1 Enzyme Immobilization

Some food processing methods employ enzymes to modify food components for improving flavor, texture, and nutritional quality. The latest methods employ enzymes that are immobilized using NPs for thorough dispersal around food matrices to increase enzyme activity (Thangavel 2014). For example, SiO_2 NPs are used for triacylglycerol lipase immobilization in olive oil. The SiO_2 NPs, with reactive aldehyde groups, covalently bind to a porcine triacylglycerol lipase and help in hydrolyzing olive oil. The enzyme immobilization also helps in improving stability, adaptability, and reusability of food products (Bai et al. 2006). Similarly, magnetite NPs are used to immobilize lipase (Lee et al. 2009). In addition, polyacrylonitrile (PAN) nanofibrous membranes were used for immobilizing *Candida rugosa* lipase for the hydrolysis of soybean oil (Li and Wu 2009).

6.3.1.2 Nanofrying

The US-based Oilfresh Corporation is marketing a new nanoceramic product with several proclaimed benefits. The benefit includes reduced use of oil in restaurants and in fast food due to the larger surface area of the product (Momin et al. 2013).

6.3.1.3 Nanofiltration

Nanofiltration is a cost-effective separation method for product separation that stands somewhere between reverse osmosis and ultrafiltration. It separates substances less than 10 Å and removes divalent and multivalent ions. Nanofiltration is commercially used for desalination of seawater, concentration of juices (Warczok et al. 2004), demineralization, and removal of color from water. The technology is also applicable in wastewater treatment, water purification, and cheese making (Hussain and Al-Rawajfeh 2009). For example, the commercial cottage cheese whey making suffers from acidity, dilution, and saltiness. Nanofiltration of cottage cheese whey effectively concentrates solids by removing sodium, potassium salts, and some acids (Nguyen et al. 2003). Another commercial example of nanofiltration is of lactic acid separation, which is an important product of the food industry that requires costly purification system. Lactic acid enrichment with inorganic nanofiltration and molecular sieving membranes by pervaporation was achieved at much lower costs than vacuum evaporation and electrodialysis (Duke et al. 2008).

6.3.1.4 Nanolamination

Food is spoiled by the accumulation of moisture, gases, and lipids, which can be protected through nanolamination. Nanolaminates are thin, harmless protective films synthesized from polysaccharides, proteins, and lipids. These are barrier against carbon dioxide and oxygen that preserves food, improves food texture, flavor, and color. In addition nanolaminations enhance the level of nutrients and antioxidants in the food (Ali et al. 2014). For example, lipid-based nanolaminates are good barriers against moisture, while polysaccharide- and protein-based films protect from oxygen- and carbon dioxide-based damage (Ravichandran 2010).

6.3.2 Food Packaging

The packaging increases the shelf life of food by reducing microbial spoilage, decreasing gas and moisture exchange. Many commercial NP-based food packaging products are available and discussed further.

6.3.2.1 Active Packaging

The NP-based active packaging provides antimicrobial, antioxidation, and moisture scavenging properties to the food items (Yildirim et al. 2017). The main action of active packaging NPs lies in the inhibition of the microbial growth that leads to the spoiling of food (Chellaram et al. 2014).

The chitosan films are used in food preservation due to antimicrobial and non-toxic properties (Tripathi et al. 2008). Nanoclay is used in the packaging of food, beer bottles, carbonated drinks, and thermoformed containers. The packaged nano-clay acts as a gas barrier that keeps oxygen-sensitive foods fresh for longer time spans. The polyethylene terephthalate (PET) beer bottles use nanoclays produced by Nanocor® which enhance beer storage time from 11 weeks to about 30 weeks (Silvestre et al. 2011). The Ag NPs combined with hydroxypropyl methylcellulose (HPMC) matrix provide antimicrobial property, which is internally coated on food packaging material (Moura et al. 2012). In addition, the ZnO NPs are also used in food packaging as they are safe and hold antimicrobial activity (Espitia et al. 2012). Also, allyl isothiocyanate (AIT) and carbon nanotubes (CNT) are utilized as antimicrobial film for the packaging of shredded and cooked chicken meat. The diffusion of the AIT from the film in the packaged chicken decreases the microbial contamination, which allows controlled oxidation and reduces color change (Dias et al. 2013). In another study, the antimicrobial activity of ZnO NP/CS film provided about 1.5- to two-fold increase in the antimicrobial activity against *B. subtilis* and *E. coli* (Priyadarshi and Negi 2016). Durethan is a transparent plastic film, which contains clay NPs that block oxygen, carbon dioxide, and moisture to keep the food fresh. The use of nanoclay in durethan preparation makes the plastic lighter, stronger, and heat resistant (Davari et al. 2017). The Ag NPs synthesized by crosslinking with trisodium citrate enhanced shelf life of the grapefruit. The Ag NPs filled with hydroxypropyl methylcellulose (HPMC) and xanthan films show good antibacterial activity with decreased decay index (Kothari and Setia 2017). A variety of NP-reinforced polymers termed as nanocomposites are used in the food packaging industry (Momin et al. 2013). For example, the composite of Ag and ZnO NPs enhances the shelf life of fresh juice (Emamifar et al. 2010).

Fruits and vegetables degrade faster after harvest due to ethylene production. Therefore, ethylene adsorbent powders are added in packaging to maintain the freshness of food and food products. Alkali-treated halloysite nanotubes have the highest ethylene adsorption capacity because the treatment increases the nanotube pore size (Gaikwad et al. 2018). Polyethylene nanocomposite films containing combinations of organoclay (OC) NP are also used in antimicrobial packaging of food that are effective against *Escherichia coli* (Fasihnia et al. 2017). Sarwar et al. (2017) found the antimicrobial activity of PVA/nanocellulose/Ag nanocomposite films against *Staphylococcus aureus* (MRSA) and *E. coli* (DH5-α). The developed film was nontoxic and possessed high thermal stability and high mechanical strength.

6.3.2.2 Biodegradable Packaging

The use of nondegradable plastics change the soil nature and causes the accumulation of toxic gases in the atmosphere, which is responsible for environmental pollution and global warming. To counter the problem associated with nondegradable plastics, the production of biodegradable plastics has been initiated. However, the biodegradable plastic production at a commercial level is a costly affair, and also they are permeable to water and gas permeable and lack mechanical strength.

The newer methods employ natural or synthetic NPs to rectify the defects of biodegradable plastics because NP-based bioplastics have advantages like biodegradability and higher mechanical strength (Chellaram et al. 2014). For example, biodegradable starch/clay nanocomposite films hold higher mechanical strength and are used in food packaging (Avella et al. 2005). In addition, the incorporation of NPs in bioplastics impart antimicrobial nature to them. For example, laser-generated Cu NPs embedded in a biodegradable polymer matrix (polylactic acid) equip them with antibacterial nature, for better food storage capability (Longano et al. 2012). A biodegradable material, poly(lactic acid) used in food packaging, provides higher mechanical strength and antimicrobial activity (Gonzalez and Igarzabal 2013). Also, specialized bioplastics are available for thermal-sensitive foods. For example, pectin-TiO_2 nanocomposite aerogels are available for packaging thermosensitive food that lowers the thermal conductivity to food. The presence of TiO_2 enhances thermal stability and provides antimicrobial activity under UV light and dark conditions (Nesi et al. 2018).

6.3.2.3 Smart Packaging

In smart packaging, the nanosensors are used for detecting microbial and biochemical changes and release of antimicrobials, antioxidants, and enzymes in food. The smart food spoilage detection is based on the concept that unpleasant odors are generated from food and drinks when contaminated with bacteria. A chemical sensor system called electronic nose senses the odor released from spoiled food (Casalinuovo et al. 2006). In a more specified example, Concina et al. (2009) used an electronic nose to detect the volatile compounds released from spoiled tomato. Electronic tongues, based on amperometric sensors, are also in use for food analysis (Scampicchio et al. 2008). In addition, surface-enhanced Raman scattering (SERS) sensors made up of a graphene and Ag nanocomposite was developed to detect the prohibited color additives in food (Xie et al. 2012).

6.3.3 Nanocoating

Wax coating is widely used for improving the shelf life of apples and cheeses. Coating delays senescence and protects the fruit from decay (Ghosh et al. 2017). The recent development of nanoscale edible coatings was made possible through nanotechnology. Edible coatings and films are used on various foods including fruits, vegetables, meats, cheese, and bakery products (Momin et al. 2013). Nanocomposite edible coating is used for increasing the shelf life of olive (Ghosh et al. 2017).

6.3.4 Nanoadditives

Nanoadditives such as vitamins, antimicrobials, antioxidants, and preservatives are used to enhance taste, absorption, and bioavailability of nutrients (Momin et al. 2013). For example, food additive nanoscale silica powder increases the shelf life and bioavailability of specific nutrients (Canham 2007). The NPs containing calcium show effective absorption properties (Jeon and Lee 2009). The TiO_2 or E-171, a colored additive, produces a nontransparent white film that shows shielding against UV light (Latva-Nirva et al. 2009). The nanocrystalline powder of lycopene and resveratrol increases the absorption and bioavailability of nutrients in the body (Hsieh 2010). Nanoadditive, acyl ascorbates, synthesized through lipase condensation of ascorbic acid with fatty acids is used as antioxidant as it inhibits the oxidation of polyunsaturated fatty acid (Sharma and Pathak 2010). Aquasol preservative manufactured by AquaNova contains nanoscale micelle that increases absorption of nutritional additives and enhanced the preservation of food. Bioral™ omega-3 fatty acid nanocochleates, manufactured by BioDelivery Sciences International, is added to cakes, muffins, pasta, soups, cookies, cereals, chips, and confectionery. Synthetic lycopene, manufactured by BASF AG and DSM Nutritional Products Ltd., is an antioxidant and used in soft drinks, juices, breakfast cereals, instant soups, salad dressings, and yogurt (Kirdar 2015). Nisin-loaded chitosan-monomethyl fumaric acid (CM-N) NPs used as food additive show antibacterial activity against food-borne pathogens in orange juice (Khan et al. 2017).

6.3.5 Nanoencapsulation

Nanoencapsulation is the process to encapsulate substances at the nanoscale range (Lopez et al. 2006). The benefit of nanoencapsulation includes increased bioavailability of nutrients, preservation of the ingredients and additives during processing, removal of unpleasant tastes and flavors, and controlled release of additives (Chaudhry et al. 2010). Lipid-based nanoencapsulation systems such as nanoliposomes, archaeosomes, and nanocochleates can protect antioxidants from degradation and increase their solubility and bioavailability (Mozafari et al. 2006). Functional bread has been enriched with nanoencapsulated omega-3 fatty acids. Encapsulation decreased lipid oxidation during baking and reduces the formation of acrylamide and hydroxymethylfurfural (HMF) in bread (Gokmen et al. 2011). Donsi et al. (2011) used the high-pressure homogenization (HPH) method to prepare sunflower oil or oil-in-water nanoemulsions. A terpene and D-limonene mixture was encapsulated into sunflower oil or essential oil-in-water nanoemulsions. Encapsulation of terpenes and D-limonene inhibits the growth of microbes. Isolated lactoferrin from camel milk is encapsulated using calcium alginate. Calcium alginate nanocapsules help to control lactoferrin digestion (Raei et al. 2015).

Curcumin is a medically important compound produced by *Curcuma longa* plants. However, they are lipophilic in nature and unstable in gastrointestinal fluids. It has been shown that encapsulation of curcumin in nanoform enhances their antioxidant properties with increased absorption rate during digestion (Rao and Khanum 2016). In addition, nanoencapsulation could be used to encapsulate fish oils that deteriorate rapidly otherwise. As a result, nanoliposome system was developed to encapsulate fish oil (Ghorbanzade et al. 2017).

6.3.6 Nanoemulsions

The emulsion is a mixture of two immiscible liquid phases in which one phase disperses as droplets. Nanoemulsions consist of nanosize oil droplets of 10–100 nm. Nanoemulsions can be produced using high- and low-energy methods. A high-energy method uses mechanical devices (high-pressure valve homogenizers, microfluidizers, and sonication) to generate intense disruptive forces that breakdown the oil and water phases to form oil droplets. A low-energy method is based on the spontaneous formation of oil droplets within mixed oil-in-water emulsifier systems (Silva et al. 2012).

6.3.7 Nutraceutical Delivery

The major problem of the nutraceutical industry lies with the active components that poorly dissolve in oil or water, thus finding it difficult to reach target sites, thereby resulting in low bioavailability of nutrients (Putheti 2015). Nanobiotechnology employs a new strategy for the efficient delivery of nutraceuticals through nanoencapsulation. For example, nanoencapsulation of nutraceuticals is most commonly done using hydrophobins (Hyd), a small cysteine-rich amphipathic protein. The Hyd proteins bind to hydrophobic materials like vitamin D3 (VD3) with high affinity and protect the vitamin against degradation. The Hyd protein also acts as nanovehicle of hydrophobic nutraceuticals (Israeli-Lev et al. 2014). Dual nutraceutical nanohybrids consisting of folic acid and calcium were synthesized based on layered double hydroxide structure through exfoliation reassembly hybridization method. The report shows that the use of nutraceutical nanohybrids increases contents of essential nutrients in the human body (Kim and Oh 2016).

6.3.8 Nanobiosensor for Detection of Food Pathogen and Other Contaminants

Traditionally, the food pathogens are detected through microbial culture observation, PCR amplification, and immunology-based method (ELISA). Recently, many NP-based biosensor systems have been developed that are fast, inexpensive and require a little experience. Magnetic NPs coated with antibody (anti-AFM_1) are used to separate the bound and unbound fractions of the component. Nanosensors successfully detect a minute change in the food color and gases released due to spoilage (Pradhan et al. 2015).

Biosensor-based methods involve detection of foodborne pathogen antibodies bound to NPs (Popov et al. 2010). An electronic tongue or nose consisting of an array of nanosensors was developed to detect the food spoilage through signals of gases released by food items (Garcia et al. 2006). Several cited examples have used nanosensors for pathogen detection. For example, nanobiosensors were used to detect food pathogens such as *Staphylococcus* spp., *Bacillus* spp., *Clostridium* spp., *Shigella*, and *E. coli* (Otles and Yalcin 2010). A gold NP-based biosensor with graphite-epoxy composite electrodes was used for the identification of *Salmonella* IS200 (Oliveira Marques et al. 2009). Streptavidin-coated magnetic NPs were used for detecting foodborne pathogens like *Escherichia coli* O157: H7, *Salmonella enterica*, *Vibrio cholera*, and *Campylobacter jejuni* (Song et al. 2013). Shelby et al. (2017) used magneto-fluorescent nanosensor for detecting foodborne pathogen *E. coli* O157: H7. Piezoelectric biosensor based on an Au NP was used for detecting *Escherichia coli* O157: H7 in apple juice, milk, and ground beef. An Au/Si hetero-nanorod biosensor was used to detect *Salmonella* sp. based on the fluorescence (Fu et al. 2008). Electrically active polyaniline-coated magnetic (EAPM) NP-based biosensor was used to detect endospores of *Bacillus anthracis* in food samples (Pal and Alocilja 2009).

Nanosensors find huge application in toxin detection from various products. For example, an electrochemical immunosensor was used for detecting aflatoxin M_1 (AFM_1) in milk. The sensor was based on a competitive immunoassay in which the enzyme horseradish peroxidase (HRP) was used as a tag (Paniel et al. 2010). Optical biosensor based on competitive dispersion of gold nanorods (GNRs) detect aflatoxin B1 (AFB_1) in food products (Xu et al. 2013). The Au NP-based aptasensor was used for detecting AFB_1 in food samples (Hosseini et al. 2015). The use of Au NP in biosensor amplifies the signal frequency change due to the relatively large mass of the NPs (Chen et al. 2008). An optical carbon nanotube (CNT) immunosensor detects *Staphylococcal* enterotoxin B (SEB) in food. The deployment of CNT has lowered the detection limit of SEB and increased the immunosensor sensitivity at least sixfold (Yang et al. 2008).

Recently, the aptamer-based biosensors are under the course of development. Aptamers are single-stranded nucleic acids (DNA and RNA) or peptides that bind to the targeted molecules with high affinity and specificity (Song et al. 2012). In a classical example, luminescent assay, based on aptamer sensors, was used for detecting toxins in food (McKeague et al. 2011). In addition, aptamer-based nanosensors successfully detect acetamiprid pesticides in food (Verdian 2017). Besides aptamer biosensors, NP-based fluorescence resonance energy transfer methods are also used for detecting organophosphorus pesticides in food samples (Long et al. 2015).

6.4 Nanoparticles: Risks and Regulations

Besides various advantages of nanobiotechnology in the food industry, some safety issues are associated with the use of NPs. Several vivid nanotechnological application can impart accidental effects on plants and animals, thus need a thorough evaluation. NPs may accumulate within body organs and tissues due to smaller dimensions (Savolainen et al. 2010). Animals may inhale NPs into the lungs, resulting in severe disorders. The NPs may deposit on leaves and floral plant parts to create a toxic layer that prevents pollen tube penetration on stigma. They may also affect the translocation of water and minerals (Tarafdar 2015). Exposure of human endothelial cells with NPs could lead to cytotoxicity, genotoxicity, and dysfunction of nitric oxide signaling (Cao 2018). The NPs can enter the human body through the lungs, intestinal tract, and skin. Skin contact with toxic NPs and their inhalation are major risks in agriculture practices (Hoet et al. 2004). Inhaled NPs may reach the brain to cause neurodegeneration (Win-Shwe and Fujimaki 2011). Some studies reported the neurotoxic nature of NPs (Wu et al. 2013; Chin-Chan et al. 2015; Coccini et al. 2015; Migliore et al. 2015).

As a result, special regulations are required for NP preparation and application in agriculture and food industries (Amenta et al. 2015). In 2011, the guidance draft on risk assessment for nanotechnology use in food, feed applications, and pesticides was issued by the European Food Safety Authority (EFSA Scientific Committee). Environmental Protection Agency (EPA), USA, issued a proposal to use Sect. 6(a)(2) of the Federal Insecticide, Fungicide, and Rodenticide Act (FIFRA) to obtain information on the use of nanomaterials in pesticide products and its potential effect on humans or the environment (EFSA Scientific Committee 2011). As a result, National Institute for Occupational Safety and Health (NIOSH), USA, issued warning against occupational exposure to carbon nanotubes and nanofibers during research and development (Howard 2013).

6.5 Conclusion

Nanobiotechnology involves the use of nanotechnology to modify living organisms and to enable the combination of biological and nonbiological materials. Nanobiotechnology has multiple applications in agriculture, food, and other sectors. The NP-based herbicides, pesticides, fertilizers, fungicides, and insecticides are needed for improving the crop productivity in an eco-friendly way. In seed science, nanotechnology is used to enhance seed germination. The role of nanobiosensors has also been proved useful in pathogen diagnosis, pesticide residue detection, and food contaminant determination. The nanobiotechnology application in food includes food processing (improving texture, color and shelf life), packaging (antimicrobial, oxygen and aroma scavenger), application of nanoadditives, nanoencapsulation, and controlled nutraceuticals delivery at the time of digestion. The

NPs have been effectively used in developing food storage packaging materials that can decrease pathogenic growth of stored food. Nanobiosensors are also used to detect the contamination of foodborne pathogens, toxic substance, and pesticide in food. However, extensive studies are required to understand the NP-associated toxicity and involved mechanism to avoid application-based harmful effects on the environment and human health.

References

Abd-Elsalam KA, Alghuthaymi MA (2015) Nanobiofungicides: are they the next-generation of fungicides? J Nanotech Mater Sci 2:38–40

Abdel-Aziz HMM, Hasaneen MNA, Omer AM (2016) Nano chitosan-NPK fertilizer enhances the growth and productivity of wheat plants grown in sandy soil. Span J Agric Res 14:e0902. https://doi.org/10.5424/sjar/2016141-8205

Abduz Zahir A, Bagavan A, Kamaraj C, Elango G, Abdul Rahuman A (2012) Efficacy of plant-mediated synthesized silver nanoparticles against *Sitophilus oryzae*. J Biopest 5:95–102

Adhikari T, Kundu S, Rao AS (2013) Impact of SiO_2 and Mo nanoparticles on seed germination of rice (*Oryza sativa* L.). Int J Agric Food Sci Technol 4:809–816

Ali MA, Rehman I, Iqbal A, Din S, Rao AQ, Latif A, Samiullah TR, Azam S, Husnain T (2014) Nanotechnology, a new frontier in agriculture. Adv Life Sci 1:129–138

Alsaeedi A, El-Ramady H, Alshaal T, El-Garawani M, Elhawat N, Al-Otaibi A (2018) Exogenous nanosilica improves germination and growth of cucumber by maintaining K^+/na^+ ratio under elevated na^+ stress. Plant Physiol Biochem 125:164–171

Amenta V, Aschberger K, Arena M, Bouwmeester H, Moniz FB, Brandhoff P, Gottardo S, Marvin HJP, Mech A, Pesudo LQ, Rauscher H, Schoonjans R, Vettori MV, Weigel S, Peters RJ (2015) Regulatory aspects of nanotechnology in the Agri/feed/food sector in EU and non-EU countries. Regul Toxicol Pharmacol 73:463–476

Ariffin SAB, Adam T, Hashim U, Faridah S, Zamri I, Uda MNA (2014) Plant diseases detection using nanowire as biosensor transducer. Adv Mater Res 832:113–117

Ardekani MRS, Abdin MZ, Nasrullah N, Samim M (2014) Calcium phosphate nanoparticles a novel non-viral gene delivery system for genetic transformation of tobacco. International journal of pharmacy and pharmaceutical. Science 6:605–609

Avella M, De Vlieger JJ, Errico ME, Fischer S, Vacca P, Volpe MG (2005) Biodegradable starch/clay nanocomposite films for food packaging applications. Food Chem 93:467–474

Bai YX, Li YF, Yang Y, Yi LX (2006) Covalent immobilization of triacylglycerol lipase onto functionalized nanoscale SiO_2 spheres. Process Biochem 41:770–777

Barik TK, Sahu B, Swain V (2008) Nanosilica – from medicine to pest control. Parasitol Res 103(2):253–258

Bhattacharyya A, Duraisamy P, Govindarajan M, Buhroo AA, Prasad R (2016) Nanobiofungicides: Emerging trend in insect pest control. In: Advances and Applications through Fungal Nanobiotechnology (ed. Prasad R), Springer International Publishing Switzerland 307–319

Bottoms M, Emerson SH (2013) Chemistry, fertilizer, and the environment. California Foundation for Agriculture in the Classroom pp 1–98

Buteler M, Sofie SW, Weaver DK, Driscoll D, Muretta J, Stadler T (2015) Development of nanoalumina dust as insecticide against *Sitophilus oryzae* and *Rhyzopertha dominica*. Int J Pest Manag 61:80–89

Canham LT (2007) Nanoscale semiconducting silicon as nutritional food additive. Nanotech 18:1–6

Cao Y (2018) The toxicity of nanoparticles to human endothelial cells. Adv Exp Med Biol 1048:59–69
Casalinuovo IA, Pierro DD, Coletta M, Francesco PD (2006) Application of electronic noses for disease diagnosis and food spoilage detection. Sensors 6:1428–1439
Chang Y, Lin Y, Xiao G, Chiu T, Hu C (2016) A highly selective and sensitive nanosensor for the detection of glyphosate. Talanta 161:94–98
Chartuprayoon N, Rheem Y, Ng JCK, Nam J, Chen W, Myung NV (2013) Polypyrrole nanoribbon based chemiresistive immunosensors for viral plant pathogen detection. Anal Methods 5:3497–3502
Chaudhry Q, Watkins R, Castle L (2010) Nanotechnologies in the food. In: Chaudhry Q, Castle L, Watkins R (eds) New opportunities, new questions, new concerns. Royal Society of Chemistry, UK, pp 1–17
Chin-Chan M, Navarro-Yepes J, Quintanilla-Vega B (2015) Environmental pollutants as risk factors for neurodegenerative disorders: Alzheimer and Parkinson diseases. Front Cell Neurosci 9:1–22
Chen SH, Wu VC, Chuang YC (2008) Using oligonucleotide-functionalized au nanoparticles to rapidly detect foodborne pathogens on a piezoelectric biosensor. J Microbiol Methods 73:7–17
Chellaram C, Murugaboopathi G, John AA, Sivakumar R, Ganesan S, Krithika S, Priya G (2014) Significance of nanotechnology in food industry. APCBEE Procedia 8:109–113
Coccini T, Grandi S, Lonati D, Locatelli C, Simone UD (2015) Comparative cellular toxicity of titanium dioxide nanoparticles 3 on human astrocyte and neuronal cells after acute and 4 prolonged exposure. Neurotoxicology 48:77–89
Concina I, Falasconi M, Gobbi E, Bianchi F, Musci M, Mattarozzi M, Pardo M, Mangia A, Careri M, Sberveglieri G (2009) Early detection of microbial contamination in processed tomatoes by electronic nose. Food Cont 20:873–880
Dasgupta N, Ranjan S, Mundekkad D, Ramalingam C, Shanker R, Kumar A (2015) Nanotechnology in agro-food: from field to plate. Food Res Int 69:381–400
Davari MR, Kazazi SB, Pivehzhani OA (2017) Nanomaterials: implications on agroecosystem. In: Prasad R, Kumar M, Kumar V (eds) Nanotechnology: an agricultural paradigm. Springer Nature, Singapore, pp 59–71
Debnath N, Das S, Seth D, Chandra R, Bhattacharya SC, Goswami A (2011) Entomotoxic effect of silica nanoparticles against *Sitophilus oryzae* (L.). J Pest Sci 84:99–105
Dehkourdi EH, Mosavi M (2013) Effect of anatase nanoparticles (TiO_2) on parsley seed germination (*Petroselinum crispum*) *in vitro*. Biol Trace Elem Res 155:283–286
Dias MV, Soares NDFF, Borges SV, Sousa MMD, Nunes CA, Oliveira IRND, Medeiros EAA (2013) Use of allyl isothiocyanate and carbon nanotubes in an antimicrobial film to package shredded, cooked chicken meat. Food Chem 141:3160–3166
Donsì F, Annunziata M, Sessa M, Ferrari G (2011) Nanoencapsulation of essential oils to enhance their antimicrobial activity in foods. LWT – Food Sci Tech 44:1908–1914
Duke MC, Lim A, Luz SC, Nielsen L (2008) Lactic acid enrichment with inorganic nanofiltration and molecular sieving membranes by pervaporation. Food Bioprod Proc 86:290–295
Durga Devi G, Murugan K, Panneer Selvam C (2014) Green synthesis of silver nanoparticles using *Euphorbia hirta* (Euphorbiaceae) leaf extract against crop pest of cotton bollworm, *Helicoverpa armigera* (Lepidoptera: Noctuidae). J Biopest 7:54–66
El-bendary HM, El-Helaly AA (2013) First record nanotechnology in agricultural: silica nanoparticles a potential new insecticide for pest control. App Sci Report 4:241–246
Elrahman SHA, Mostafa MAM (2015) Applications of nanotechnology in agriculture: an overview. Egyptian J Soil Sci 55:197–214
Emamifar A, Kadivar M, Shahedi M, Soleimanian-Zad S (2010) Evaluation of nanocomposite packaging containing ag and ZnO on shelf life of fresh orange juice. Innovative Food Sci Emerg Technol 11:742–748
Espitia PJP, Soares NDFF, Coimbra JSDR, Andrade NJD, Cruz RS, Medeiros EAA (2012) Zinc oxide nanoparticles: synthesis, antimicrobial activity and food packaging applications. Food Bioprocess Technol 5:1447–1464

EFSA Scientific Committee (2011) Guidance on the risk assessment of the application of nanoscience and nanotechnologies in the food and feed chain. EFSA J 9:1–36
Fang Y, Umasankar Y, Ramasamy RP (2014) Electrochemical detection of p-ethylguaiacol, a fungi infected fruit volatile using metal oxide nanoparticles. Analyst 139:3804–3810
Fasihnia SH, Peighambardoust SH, Peighambardoust SJ (2017) Nanocomposite films containing organoclay nanoparticles as an antimicrobial (active) packaging for potential food application. J Food Process Preserv e13488:1–10
Feizi H, Moghaddam PR, Shahtahmassebi N, Fotovat A (2012) Impact of bulk and nanosized titanium dioxide (TiO_2) on wheat seed germination and seedling growth. Biol Trace Elem Res 146:101–106
Fletcher J, Bender C, Budowle B, Cobb WT, Gold SE, Ishimaru CA, Luster D, Melcher U, Murch R, Scherm H, Seem RC, Sherwood JL, Sobral BW, Tolin SA (2006) Plant pathogen forensics: capabilities, needs, and recommendations. Microbiol Mol Biol Rev 70:450–471
Fu J, Park B, Siragusa G, Jones L, Tripp R, Zhao Y, Cho Y (2008) An au/Si hetero-nanorod-based biosensor for *Salmonella* detection. Nanotech 19:1–7
Gaikwad KK, Singh S, Lee YS (2018) High adsorption of ethylene by alkali-treated halloysite nanotubes for food-packaging applications. Environ Chem Letters 16:1055–1062
Garcia M, Aleixandre M, Gutiérrez J, Horrillo MC (2006) Electronic nose for wine discrimination. Sensors Actuat 113:911–916
Gopinath K, Gowri S, Karthika V, Arumugam A (2014) Green synthesis of gold nanoparticles from fruit extract of *Terminalia arjuna*, for the enhanced seed germination activity of *Gloriosa superba*. J Nanostr Chem 4:115–126
Ghorbanzade T, Jafari SM, Akhavan S, Hadavi R (2017) Nano-encapsulation of fish oil in nano-liposomes and its application in fortification of yogurt. Food Chem 216:146–152
Ghormade V, Deshpande MV, Paknikar KM (2011) Perspective for nano-biotechnology enabled protection and nutrition of plants. Biotech Adv 29:792–803
Ghosh A, Dey K, Mani A, Dey AN, Bauri FK (2017) Implication of nanocomposite edible coating for shelf life extension of Indian olive (*Elaeocarpus floribundus* Blume). Curr J App Sci Tech 22:1–8
Gokmen V, Mogol BA, Lumaga RB, Fogliano V, Kaplun Z, Shimoni E (2011) Development of functional bread containing nanoencapsulated omega-3 fatty acids. J Food Eng 105:585–591
Gonzalez A, Igarzabal CIA (2013) Soy protein-poly(lactic acid) bilayer films as biodegradable material for active food packaging. Food Hydrocoll 33:289–296
Grillo R, Pereira AE, Nishisaka CS, de Lima R, Oehlke K, Greiner R, Fraceto LF (2014) Chitosan/tripolyphosphate nanoparticles loaded with paraquat herbicide: an environmentally safer alternative for weed control. J Hazard Materials 278:163–171
Hafeez A, Razzaq A, Mahmood T, Jhanzab HM (2015) Potential of copper nanoparticles to increase growth and yield of wheat. J Nanosc Advanced Tech 1:6–11
He L, Liu Y, Mustapha A, Lin M (2011) Antifungal activity of zinc oxide nanoparticles against *Botrytis cinerea* and *Penicillium expansum*. Microbiol Res 166:207–215
Hoet PHM, Brüske-Hohlfeld I, Salata OV (2004) Nanoparticles–known and unknown health risks. J Nanobiotechnol 2:1–15
Hosseini M, Khabbaz H, Dadmehr M, Ganjali MR, Mohamadneja J (2015) Aptamer-based colorimetric and chemiluminescence detection of aflatoxin B1 in foods samples. Acta Chim Slov 62:721–728
Howard J (2013) Occupational exposure to carbon nanotubes and nanofibers. https://www.cdc.gov/niosh/docs/2013-145/pdfs/. Accessed 20 June 2018
Hsieh KL (2010) Lycopene and resveratrol dietary supplement. US Patent WO/2010/132021, 18 Nov 2010
Hussain AA, Al-Rawajfeh AE (2009) Recent patents of nanofiltration applications in oil processing, desalination, wastewater and food industries. Recent Patents Chemical Eng 2:51–66
Hussain M, Raja NI, Mashwani ZUR, Iqbal M, Sabir S, Yasmeen F (2017) In vitro seed germination and biochemical profiling of *Artemisia absinthium* exposed to various metallic nanoparticles. 3Biotech 7:101–108

Ing LY, Zin NM, Sarwar A, Katas H (2012) Antifungal activity of chitosan nanoparticles and correlation with their physical properties. Int J Biomat 632698. https://doi.org/10.1155/2012/632698

Ismail M, Prasad R, Ibrahim AIM, Ahmed ISA (2017) Modern prospects of nanotechnology in plant pathology. In: Nanotechnology (eds. Prasad R, Kumar M, Kumar V), Springer Nature Singapore Pte Ltd. 305–317

Jamdagni P, Rana JS, Khatri P, Nehra K (2018) Comparative account of antifungal activity of green and chemically synthesized zinc oxide nanoparticles in combination with agricultural fungicides. Int J Nano Dimen 9:198–208

Jampílek J, králová K (2015) Application of nanotechnology in agriculture and food industry, its prospects and risks. Ecol Chem Eng S22:321–361

Jeon JS, Lee HS (2009) Nano-particles containing calcium and method for preparing the same. US Patent WO/2009/011520, 22 Jan 2009

Jo Y, Kim BH, Jung G (2009) Antifungal activity of silver ions and nanoparticles on phytopathogenic fungi. Plant Dis 93:1037–1043

Kanhed P, Birla S, Gaikwad S, Gade A, Seabra AB, Rubilar O, Duran N, Rai M (2014) In vitro antifungal efficacy of copper nanoparticles against selected crop pathogenic. Fungi Mater Lett 115:13–17

Kashyap PL, Rai P, Sharma S, Chakdar H, Kumar S, Pandiyan K, Srivastava AK (2016) Nanotechnology for the detection and diagnosis of plant pathogens. In: Ranjan S et al (eds) Nanoscience in food and agriculture 2, sustainable agriculture reviews 21. Springer, Basel pp 253–276

Khan I, Tango CN, Sumaira M, Deog-Hwan O (2017) Evaluation of nisin-loaded chitosan-monomethyl fumaric acid nanoparticles as a direct food additive. Carbohydrate Poly 184:100–107

Khalifa NS, Hasaneen MN (2018) The effect of chitosan–PMAA–NPK nanofertilizer on *Pisum sativum* plants. 3Biotech 8:193–205

Khodakovskaya M, Dervishi E, Mahmood M, Xu Y, Li Z, Watanabe F, Biris AS (2009) Carbon nanotubes are able to penetrate plant seed coat and dramatically affect seed germination and plant growth. ACS Nano 3:3221–3227

Khooshe-Bast SN, Ghaffari-moghaddam M, Mirshekar A (2016) Insecticidal effects of zinc oxide nanoparticles and *Beauveria bassiana* TS11 on *Trialeurodes vaporariorum* (Westwood, 1856) (Hemiptera: Aleyrodidae). Acta Agri Slovenica 107:299–309

Kim T, Oh J (2016) Dual nutraceutical nanohybrids of folic acid and calcium containing layered double hydroxides. J Solid State Chem 233:125–132

Kim KJ, Sung WS, Suh BK, Moon SK, Choi JS, Kim JG, Lee DG (2009) Antifungal activity and mode of action of silver nano-particles on *Candida albicans*. Biometals 22:235–242

Kim SW, Jung JH, Lamsal K, Kim YS, Min JS, Lee YS (2012) Antifungal effects of silver nanoparticles (AgNPs) against various plant pathogenic fungi. Mycobiology 40:53–58

Kirdar SS (2015) ISITES conference. Valencia, pp 1517–1527

Ko KS, Koh DC, Kong IC (2017) Evaluation of the effects of nanoparticle mixtures on *Brassica* seed germination and bacterial bioluminescence activity based on the theory of probability. Nano 7:344–354

Kothari P, Setia H (2017) Silver nanoparticle filled HPMC and xanthan films for food packaging and safety. Indian J Sci Tech 10:1–6

Kumar GD, Natarajan N, Nakkeeran S (2016) Antifungal activity of nanofungicide Trifloxystrobin 25% + Tebuconazole 50% against *Macrophomina phaseolina*. African J Microbiol Res 10:100–105

Lin HY, Huang CH, Lu SH, Kuo IT, Chau LK (2014) Direct detection of orchid viruses using nanorod-based fiber optic particle plasmon resonance immunosensor. Biosens Bioelectron 51:371–378

Lamsal K, Kim SW, Jung JH, Kim YS, Kim KS, Lee YS (2011) Inhibition effects of silver nanoparticles against powdery mildews on cucumber and pumpkin. Mycobiol 39:26–32

Latva-Nirva, E, Dahms G, Jung A, Fiebrig B (2009) Ultrafine titanium dioxide nanoparticles and dispersions thereof. US Patent WO 2009/141499 Al, 26 Nov 2009

Le Israeli-Lev G, Livney YD (2014) Self-assembly of hydrophobin and its coassembly with hydrophobic nutraceuticals in aqueous solutions: towards application as delivery systems. Food Hydrocoll 35:28–35

Lee DG, Ponvel KM, Kim M, Hwang S, Ahn IK, Lee CH (2009) Immobilization of lipase on hydrophobic nano-sized magnetite particles. J Mol Catal B Enzym 57:62–66

Lopez A, Gavara R, Lagaron J (2006) Bioactive packaging: turning foods into healthier foods through biomaterials. Trends Food Sci Technol 17:567–575

Li FS, Wu WT (2009) Lipase-immobilized electrospun PAN nanofibrous membranes for soybean oil hydrolysis. Biochem Eng J 45:48–53

Liu R, Lal R (2014) Synthetic apatite nanoparticles as a phosphorus fertilizer for soybean (*Glycine max*). Sci Reports 4:5686–5691

Long Q, Li H, Zhang Y, Yao S (2015) Upconversion nanoparticle-based fluorescence resonance energy transfer assay for organophosphorus pesticides. Biosens Bioelectron 68:168–174

Longano D, Ditaranto N, Cioffi N, Niso FD, Sibillano T, Ancona A, Conte A, Del Nobile MA, Sabbatini L, Torsi L (2012) Analytical characterization of laser-generated copper nanoparticles for antibacterial composite food packaging. Anal Bioanal Chem 403:1179–1186

Mahmoodzadeh H, Nabavi M, Kashefi H (2013) Effect of nanoscale titanium dioxide particles on the germination and growth of canola (*Brassica napus*). J Ornam Hort Plants 3:25–32

Malaikozhundan B, Vaseeharan B, Vijayakumar S, Thangaraj MP (2017) *Bacillus thuringiensis* coated zinc oxide nanoparticle and its biopesticidal effects on the pulse beetle, *Callosobruchus maculatus*. J Photochem Photobiol B Biol 174:306–314

Manikandan A, Subramanian KS (2016) Evaluation of zeolite based nitrogen nano-fertilizers on maize growth, yield and quality on inceptisols and alfisols. Int J Plant Soil Sci 9:1–9

Manimaran M (2015) A review on nanotechnology and its implications in agriculture and food industry. Asian J Plant Sci Res 5:13–15

Martin-Ortigosa S, Peterson DJ, Valenstein JS, Lin VSY, Trewyn BG, Lyznik LA, Wang K (2014) Mesoporous silica nanoparticle-mediated intracellular Cre protein delivery for maize genome editing via loxP site excision. Plant Physiol 164:537–547

McKeague MK, Giamberardino A, DeRosa MC (2011) Advances in aptamer based biosensors for food safety. In: Somerset V (ed) Environmental biosensors. InTech, Croatia, pp 17–42

Mehrazar E, Rahaie M, Rahaie S (2015) Application of nanoparticles for pesticides, herbicides, fertilisers and animals feed management. Int J Nanoparticles 8:1–19

Migliore L, Uboldi C, Di Bucchianico S, Coppede F (2015) Nanomaterials and neurodegeneration. Environ Mol Mutagen 56:149–170

Mishra S, Singh BR, Singh A, Keswani C, Naqvi AH, Singh HB (2014) Biofabricated silver nanoparticles act as a strong fungicide against *Bipolaris sorokiniana* causing spot blotch disease in wheat. PLoS One 9:e97881. https://doi.org/10.1371/journal.pone.0097881

Moghadam A, Vattani H, Baghaei N, Keshavarz N (2012) Effect of different levels of fertilizer nano-iron chelates on growth and yield characteristics of two varieties of spinach (*Spinacia oleracea* L.): Varamin 88 and Viroflay. Res J App Sci Eng Tech 4:4813–4818

Momin JK, Jayakumar C, Prajapati JB (2013) Potential of nanotechnology in functional foods. Emirates J Food Agricult 25:10–19

Moura MRD, Mattoso LHC, Zucolotto V (2012) Development of cellulose-based bactericidal nanocomposites containing silver nanoparticles and their use as active food packaging. J Food Eng 109:520–524

Mozafari MR, Flanagan J, Matia-Merino L, Awati A, Omri A, Suntres ZE, Singh H (2006) Recent trends in the lipid-based nanoencapsulation of antioxidants and their role in foods. J Sci Food Agri 86:2038–2045

Mukhopadhyay SS (2014) Nanotechnology in agriculture: prospects and constraints. Nanotech Sci App 7:63–71

Naderi MR, Danesh-Shahraki A (2013) Nanofertilizers and their roles in sustainable agriculture. Intl J Agri Crop Sci 5:2229–2232

Nesi A, Gordi M, Davidovi S, Radovanovic Z, Nedeljkovi J, Smirnova I, Gurikov P (2018) Pectin-based nanocomposite aerogels for potential insulated food packaging application. Carbohydrate Poly 195:128–135
Nguyen M, Reynolds N, Vigneswaran S (2003) By-product recovery from cottage cheese production by nanofiltration. J Clean Prod 11:803–807
Niemeyer CM, Doz P (2001) Nanoparticles, proteins, and nucleic acids: biotechnology meets materials science. Angewandte Chemie Int Ed 40:4128–4158
Oliveira Marques PRB, Lermo A, Campoy S, Yamanaka H, Barbe J, Alegret S, Pividori MI (2009) Double-tagging polymerase chain reaction with a thiolated primer and electrochemical genosensing based on gold nanocomposite sensor for food safety. Anal Chem 81:1332–1339
Oskam G (2006) Metal oxide nanoparticles: synthesis, characterization and application. J SolGel Sci Technol 37:161–164
Otles S, Yalcin B (2010) Nano-biosensors as new tool for detection of food quality and safety. LogForum 6:67–70
Ouda SM (2014) Antifungal activity of silver and copper nanoparticles on two plant pathogens, *Alternaria alternata* and *Botrytis cinerea*. Res J Microbiol 9:34–42
Pal S, Alocilja EC (2009) Electrically active polyaniline coated magnetic (EAPM) nanoparticle as novel transducer in biosensor for detection of *Bacillus anthracis* spores in food samples. Biosens Bioelectron 24:1437–1444
Panacek A, Kolar M, Vecerova R, Prucek R, Soukupova J, Krystof V, Hamal P, Zboril R, Kvıtek L (2009) Antifungal activity of silver nanoparticles against *Candida* spp. Biomaterials 30:6333–6340
Paniel N, Radoi A, Marty J (2010) Development of an electrochemical biosensor for the detection of aflatoxin M_1 in milk. Sensors 10:9439–9448
Parveen A, Rao S (2014) Effect of nanosilver on seed germination and seedling growth in *Pennisetum glaucum*. J Clust Sci 26:693–701
Pereira AES, Grillo R, Mello NFS, Rosa AH, Fraceto LF (2014) Application of poly(epsilon-caprolactone) nanoparticles containing atrazine herbicide as an alternative technique to control weeds and reduce damage to the environment. J Hazard Mat 268:207–215
Perlatti B, Bergo PLD, da Silva MFDGF, Fernandes JB, Forim MR (2013) Polymeric nanoparticle-based insecticides: a controlled release purpose for agrochemicals. In: Trdan S (ed) Insecticides - development of safer and more effective technologies, InTech, Croatia, pp 523–550
Pham DC, Nguyen TH, Ngoc UTH, Le NTT, Tran TV, Nguyen DH (2018) Preparation, characterization and antifungal properties of chitosan-silver nanoparticles synergize fungicide against *Pyricularia oryzae*. J Nanosc Nanotech 18:1–7
Popov KI, Filippov AN, Khurshudyan SA (2010) Food nanotechnologies. Russian J Gen Chem 80:630–642
Pradhan N, Singh S, Ojha N, Shrivastava A, Barla A, Rai V, Bose S (2015) Facets of nanotechnology as seen in food processing, packaging, and preservation industry. Biomed Res Int 365672. https://doi.org/10.1155/2015/365672
Prasad R, Kumar V, Prasad KS (2014) Nanotechnology in sustainable agriculture: present concerns and future aspects. Afr J Biotechnol 13(6):705–713
Prasad R, Bhattacharyya A, Nguyen QD (2017a) Nanotechnology in sustainable agriculture: Recent developments, challenges, and perspectives. Front Microbiol 8:1014. doi: https://doi.org/10.3389/fmicb.2017.01014
Prasad R, Kumar V, Kumar M (2017b) Nanotechnology: Food and Environmental Paradigm. Springer Nature Singapore Pte Ltd. (ISBN 978-981-10-4678-0)
Priyadarshi R, Negi YS (2016) Effect of varying filler concentration on zinc oxide nanoparticle embedded chitosan films as potential food packaging material. J Polym Environ 25:1087–1098
Puoci F, Lemma F, Spizzirri UG, Cirillo G, Curcio M, Picci N (2008) Polymer in agriculture: a review. Am J Agri Biol Sci 3:299–314
Putheti S (2015) Application of nanotechnology in food, nutraceuticals and pharmaceuticals. e-J Sci Tech 2:17–23

Rad F, Mohsenifar A, Tabatabaei M, Safarnejad MR, Shahryari F, Safarpour H, Foroutan A, Mardi M, Davoudi D, Fotokian M (2012) Detection of *Candidatus* Phytoplasma aurantifolia with a quantum dots FRET-based biosensor. J Plant Pathol 94:525–534

Raei M, Rajabzadeh G, Zibaei S, Jafari SM, Sani AA (2015) Nano-encapsulation of isolated lactoferrin from camel milk by calcium alginate and evaluation of its release. Int J Biol Macromol 79:669–673

Ragaei M, Sabry AH (2014) Nanotechnology for insect pest control. Int J Sci Environ Tech 3:528–545

Rao PJ, Khanum H (2016) A green chemistry approach for nanoencapsulation of bioactive compound curcumin. LWT - Food Sci Tech 65:695–702

Ravichandran R (2010) Nanotechnology applications in food and food processing: innovative green approaches, opportunities and uncertainties for global market. Int J Green Nanotech 1:P72–P96

Rouhani M, Samih MA, Kalantari S (2012) Insecticide effect of silver and zinc nanoparticles against *Aphis nerii* Boyer de Fonscolombe (Hemiptera: Aphididae). Chilean J Agri Res 72:590–594

Rui M, Ma C, Hao Y, Guo J, Rui Y, Tang X, Zhao X, Fan X, Zhang Z, Hou T, Zhu S (2016) Iron oxide nanoparticles as a potential iron fertilizer for peanut (*Arachis hypogaea*). Front Plant Sci 7:815–824

Saharan V, Sharma G, Yadav M, Choudhary MK, Sharma SS, Pal A, Raliya R, Biswas P (2015) Synthesis and *in vitro* antifungal efficacy of cu–chitosan nanoparticles against pathogenic fungi of tomato. Int J Biol Macromol 75:346–353

Sahayaraj K, Madasamy M, Radhika SA (2016) Insecticidal activity of bio-silver and gold nanoparticles against *Pericallia ricini* fab. (Lepidaptera: Archidae). J Biopest 9:63–72

Sahoo D, Mandal A, Mitra T, Chakraborty K, Bardhan M, Dasgupta AK (2018) Nanosensing of pesticides by zinc oxide quantum dot: an optical and electrochemical approach for the detection of pesticides in water. J Agric Food Chem 66:414–423

Sarwar MS, Niazi MBK, Jahan Z, Ahmad T, Hussain A (2017) Preparation and characterization of PVA/nanocellulose/ag nanocomposite films for antimicrobial food packaging. Carbohydrate Poly 184:453–464

Savithramma N, Ankanna S, Bhumi G (2012) Effect of nanoparticles on seed germination and seedling growth of *Boswellia ovalifoliolata* – an endemic and endangered medicinal tree. Taxon Nano Vision 2:61–68

Savolainen K, Pylkkänen L, Norppa H, Falck G, Lindberg H, Tuomi T, Vippola M, Alenius H, Hämeri K, Koivisto J, Brouwer D, Mark D, Bard D, Berges M, Jankowska E, Posniak M, Farmer P, Singh R, Krombach F, Bihari P, Kasper G, Seipenbusch M (2010) Nanotechnologies, engineered nanomaterials and occupational health and safety – a review. Saf Sci 48:957–963

Scampicchio M, Ballabio D, Arecchi A, Cosio SM, Mannino S (2008) Amperometric electronic tongue for food analysis. Microchim Acta 163:11–21

Scrinis G, Lyons K (2007) THE emerging nano-corporate paradigm: nanotechnology and the transformation of nature, food and Agri-food systems. Int J Socio Food Agri 15:22–44

Singh NB, Amist N, Yadav K, Singh D, Pandey JK, Singh SC (2013) Zinc oxide nanoparticles as fertilizer for the germination, growth and metabolism of vegetable crops. J Nanoeng Nanomanufacturing 3:353–364

Singh T, Shukla S, Kumar P, Wahla V, Bajpai VK, Rather IA (2017) Application of nanotechnology in food science: perception and overview. Front Microbiol 8:1–7

Shahrekizad M, Ahangara AG, Mirb N (2015) EDTA-coated Fe_3O_4 nanoparticles: a novel biocompatible fertilizer for improving agronomic traits of sunflower (*Helianthus annuus*). J Nanostruct 5:117–127

Shaker AM, Zaki AH, Abdel-Rahim EF, Khedr MH (2016) Novel CuO nanoparticles for pest management and pesticides photodegradation. Adv Environ Biol 10:274–283

Sharma R, Pathak Y (2010) Acyl ascorbate in enzymatic synthesis: industrial uses as a food nanoadditive. Nanotech 1:79–82

Sharon M, Choudhary AK, Kumar R (2010) Nanotechnology in agricultural diseases and food safety. J Phytology 2:83–92
Shelby T, Sulthana S, McAfee J, Banerjee T, Santra S (2017) Foodborne pathogen screening using magneto-fluorescent nanosensor: rapid detection of *E. coli* O157: H7. J Vis Exp 127:e55821. https://doi.org/10.3791/55821
Siddiqui MH, Al-Whaibi MH (2014) Role of nano-SiO_2 in germination of tomato (*Lycopersicum esculentum* seeds mill.). Saudi J Bio Sci 21:13–17
Silva AT, Nguyen A, Ye C, Verchot J, Moon JH (2010) Conjugated polymer nanoparticles for effective siRNA delivery to tobacco BY-2 protoplasts. BMC Plant Biol 10:291–304
Silva HD, Cerqueira MA, Vicente AA (2012) Nanoemulsions for food applications: development and characterization. Food Bioprocess Technol 5:854–867
Silvestre C, Duraccio D, Cimmino S (2011) Food packaging based on polymer nanomaterials. Prog Poly Sci 36:1766–1782
Siva GV, Benita LFJ (2016) Iron oxide nanoparticles promotes agronomic traits of ginger (*Zingiber officinale* Rosc.). Int J Adv Res Biol Sci 3:230–237
Song K, Lee S, Ban C (2012) Aptamers and their biological applications. Sens 12:612–631
Song L, Hongna L, Yan D, Lin L, Nongyue H (2013) Development of a magnetic nanoparticles microarray for simultaneous and simple detection of foodborne pathogens. J Biomed Nanotech 9:1254–1260
Sousa GFM, Gomes DG, Campos EVR, Oliveira JL, Fraceto LF, Stolf-Moreira R, Oliveira HC (2018) Post-emergence herbicidal activity of nanoatrazine against susceptible weeds. Front Environ Sci 6:1–6
Srivastava G, Das CK, Das A, Singh SK, Roy M, Kim H, Sethy N, Kumar A, Sharma RK, Singh SK, Philip D, Das M (2014) Seed treatment with iron pyrite (FeS2) nanoparticles increases the production of spinach. RSC Adv 4:58495–58504
Subramanian KS, Tarafdar JC (2011) Prospects of nanotechnology in Indian farming. Indian J Agri Sci 81:887–893
Sun X, Liu B, Xia K (2011) A sensitive and regenerable biosensor for organophosphate pesticide based on self-assembled multilayer film with CdTe as fluorescence probe. Luminescence 26:616–621
Tahira MA, Bajwa SZ, Mansoor S, Briddon RW, Khan WS, Scheffler BE, Amin I (2018) Evaluation of carbon nanotube based copper nanoparticle composite for the efficient detection of agroviruses. J Hazard Mater 346:27–35
Tarafdar JC (2015) Nanoparticle production, characterization and its application to horticultural crops. In: Aishwath OP, Singh B, Dubey PN, Mishra BK (eds) Winter School on "utilization of degraded land and soil through horticultural crops for agricultural productivity and environmental quality". NRCSS, Ajmer, Rajasthan, pp 222–229
Thangavel G (2014) Nanotechnology in food industry – a review. Int J Chem Tech 6:4096–4101
Torney FO, Trewyn BG, Lin VSY, Wang K (2007) Mesoporous silica nanoparticles deliver DNA and chemicals into plants. Nature Nanotech 2:295–300
Tripathi S, Mehrotra GK, Dutta PK (2008) Chitosan based antimicrobial films for food packaging applications. E-Polymers 93:1–7
Upadhyaya H, Roy H, Shome S, Tewari S, Bhattacharya MK, Panda SK (2017) Physiological impact of zinc nanoparticle on germination of rice (*Oryza sativa* L) seed. J Plant Sci Phytopathol 1:62–70
Vani C, Brindhaa U (2013) Silica nanoparticles as nanocides against *Corcyra cephalonica* (S.), the stored grain pest. Int J Pharm Biol Sci 4:1108–1118
Verdian A (2017) Apta-nanosensors for detection and quantitative determination of acetamiprid – a pesticide residue in food and environment. Talanta 176:456–464
Vimala V, Clarke SK, Urvinder Kaur S (2016) Pesticides detection using acetylcholinesterase nanobiosensor. Biosens J 5:1–4
Wang M, Li Z (2008) Nano-composite ZrO2/au film electrode for voltammetric detection of parathion. Sensors Actuators B Chem 133:607–612

Wang SL, Nguyen AD (2018) Effects of Zn/B nanofertilizer on biophysical characteristics and growth of coffee seedlings in a greenhouse. Res Chem Intermed 44:4889–4901

Wang Z, Wei F, Liu SY, Xu Q, Huang J-Y, Dong XY, Hua JH, Yang Q, Zhao YD, Chen H (2010) Electrocatalytic oxidation of phytohormone salicylic acid at copper nanoparticles-modified gold electrode and its detection in oilseed rape infected with fungal pathogen *Sclerotinia sclerotiorum*. Talanta 80:1277–1280

Wani AH, Shah MA (2012) A unique and profound effect of MgO and ZnO nanoparticles on some plant pathogenic fungi. J App Pharm Sci 02:40–44

Warczok J, Ferrando M, Lopez F, Guell C (2004) Concentration of apple and pear juices by nanofiltration at low pressures. J Food Eng 63:63–70

Win-Shwe T, Fujimaki H (2011) Nanoparticles and neurotoxicity. Int J Mol Sci 12:6267–6280

Wu J, Ding T, Sun J (2013) Neurotoxic potential of iron oxide nanoparticles in the rat brain striatum and hippocampus. Neurotoxicology 34:243–253

Xie Y, Li Y, Niu L, Wang H, Qian H, Yao W (2012) A novel surface-enhanced Raman scattering sensor to detect prohibited colorants in food by graphene/silver nanocomposite. Talanta 100:32–37

Xu X, Liu X, Li Y, Ying Y (2013) A simple and rapid optical biosensor for detection of aflatoxin B1 based on competitive dispersion of gold nanorods. Biosens Bioelectron 47:361–367

Yan X, Song Y, Zhu C, Li H, Du D, Su X, Lin Y (2018) MnO_2 nanosheet-carbon dots sensing platform for sensitive detection of organophosphorus pesticides. Anal Chem 90:2618–2624

Yang M, Kostov Y, Rasooly A (2008) Carbon nanotubes based optical immunodetection of staphylococcal enterotoxin B (SEB) in food. Int J Food Microbiol 127:78–83

Yang S, Wu T, Zhao X, Li X, Tan W (2014) The optical property of core-shell nanosensors and detection of atrazine based on localized surface plasmon resonance (LSPR) sensing. Sensors 14:13273–13284

Yao KS, Li SJ, Tzeng KC, Cheng TC, Chang CY, Chiu CY, Liao CY, Hsu JJ, Lin ZP (2009) Fluorescence silica nanoprobe as a biomarker for rapid detection of plant pathogens. Adv Materials Res 79-82:513–516

Yassen A, Abdallah E, Gaballah M, Zaghloul S (2017) Role of silicon dioxide nano fertilizer in mitigating salt stress on growth, yield and chemical composition of cucumber (*Cucumis sativus* L.). Int J Agri Res 12:130–135

Yildirim S, Röcker B, Pettersen MK, Nilsen-Nygaard J, Ayhan Z, Rutkaite R, Radusin T, Suminska P, Marcos B, Coma V (2017) Active packaging applications for food. Comp Rev Food Sci Food Safety 17:165–199

Zacco E, Pividori MI, Alegret S (2006) Electrochemical magnetoimmunosensing strategy for the detection of pesticides residues. Anal Chem 78:1780–1788

Zhao W, Ge P, Xu J, Chen H (2009) Selective detection of hypertoxic organophosphates pesticides via PDMS composite based acetylcholinesterase-inhibition biosensor. Environ Sci Technol 43:6724–6729

Ziaee M, Ganji Z (2016) Insecticidal efficacy of silica nanoparticles against *Rhyzopertha dominica* F. And *Tribolium confusum* Jacquelin du Val. J Plant Prot Res 56:250–255

Chapter 7
Application of Nanotechnology for Sustainable Crop Production Systems

Akbar Hossain, Rout George Kerry, Muhammad Farooq, Nawfel Abdullah, and M. Tofazzal Islam

Contents

A. Hossain (✉)
Bangladesh Wheat and Maize Research Institute, Dinajpur, Bangladesh

R. G. Kerry
PG Department of Biotechnology, Utkal University, Bhubaneswar, Odisha, India

M. Farooq
Department of Crop Sciences, College of Agricultural and Marine Sciences, Sultan Qaboos University, Muscat, Oman

Department of Agronomy, University of Agriculture, Faisalabad, Pakistan

The UWA Institute of Agriculture and School of Agriculture and Environment, The University of Western Australia, Perth, WA, Australia

N. Abdullah
Australian Institute of Innovative Materials (AIIM), University of Wollongong, North Wollongong, NSW, Australia

M. Tofazzal Islam
Institute of Biotechnology and Genetic Engineering (IBGE), Bangabandhu Sheikh Mujibur Rahman Agricultural University, Gazipur, Bangladesh

D. Thangadurai et al. (eds.), *Nanotechnology for Food, Agriculture, and Environment*, Nanotechnology in the Life Sciences,
https://doi.org/10.1007/978-3-030-31938-0_7

7.1 Introduction

Nanotechnology is the science of manipulation, control, precision placement, modelling and integration of nanoscale (1–100 nm) substances to form structures, components, devices and systems having new problem-solving properties and functions (Halford 2005; US EPA 2007; Sargent Jr. 2011; Jin 2012). However, the simultaneously colloidal particulate nanoscale (10 to 1000 nm) size can be considered as nanoparticles for application in agriculture and allied disciplines (Nakache et al. 1999; US Department of Agriculture 2002). The versatile use of nanoparticles is linked with their high surface areas and unique physicochemical properties. Therefore, synthesis of a certain dimension of nanoparticles of metallic and non-metallic compounds/atoms can be used for a certain purpose. Nanomaterials containing an array of nanosized pores have large surface area-to-volume ratios. They have been actively researched and developed due to their potential as catalytic and adsorption materials, offering sites for novel chemical reactions. For example, recently Jiang et al. (2017) synthesized mesoporous rhodium nanoparticles that are two to threefold more efficient as catalytic converter for vehicle (Figs. 7.1 and 7.2).

The European Commission declared the nanotechnology as one of its six "vital empowering technologies" in several industrial sectors (Parisi et al. 2015), such as biotechnology, medicine, electronics, material science and energy sectors, among others (Iqbal et al. 2017), whereas Hornyak et al. (2008) opined that nanotechnology is a convergent and enabling horizontal technology that cuts across all vertical industrial sectors such as agriculture, biotechnology, chemical, recreation/sports, telecommunication, computers/data storage, construction, transportation, health care/pharmaceuticals, aerospace/defence and energy.

It has become a robust technology for the sustainability of agriculture (Sastry et al. 2011; Mukhopadhyay 2014; Prasad et al. 2014). Although the agricultural sector has benefited from many other modern technological innovations, such as hybrid crop cultivars, synthetic fertilizers, genetic engineering and gene editing through biotechnological innovation, the real promise of nanotechnology in agricultural development is yet to be realized at large (Parisi et al. 2015). The potential of application of nanotechnology in climate-resilient and smart agriculture is manifold. Nanomaterials can help reduce the amount of sprayed chemical products by smart delivery of active ingredients, minimize nutrient losses in fertilization (Gogos et al. 2012) and increase yields through optimized water and nutrient management.

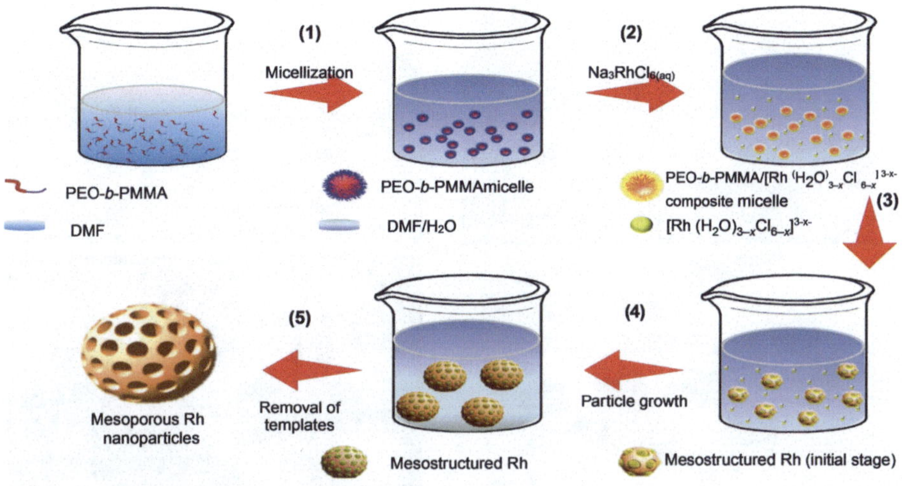

Fig. 7.1 An illustration describing the formation mechanism for mesoporous Rh nanostructures. Mesoporous Rh nanostructures form via chemical reduction on self-assemble polymeric PEO-b-PMMA micelle templates. The PEO-b-PMMA micelles function as a soft template, Na_3RhCl_6 as the Rh precursor, AA as the reducing agent, and DMF/H_2O as the mixed solvent, respectively. The synthetic process has five steps: (1) addition of water causes the PEO-b-PMMA to self-assemble into spherical micelles with a PMMA core and a PEO shell; (2) Na_3RhCl_6 source is dissolved into Na^b and $[Rh(H_2O)_{3-x}Cl_{6-x}]^{(3-x)-}$, and then the aqua complexes interact with the PEO moieties via hydrogen bonding between the PEO and aqua complexes (i.e. the formation of PEO-b-PMMA/ $[Rh(H_2O)_{3-x}Cl_{6-x}]^{(3-x)-}$ composite micelles); (3) the Rh species are reduced to form solid Rh nuclei, (4) which coalesce and further grow into mesoporous Rh nanostructures over; and (5) the templates are removed by a solvent extraction method. (Adapted from Jiang et al. 2017 with permission)

The smart use of natural resources such as water, nutrients and chemicals in agricultural sectors by using nanosensors is cost-effective and environment-friendly (Jie et al. 2013; Prasad et al. 2017). Nanomaterials and global positioning systems with satellite imaging, of fields, can help farmers in detecting crop pests and to precisely know the status of stresses such as drought, salinity, flood and soil nutrient deficiencies. These smart materials can detect the presence of viruses (Jhanzab et al. 2015; Manjunatha et al. 2016; Wang and Xie 2018) and other microbial pathogens; increase the use efficiency of water, fertilizer and other agricultural inputs; and thus contribute to reduce the environmental pollution (Prasad et al. 2014; Jhanzab et al. 2015). Nanomaterials, e.g. zeolites and nano-clays, can improve the retention of water or liquid agrochemicals in the soil for their slow release to the plants (Duhan et al. 2017). Nanotechnology-led innovations are also being used in plant improvement and genomic transformation programmes (Torney et al. 2007).

Fertilizers are very important in crop production and contribute 40–50% total yield of field crops (Das and Mandal 2015). However, fertilizer use efficiency seldom increases by 30% in most of the cases. Overuse of fertilizers may cause water and air pollution (Smith 2003; Smith et al. 2006; Bhateria and Jain 2016). Use of

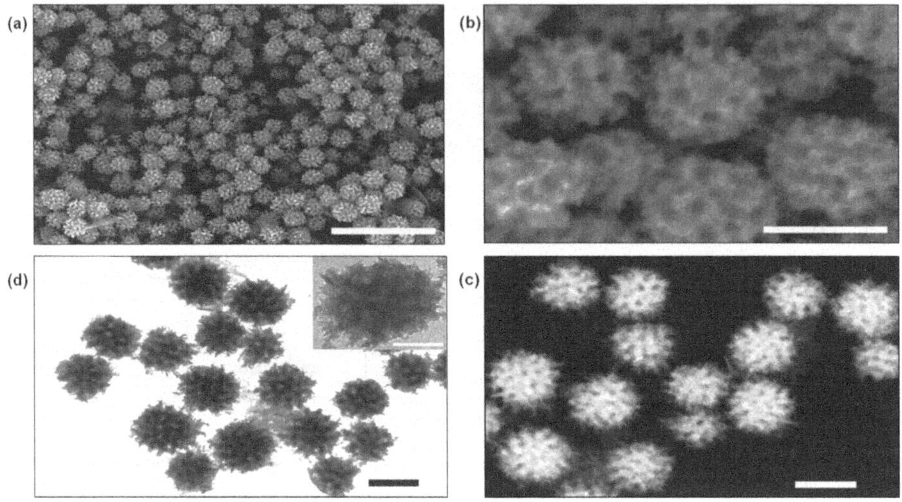

Fig. 7.2 Structural characterization of mesoporous Rh nanoparticles. (**a**) Low-magnification SEM micrograph (scale bar, 500 nm), (**b**) high-magnification SEM micrograph (scale bar, 100 nm), (**c**) TEM micrograph (scale bar, 100 nm), and (**d**) HAADF-STEM micrograph (scale bar, 100 nm) of mesoporous Rh nanoparticles. The inset in c is a high-magnification TEM micrograph of a single mesoporous Rh nanoparticle (scale bar, 50 nm). (Adapted from Jiang et al. 2017 with permission)

nanotechnology for slow release nanofertilizers seems an attractive alternative in changing climate and can also help decrease nitrogen losses through leaching, emissions and long-term soil absorption.

Application of nanoparticles has great scope in developing environmentally sound and cost-effective methods for controlling the plant pathogens (Zargar et al. 2011; Ahmed et al. 2016), insect pests (Park et al. 2006; Scrinis and Lyons 2007; Suman et al. 2010) and weeds (De et al. 2014; Chhipa and Joshi 2016; Pandey et al. 2016). For example, use of nano-silver compounds provided effective control of different fungal pathogens than the commercial fungicides (Ouda 2014). Nanoencapsulation, of insecticides, with nanoparticles may help efficient plant absorption for effective control against insect pests (Park et al. 2006; Scrinis and Lyons 2007; Suman et al. 2010) and weeds (Ali et al. 2014; De et al. 2014; Kumar et al. 2017). Nanocapsules, nanoparticles, nanoemulsions and viral capsids act as smooth carriage systems for controlling disease, insect pest and weed in association with plants (Dimetry and Hussein 2016).

Therefore, the application of nanotechnology is one of the emerging technologies for sustainability of crop production systems to meet the growing food demand of increasing population in the world. In this chapter, recent progress and potential applications of nanotechnology for sustainable crop production systems are discussed. Challenges and constraints in the use of nanotechnology have also been described.

7.2 Nanomaterials for Use in Crop Production Systems

Agonizing deeds of mankind to selfishly deracinate their wants has left Mother Nature under excruciation. Forgetting the consequences that will raise serious issues of global warming and unleash the tears of its weeping in the form of climate change leads to startled natural calamities for which the mankind had never considered for preparation.

A crop production system in general is a precision agricultural approach, where food and fibre produced are concurrently profitable; uses on-farm affordable resources without hampering biodiversity; conserves quality of products, dynamic nature of soil and systemic nutrient density of the available water; and supports energetic rural community (Walters et al. 2016; Duhan et al. 2017). This eco-friendly crop production system could be further supported by the advanced application of nanotechnology-based approach (Duhan et al. 2017). Currently nanotechnological approaches have stigmatized the concept that something is impossible and beyond the reach of mankind. Therefore, this technology has now touched every aspect of life starting from medical to agricultural industry. Among the three most abundantly explored fields, agriculture stands out based on its major requirements to sustain its necessity. Present advancement made in the field of agriculture may majorly improve two basic aspects of agriculture that is soil and productivity apart from universally required water. Nanomaterials for the improvement and sustenance of soil and improvement of crop are generally either of organic (chitosan, polyacrylic acid, clay or zeolite), inorganic (Fe, Zn, SiO_2, TiO_2) or both hybrid origins (polymer-encapsulated carbon nanotubes, nanodiamonds, graphene) (DeRosa et al. 2010; Duhan et al. 2017; Morales-Díaz et al. 2017). These nanomaterials, which are used for the preparation of various forms of nano-based agricultural tools to maintain soil dynamic nature of the soil as well as to improve crop productivity by nanofertilizer and further to protect the crops from biotic stress by nanopesticides, have seriously made an impeccable impact in the sustainable improvement of agricultural research (Chhipa 2017).

7.2.1 *Nanofertilizers*

The chemical or natural substances, applied to improve soil fertility, are known as fertilizers/biofertilizers (natural) (Dong et al. 2012; Bhardwaj et al. 2014). Nanofertilizer and nanobiofertilizers contain synthetic and natural substances, respectively, and enhance the soil fertility as well as bioavailability more efficiently than the conventional fertilizers (Chhipa 2017). However, to be declared as nanofertilizer, the individual size of the molecules, in dispersed and/or aggregated state, should be $\leq$100 nm, and the bulk size should be closer to 100 nm, and the nanoproduct should be stable, i.e. the nanofertilizer should retain its nanoscale dimension and intactness during pre- and post-interaction with the soil or the crop.

However, the rate of dissolution and bioactivity are inevitably influenced by the shape. The design of the nanofertilizer where the products are functionalized with certain target agents and/or for control release agents to develop a hybrid product (Dimkpa and Bindraban 2017).

Among the abovementioned features of nanofertilizer, the most spectacular application is the optimization of nutrient use efficiency (NUE), control release of the active substances and no effect on soil biodiversity (Sempeho et al. 2014). These nanofertilizers, based on their biodistribution and nutrient composition, by the plant may be subcategorized into macronutrient nanofertilizers, micronutrient nanofertilizers and nanoparticulate fertilizers (Chhipa 2017).

7.2.1.1 Macronutrient Nanofertilizers

The demand for macronutrients such as calcium (Ca), phosphorus (P), potassium (K), magnesium (Mg) nitrogen (N) and sulphur (S) has increased with the increase in pressure for more food production (Chhipa 2017). This increased demand for food is expected to further increase (FAO 2017). However, application of conventional fertilizer at higher rates, to achieve the target food production level, would be catastrophic. Acclimatization of the soil, resulting from agrochemical residue accumulation, may become an unalterable future. This may further pollute the running water along with it. But alternatively, the situation could be encountered by high-volume-to-surface ratio of nanoparticles, whose increased efficiency to withhold increases number of macronutrients and itself being so less in amount (Liu and Lal 2015; Ditta et al. 2015). For example, a nanofertilizer of Ca- and P-hydroxyapatite nanoparticles was prepared by Liu and Lal (2015) that showed higher increment in soybean seed yield in comparison to conventional fertilizer. Kottegoda et al. (2017) later developed another biocompatible urea functionalized hydroxyapatite nanoparticle which is also a rich source of phosphorus with controlled release of nitrogen. However, nanohybrids of this kind so far were only developed for biomedical applications.

7.2.1.2 Micronutrient Nanofertilizers

The nutrients that are required by the plants in less quantity but are equally important for the plant metabolism such as boron (B), chloride (Cl), copper (Cu), iron (Fe), manganese (Mn), molybdenum (Mo), nickel (Ni) and zinc (Zn) are termed as micronutrients. Zinc oxide (ZnO) nanoparticles, as nanofertilizer, improved the germination, sugar and protein contents and antioxidant activity in cabbage (*Brassica oleracea* var. *capitata* cv. Acre and *B. oleracea* var. *botrytis*) and tomato (*Lycopersicon esculentum* L. cv. Pusa ruby) (Singh et al. 2013). Askary and co-workers (Askary et al. 2016) investigated the impact of nano-iron fertilizer on the physiology of Madagascar periwinkle (*Catharanthus roseus*) with varying concentrations and found that plant growth, pigment composition and total protein contents were significantly improved (Askary et al. 2016). Similarly, Mahmoodi et al. (2017) evaluated the impact of nano-urea and nano-iron fertilizer along with conventional fertilizer on borage (*Borago officinalis* L.) and concluded that only nano-urea improved plant growth and essential oil yield, whereas nano-iron fertilizer increased number of seeds in plants. But all fertilizers, nano-urea and nano-iron fertilizer and conventional fertilizer, were equally effective in improving plant height.

7.2.1.3 Nanoparticulated Fertilizers

Nanoparticulated fertilizers are different from nanoparticulated systems used for the delivery of essential bioactive compounds to the plant directly or indirectly. In case of fertilizers, these nanoparticulated systems may be bases on carbon, silica, based or other organic polymers. Lahiani et al. (2013) reported that application of multi-walled carbon nanotubes (MWCNTs) improved the seed germination and plant growth and development in barley (*Hordeum vulgare*), soybean (*Glycine max*) and maize (*Zea mays*). Abdel-Aziz et al. (2016) evaluated the efficiency of chitosan nanoparticle (polymeric nanoparticle) encapsulating N, P and K for foliar uptake by wheat (*Triticum aestivum*) grown in sandy soil and found improvement in the harvest index, crop index and mobilization index by nanoparticulated fertilizer in comparison to control, non-fertilized and normal fertilized NPK. McGehee et al. (2017) also reported yield enhancement in hydroponically grown tomato with the application of MWCNT improves the production of fruit.

7.2.2 *Nanopesticides*

In general, a pesticide should be selectively toxic in its nature to the biotic stressors of the crop plants without any effect on plant itself and the symbiotic flora, fauna as well as humans (Hassall 1965). But the extensive use of these pesticides has led to the serious public concerns and environmental pollution. Bioaccumulation is one of the major concerns of the current generation, which emerged due to biomagnifying of persistent organic pollutants (POPs) soon to be considered as persistent toxic substances (PTS) (Kutz et al. 1991). Development of resistance in the target pests' poses another threat for the continuous use of conventional pesticides (Chareonviriyaphap et al. 2013; Soko et al. 2015). These limitations could be overthrown by the exploitation of nanotechnology in engineering various nano-based pesticides or nanopesticides (Chhipa 2017). Some of the commonly used nanopesticides include nanofungicides, nanoherbicides, nanomolluscicides, nanonematicides and nanoparticulated growth regulators (Guang et al. 2013; Cromwell et al. 2014; Oliveira et al. 2015; Pereira et al. 2017; Antonoglou et al. 2018).

The current interest in the use of nanoparticles as agricultural pesticides is based on the fact that these enhance the efficacy and environment and reduce the cost (Zhao et al. 2017; Fig. 7.3). They also noted that for increasing the formulation

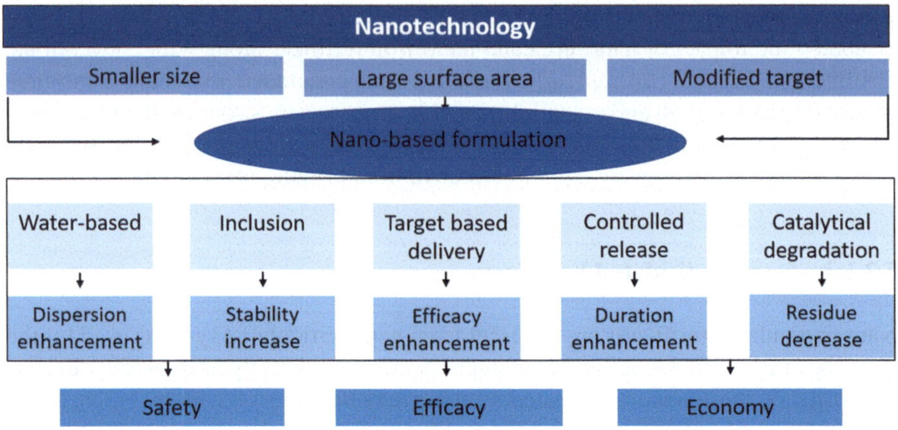

Fig. 7.3 The prospects of nanotechnology where nanopesticides are part. (Adapted from Zhao et al. 2017 with permission)

properties, including water dispersion, chemical stability, targeting adhesion, permeability and controlled release, nanosuspensions, nanoemulsions, nanospheres, nanomicelles and nanocapsules showed great potential, and these are also environmentally sound and eco-friendly (Zhao et al. 2017; Prasad et al. 2019; Fig. 7.4).

Achieving and maintaining environmental sustainability are among the sustainable developmental goals, where efforts to save the environment have been on the spotlight. The use of nanomaterials as agricultural pesticides is based on the underlying nanotechnology that helps to develop systems that protect the environment and plants from hazards by the nano-chemicals. This technology may help attain these objectives through nanocarriers and nanodevices to aid in controlled release and delivery of the content to the targeted site (Huang et al. 2007; Sarlak et al. 2014; Mehrazar et al. 2015). Nanotechnology helps develop nano-based pesticides with controlled release mechanisms (Khot et al. 2012; Mukhopadhyay 2014). The use of nanoparticles in the development of agricultural pesticides may help decrease the residual life in the environment (Karn et al. 2009; Mukhopadhyay 2014; Rani et al. 2017). The most common agricultural nanoparticles are the surfactants and the organic ligands, but various limitations have led to the exploration of other base elements, especially silver (Elek et al. 2010; Du et al. 2011). According to Singh et al. (2013), silica, carbon, silver and aluminium silicate are the potential base ingredients used for the development of nano-pesticides. Use of nanoparticles in pesticide, with less residual life, is illustrated in Fig. 7.5 (Nuruzzaman et al. 2016).

The cellulosic paper and paper-based products are ubiquitous in our daily life. These cellulosic materials may serve as vector of transmission of numerous infectious human diseases (Angelakis et al. 2014). Nano-coating of these useful materials may yield antimicrobial cellulosic paper for safe use and improved packaging materials to enhance shelf life of fruit, vegetables and other agricultural produce. Recently, Islam and co-workers immobilized silver nanoparticles (AgNPs) onto

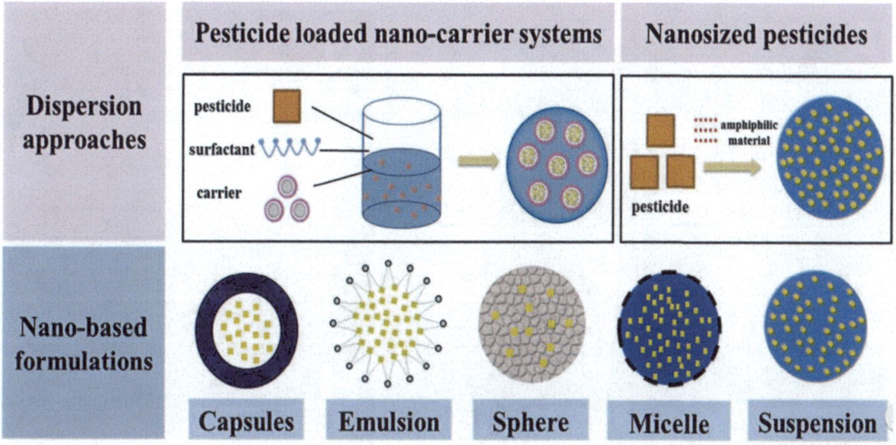

Fig. 7.4 Water-based dispersion pesticide nanoformulation increases the formulation properties. (Adapted from Zhao et al. 2017 with permission)

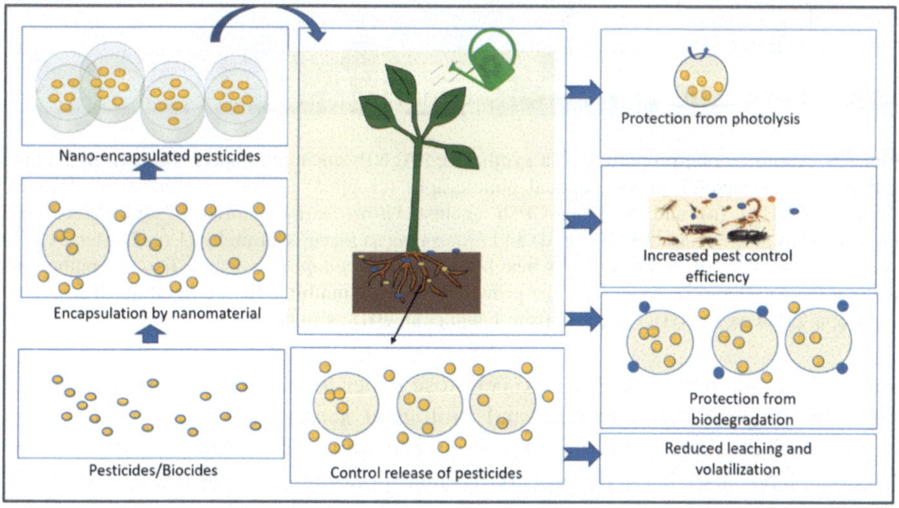

Fig. 7.5 Nanoparticles enable pesticides to reach the targeted areas without biodegrading prematurely. (Adapted from Nuruzzaman et al. 2016 with permission)

dopamine-modified cellulose paper (Fig. 7.6). The AgNP-loaded paper showed excellent antimicrobial activities against some antibiotic-resistant virulent fish pathogenic bacteria such as *Vibrio parahaemolyticus* strains 2A1 and 2A2, *Enterococcus faecalis* strains F1B1 and EF11, *Serratia marcescens* 4 V3 and a disastrous fungal phytopathogen *Magnaporthe oryzae Triticum*. Owing to the high adhesion of the dopamine moiety to nearly all inorganic substances, the novel approach of the preparation of fabricated Ag-DOPA-CP established by Islam et al. (2018; Fig. 7.6) might be extendable to robust immobilization of any other metals

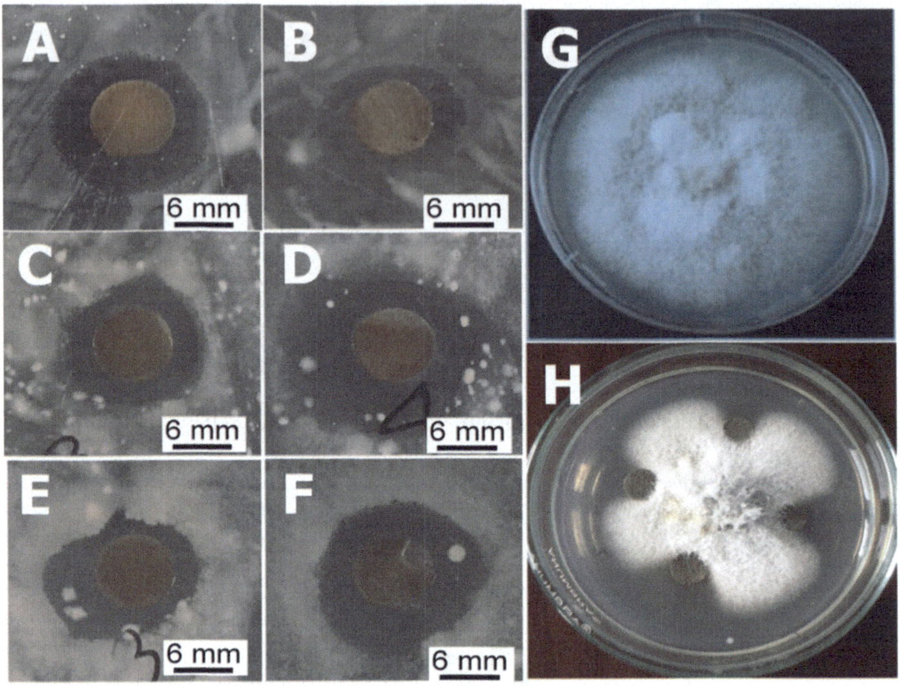

Fig. 7.6 Antimicrobial properties of a synthesized AgNPs anchored papers. (**a–f**) Antibacterial activity of Ag-Dopa-CP papers: optical photographs (cropped) of zone of inhibition (ZOI) of Ag-Dopa-CP15m (**a**) and Ag-Dopa-CP8h against *Vibrio parahaemolyticus* strain 2A1 (**b**), *Enterococcus faecalis* strain F1B1 (**c**, **d**) and *Enterococcus faecalis* strain FF11 (**e**, **f**) after 24 h of incubation at 28 °C. (**g**, **h**) Growth of wheat blast fungus *Magnaporthe oryzae Triticum* pathotype without the Ag-Dopa-CP papers (**g**); the growth is strongly inhibited by the Ag-DopaCP15m and Ag-Dopa-CP8h papers (H). (Adapted from Islam et al. 2018 with permission)

and metal oxide nanoparticles onto cellulose paper to explore new applications in numerous paper-based chemical and biological sensors (Ge et al. 2017). The daylight-driven rechargeable antibacterial and antiviral nanofibrous membranes (RNMs) have also been developed by incorporating daylight-active chemicals which can efficiently produce reactive oxygen species (ROS) for controlling bacterial and viral phytopathogens both under dim light and dark conditions (Si et al. 2018). These materials may be used for limited transmission of plant diseases through agricultural equipment for controlling bacterial and viral phytopathogens.

7.2.2.1 Nanofungicides

The fungicides such as cupric salt, dicarboximides, dithiocarbamates, dinitrophenoles, tiabendazoles, triazoles, thiocarbamates and organotin compounds are effective in controlling fungi and protecting fungal diseases. The target specificity of most of the fungicides has yet not been properly determined. However, the mode of action and effect of fungicides on components of the targeted cell membrane, protein translation, DNA replication, signal cascade, metabolic pathways and the phases of cell cycle are known (Yang et al. 2011). The target specificity could be greatly accelerated by amalgamating with nanotechnology. This amalgamation laid the formulation of nanofungicides. Kumar et al. (2016) formulated a nanoform of a commercially available fungicide against *Macrophomina phaseolina* containing 25% trifloxystrobin and 50% tubuconazole. The formulated nanoform had broad spectrum action with an enhanced antifungal activity including malformed hyphae, lysis of the hyphae and malformation of sclerotia in *Macrophomina phaseolina*. Antonoglou et al. (2018) prepared inexpensive Cu-Zn bimetallic nanoparticles against *Saccharomyces cerevisiae* and evaluated their phytotoxicity in phosystem II activity of tomato through chlorophyll fluorescence imaging. They found that the bimetallic nanoformulation had negligible effect on photosystem II of tomato leaf but showed effective antifungal activity against *S. cerevisiae*.

7.2.2.2 Nanoherbicides

Herbicides are substances used in agriculture and landscapes to control unwanted plants. For instance, some of nonselective conventional herbicides include S-ethyl-N, N-dipropyl thiocarbamate, glyphosate, simazine, 4-(2,4-dichlorophenoxy) butyric acid, bromoxynil, diuron, terbacil and hexazinone, whereas 4-dichlorophenoxyacetic acid, paraquat and glyphosate may be categorized as selective herbicides (Rolando et al. 2013). Now the advanced form of herbicides, the nanoherbicides, is slowly taking over the conventional herbicides in laboratory research. For example, a polymeric nanoformulation herbicide consisting of poly(epsilon-caprolactone) and chitosan/tripolyphosphate encapsulating atrazine/paraquat was developed (Grillo et al. 2015). The cytotoxicity and genotoxicity were also determined which showed that the nanoformulation was less toxic than the free herbicides and were much useful in agricultural practices. Similarly, Oliveira et al. (2015) developed poly(epsilon-caprolactone) polymeric nanoparticles loaded with atrazine and determined the post-emergence herbicidal efficiency using mustard greens (*Brassica juncea*). The nanoencapsulated atrazine was more efficient as post-emergence herbicides when applied at lower rate.

7.2.2.3 Nanomolluscicides

To counteract the invasion and damage caused by different species of snails and slugs, various molluscicides including metaldehyde, methiocarb and ferrous phosphate are being used (Abubakar et al. 2017). These commercially available conventional molluscicides have certain limitations such as cost-deficit and nonspecific adverse effect which is the primary concern of the present research and innovation (Coelho and Caldeira 2016). With the advancement of nanotechnology, the hope of developing cost-effective and target-specific nanomolluscicides is no more a dream. Guang et al. (2013) evaluated the toxic effect of green synthesized silver nanoparticles on *Oncomelania hupensis* snails and crucian seedlings. The nanoparticles were toxic to the snails but had negligible toxicity towards crucian seedlings. Omobhude et al. (2017) developed a molluscicidal polymeric nanoparticle of polylactic acid encapsulating curcumine and nisin and tested it against *Biomphalaria pfeifferi*. The nanoformulation was a potent molluscicide in proportion to various stages of development in *B. pfeifferi*.

7.2.2.4 Nanonematicides

There is an urgent need for a novel alternative for the conventional nematicides such as aldicarb, fosthiazate, fenamiphos and oxamyl compounds used to control phytonematodes (Ntalli and Caboni 2012). The toxicity of these nematicides to both human and the environment might also increase due to the increase in expected use (Sánchez-Moreno et al. 2009). The coherent demand for an alternative form of nematicide seems to be positively achieved by an advanced exploration of nanotechnology. The nanonematicide is at the point-blank range of the possible advancement needed to satisfy the urgent demand for novel alternative nematicides. For example, Cromwell et al. (2014) determined the potency of silver nanoparticles against *Maloidogyne graminis* and *M. incognita* in vitro. They demonstrated that the silver nanoparticles are efficient nematicidal. Abdellatif et al. (2016) devised a cutting-edge approach by using green silver nanoparticles as nanonematicide against *M. javanica* infested in tomato and compared the nematicidal efficiency with chemical control of *S. melongena* (Eggplant). They found that the green silver nanoparticles were more effective nematicides in comparison to chemical control of eggplant.

7.2.3 Nanomaterials for Soil Remediation

The transformation of rural areas to urban states is backed up by either establishment of small-scale industries that gradually grows into a setup of mega factories or a multinational corporation, although this transformation hugely benefits to uplift the socioeconomic status of the inhabitants but fails to sustain and conserve the biodiversity. Improper industrial debris disposal severely damages the food chain of the associate floral and faunal community due to contaminated water and soil. Soil and water on the other hand are the basic factor required for sustainable crop production. The contaminants from the water can be precipitate or filtered, but soil on the other side is much more difficult to decontaminate.

Some of the common contaminants sourced from industrial waste include heavy metals such as arsenic (As V, III), lead (Pb), copper (Cu), zinc (Zn), nickel (Ni), cadmium (Cd), radionuclides (uranium and thorium) and organic pollutants such as chlorinated organic compounds (trichloroethylene), petroleum monoaromatics, nitroaromatics and synthetic dyes (Araújoa et al. 2015). The conventional chemical and physical methods available may contribute to secondary contamination due to the use of high quantity of these remediating agents which further poses a limitation for their use (Watson 1996; Yao et al. 2012). On the other hand, the eco-friendly microbial-based soil remediation requires relatively high cost, but outcome may be uncertain (Sharma and Reddy 2004). These drawbacks of conventional chemical-, physical- and microbial-based methods could be eliminated by the utilization of nano-based approaches such as nanofertilizers, nanobiosensors and different nanoremediation processes (Rai et al. 2012; Rizwan et al. 2014; Raliya et al. 2017).

Nanofertilizer, as mentioned above, has immensely influenced the prospect of sustainable agriculture to a greater extent. The serviceability of this novel form of nano-based fertilizer is seen in the form of macronutrient nanofertilizers, micronutrient nanofertilizers and nanoparticulate fertilizers (Chhipa 2017). Under certain nutrient-deficit condition, nanofertilizer of Ca and P hydroxyapatite nanoparticles could be a useful mode of nutrient supply and influence soybean seed yield, and other place nano-urea and nano-iron fertilizer improved growth traits and increased production of essential oil yield of borage (*Borago officinalis* L.) (Liu and Lal 2015; Mahmoodi et al. 2017). A digital/analytical portable device possessing at least a single sensing unit in the nano-dimension which is engineered for monitoring physicochemical variations at locations complicated to reach is regarded as nanosensors (Fraceto et al. 2016). If these digital/analytical portable devices are conjugated to a specific structural feature of single biomolecules such as nucleic acid, protein or carbohydrate, then the devices may also be regarded as nanobiosensors (Yuce and Kurt 2017). Nanobiosensors on the other hand are an exceptional tool for determining the contaminants at a minimum concentration. A nanobiosensor based on grapheme nanoribbon could be used for detection of Mn impurities (Enciu et al. 2014). Salinas et al. (2014) developed a nanosensor for the detection of Ar (argon) by the help of self-assembled monolayers on gold substrate type (III) in an aqueous medium. Nanoremediation is mostly focused on mitigating the impact of harmful contaminants on the environment (Prasad et al. 2017). Specific organic and inorganic compounds such as natural short-ordered aluminosilicate, surface of titanium oxide and humic acids can be coupled with Ni through multiwalled carbon nanotube and serve as an efficient nanobioremediation tool in sustainable agricultural system (Raliya et al. 2013; Prasad et al. 2017).

Industrial wastewater is one of the major sources of contamination of the crop field and waterbodies (Carpenter et al. 1998; Xing et al. 2001; Krishna et al. 2009). Nanoporous carbon has a myriad of application ranging from wastewater purification to gas separation to energy conversion and storage (Diallo and Brinker 2011). Recently, Khan et al. (2018) developed an economically viable and environment-friendly novel strategy for the preparation of nanoporous carbon (NC) from jute. The NC prepared at 800 °C resulted in a high surface area (981 m^2 g^{-1}) associated with the retention of the original fibrous shapes. Industrial application of this novel technology in conversion of jute fibre to the NC would lead to a re-emergence of jute as a black gold in the regional economy.

7.2.4 Nanomaterials for Crop Improvement

Improvement of crops and its production is inversely proportional to the biotic and abiotic stresses. The synthetic pesticides and growth regulators used as a conventional means to mitigate biotic and abiotic stresses further pose additional jeopardy to the sustenance of agriculture. Nanotechnology as previously mentioned can be a verified asset in these conditions. Nanopesticides in the form of nanofungicides, nanoherbicides, nanomolluscicides and nanonematicides are previously been described with example (Guang et al. 2013; Cromwell et al. 2014; Oliveira et al. 2015; Antonoglou et al. 2018). These nano-based products are equipped with a common component that is a nano-based carrier or nanocarrier. Basically, the nanocarriers are categorized as soft and hard nanoparticles (Kah et al. 2018). Polymers, solid lipids and liposomes are considered as soft nanoparticles, and carbon-, silica- and titanium-based nanoparticles and other metallic nanoparticles are considered as hard nanoparticles (Mukhopadhyay 2014; Siafaka et al. 2016; Kah et al. 2018).

In a broad-spectrum nanomaterial-based crop, improvement can also be visualized in various applications such as improved germination of seed and growth of plants, protection of plants from pathogens and precise detection of phytopathogens (Khot et al. 2012; Ismail et al. 2017). Varying doses of silver nanoparticles significantly enhance the rate of seed germination of maize, *Citrullus lanatus* [Thunb.] Matsum. & Nakai (watermelon) and *Cucurbita pepo* L. (pumpkin) by showing slight toxic effect in root elongation of maize (Almutairi and Alharbi 2015). The germination of *Lactuca sativa* (lettuce) seed is also enhanced by the single-capillary electrospraying of TiO_2 nanoparticles (Wu et al. 2014). The nanopriming of *Oryza sativa* L. cv. KDML105 (rice) by a biocompatible silver nanoparticle produced using *Citrus hystrix* D.C. (Kaffir lime) leaf extract significantly improves seed germination and α-amylase activity associated with seedling growth and stimulates upregulation of aquaporin genes (Mahakham et al. 2017). Protection of plants from pathogens have been described above where nanoform of fungicide containing trifloxystrobin and tubuconazole against *Macrophomina phaseolina* and Cu-Zn bimetallic nanoparticles against *Saccharomyces cerevisiae* (Kumar et al. 2016; Antonoglou et al. 2018). Silver nanoparticulated molluscicide against *Oncomelania hupensis* snails and polylactic acid encapsulating curcumin-nisin against *Biomphalaria pfeifferi* (Guang et al. 2013; Omobhude et al. 2017). Crop production could further be enhanced by an efficient detection of the disease. Nanotechnology-based approaches could be the useful tools for this important task. Nano-based biosensors, nanoparticulated systems, nanoimaging and nanopore DNA sequencing tools have tremendously enhanced accuracy, persistence and time duration of pathogen detection and upgradation of high-throughput instrumentation for quality detection of plant pathogens (Khiyami et al. 2014). Recently, quantum dots having versatile surface chemistry, bright fluorescence and negligible photobleaching effect could easily cross plant cell wall pores and might be useful in detection of biotic and abiotic stresses (Wu et al. 2017).

7.3 Effect of Nanomaterials on Soil, Water and Environmental Health and Crop Productivity

7.3.1 Soil, Water and Environmental Health

To meet the growing demand of food for increasing population under the limiting natural resource, agencies and research institutes involved in food security (FAO 2009) are encouraging "sustainable crop intensification". Among the emerging technologies for addressing new challenges in crop production under the changing climate, nanotechnology is emerging as a potential frontier technology to promote sustainable productivity in agriculture. It can be used for remediation of deteriorated environment and also precisely detection of environmental pollutants (Neethirajan and Jayas 2011). Application of nanomaterials such as nanofertilizers, nanopesticides, nanocarriers, nanosensors, nano-packaging and nano-chips is revolutionized in the current agricultural production system. Nanotechnology could promote precision and smooth agriculture and significantly reduce production and postharvest losses in crop production system (Gruere et al. 2013). It is environmentally sound as it can reduce the bulk use of pesticide by adding nano-silver particles to pesticides. The use of nanometal oxides to target soil pathogens and adding metallic nanomaterials such as nanosilicon to increase water uptake efficiency in plants are considered highly promising nanotechnology in agriculture (Servin et al. 2015). The use of synthetic clay nanomineral and zinc oxide nanoparticles for arsenic removal seems useful in water purification device (Clare e al. 2011). The water to be filtered is percolated through a column of hydrotalcite (synthetic clay mineral) (Prasad et al. 2014). The development of a DNA-based nanobiosensor in a polymer to coat fertilizers would release only as much fertilizer as "demanded" by plant root ionic signals (Gautam et al. 2014; Prasad et al. 2014). Similarly, nanoscale zerovalent iron could be remediated pollutants from soil or groundwater and is the most widely used nanomaterials for water purification in the world (Li et al. 2006). Other potential nanomaterials for remediation include nanoscale zeolites, metal oxides, carbon nanotubes and fibres, enzymes, various noble metals (mainly as bimetallic nanoparticles) and titanium dioxide. Nanoparticle filters such as dichorodiphenyltrichloroethane (DDT), endosulfan, malathion and chlor-pyrifos) can be used to remove organic particles and pesticides from water (Pradeep 2012; Bootharaju and Pradeep 2012; Ul-Islam et al. 2012; Thatai et al. 2014).

However, exceptionally, nanoparticles might cause some impulsive contrary effects on agricultural productivity, soil health and ecosystem (Brumfiel 2003; Service 2003; Zhang 2003; Kelly 2004). The applied nanoparticles will gather over time in soils and water environment (Boxall et al. 2007; Gottschalk et al. 2009). Some nanoparticles such as fertilizers and pesticides may flow into food chain through the environmental systems that threaten human health and ecosystem (Zhao et al. 2017). Hence, for avoiding the risk, investigation might be established on safety and risk assessments of nanopesticides. Investigation related to the toxic effect of nanoparticles and the interaction between nanoparticles and plants will provide a theoretical basis of knowledge for the development of nanopesticides for sustainable implementation of nanotechnology in agriculture (Zhao et al. 2017), while nanopesticides can accelerate the catalytic degradation of toxic residues as well as reduce the pesticides residual effect on the soil and water environment through introducing biodegradable carriers' material and adding photocatalysts (Pierluigi et al. 2003; Zhao et al. 2017; Fig. 7.7).

7.3.2 Crop Productivity

Efficient crop productivity is a direct result of proper management of plant growth under diverse physiological and biological stresses. Nanomaterials may have strong impact on the management and sustenance of plant growth under biotic and abiotic stresses. The effect of applied nanomaterials varies based on stress condition, quantity and the type of nanometerials used (Siddiqui et al. 2015). For example, soft/organic nanomaterials such as polymer and solid lipid nanoparticles and metallic nanoparticles and carbon-, silica- and titanium-based nanoparticles are considered hard/inorganic nanoparticles being evaluated for their use in sustainable agriculture (Kah et al. 2018).

The insecticidal activity of polycaprolactone and polylactic acid nanosphere loaded with ethiprole had better penetration in comparison to reference formulation (Boehm et al. 2003). With growing knowledge of nanoscience, stimulation and

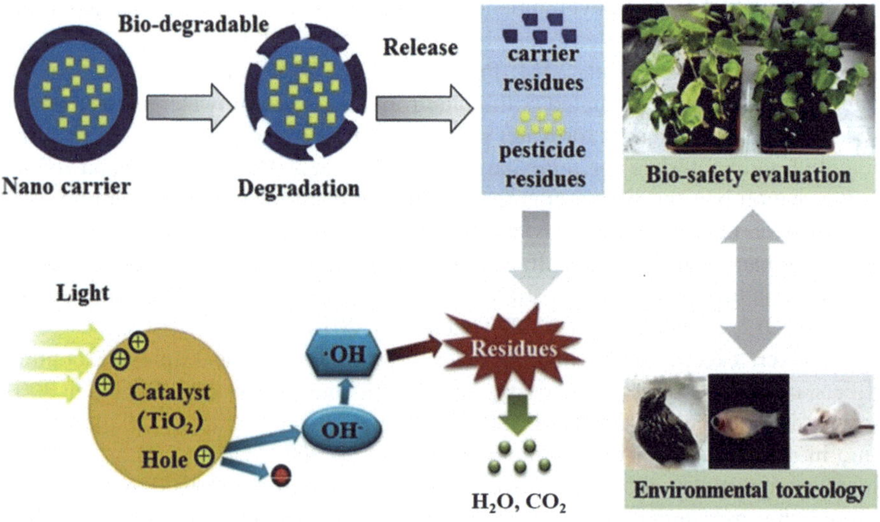

Fig. 7.7 Nanopesticides accelerate the catalytic degradation of toxic residues as a biosafety. (Adapted from Zhao et al. 2017 with permission)

control of nanoparticulated system based on the targets are quite possible (Gogos et al. 2012). While reviewing the controlled release efficiency of imidacloprid (1-(6 chloro-3-pyridinyl methyl)-N-nitro imidazolidin-2-ylideneamine) encapsulated in polyethylene glycol, Sekhon (2014) inferred that these were quite effective against *Melanagromyza sojae* Zehntmer and *Bemisia tabaci* Gennadius. Lipid-based nanometerials can also be effective in sustainable crop production systems (Naseri et al. 2015). Two herbicides, namely, atrazine and simazine, encapsulated by solid lipid nanomaterial were effective against wild radish (*Raphanus raphanistrum*) in maize with no adverse effect on the crop (de Oliveira et al. 2015).

Hard/inorganic nanoparticles are equally important in crop production systems. The silicon dioxide (SiO_2) nanoparticles can help improve seed germination, plant growth and yield formation in tomato, Olgan larch (*Larix olgensis*) and maize. These nanoparticles also stimulated the seed germination of tomato and squash and antioxidant system (Siddiqui et al. 2015). Carbon-based nanomaterials are more extensively used for disease treatment and diagnostics in animals. But recently they are also being used in crop production. Many molecules such as nano-onions, nanohorns, nano-cones, nanodots, nanotubes, nano-beads, nanofibres, nanodiamonds, fullerenes and graphene are included under the family of carbon nanomaterials (Mukherjee et al. 2016). The interaction of nanomaterials with bioactive compounds results in altered gene expression, DNA damage and increased oxidative damage (Zaytseva and Neumann 2016). However, certain carbon-based materials such as *s*-multiwalled carbon nanotube (*s*MWCNT) may improve seed germination by induction of perforation of the seed coat in *Salvia macrosiphon*, peppers (*Capsicum annuum*) and tall fescue (*Festuca arundinacea*) (Pourkhaloee et al. 2011; Zaytseva and Neumann 2016). Thus, application of nanomaterials has great scope for improvement in crop productivity; however, some nanomaterials may have toxic effects on nontargeted organisms. Therefore, a cautious selection of nanomaterials with specific action is needed for a sustainable crop production.

7.4 Constraints in the Use of Nanotechnology in Crop Production Systems

Every new technology has some limitations of its use. One of the constraints of using nanomaterials in agriculture is nanotoxicity. Although many nanomaterials have shown high promise as nanopesticides, a large proportion of them showed toxicity to the nontargeted organisms. The fate and dispersion/transport of synthesized nanoparticles in the environment critically depend on plants which are an indispensable constituent of our ecosystem (Xingmao et al. 2010). It has also been speculated that plant uptake and bioaccumulation of nanoparticles may also alter gene expression, DNA damage and increased oxidative damage (Xingmao et al. 2010; Zaytseva and Neumann 2016). Therefore, specificity of the activity of nanomaterials is a prerequisite of its larger use in agriculture and environment.

The probable mechanism of emergence of nanotoxicity in plants may initiate from low degradation rate and prolonged adherence of the nanoparticle to the shoot through nanopesticides or root through nanofertilizer. In root, the nanoparticles first must reach vascular tissue (xylem) making their way through thick cell wall and plasma membrane of epidermal layer in order to be properly dispersed throughout plant system (Xingmao et al. 2010). One of the important foci of research is to understand the mode of action of the nanomaterials and explore their effects on nontargeted organisms. Another important focus of research is the application method of nanoparticles as pesticide which is still challenging and could be supplemented with a sophisticated method for developing broad spectrum nanocarriers. Active ingredients with a wide range of efficiency determine intrinsic factors like rate degradation of nanocarriers in the environment. This could determine the rate of releasing of active ingredients and availability of active ingredients for their pesticidal effect (Kah et al. 2018).

Nanofertilizer as macronutrient nanofertilizers, micronutrient nanofertilizers and nanoparticulate fertilizers are currently experimented for their use, and some of them are already in the market (Chhipa 2017). Recently, it has been observed that nanofertilizer shows low mobility and low rate of degradability (Kah et al. 2018). This stagnancy may further lead to deposition of the nano-based micronutrients, the excess of which may lead to nanoaccumulation and nanotoxicity (Chhipa 2017; Kah et al. 2018). These constraints could be mitigated when more advanced form of nanoformulations, like functionalized or tuned nanocarriers, will be used. These carriers could be triggered by control release of the nanoparticulated systems which is achieved either by intracellular (pH, ATP, GSH, enzyme, glucose and H_2O_2) or exogenous (temperature, light, magnetic field, ultrasound and electricity) stimuli (Song et al. 2017). These forms of nanocarriers are already more extensively used and evaluated in the biomedical sciences for various serious and life-threatening diseases. By regulating and modulating certain synthesis parameters, these nanocarriers might be used for sustainable agricultural developments. But the agricultural research using these concepts is scanty; therefore, further detailed and extensive research is essential requirement to develop more efficient nanoformulation for sustainable agricultural development.

7.5 Conclusion and Future Research Thrusts

Nanotechnology has contributed a lot in electronics, energy generation, drug delivery, and disease diagnosis. This technology has great scope and application for sustainable climate resilient and smart crop production systems as well. The nanomaterials can be used to develop nanofertilizers, nanopesticides, nanocarriers, nanosensors, nano-packaging and nano-chips to improve the crop productivity and resource use efficiency through significant reduction in production cost and postharvest losses. The use of nanomaterials may help reduce the amount of sprayed agrochemicals by smart delivery of active ingredients, minimize nutrient losses in fertilization and increase yields through optimized water and nutrient management. Using nanosensors in the application of water, nutrients and chemicals may help further reduce production cost and environmental concerns. Nanomaterials may also be used for the improvement of soil water retention. Nanotechnology-led innovations are also being used in plant improvement and genomic transformation programs. Nonetheless, like other synthetic agrochemicals, unwise and indiscriminate use of nanomaterials may have deleterious impacts on the environment. Therefore, research-based, wise application of nanotechnology is needed for sustainability of environment and crop production systems.

Acknowledgements We are thankful to the Government of Bangladesh for partially funding this work through a HEQEP CPSF#2071 to the Department of Biotechnology of Bangabandhu Sheikh Mujibur Rahman Agricultural University, Bangladesh.

References

Abdel-Aziz HMM, Hasaneen MNA, Omer AM (2016) Nano chitosan-NPK fertilizer enhances the growth and productivity of wheat plants grown in sandy soil. Span J Agric Res 14(1):e0902. https://doi.org/10.5424/sjar/2016141-8205

Abdellatif KF, Abdelfattah RH, El-Ansary MSM (2016) Green nanoparticles engineering on root-knot nematode infecting eggplants and their effect on plant DNA modification. Iran J Biotechnol 14(4):250–259

Abubakar A, Bala AY, Singh K (2017) Plant molluscicides and their modes of action: a review. Int J Sci Res Technol 2(1):37–49

Ahmed S, Ahmad M, Swami BL, Ikram S (2016) A review on plants extract mediated synthesis of silver nanoparticles for antimicrobial applications: a green expertise. J Adv Res 7(1):17–28

Ali MA, Rehman I, Iqbal A, Din S, Rao AQ, Latif A, Samiullah TR, Azam S, Husnain T (2014) Nanotechnology, a new frontier in agriculture. Adv Life Sci 1(3):129–138

Almutairi ZM, Alharbi A (2015) Effect of silver nanoparticles on seed germination of crop plants. Int J Nucl Quantum Engg 9(6):594–598

Angelakis E, Azhar EI, Bibi F, Yasir M, Al-Ghamdi AK, Ashshi AM, Elshemi AG, Raoult D (2014) Paper money and coins as 772 potential vectors of transmissible disease. Future Microbiol 9(773):249–261

Antonoglou O, Moustaka J, Adamakis ID, Sperdouli I, Pantazaki AA, Moustakas M, Dendrinou-Samara C (2018) Nanobrass CuZn nanoparticles as foliar spray nonphytotoxic fungicides. ACS Appl Mater Interfaces 10(5):4450–4461

Araújoa R, Castrob ACM, Fiúza A (2015) The use of nanoparticles in soil and water remediation processes. Mater Today: Proceed 2:315–320

Askary M, Amirjani MR, Saberi T (2016) Comparison of the effects of nano-iron fertilizer with iron-chelate on growth parameters and some biochemical properties of *Catharanthus roseus*. J Plant Nutr 40(7):974–982

Bhardwaj D, Ansari MW, Sahoo RK, Tuteja N (2014) Biofertilizers function as key player in sustainable agriculture by improving soil fertility, plant tolerance and crop productivity. Microbiol Cell Fact 13:66. https://doi.org/10.1186/1475-2859-13-66

Bhateria R, Jain D (2016) Water quality assessment of lake water: a review. Sust Water Resour Manage 2(2):161–173

Boehm AL, Martinon I, Zerrouk R, Rump E, Fessi H (2003) Nanoprecipitation technique for the encapsulation of agrochemical active ingredients. J Microencapsul 20(4):433–441

Bootharaju MS, Pradeep T (2012) Understanding the degradation pathway of the pesticide, chlorpyrifos by noble metal nanoparticles. Langmuir 28(5):2671–2679

Boxall AB, Tiede K, Chaudhry Q (2007) Engineered nanomaterials in soils and water: how do they behave and could they pose a risk to human health? Nanomedicine 2:919–927

Brumfiel G (2003) Nanotechnology: a little knowledge. Nature 424:246–248

Carpenter SR, Caraco NF, Correll DL, Howarth RW, Sharpley AN, Smith VH (1998) Nonpoint pollution of surface waters with phosphorus and nitrogen. Ecol Appl 8(3):559–568

Chareonviriyaphap T, Bangs MJ, Suwonkerd W, Kongmee M, Corbel V, Ngoen-Klan R (2013) Review of insecticide resistance and behavioural avoidance of vectors of human diseases in Thailand. Parasitol Vectors 6:280. https://doi.org/10.1186/1756-3305-6-280

Chhipa H (2017) Nanofertilizers and nanopesticides for agriculture. Environ Chem Lett 15(1):15–22

Chhipa H, Joshi P (2016) Nanofertilisers, nanopesticides and nanosensors in agriculture. In: Ranjan S, Dasgupta N, Lichtfouse E (eds) Nanoscience in food and agriculture, vol 1. Sustainable Agriculture Reviews. Springer, Cham, pp 247–282

Coelho P, Caldeira R (2016) Critical analysis of molluscicide application in schistosomiasis control programs in Brazil. Infect Dis Poverty 5:57. https://doi.org/10.1186/s40249-016-0153-6

Clare N, Linda A, Guillaume G (2011) Agricultural, food, and water nanotechnologies for the poor: opportunities, constraints, and role of the consultative group on international agricultural research. The International Food Policy Research Institute, Washington, DC

Cromwell WA, Yang J, Starr JL, Jo YK (2014) Nematicidal effects of silver nanoparticles on root-knot nematode in bermudagrass. J Nematol 46(3):261–266

Das DK, Mandal M (2015) Advanced technology of fertilizer uses for crop production. In: Sinha S, Pant KK, Bajpai S (eds) Fertilizer technology, vol 1. Synthesis. Studium Press LLC, Houston, pp 19–67

De A, Bose R, Kumar A, Mozumdar S (2014) Targeted delivery of pesticides using biodegradable polymeric nanoparticles. Springer, New Delhi

DeRosa MC, Monreal C, Schnitzer M, Walsh R, Sultan Y (2010) Nanotechnology in fertilizers. Nat Nanotechnol 5(2):91. https://doi.org/10.1038/nnano.2010.2

Diallo M, Brinker CJ (2011) Nanotechnology for sustainability: environment, water, food, minerals, and climate. In: Roco MC, Hersam MC, Mirkin CA (eds) Nanotechnology research directions for societal needs in 2020. Springer, Dordrecht, pp 221–259

Dimetry NZ, Hussein HM (2016) Role of nanotechnology in agriculture with special reference to pest control. Int J Pharm Technol Res 9:121–144

Dimkpa CO, Bindraban PS (2017) Nanofertilizers: new products for the industry? J Agric Food Chem 66(26):6462–6473

Ditta A, Arsha M, Ibrahim M (2015) Nanoparticles in sustainable agricultural crop production: applications and perspectives. In: Siddiqui M, Al-Whaibi M, Mohammad F (eds) Nanotechnology and plant sciences. Springer, Cham, pp 55–75

Dong W, Zhang X, Wang H, Dai X, Sun X, Qiu W, Yang F (2012) Effect of different fertilizer application on the soil fertility of paddy soils in red soil region of southern China. PLoS One 7(9):e44504. https://doi.org/10.1371/journal.pone.0044504

Du W, Sun Y, Ji R, Zhu J, Wu J, Guo H (2011) TiO_2 and ZnO nanoparticles negatively affect wheat growth and soil enzyme activities in agricultural soil. J Environ Monitor 13(4):822–828

Duhan JS, Kumar R, Kumar N, Kaur P, Nehra K, Duhan S (2017) Nanotechnology: the new perspective in precision agriculture. Biotechnol Rep 15:11–23

Elek N, Hoffman R, Raviv U, Resh R, Ishaaya I, Magdassi S (2010) Novaluron nanoparticles: formation and potential use in controlling agricultural insect pests. Coll Surfaces A: Physicochem Engg Aspects 372(1–3):66–72

Enciu D, Toader A, Ursu I (2014) Magnetic field nanosensor based on Mn impurities. Incas Bulletin 6(2):51–60

FAO (2017) The future of food and agriculture – trends and challenges. Food and Agriculture Organization, Rome

FAO (Food and Agriculture Organization) (2009). Sustainable crop production intensification. Available at http://www.fao.org/agriculture/crops/core-themes/theme/spi/scpi-home/framework/sustainable-intensification-in-fao/en/. Accessed 27 July 2018

Fraceto LF, Grillo R, de Medeiros GA, Scognamiglio V, Rea G, Bartolucci C (2016) Nanotechnology in agriculture: which innovation potential does it have? Front Environ Sci 4:20. https://doi.org/10.3389/fenvs.2016.00020

Gautam RK, Mudhoo A, Lofrano G, Chattopadhyaya MC (2014) Biomass-derived biosorbents for metal ions sequestration: adsorbent modification and activation methods and adsorbent regeneration. J Environ Chem Engg 2(1):239–259

Ge S, Zhang L, Zhang Y, Lan F, Yan M, Yu J (2017) Nanomaterials-modified cellulose paper as a platform for biosensing applications. Nanoscale 9(13):4366–4382

Gogos A, Knauer K, Bucheli TD (2012) Nanomaterials in plant protection and fertilization: current state, foreseen applications, and research priorities. J Agric Food Chem 60(39):9781–9792

Gottschalk F, Sonderer T, Scholz RW, Nowack B (2009) Modelled environmental concentrations of engineered nanomaterials (TiO_2, ZnO, ag, CNT, fullerenes) for different regions. Environ Sci Technol 43:9216–9222

Grillo R, Oliveira HC, Lima R, Fraceto LF (2015) Polymeric nanoparticles as carrier systems for herbicides. J Nanomed Nanotechnol 6(4). https://doi.org/10.4172/2157-7439.S1.022

Gruere G, Narrod C, Abbott L (2013) Agriculture, food, and water nanotechnologies for the poor opportunities and constraints. International Food Policy Research Institute (IFPRI), Washington, DC

Guang XY, Wang JJ, He ZG, Chen GX, Ding L, Dai JJ, Yang XH (2013) Molluscicidal effects of nano-silver biological molluscicide and niclosamide. Zhongguo Xue Xi Chong Bing Fang Zhi Za Zhi 25(5):503–505

Hassall KA (1965) Pesticides: their properties, uses and disadvantages: part I: general introduction; insecticides and related compounds. British Vet J 121(3):105–118

Halford B (2005) Nano dictionary. Chem Engg News 83(15):31

Hornyak GL, Dutta J, Tibbals HF, Rao AK (2008) Introduction to nanoscience and nanotechnology. CRC Press, Boca Raton, p 1640. https://doi.org/10.1201/9781420047806

Huang J, Li Q, Sun D, Lu Y, Su Y, Yang X, Wang H, Wang Y, Shao W, He N, Hong J (2007) Biosynthesis of silver and gold nanoparticles by novel sundried *Cinnamomum camphora* leaf. Nanotechnology 18(10):105104. https://doi.org/10.1088/0957-4484/18/10/105104

Iqbal M, Li C, Jiang B, Hossain MSA, Islam MT, Henzie J, Yamauchi Y (2017) Tethering mesoporous Pd nanoparticles to reduced graphene oxide sheets forms highly efficient electrooxidation catalysts. J Mater Chem A 5:21249–21256

Ismail M, Prasad R, Ibrahim AIM, Ahmed ISA (2017) Modern prospects of nanotechnology in plant pathology. In: Nanotechnology (eds. Prasad R, Kumar M, Kumar V), Springer Nature Singapore Pte Ltd. 305–317

Islam MS, Akter N, Rahman MM, Shi C, Islam MT, Zeng H, Azam MS (2018) Mussel-inspired immobilization of silver nanoparticles toward antimicrobial cellulose paper. ACS Sustain Chem Engg 6(7):9178–9188

Jin R (2012) The impacts of nanotechnology on catalysis by precious metal nanoparticles. Nanotechnol Rev 1(1):31–56

Jhanzab HM, Razzaq A, Jilani G, Rehman A, Hafeez A, Yasmeen F (2015) Silver nano-particles enhance the growth, yield and nutrient use efficiency of wheat. Int J Agron Agri Res 7(1):15–22

Jiang B, Li C, Dag O, Abe H, Takei T, Imai T, Hossain MSA, Islam MT, Wood K, Henzie J, Yamauchi Y (2017) Mesoporous metallic rhodium nanoparticles. Nat Commun 8:15581. https://doi.org/10.1038/ncomms15581

Jie H, Jose RP, Jorge LG (2013) Nanomaterials in agricultural production: benefits and possible threats? Sustainable nanotechnology and the environment: advances and achievements. American Chemical Society, USA, pp 73–90

Kah M, Kookana RS, Gogos A, Bucheli TD (2018) A critical evaluation of nanopesticides and nanofertilizers against their conventional analogues. Nat Nanotechnol 13:677–684

Karn B, Kuiken T, Otto M (2009) Nanotechnology and in situ remediation: a review of the benefits and potential risks. Environ Health Persp 117(12):1813. https://doi.org/10.1289/ehp.0900793
Khan JH, Lin J, Young C, Matsagar BM, Wu KC, Dhepe PL, Islam MT, Rahman M, Shrestha LK, Alshehri SM, Ahamad T (2018) High surface area nanoporous carbon derived from Bangladeshi jute. Mater Chem Phys 216(1):491–495
Khiyami MA, Almoammar H, Awad YM, Alghuthaymi MA, Abd-Elsalam KA (2014) Plant pathogen nanodiagnostic techniques: forthcoming changes? Biotechnol Biotechnol Equip 28(5):775–785
Khot LR, Sankaran S, Maja JM, Ehsani R, Schuster EW (2012) Applications of nanomaterials in agricultural production and crop protection: a review. Crop Prot 35:64–70
Kelly KL (2004) Nanotechnology grows up. Science 304:1732–17345
Kottegoda N, Sandaruwan C, Priyadarshana G, Siriwardhana A, Rathnayake UA, Arachchige DMB, Kumarasinghe AR, Dahanayake D, Karunaratne V, Amaratunga GA (2017) Urea-hydroxyapatite nanohybrids for slow release of nitrogen. ACS Nano 11(2):1214–1221
Krishna AK, Satyanarayanan M, Govil PK (2009) Assessment of heavy metal pollution in water using multivariate statistical techniques in an industrial area: a case study from Patancheru, Medak District, Andhra Pradesh, India. J Hazard Mater 167(1–3):366–373
Kumar GD, Natarajan N, Nakkeeran S (2016) Antifungal activity of nanofungicide Trifloxystrobin 25% + Tebuconazole 50% against *Macrophomina phaseolina*. Afr J Microbiol Res 10(4):100–105
Kumar M, Shamsi TN, Parveen R, Fatima S (2017) Application of nanotechnology in enhancement of crop productivity and integrated pest management. In: Prasad R, Kumar M, Kumar V (eds) Nanotechnology. Springer, Singapore, pp 361–371
Kutz FW, Wood PH, Bottimore DP (1991) Organochlorine pesticides and polychlorinated biphenyls in human adipose tissue. Rev Environ Contam Toxicol 120:1–82
Lahiani MH, Dervishi E, Chen J, Nima Z, Gaume A, Biris AS, Khodakovskaya MV (2013) Impact of carbon nanotube exposure to seeds of valuable crops. ACS Appl Mater Interf 5(16):7965–7973
Li XQ, Elliott DW, Zhang WX (2006) Zero-valent iron nanoparticles for abatement of environmental pollutants: materials and engineering aspects. Crit Rev Solid State Mater Sci 31(4):111–122
Liu R, Lal R (2015) Potentials of engineered nanoparticles as fertilizers for increasing agronomic productions. Sci Total Environ 514:131–139
Mahakham W, Sarmah AK, Maensiri S, Theerakulpisut P (2017) Nanopriming technology for enhancing germination and starch metabolism of aged rice seeds using phytosynthesized silver nanoparticles. Sci Rep 7:8263. https://doi.org/10.1038/s41598-017-08669-5
Mahmoodi P, Yarnia M, Amirnia R, Tarinejad A, Mahmoodi H (2017) Comparison of the effect of nano urea and nono iron fertilizers with common chemical fertilizers on some growth traits and essential oil production of *Borago officinalis* L. Dairy Vet Sci J 2(2):555585. https://doi.org/10.19080/JDVS.2017.02.555585
Manjunatha SB, Biradar DP, Aladakatti YR (2016) Nanotechnology and its applications in agriculture: a review. J Farm Sci 29(1):1–13
McGehee DL, Lahiani MH, Irin F, Green MJ, Khodakovskaya MV (2017) Multiwalled carbon nanotubes dramatically affect the fruit metabolome of exposed tomato plants. ACS Appl Mater Interfaces 9(38):32430–32435
Mehrazar E, Rahaie M, Rahaie S (2015) Application of nanoparticles for pesticides, herbicides, fertilisers and animals feed management. Int J Nanopart 8(1):1–9
Morales-Díaz AB, Ortega-Ortíz H, Juárez-Maldonado A, Cadenas-Pliego G, González-Morales S, Benavides-Mendoza A (2017) Application of nanoelements in plant nutrition and its impact in ecosystems. Adv Nat Sci Nanosci Nanotechnol 8:13001. https://doi.org/10.1088/2043-6254/8/1/013001
Mukhopadhyay SS (2014) Nanotechnology in agriculture: prospects and constraints. Nanotechnol Sci Appl 7:63–71

Mukherjee A, Majumdar S, Servin AD, Pagano L, Dhankher OP, White JC (2016) Carbon nanomaterials in agriculture: a critical review. Front Plant Sci 7:172. https://doi.org/10.3389/fpls.2016.00172

Nakache E, Poulain N, Candau F, Orecchioni AM, Irache JM (1999) Biopolymer and polymer nanoparticles and their biomedical applications. In: Nalwa HS (ed) Handbook of nanostructured materials and nanotechnology. Academic Press, New York, p 3461

Naseri N, Valizadeh H, Zakeri-Milani P (2015) Solid lipid nanoparticles and nanostructured lipid carriers: structure, preparation and application. Adv Pharm Bull 5(3):305–313

Neethirajan S, Jayas DS (2011) Nanotechnology for the food and bioprocessing industries. Food Bioproc Technol 4(1):39–47

Ntalli NG, Caboni P (2012) Botanical nematicides: a review. J Agric Food Chem 60(40):9929–9940

Nuruzzaman M, Rahman MM, Liu Y, Naidu R (2016) Nanoencapsulation, nano-guard for pesticides: a new window for safe application. J Agric Food Chem 64(7):1447–1483

de Oliveira JL, Campos EVR, da Silva CMG, Pasquoto T, Lima R, Fraceto LF (2015) Solid lipid nanoparticles co-loaded with simazine and atrazine: preparation, characterization, and evaluation of herbicidal activity. J Agric Food Chem 63(2):422–432

Oliveira HC, Stolf-Moreira R, Martinez CBR, Grillo R, de Jesus MB, Fraceto LF (2015) Nanoencapsulation enhances the post-emergence herbicidal activity of atrazine against mustard plants. PLoS One 10(7):e0132971. https://doi.org/10.1371/journal.pone.0132971

Omobhude ME, Morenikeji OA, Oyeyemi OT (2017) Molluscicidal activities of curcumin-nisin polylactic acid nanoparticle on *Biomphalaria pfeifferi*. PLoS Negl Trop Dis 11(8):e0005855. https://doi.org/10.1371/journal.pntd.0005855

Ouda SM (2014) Antifungal activity of silver and copper nanoparticles on two plant pathogens, *Alternaria alternata* and *Botrytis cinerea*. Res J Microbiol 9(1):34–42

Pandey S, Giri K, Kumar R, Mishra G, Rishi RR (2016) Nanopesticides: opportunities in crop protection and associated environmental risks. Proceed Nat Acad Sci India Sec B Biol Sci 2016:1–22

Parisi C, Vigani M, Rodríguez-Cerezo E (2015) Agricultural nanotechnologies: what are the current possibilities? Nano Today 10(2):124–127

Park HJ, Kim SH, Kim HJ, Choi SH (2006) A new composition of nanosized silica-silver for control of various plant diseases. Plant Pathol J 22:25–34

Pereira AES, Sandoval-Herrera IE, Zavala-Betancourt SA, Oliveira HC, Ledezma-Pérez AS, Romero J, Fraceto LF (2017) γ-Polyglutamic acid/chitosan nanoparticles for the plant growth regulator gibberellic acid: characterization and evaluation of biological activity. Carbohydr Polym 157:1862–1873

Pierluigi C, Robert ES, John EC (2003) Phenylpyrazole insecticide photochemistry, metabolism and GABAergic action: ethiprole compared with fipronil. J Agri Food Chem 51:7055–7061

Pourkhaloee A, Haghighi M, Saharkhiz MJ, Jouzi H, Doroodmand MM (2011) Carbon nanotubes can promote seed germination via seed coat penetration. Seed Technol 33:155–169

Pradeep T (2012) Noble metal nanoparticles for water purification: a critical review. Thin Solid Films 517(24):6441–6478

Prasad R, Kumar V, Prasad KS (2014) Nanotechnology in sustainable agriculture: present concerns and future aspects. Afr J Biotechnol 13(6):705–713

Prasad R, Bhattacharyya A, Nguyen QD (2017) Nanotechnology in sustainable agriculture: recent developments, challenges, and perspectives. Front Microbiol 8:1014. https://doi.org/10.3389/fmicb.2017.01014

Prasad R, Kumar V, Kumar M, Choudhary D (2019) Nanobiotechnology in Bioformulations. Springer International Publishing (ISBN 978-3-030-17061-5) https://www.springer.com/gp/book/9783030170608

Rai V, Acharya S, Dey N (2012) Implications of nanobiosensors in agriculture. J Biomater Nanobiotechnol 3:315–324

Rani M, Shanker U, Jassal V (2017) Recent strategies for removal and degradation of persistent and toxic organochlorine pesticides using nanoparticles: a review. J Environ Manag 190:208–2022

Raliya R, Tarafdar JC, Gulecha K, Choudhary K, Ram R, Mal P, Saran RP (2013) Scope of nanoscience and nanotechnology in agriculture. J Appl Biol Biotechnol 1:41–44
Raliya R, Saharan V, Dimkpa C, Biswas P (2017) Nanofertilizer for precision and sustainable agriculture: current state and future perspectives. J Agric Food Chem 66(26):6487–6503
Rizwan MD, Singh M, Mitra CK, Morve RK (2014) Eco-friendly application of nanomaterials: Nanobioremediation. J Nanopart 431787(7). https://doi.org/10.1155/2014/431787
Rolando CA, Garrett LG, Baillie BR, Wat MS (2013) A survey of herbicide use and a review of environmental fate in New Zealand planted forests. New Zealand J Forest Sci 43:17. https://doi.org/10.1186/1179-5395-43-17
Sánchez-Moreno S, Alonso-Prados E, Alonso-Prados JL, García-Baudín JM (2009) Multivariate analysis of toxicological and environmental properties of soil nematicides. Pest Manag Sci 65(1):82–92
Sargent Jr JF (2011) Nanotechnology and environmental health and safety: issues for consideration. CRS Report for Congress, Congressional Research Service, 7-5700, www.crs.gov, RL34614, p 37
Sarlak N, Taherifar A, Salehi F (2014) Synthesis of nanopesticides by encapsulating pesticide nanoparticles using functionalized carbon nanotubes and application of new nanocomposite for plant disease treatment. J Agric Food Chem 62(21):4833–4838
Salinas S, Mosquera N, Yate L, Coy E, Yamhure G, González E (2014) Surface plasmon resonance nanosensor for the detection of arsenic in water. Sens Transducers 183(12):97–102
Sastry RK, Rashmi HB, Rao NH (2011) Nanotechnology for enhancing food security in India. Food Policy 36(3):391–400
Scrinis G, Lyons K (2007) The emerging nano-corporate paradigm: nanotechnology and the transformation of nature, food and Agri-food systems. J Sci Food Agric 15:22–44
Sekhon BS (2014) Nanotechnology in Agri-food production: an overview. Nanotechnol Sci Appl 7:31–53
Sempeho SI, Kim HT, Mubofu E, Hilonga A (2014) Meticulous overview on the controlled release fertilizers. Adv Chem 363071:16. https://doi.org/10.1155/2014/363071
Servin A, Elmer W, Mukherjee A, De La Torre-Roche R, Hamdi H, White JC, Bindraban P, Dimkpa C (2015) A review of the use of engineered nanomaterials to suppress plant disease and enhance crop yield. J Nanopart Res 117(2):92. https://doi.org/10.1007/s11051-015-2907-7
Service RF (2003) Nanomaterials show signs of toxicity. Science 300:243
Singh NB, Amist N, Yadav K, Singh D, Pandey JK, Singh SC (2013) Zinc oxide nanoparticles as fertilizer for the germination, growth and metabolism of vegetable crops. J Nanoeng Nanomanuf 3:353–364
Sharma HD, Reddy KR (2004) Geo-environmental engineering: site remediation, waste containment, and emerging waste management technologies. Wiley, Hoboken
Si Y, Zhang Z, Wu W, Fu Q, Huang K, Nitin N, Ding B, Sun G (2018) Daylight-driven rechargeable antibacterial and antiviral nanofibrous membranes for bioprotective applications. Sci Adv 4(3):eaar5931. https://doi.org/10.1126/sciadv.aar5931
Siafaka PI, Okur NU, Karavas E, Bikiaris DN (2016) Surface modified multifunctional and stimuli responsive nanoparticles for drug targeting: current status and uses. Int J Mol Sci 17(9):1440. https://doi.org/10.3390/ijms17091440
Siddiqui MH, Al-Whaibi MH, Firoz M, Al-Khaishany MY (2015) Role of nanoparticles in plants. In: Siddiqui MH, Al-Whaibi MH, Firoz M (eds) Nanotechnology and plant sciences. Springer, Switzerland, p 303
Smith VH (2003) Eutrophication of freshwater and coastal marine ecosystems a global problem. Environ Sci Pollut Res 10(2):126–139
Smith VH, Joye SB, Howarth RW (2006) Eutrophication of freshwater and marine ecosystems. Limnol Oceanogr-Meth 51(1–2):351–355
Soko W, Chimbari MJ, Mukaratirwa S (2015) Insecticide resistance in malaria-transmitting mosquitoes in Zimbabwe: a review. Infect Dis Poverty 4:46. https://doi.org/10.1186/s40249-015-0076-7

Song Y, Li Y, Xu Q, Liu Z (2017) Mesoporous silica nanoparticles for stimuli-responsive controlled drug delivery: advances, challenges, and outlook. Int J Nanomedicine 12:87–110

Suman PR, Jain VK, Varma A (2010) Role of nanomaterials in symbiotic fungus growth enhancement. Curr Sci 99:1189–1191

Thatai S, Khurana P, Boken J, Prasad S, Kumar D (2014) Nanoparticles and core–shell nanocomposite based new generation water remediation materials and analytical techniques: a review. Microchem J 116:62–76

Torney F, Trewyn BG, Lin VSY, Wang K (2007) Mesoporous silica nanoparticles deliver DNA and chemicals into plants. Nat Nanotechnol 2(5):295. https://doi.org/10.1038/nnano.2007.108

Ul-Islam M, Khan T, Park JK (2012) Nanoreinforced bacterial cellulose–montmorillonite composites for biomedical applications. Carbohydr Polym 4:1189–1197

US (EPA) Environmental Protection Agency (2007) Nanotechnology white paper. EPA, Washington, DC. http://www.epa.gov/osainter/pdfs/nanotech/epa-nanotechnologywhitepaper-0207.pdf. Accessed 26 June 2018

US Department of Agriculture (2002) Nanoscale science and engineering for agriculture and food systems. United States Department of Agriculture, National Planning Workshop, November 18–19, 2002, Washington, DC

Wang X, Xie H (2018) A review on applications of remote sensing and geographic information systems (GIS) in water resources and flood risk management. Water 10(5):608. https://doi.org/10.3390/w10050608

Watson JG (1996) Physical/chemical treatment of organically contaminated soils and sediments. J Air Waste Manage Assoc 46(10):993–1003

Walters JP, Archer DW, Sassenrath GF, Hendrickson JR, Hanson JD, Halloran JM, Vadas P, Alarcon VJ (2016) Exploring agricultural production systems and their fundamental components with system dynamics modelling. Ecol Model 333(10):51–65

Wu SG, Huang L, Head J, Ball M, Tang YJ, Chen D-R (2014) Electrospray facilitates the germination of plant seeds. Aerosol Air Qual Res 14:632–641

Wu H, Santana I, Dansie J, Giraldo JP (2017) *In vivo* delivery of nanoparticles into plant leaves. Curr Protoc Chem Biol 9(4):269–284

Xing G, Cao Y, Shi S, Sun G, Du L, Zhu J (2001) N pollution sources and denitrification in water bodies in Taihu Lake region. Sci China Ser B: Chem 44(3):304–314

Xingmao M, Geiser-Lee J, Deng Y, Kolmakov A (2010) Interactions between engineered nanoparticles ENPs and plants: Phytotoxicity, uptake and accumulation. Sci Total Environ 408:3053–3061

Yang C, Hamel C, Vujanovic V, Gan Y (2011) Fungicide: modes of action and possible impact on non-target microorganisms. ISRN Ecol 130289:8. https://doi.org/10.5402/2011/130289

Yao Z, Li J, Xie H, Yu C (2012) Review on remediation technologies of soil contaminated by heavy metals. Procedia Environ Sci 16:722–729

Yuce M, Kurt H (2017) How to make nanobiosensors: surface modification and characterisation of nanomaterials for biosensing applications. RSC Adv 7:49386. https://doi.org/10.1039/C7RA10479K

Zargar M, Hamid AA, Bakar FA, Shamsudin MN, Shameli K, Jahanshiri F (2011) Green synthesis and antibacterial effect of silver nanoparticles using *Vitex negundo* L. Molecules 16(8):6667–6676

Zaytseva O, Neumann G (2016) Carbon nanomaterials: production, impact on plant development, agricultural and environmental applications. Chem Biol Technol Agric 3:17. https://doi.org/10.1186/s40538-016-0070-8

Zhang W (2003) Environmental technologies at the nanoscale. Environ Sci Technol 7:103–108

Zhao X, Cui H, Wang Y, Sun C, Cui B, Zeng Z (2017) Development strategies and prospects of nano-based smart pesticide formulation. J Agric Food Chem 66(26):6504–6512

Chapter 8
Nanoparticles from Endophytic Fungi and Their Efficacy in Biological Control

B. Shankar Naik

Contents

8.1 Introduction

Resistance to antimicrobial agents in different microorganisms has attracted the attention of researchers toward a better alternative agent to control pathogens causing disease to plants, animals, and humans. Nanotechnology is an emerging area that has demonstrated multiple applications in optoelectronics, textiles, medicine, agriculture, and also ecosystem management (Nanda and Majeed 2014). Nanoparticles have many unique properties that make them suitable for their use in various applications. Nanomaterials have been proven better class of agents with antibacterial, antiviral, antiprotozoal, antifungal, larvicidal, acaricidal, anthelmintic, and antitumor activities (Ortega et al. 2015; Yadav et al. 2015). Several methods mainly chemical, physical, and biological methods are commonly used for the fabrication of metal nanoparticles. Chemical methods are employed because of large production in relatively short time and control on size and distribution (He et al. 2015). Apart from these advantages, chemical methods

B. Shankar Naik (✉)
Department of Biology, Government Science College, Basavanahalli Extension, Chikmagalur, Karnataka, India

D. Thangadurai et al. (eds.), *Nanotechnology for Food, Agriculture, and Environment*, Nanotechnology in the Life Sciences,
https://doi.org/10.1007/978-3-030-31938-0_8

employ toxic chemicals and intensive energy and produce hazardous wastes that are major risks to the environment.

In physical methods, sputter deposition, laser ablation or cluster beam deposition, and microwave-assisted synthesis are used for the synthesis of metal nanoparticles. Involvement of high temperature, radiations, and pressure associated with these methods is posing threat to both organisms and the environment (Alzahrani et al. 2015). Biological systems, i.e., use of microbes (bacteria, fungi, algae, cyanobacteria, and myxobacteria) (Prasad et al. 2016) and plants, are effectively being used for either intra- or extracellular synthesis of different metal nanoparticles (Kar et al. 2014; Singh et al. 2015; Ahmed et al. 2016). Among these, fungi are considered to be a better microbial system because of their ubiquitous distribution and role in the synthesis of metal nanoparticles (Kar et al. 2014; Prasad 2016, 2017). Hence, the nanobiotechnology involving fungi now known to be myconanobiotechnology emerged as a better biological alternative to physical and chemical methods for synthesis and fabrication of nanoparticles (Rai et al. 2009; Prasad et al. 2018).

Physical and chemical methods for synthesis of nanoparticles involve chemical reduction of metal ions in aqueous solutions with or without the use of stabilizing agents (Rai et al. 2008; Sharma et al. 2009). Intensive use of a large amount of heat and energy, elevated temperature, and toxic chemicals are a threat to organisms and also the environment. The synthesis of nanoparticles by microbes and plants which is also referred to as green nanotechnology has a potential application such as therapeutics, biosensors, and bionanocatalysts (Rai et al. 2009; Narayanan and Sakthivel 2010). Fungi are non-phototrophic eukaryotic microorganisms consisting of the rigid cell wall (Bowman and Free 2011), being chemoautotrophic requiring simple nutritional requirements (Holan and Volesky 1995). They grow on land and dead organic matter and also as mutualists and parasites. Fungi secrete several enzymes extracellularly and digest organic matter (Madigan et al. 2006). Fungal cell wall provides the cell resistance to extreme osmotic pressure and environmental stresses due to their unique chemical composition in their cell wall (Duran and Nombela 2004; Bowman and Free 2011). The fungal cell wall comprised of glycoproteins and polysaccharides cross-linked together in a complex network provides rigid and more resistance cover for the cell (Bowman and Free 2011). Fungi secrete a large number of enzymes required for the synthesis and yield of nanoparticles (Birla et al. 2009; Narayanan and Sakthivel 2010; Yadav et al. 2015; Abdel-Aziz et al. 2018).

8.2 Progress in Microfabrication of Metal Nanoparticles

Nanoparticles can be synthesized intra- and extracellularly from fungi (Li et al. 2012). In intracellular synthesis, fungal biomass is treated with a metal salt solution and incubated for 24 h in dark. Nanoparticles synthesized using intracellular method are smaller in size compared to nanoparticles synthesized through extracellular method (Narayanan and Sakthivel 2010). Downstream processing becomes difficult and costly as synthesis takes place inside the cell (Zhang et al. 2011;

Dhillon et al. 2012). In extracellular nanoparticle synthesis in fungi, the filtrate is treated with a metal salt solution and incubated for 24 h in the dark (Mukherjee et al. 2008; Nanda and Majeed 2014). This method does not require extensive downstream processing which offers easier and cost-effective synthesis (Devi and Joshi 2012).

8.3 Mechanism of Mycosynthesis of Metal Nanoparticles

Fungi produce several extracellular metabolites when exposed to different environmental stresses such as toxic substances (metallic ions), predators, and temperature variations (Mehra and Winge 1991). During the synthesis of metal nanoparticles by a fungus, the fungal mycelium is exposed to the metal salt solution which prompts the fungi to release enzymes/secondary metabolites. In this process, the toxic metal ions are reduced to the nontoxic metallic solid nanoparticles through the catalytic effect of the extracellular enzyme and metabolites of the fungus (Vahabi et al. 2011; Yadav et al. 2015). Ahmed et al. (2003) reported that NADH-dependent reductase was responsible for the reduction of Ag ions and the formation of silver nanoparticles by *Fusarium oxysporum* based on protein assay.

NADPH–nitrate reductase and anthraquinone were known responsible for the synthesis of silver nanoparticles (Duran et al. 2011). The formation of silver nanoparticles requires reduction of NADPH to NADP (Kumar et al. 2007). They hypothesized that hydroxyquinoline probably acts as an electron shuttle transferring the electron generated during the reduction of nitrate to Ag^+ ions converting them to Ag^0. Ingle et al. (2008) supported the hypothetical mechanism where the cofactor NADH and nitrate reductase enzyme were responsible for the synthesis of silver nanoparticles from *Fusarium acuminatum*. Mukherjee et al. (2008) suggested the Michaelis Menten type of mechanism for the synthesis of silver nanoparticles. In this mechanism during the initial phase when the concentration of silver nitrate is higher, the reaction is rather slow, and as the reaction proceeds, the concentrate of silver nitrate lowers down considerably. They propose cysteine (-SH bonds) undergoes dehydrogenation on reaction with silver nitrate to produce silver nanoparticles and free amino acid groups possibly serve as a capping for silver nanoparticles. Das et al. (2009) hypothesized that surface-bound protein molecules acted as reducing and stabilizing agent. In their study, the fungal culture containing $AuCl_4$ (auric chloride) confirmed the presence of amide I, II, and III groups and disappearance of carbonyl groups present in the mycelia by FTIR spectra. Stabilization of nanoparticles by fungal protein and the reduction of silver ions by amide I and II groups during the synthesis of nanoparticles by *Coriolus versicolor* were reported by Sanghi and Verma (2009).

Two-step hypothetical mechanism for the synthesis of silver nanoparticles was reported by Jain et al. (2011). In the first step, reduction of silver ions to silver nanoparticles takes place by a 32KDa protein which might be a reductase secreted by *Aspergillus flavus*. In the second step, stabilization was conferred by 32KDa proteins capping silver nanoparticles. In another study, the presence of two extracellular proteins with molecular weight of 24 and 28 kDa is responsible for zirconium oxide nanoparticles by *Fusarium oxysporum* (Bansal et al. 2004). Chan and Mashitah (2012) supposed that reduction of silver nanoparticles was possibly due to the presence of diketone compound which was confirmed by GC-MS analysis while working on three different macrofungi (*Pycnoporus sanguineus*, *Schizophyllum commune*, and *Lentinus sajor-caju*). Li et al. (2012) inferred that NADH-dependent reductase might be responsible for silver nanoparticles produced by *Aspergillus terreus*. Kashyap et al. (2013) indicated that reduction of NADPH to $NADP^+$ is essential for the synthesis of silver nanoparticles. Here, hydroquinone acts as electron shuttle transferring the electrons generated during the reduction of nitrate to Ag^+ ions and converting them to Ag^0.

8.4 Biomedical Applications of Nanoparticles

The metal nanoparticles especially silver, gold, and platinum have huge biomedical applications (Yadav et al. 2015). Silver nanoparticles have shown to be strong antimicrobial potential against bacteria, viruses, and fungi. Several researchers reported the antimicrobial activity of silver nanoparticles against both Gram-positive and Gram-negative bacteria (Rai et al. 2009; Aziz et al. 2014, 2015, 2016). Silver nanoparticles have also shown the promising result for control and prevention of viral diseases. These particles have inhibitory activity against influenza A H1N1 virus which was proved through hemagglutination inhibition test and embryo inoculation assays (Xiang et al. 2011). The silver nanoparticles in concentration of 24 μg/ml significantly reported inhibiting HIV infection in CD4 MT2 cells and cMAG HIV-1 receptor cells (Elechiguerra et al. 2005). These nanoparticles have shown to be effective against herpes simplex virus type 1 and 2 (HSV1 and 2) and human parainfluenza virus type 3 (HPIV-3) (Chen et al. 2013; Hu et al. 2014).

Antifungal activity of silver nanoparticles was reported (Gajbhiye et al. 2009; Kim et al. 2012). Silver nanoparticles are used as biocides to prevent infections (Franci et al. 2015; Rai et al. 2015), water disinfectant (Bhattacharya and Mukherjee 2008), silver dressings, surgical masks, gloves, and catheters (Rai et al. 2009). Gold nanoparticles also have tremendous biomedical applications. They can be used for the diagnosis and treatment of cancer, AIDS, tuberculois, and other diseases through dry delivery (Johnson et al. 2013).

8.5 Applications of Nanoparticles in Agriculture

Nanoparticles have potential application in agriculture (Goel 2015). They can serve as nano-pesticides, nanoinsecticides, and nanofertilizers and are also useful for developing nano-biosensors used for the preparation of devices which can be applied in precision farming. Nano-biosensors can be effectively used for sensing a wide variety of fertilizers, herbicides, pesticides insecticides, pathogen, moisture, and soil pH (Rai et al. 2012). Precision farming involves system controller for each growth factor such as nutrition, light, and temperature. This system should have information from planting to harvest time which can be controlled by satellite systems so that farmers can be able to know the best time to invest. Nanoparticles can be effectively used as nanofungicides against a large number of fungal pathogens which include species of *Fusarium*, *Phoma*, *Aspergillus*, *Phytophthora*, and *Phyllosticta* (Singh et al. 2015). Nanoparticles can be effectively used as nanofertilizers which could be a better alternative to chemical fertilizers. It is an eco-friendly approach resulting in an increase in nutrient efficiency, reducing soil toxicity. Corradini et al. (2010) proposed the incorporation of chemical fertilizers with chitosan nanoparticles for slow and constant release of inadequate amount in the soil. Naturally occurring minerals such as nano-clays and zeolites (Chinnamuthu and Bhoopathi 2009), urea-fertilized zeolite chips (Millan et al. 2008), ammonium-charged zeolite, and potash fertilizer with a coating of wax and plaster of Paris (Subbarao et al. 2013) help in the slow release of fertilizers and minimizing fertilizer loss.

Nanoparticles can also be used effectively as nano-pesticides (Barik et al. 2008; Bhattacharyya et al. 2016). Similarly, Wang et al. (2007) reported that nanoemulsions can be used against various insect pests in agriculture. Nanosilica can be used as nano-pesticide to control insect pests (Barik et al. 2008). According to them, nanoparticles get absorbed into the cuticular lipids by physisorption and thereby cause the death of insect pests (Barik et al. 2008). Application of different kinds of nanoparticles, viz., silver nanoparticles, aluminum oxide, zinc oxide, and titanium rice weevil, helps eliminate grasseric disease in silkworm (*Bombyx mori*) caused by *Sitophilus oryzae* and baculovirus from *Bombyx mori* (Goswami et al. 2010).

Nanotechnology has great prospects in maintaining the clean environment. Metal nanoparticles play a vital role in absorbing water contaminants, for example, iron nanoparticles transform and detoxify environmental pollutants such as chlorinated organic solvents, pesticides, and polychlorinated biphenyls (Zhang 2003). Groundwater can be made potable by the use of zero-valent iron nanoparticles (Rajan 2011). Magnetic nanoparticles are effective in removing mercury ions (Farukh et al. 2013) and chromium ions (Telling et al. 2009) from contaminated water. Magnetic iron oxide nanoparticles absorb arsenic and are highly biocompatible and can be easily degraded in the environment. Photocatalytic degradation can be achieved through TiO_2 nanoparticles against pollutants such as phenols, volatile compounds, and dyes (Chong et al. 2010).

Nanotechnology is the branch of research involving novel strategies for the design synthesis, manipulation, and application of particles with a dimension smaller than 100 nm called as nanoparticles. The extreme small size and high surface area-to-volume ratio made nanoparticles have extraordinary physicochemical properties. These properties make them applicable in different sectors, viz., electronics, biomedicine, agriculture, and environmental remediation (Chong et al. 2010). Metallic nanoparticles including gold, silver, iron, zinc, and metal oxide nanoparticles have shown great potential in terms of biomedical applications (Bhattacharya and Mukherjee 2008). Gold and cerium oxide nanoparticles have shown to be effective against the treatment of tumors and anti-inflammation (Bhattacharya and Mukherjee 2008).

8.6 Methods of Synthesis of Nanoparticles

The nanoparticles of a wide range of materials can be prepared by a number of methods which are categorized into three main groups: chemical, physical, and biological methods. These methods follow either a top-down approach or bottom-up approach. In the top-down approach, the bulk materials are mechanically grinded, and the resulting nanosized particles are stabilized by the addition of colloidal stabilizing agents, whereas in the bottom-up approach, the bulk metals are reduced by electrochemical methods (Amulyavichus et al. 1998). The physicochemical methods are usually associated with certain disadvantages such as the requirement of expensive equipment and use of toxic-reducing agents like sodium borohydride and N,N-dimethylformamide which produce hazardous effects on the environment and health.

The physicochemical methods involved in the production of metal-based nanoparticles mainly include chemical reduction, thermal treatment, irradiation, and laser ablation (Sun and Luo 2005; Peterson et al. 2007). Hence, biological methods gain interest among the researchers. Biosynthetic approach involves the use of biological entities including plant and plant products, algae, fungi, yeast, bacteria, and viruses for the synthesis of nanoparticles (Sunkar and Nachiyar 2012; Prasad et al. 2018a). The nanoparticles may be synthesized intracellularly or extracellularly by uni- or multicellular organisms. This technology led to the development of a new branch of biotechnology called green nanotechnology or nanobiotechnology.

Silver nanoparticle has great potential in biomedical applications. Biosynthetic methods come under the bottom-up approach wherein the silver nanoparticles are produced by reduction via enzymes and other metabolites secreted by biological agents. The use of silver nanoparticles has emerged as the most promising approach for overcoming the antibiotic resistance of microorganisms. Silver is advantageous over other metals since it exhibits higher toxicity to a broad spectrum of microorganisms and lower toxicity to mammalian cells (Sharma et al. 2013). The mechanism of antimicrobial activity of silver nanoparticles involves penetration of silver nanoparticles inside the cells and disturbs the functions of a cell membrane such as permeability and respiration. They also inhibit the functioning of enzyme machinery and phosphorus-containing compounds such as DNA which leads to cell death (Sondi and Sondi 2004; Aziz et al. 2015, 2016, 2019). Recently, synthesis of nanoparticles from endophytic fungi and examination of the efficacy of the particles against pathogens gain much interest (Bharathidasan and Panneerselvam 2012).

8.7 Silver Nanoparticles in Clinical Medicine

Over the past decade, silver has been used for the treatment of a variety of diseases, viz., epilepsy, general infections, and leg ulcers. Silver nanoparticles also have applications in the healing of surgical wounds and reducing postoperative infections; pencils containing silver nitrate mitigated with potassium nitrate were used for wart removal and ulcer debridement (Klasen 2000). Surface-enhanced Raman spectroscopy (SERS) has emerged as a powerful analytical tool for the diagnosis of respiratory infections. It used silver nanorods to enhance the vibrational spectra of molecules adsorbed to the surface that extends the possibilities of vibrational spectroscopy. Recently, four strains of the respiratory syncytial virus (RSV) were readily detected as very low detection limits by using SERS composed of silver nanorods (Shanmukh et al. 2008). In another diagnostic study, Au-Ag nanorods were used as a nanoplatform for multivalent binding by multiple aptamers in cancer cell recognition. Here, the molecular assembly of aptamers on the nanorods was shown to lead to a 26-fold higher affinity than the original aptamer probes (Huang et al. 2008). RT-PCR studies revealed that silver nanoparticles have anti-inflammatory properties. Reports on burn cases revealed that silver nanoparticle-treated animals showed lower levels of pro-inflammatory cytokine IL-6. Similarly, mRNA levels of IL-10, an anti-inflammatory cytokine, remain higher in the silver nanoparticle-treated cells (Wong et al. 2009).

Silver is being used for wound dressings (Fong and Wood 2006). Acticoat is the first commercially approved dressing kit made up of two layers of polyamide ester membranes covered with nanocrystalline silver ions which have the lowest MIC and MBC values. It has the strongest antibacterial activity against five bacteria. It has been also reported that the frequency of burn wound sepsis and sec bacteremia was lower in patients treated with silver nanoparticles (Sibbald et al. 2007). Silver-impregnated central venous catheters have been developed for clinical use on which silver ions are bonded with an inert ceramic zeolite (Mermel 2000). InterGard Silver has been shown to a low rate of graft infection (Ricco 2006). Silver-impregnated ventricular catheters have resulted in increased clinical efficacy in the treatment of neurological and neurosurgical patients requiring external SF drainage (Galiano et al. 2008). Silver nanoparticles could play a critical role in orthopedics. Bone cement loaded with nanosilver (Alt et al. 2004) has shown high antibacterial activity against methicillin-resistant *Staphylococcus aureus* (MRSA). Here nanoparticles were not found to be cytotoxic against osteoblasts grown in vitro (Yadav et al. 2015). Silver nanoparticle-coated polypropylene mesh can decrease the prosthetic infection rate and the host inflammatory response in the clinical setting (Cohen et al. 2007). It has also been shown that silver nanoparticles not only possess antiretroviral activities but also inhibit HI-1 replication, and these nanoparticles did not show acute cytotoxicity to either Hut/r5 cells or the normal peripheral blood mononuclear cells (Sun et al. 2005). Silver nanoparticles are known to be effective in the treatment of thrombotic disorders by causing inhibition of platelet functional responses like aggregation, secretion, and adhesion (Shrivastava et al. 2009).

Nanoparticles from mycoendophytes possess unique physical and chemical properties and can be used for the synthesis of stable and bioactive AgNPs (Thakkar et al. 2010; Rathna et al. 2013). A number of mechanism have been proposed for the synthesis of metal nanoparticles in biological systems in which bioreduction is a key step (Devi et al. 2014; Balakumaran et al. 2015). The proposed mechanism for the synthesis of AgNPs from *Fusarium oxysporum* was based on the presence of an NADH-dependent reductase responsible for the reduction Ag ions and subsequent formation of silver nanoparticles (Kitching et al. 2015). Reduction of silver ions may be caused by electron transfer from the NADH by NADH-dependent reductase as an electron carrier, and the electron-deficient silver ions (Ag^+) accept the electrons and are reduced to silver neutral (Ag^0) leading to the formation of silver nanoparticles (Vago et al. 2016). It was reported that the biosynthesis of AgNPs occurred in the presence of anthraquinone or hydroxyquinoline and NADPH nitrate reductase (Ahmad et al. 2003; Li et al. 2012). The reduction of NADPH to NADP+ is required in this process, and electrons generated during the reduction of silver ions to Ag0 were donated from both quinine or hydroxyquinoline and NADPH. Another hypothetical mechanism is that the cofactor NADH and nitrate reductase are involved in the synthesis of silver nanoparticles (Balakumaran et al. 2015). The amino acids containing –SH bonds most likely cysteine react with silver nitrate by dehydrogenation resulting in production of silver nanoparticles; meanwhile, free amino acid groups serve as capping for silver nanoparticles. The involvement of polypeptides/proteins in the bioreduction of metal ions was supported by many workers (Sanghi and Verma 2009). Das et al. (2009) studied the spectra of fungal culture containing $AuCl_4$ (auric chloride) and observed the presence of amide I, II, and III groups and the disappearance of carbonyl groups present in the mycelia. Similarly, Sanghi and Verma (2009) synthesized silver nanoparticle from *Coriolus versicolor*.

8.8 Metal Nanoparticles from Endophytic Fungi

Medically and industrially important compounds from mycoendophytes have been detailed, but attention has not been paid on the potential of endophytes for the reduction of metals to nanosize (Netala et al. 2015). A biosynthetic approach using mycoendophytes is a novel way toward the development of safe and cost-effective methods for the synthesis of nanoparticles. Here nanoparticles can be used in many antimicrobial formulations (Rahi and Parmar 2014). Endophytic fungi secrete a large number of enzymes which directly contribute to the higher productivity of metal nanoparticles (Raghupathi et al. 2011). Recently, mycoendophytes are being used for the synthesis of nanoparticles isolated from different explants such as leaves, roots, and stems of various medicinal plants (Ramalingmam et al. 2015; Raheman et al. 2011). Several genera of mycoendophytes such as *Aspergillus*, *Penicillium*, *Colletotrichum*, *Phomopsis*, *Alternaria*, *Curvularia*, and *Pestalotia* have been used for the synthesis of nanoparticles (Riddin et al. 2006; Ramalingmam et al. 2015). *Guignardia mangiferae*, *Epicoccum nigrum*, and *Amylomyces rouxii* have also been exploited for nanoparticle synthesis (Mohanpuria et al. 2008). Antimicrobial activity of silver nanoparticles produced by endophytic fungi against various human and plant pathogens was documented (Golinska et al. 2015, 2016). These bacteria include *E. coli*, *Proteus mirabilis*, *Klebsiella pneumonia*, *Pseudomonas aeruginosa*, *Salmonella typhi*, *S. typhimurium*, *Shigella dysenteriae*, *Enterobacter aerogenes*, *Citrobacter* sp., and Gram-positive bacteria such as *Enterococcus faecalis* and *S. aureus* including methicillin-resistant *Staphylococcus aureus* (MRSA), *S. epidermidis*, *Streptococcus pyrogenes*, and *Bacillus subtilis* (Rai et al. 2009; Manjunath et al. 2014). Similarly, antifungal activity of silver nano particles against *Candida albicans*, *C. tropicalis*, *C. glabrata*, *C. krusei*, *C. parapsilosis*, *Cryptococcus neoformans*, *C. gatti*, *A. flavus*, *A. fumigatus*, *F. solani*, *Sporothrix schenckii*, *trichophyton rubrum*, and *Epidermophyton floccosum* has been reported by several researchers (Rai et al. 2009; Thakkar et al. 2010; Ramalingmam et al. 2015).

Recently, application of AgNPs with a combination of certain antibiotics has shown to be very effective against bacterial pathogens (Rahi and Parmar 2014). Antibiotics such as ampicillin, rifampicin, chloramphenicol, erythromycin, kanamycin, penicillin, carbenicillin, gentamycin, and sulfamethizole in combination with AgNPs from *Pestalotia* sp. and *Curvularia lunata* were tested against *B. subtilis*, *S. aureus*, *E. coli*, *S. paratyphi*, and *S. typhi* (Raheman et al. 2011; Ramalingmam et al. 2015) (Table 8.1).

8.9 Mechanism of Antimicrobial Activity by Nanoparticles

Mechanism of silver nanoparticle action on microbial cells has been proposed. Silver nanoparticles attach to the bacterial cell membrane and disturb its normal functioning. They either accumulate in the cytoplasm or in the periplasmic region resulting in cell membrane disruption or disorganization of membranes followed by the release of the cell contents and disturbance in power function. The disorganization of the cell membrane may also cause the formation of pits and holes on cell surface leading to the loss of minerals, proteins, and genetic and other cytoplasmic materials. The disturbance of cell membranes may involve the binding of nanoparticles to sulfur-containing proteins in the bacterial cell membrane (Singh et al. 2013).

It has also been reported that silver nanoparticles specifically target pathways of protein synthesis, nucleic acid synthesis, and also the synthesis of bacterial cell wall (Jung et al. 2008). The change in the structure of bacterial membrane as well as damaging the DNA may affect the respiratory chain, cell division, and DNA replication and finally leads to cell death (Andrade et al. 2016). The silver nanoparticles when dispersed in an aqueous medium can release silver ions which may contribute to the antimicrobial activity of metal nanoparticles (Kim et al. 2009).

The activity of ZnO nanoparticles was enhanced when suspended in water. It is due to the dissolution of ZnO in water to release Zn^{2+}ions (Brayner et al. 2006). It

Table 8.1 Nanoparticles obtained from endophytic fungi and their biological activity

Endophytic fungi	Host plant	Metal	Size (nm)	Activity	Reference
Alternaria solani	*Solanum lycopersicum*	AgNps	14.8 ± 1.2	Antifungal	Hafez et al. (2016)
Alternaria sp.	*Raphanus sativus*	AgNps	4–30	Antibaterial	Singh et al. (2017)
Amylomyces rouxii	*Phoenix dactylifera*	AgNps	5–27	Antimicrobial	
Aspergillus clavatus	*Azadirachta indica*	AgNps		Antifungal, antibacterial	Verma et al. (2010)
Aspergillus conicus	*Avicennia marina*		6–12	Antibacterial	Bharathidasan and Panneerselvam (2012)
Alternaria solani	*Gloriosa superba*	AgNps	5–20	Antifungal, antibacterial	Devi et al. (2014)
Aspergillus Niger	*Centella asiatica*	AgNps	10–50	Antibacterial	Netala et al. (2015)
Aspergillus terreus	*Calotropis procera*	AgNps	16.54	Antibacterial	Li et al. (2012)
Aspergillus terreus	*Rhizophora annamalayana*	AfNps	100	Antidermatophytic	Rathna et al. (2013)
Aspergillus versicolor	*Centella asiatica*	AgNps	3–40	Antibacterial and antifungal	Netala et al. (2016)
Cladosporium sp.	*Melia dubia*	nNps	50–100	–	Vidhya and Rajgopal (2016)
Cladosporium sphaermum	Tomato	AgNps	15.1–41.2	–	Hafez et al. (2016)
Colletotrichum sp.	*Pelargonium graveolens*	GFold		–	Shivshankar et al. (2003)
Curvularia lunata	*Catharanthus roseus*	AgNps	10–50	Antibacterial, antimicrobial	Ramalingmam et al. (2015)
Epicoccum nigrum	*Phellodendrom amurense*	AgNps		Antifungal	Qian et al. (2013)
Epicoccum nigrum	*Lycopersicon esculentum*	AgNps	37	Antifungal, antioxidant	Hafez et al. (2017)
Fusarium solani	*Withania somnifera*	AgNps	10–50	Antibacterial, cytotoxic	Vijayan et al. (2016)
Fusarium sp.	*Withania somnifera*			Antibacterial	Singh et al. (2015)
Penicillium citrinum	Brown algae	Gold			Manjunath et al. (2017)
Penicillium funiculosum	*Gloriosa superba*	AgNps	5–20	Antibacterial, antifungal	Devi et al. (2014)
Penicillium nodositatum	*Randia dumetyorum*	AgNps	100	Antibacterial	Manjunath et al. (2014)

Penicillium sp.	*Curcuma longa*			Antibacterial	Singh et al. (2014)
Penicillium sp.	*Glycosmis mauritiana*	AgNps		Antiimflammatory, antioxidant, antimicrobial	Govindappa et al. (2016)
Penicillium sp.	*Centella asiatica*	AgNps	20–200	Antibacterial	Nameirakpam et al. (2012)
Penicillium sp.	*Aloe vera*		16.1–39.6	Antibacterial	Rahi and Parmar (2014)
Penicillium spinulosum	*Curcuma longa, Catharanthus roseus*	AgNps	25–30	–	Singh et al. (2016)
Pestalotia sp.	*Madhuca longifolia*	AgNps	<40	Antibacterial and antioxidant	Verma and Gond (2016)
Pestalotiopsismicrospora	*Gymnema sylvestre*	AgNps		Anticancer	Netala et al. (2016)
Trichoderma harzianum	Tomato	AgNps	12.7 ± 0.8	Antifungal	Moslamy et al. (2017)
Guignardia mangiferae	*Citrus* sp.	AgNps	5–30	Antifungal, antibacterial	Balakumaran et al. (2015)
Pestalotia sp.	*Syzygium cumuni*	AgNps	10–40	Antibacterial	Raheman et al. (2011)
Syncephalis sp.	*Sida acuta*	AgNps		Antibacterial	Rahi and Parmar (2014)

has been suggested that silver ions generated from the dissolution of silver nanoparticles become complexed with electron donor groups consisting of sulfur, oxygen, or nitrogen atoms present in thiols or phosphates on amino acids and nucleic acids (Louise et al. 2010). NP production may be affected by inactivation of the enzyme due to the denaturing of a 30S subunit of ribosomes (Chauhan et al. 2013). Thus, the obstruction of respiratory enzymes causes an elevated level of reactive oxygen species (ROS) (Onodera et al. 2015).

Biogenic synthesis of nanoparticles is eco-friendly and cost-effective. They are produced from different microbes and plants. Biomolecules can be attached to various types of surfaces via diffusion, adsorption/absorption, covalent cross-linking, and affinity interaction (Choi and Hu 2008). Stabilization of nanoparticles is achieved by adding a capping agent, which is essential to prevent aggregation of nanoparticles, increase the solubilities of the nanosystem which is an important step in decreasing the toxicity, and also can be used as a site for bioconjugation of the nanoparticles with molecules (Prabhu and Poulose 2012).

The physical and chemical characters of nanoparticles play a significant role in their effect on biological systems (De Lima et al. 2012). The nature (chemical and biogenic) and size of the nanoparticles, its surface area, and surface functionalization (capping agents) are important factors that influence biokinetics and toxicity (Prabhu and Poulose 2012). The exposure to capping agents and their presence in a system also play a key factor in in vitro exhibition of cytotoxicity of silver nanoparticles (AgNPs) (De Lima et al. 2012). Silver nanoparticles synthesized by biological methods and biocompatibility of capping agents indicated no cytotoxicity in eukaryotic cells (Donaldson 2004).

The cytotoxic effect of silver nanoparticles (AgNPs) on mammalian cells depends on dose and size of nanoparticles and also nature of treated cells (Zhang et al. 2014). It has been reported that small-sized nanoparticles are more toxic compared with the larger ones (Manjunath et al. 2014). Studies on human health have demonstrated that nanoparticles have toxic effect at the cellular, subcellular, and biomolecular level (Zhang et al. 2014). Nanoparticles are shown to be preferentially localized in mitochondria inducing structural and functional damage as well as oxidative stress (Arora et al. 2008). The functional damage of the mitochondria could affect cellular metabolic inhibition followed by a decrease in nanoparticle yield (Choi and Hu 2008). Oxidative stress is a common mechanism for cell damage induced by nanoparticles leading to oxidative damage of proteins and DNA (Onodera et al. 2015; Aziz et al. 2019). Balakumaran et al. (2015) studied endophytic fungus *Gugnardia mangiferae* for the extracellular synthesis of silver nanoparticles and in vitro biological activities. Cytotoxic effects of AgNPs showed IC50 values of 63.37, 27.54, and 23.84 μg/ml against normal African monkey kidney (vero), Hela (cervical), and F-7 (breast) cells, respectively. Cell viability assays of NIH3T3 mouse embryonic fibroblast cell line and DAB-231 human breast carcinoma cell line after treatment with mycoendophyte-derived AgNPs were carried out by Syed et al. (2013).

8.10 Conclusion and Future Perspectives

The development of resistance to antibiotics by human pathogens is a major threat in the field of pharmaceuticals and biomedicine. Nanotechnology provides an excellent platform to develop antimicrobials having potential applications in the medical field. Nanoparticle synthesis from endophytic fungi has several advantages over other methods which include highly reproducible synthesis, large-scale synthesis, higher metal tolerance, and good dispersion in the medium. Silver nanoparticles show great promise in biomedical applications. Application of silver nanoparticles in drug delivery system and diagnostics attracted researchers to study and investigate safe eco-friendly approaches using endophytic fungi for large-scale production of nanoparticles. It is clear from the present chapter that various sized monodispersed nanoparticles synthesized from fungal endophytes show antimicrobial properties against various pathogens and they can be employed as nanomedicines to treat diseases.

There is need to explore endophytic fungi from large hosts for production of monodispersed nanoparticles with defined composition. Research should be aimed at large-scale production of defined nanoparticles from endophytes and optimization of growth conditions such as pH, temperature, culture age, biomass, and salt concentration. Investigation should also be focused on exact mechanism of microbially synthesized nanoparticles and the role of reductases/biomolecules/proteins and also stabilizing agents in synthesis and characterization of nanoparticles and their applications in various fields. Efforts are also needed to minimize metal toxicity and improve biocompatibility and effectiveness.

References

Abdel-Hafez SII,Nafady NA, Abdel-Rahim IR, Shaltout AM, J Daròs JA, Mohamed MA (2016a) Assessment of protein silver nanoparticles toxicity against pathogenic Alternaria solani. 3 Biotech 6(2):1–12

Abdel-Hafez SII, Nafady NA, Abdel-Rahim IR, Shaltout AM, Mohamed MA (2016b) Biogenesis and optimisation of silver nanoparticles by the endophytic fungus *Cladosporium sphaerospermum*. Int J Nano Chem 2(1):11–19

Abdel-Hafez SII, Nafady NA, Abdel-Rahim IR, Shaltout AM, Dars JA, Mohamed MA (2017) Biosynthesis of Silver Nanoparticles Using the Compound Curvularin isolated from the Endophytic Fungus Epicoccum Nigrum: Characterization and Antifungal activity. Journal of Pharmaceutical and Applied Chemistry 3(2):135–146

Abdel-Aziz SM, Prasad R, Hamed AA, Abdelraof M (2018) Fungal nanoparticles: A novel tool for a green biotechnology? In: Fungal Nanobionics: Principles and Applications (eds. Prasad R, Kumar V, Kumar M, Wang S), Springer Singapore Pte Ltd. 61–87

Ahmad A, Senapati S, Khan MI, Kumar R, Ramani R, Srinivas V, Sastry M (2003) Intracellular synthesis of gold nanoparticles by a novel alkalotolerant actinomycete, *Rhodococcus* species. Nanotechnology 14:824–828

Ahmed S, Ahmad M, Swami BL, Ikram S (2016) A review on plant extract mediated synthesis of silver nanoparticles for antimicrobial applications: a green expertise. J Adv Res 7(1):17–28

Alt V, Bechert T, Steinrucke P, Wagener M, Siedel P, Dingeldein E, Scheddin D, Domann E, Schnettler R (2004) Nanoparticulate silver: a new antimicrobial substance for bone cement. Orthopade 33:885–892

Alzahrani E, Sharfalddin A, Alamodi M (2015) Microwave hydrothermal synthesis of ferric oxide doped with cobalt. Adv Nanopart 4:53–60

Amulyavichus A, Daugvila R, Davidonis R, Sipavichus C (1998) Study of chemical composition of nanostructural materials prepared by laser cutting of metals. Fizika Met 85:111–117

Andrade PF, de Faria AF, Oliveira SR, Goncalves MDC, Arruda MAZ (2016) Improved antibacterial activity of nanofiltration polysulfone membranes modified with silver nano particles. Water Res 81:333–342

Arora S, Jain J, Rajwade JM, Paknikar KM (2008) Cellular responses induced by silver nanoparticles: *in vitro* studies. Toxicol Lett 79:93–100

Aziz N, Fatma T, Varma A, Prasad R (2014) Biogenic synthesis of silver nanoparticles using Scenedesmus abundans and evaluation of their antibacterial activity. Journal of Nanoparticles, Article ID 689419, https://doi.org/10.1155/2014/689419

Aziz N, Faraz M, Pandey R, Sakir M, Fatma T, Varma A, Barman I, Prasad R (2015) Facile algae-derived route to biogenic silver nanoparticles: Synthesis, antibacterial and photocatalytic properties. Langmuir 31:11605–11612 DOI: https://doi.org/10.1021/acs.langmuir.5b03081

Aziz N, Pandey R, Barman I, Prasad R (2016) Leveraging the attributes of Mucor hiemalis-derived silver nanoparticles for a synergistic broad-spectrum antimicrobial platform. Front Microbiol 7:1984. doi: https://doi.org/10.3389/fmicb.2016.01984

Aziz N, Faraz M, Sherwani MA, Fatma T, Prasad R (2019) Illuminating the anticancerous efficacy of a new fungal chassis for silver nanoparticle synthesis. Front Chem 7:65. doi: https://doi.org/10.3389/fchem.2019.00065

Balakumaran MD, Ramachandran R, Kalaichelvan PT (2015) Exploitation of endophytic fungus *Guignardia mangiferae* for extracellular synthesis of silver nanoparticles and their *in vitro* biological activities. Microbiol Res 178:9–17

Bansal V, Rautray D, Ahmad A, Sastry M (2004) Biosynthesis of zirconia nanoparticles using the fungus *Fusarium oxysporum*. J Mat Chem 14:3303–3305

Barik TK, Sahu B, Swain V (2008) Nano-silica medicine to pest control. Parasitol Res 103:253–258

Bharathidasan R, Panneerselvam A (2012) Biosynthesis and characterization of silver nanoparticles using endopyhtic fungus *Aspergillus conius*, *Penicillium janthinellum,* and *Phomopsis* sp. Int J Pharm Sci Res 3:3163–3169

Bhattacharya R, mukherjee P (2008) Biological properties of naked nanoparticles. Adv Drug Del Rev 60:1289–1306

Bhattacharyya A, Duraisamy P, Govindarajan M, Buhroo AA, Prasad R (2016) Nano-biofungicides: Emerging trend in insect pest control. In: Advances and Applications through Fungal Nanobiotechnology (ed. Prasad R), Springer International Publishing Switzerland 307–319

Birla SS, Tiwari VV, Gade AK et al (2009) Fabrication of silver nanoparticles by *Phoma glomerata* and its combined effect against *Escherchia coli, Pseudomonas auruginosa* and *Staphylococcus aureus*. Lett App Microbiol 48:173–179

Bowman SM, Free JF (2011) The structure and synthesis of the fungal cell-wall. Bio Essays 28:799–808

Brayner R, Ferrari-Iliou R, Brivois N, Djediyat S, Benedetti MF, Flevet F (2006) Toxicological impact studies based on *Escherichia coli* bacteria in ultrafine ZnO nanoparticles colloidal medium. Nano Lett 6:866–870

Chan YS, Mashitah MD (2012) Instantaneous biosynthesis of silver nanoparticles by selected macrofungi. Aust J Basic Apl. Sci 6: 222–226

Chauhan R, Kumar A, Abraham J (2013) A biological approach to the synthesis of silver nanoparticles with *Streptomyces* sp. JAR1 and its antimicrobial activity. Sci Pharm 81:607–621

Chen N, Zheng Y, Yin J et al (2013) Inhibitory effects of silver nanoparticles against adenovirus type 3 in vitro. J ViroIMeth 193:470–477

Chinnamuthu CR, Boopathi PM (2009) Nanotechnology and agroecosystem. Madras Agric J 96:17–31
Choi O, Hu Z (2008) Size dependent and reactive oxygen species related nanosilver toxicity to nitrifying bacteria. Environ Sci Technol 42:4583–4588
Chong MN, Jin B, Chow CWK, Saint C (2010) Recent developments in photocatalytic water treatment technology: a review. Water Res 44:2997–3027
Cohen MS, Stern JM, Vanni AJ, Kelley RS, Baumgart E, Field D, Libertino JA, Summerhayes IC (2007) In vitro analysis of a nanocrystalline silver-coated surgical mesh. Surg Infect 8:397–403
Corradini E, de-Moura MR, Mattoso LHC (2010) A preliminary study of the incorporation of NPK fertilizer into chitosan nanoparticles. Poly Lett 4:509–515
Das SK, Das AR, Guha AK (2009) Gold nanoparticles: microbial synthesis and application in water hygiene management. Langmuir 25:8192–8199
De Lima R, Seabra AB, Duran N (2012) Silver nanoparticles: a brief review of cytotoxicity and genotoxicity of chemically and biogenically synthesized nanoparticles. J Appl Toxicol 32:867–879
Devi LS, Joshi SR (2012) Antimicrobial and synergistic effects of silver nanoparticles synthesized using soil fungi of high altitudes of eastern himalaya. Mycobiology 40:27–34
Devi LS, Bareh DA, Joshi SR (2014) Studies on biosynthesis of antimicrobial silver nanoparticles using endophytic fungi isolated from the ethno-medicinal plant *Gloriosa superba* L. Proc Natl Acad Sci India Section B Biol Sci 84:1091–1099
Dhillon GS, Brar SK, Kaur S, Verma M (2012) Green approach for nanoparticles biosynthesis by fungi: current trends and applications. Crit Rev Biotechnol 32:49–73
Donaldson K (2004) Nanotoxicology. Occup Environ Med 61:727–728
Duran A, Nombela C (2004) Fungal cell-wall biogenesis: building a dynamic interface with the environment. Microbiology 150:3099–3103
Duran N, Marcato PD, Duran M et al (2011) Mechanistic aspects in the biogenic synthesis of extracellular metal nanoparticles by peptides, bacteria, fungi and plants. App Microbiol Biotechnol 90:1609–1624
Elechiguerra JL, Burt JL, Morones JR, Camacho-Bragado A, Gao X, Lara HH, Yacaman MJ (2005) Interaction of silver nanoparticles with HIV. J Nanobiotechnol 3:1–10
EL-Moslamy SH, Elkady MF, Rezk AH, Abdel-Fattah YR (2017) Applying Taguchi design and large-scale strategy for mycosynthesis of nano-silver from endophytic *Trichoderma harzianum* SYA. F4 and its application against phytopathogens. Sci Rep 7:45297. https://doi.org/10.1038/srep45297
Farrukh A, Akram A, Ghaffar A, Yameen B (2013) Design of polymer-brush-grafted magnetic nanoparticles for highly efficient water remediation. ACS Appl Mater Interfaces 5:37843793
Fong J, Wood F (2006) Nanocrystalline silver dressings in wound management: a review. Int J Nanomedicine 1:441–449
Franci G, Falanga A, Galdiero S, Palomba L, Rai M, Morelli G, Galdiero M (2015) Silver nanoparticles as potential antibacterial agents. Molecules 20:8856–8874
Gajbhiye MB, Kesharwani JG, Ingle AP, Gade AK, Rai M (2009) Fungus mediated synthesis of silver nanoparticles and its activity against pathogenic fungi in combination of fluconazole. Nanomed 5:382–386
Galiano K, Pleifer C, Engelhardt K, Brossner G, Lackner P, Huck C, Lass-Flörl C, Obwegeser A (2008) Silver segregation and bacterial growth of intraventricular catheters impregnated with silver nanoparticles in cerebrospinal fluid drainages. Neurol Res 30:285–287
Goel A (2015) Agricultural applications of nanotechnology. J Biol Chem Res 32:260–266
Golinska P, Wypij M, Rathod S, TIKAR S, Dahm H, Rai M (2015) Synthesis of silver nanoparticles from two acidophilic strains of *Pilimelia columellifera* subsp. *pallida* and their antibacterial activities. J Basic Microbiol 56:541–556
Golinska P, Rathod D, Wypij M, Gupta I, Składanowski M, Paralikar P, Dahm H & Rai M Mycoendophytes as efficient synthesizers of bionanoparticles: nanoantimicrobials, mechanism, and cytotoxicity. Crit Rev Biotechnol 37(6):765–778

Goswami A, Roy I, Sengupta S, Debnath N (2010) Novel applications of solid and liquid formulations of nanoparticles against insect pests and pathogens. Thin Solid Films 519:1252–1257

Govindappa M, Farheen H, Chandrappa CP, Channabasava RRV, Raghavendra VB (2016) Mycosynthesis of silver nanoparticles using extract of endophytic fungi, *Penicillium* species of *Glycosmis mauritiana,* and its antioxidant, antimicrobial, anti-inflammatory and tyrokinase inhibitory activity. Adv Nat Sci Nanosci Nanotechnol 7:035014

He M, Protesescu L, Caputo R et al (2015) A general synthesis strategy for monodisperse metallic and metalloid nanoparticles (in, Ga, bi, Sb, Zn, cu, Sn and their alloys) via *in situ* formed metal long-chain amides. Chem Mater 27:635–647

Holan ZR, Volesky B (1995) Accumulation of cadmium, lead and nickel by fungal and wood biosorbents. App Biochem Biotechnol 53(2):133–146

Hu RL, Li SR, Kong FJ et al (2014) Inhibition effect of silver nanoparticles on herpes simplex virus 2. Gene Mol Res 13:7022–7028

Huang YF, Chang HT, Tan WH (2008) Cancer cell targeting using multiple aptamers conjugated on nanorods. Anal Chem 80:567–572

Ingle A, Gade A, Pierrat S et al (2008) Mycosynthesis of silver nanoparticles using the fungus *Fusarium acuminatum* and its activity against some human pathogenic bacteria. Curr Nanosci 4:141–144

Jain N, Bhargava A, Majumdar S et al (2011) Extracellular biosynthesis and characterization of silver nanoparticles using *Aspergillus flavus* NJP08: a mechanism perspective. Nanoscale 3:635–641

Johnson R, Sabnis N, McConathy WJ, Lacko AG (2013) The potential role of nanotechnology in therapeutic approaches for triple negative breast cancer. Pharmaceutics 5:353–370

Jung WK, Koo HC, Kim KW et al (2008) Antibacterial activity and mechanism of action of the silver ion in *Staphylococcus aureus* and *Escherichia coli*. Appl Environ Microbiol 74:2171–2178

Kar P, Murmu S, Saha S et al (2014) Anthelmintic efficacy of gold nanoparticles derived from a phytopathogenic fungus, *Nigrospora oryzae*. PLoS One 9:84693

Kashyap PL, Kumar S, Srivastava AK, Sharma AK (2013) Myconanotechnology in agriculture: a perspective. World J Microbiol Biotechnol 29:191–207

Kim S, Choi JE, Cho J et al (2009) Oxidative stress-dependent toxicity of silver nanoparticles in human hepatomacells. Toxicol In Vitro 23:1076–1084

Kim SW, Jung JH, Lamsal K et al (2012) Antifungal effects of silver nanoparticles (AgNPs) against various plant pathogenic fungi. Mycobiology 40:53–58

Kitching M, Ramani M, Marsili E (2015) Fungal biosynthesis of gold nanoparticles: mechanism and scale up. Microb Biotechnol 8:904–917

Klasen HJ (2000) A historical review of the use of silver in the treatment of burns II. Renewed interest for silver. Burns 26:131–138

Kumar SA, Abyaneh MK, Gosavi SW et al (2007) Nitrate reductase-mediated synthesis of silver nanoparticles from $AgNO_3$. Biotechnol Lett 29:439–445

Li G, He D, Qian Y et al (2012) Fungus mediated green synthesis of silver nanoparticles using *Aspergillus terreus*. Int J Mol Sci 13:466–476

Louise R, Zhou X, Tang L et al (2010) Immobilization of living bacteria for AFM imaging under physiological conditions. Ultramicroscopy 110:1349–1357

Madigan MT, Martinko JM, Brock TD (2006) Brock biology of microorganisms. Prentice Hall, Upper Saddle River, p 992

Manjunath HM, Joshi CG, Raju NG (2014) Biogenic synthesis of silver nano particles using endophytic fungi *Penicillium nodositatum* and its antibacterial activity. J Chem Pharm Res 6:112–117

Manjunath HM, Joshi CG, Raju NG (2017) Biofabrication of gold nanoparticles using marine endophytic fungus – *Penicillium citrinum*. IET Nano biotechnol 11(1):40–44

Mehra RK, Winge DR (1991) Metal ion resistance in fungi: molecular mechanisms and their regulated expression. J Cell Biochem 45:30–40

Mermel LA (2000) Prevention of intravascular catheter-related infections. Ann Intern Med 132:391–402

Millan G, Agosto F, Vazquez M (2008) Use of clinoptilolite as a carrier for nitrogen fertilizers in soils of the Pampean regions of Argentina. Cien Inv Agr 35:293–302

Mohanpuria P, Rana KN, Yadav SK (2008) Biosynthesis of nanoparticles: technological concepts and future applications. J Nanopart Res 10:507–517

Mukherjee P, Roy M, Mandal BP et al (2008) Green synthesis of highly stabilized nanocrystalline silver particles by a non-pathogenic and agriculturally important fungus *T. asperellum*. Nanotechnology 19:103–110

Nanda A, Majeed S (2014) Enhanced antibacterial efficacy of biosynthesized AgNPs from *Penicillium glabrum* (MTCC1985) pooled with different drugs. Int J Pharm Tech Res 6:217–223

Narayanan KB, Sakthivel N (2010) Biological synthesis of metal nanoparticles by microbes. Adv Colloid Interf Sci 156:1–13

Netala VR, Bobbu P, Ghosh SB et al (2015) Endophytic fungal assisted synthesis of silver nanoparticles, characterization, and antimicrobial activity. Asian J Pharm Clin Res 8:113–116

Netala VR, Kotakadi VS, Bobbu P, Gaddam SA, Tartte V (2016) Endophytic fungal isolate mediated biosynthesis of silver nanoparticles and their free radical scavenging activity and anti microbial studies. 3Biotech 6(2):132

Nameirakpam ND, Dheeban SP, Sutha S (2012) Biomimetic synthesis of silver nanoparticles from an endophytic fungus and their antimicrobial efficacy. Int J Biomed Adv Res 3(5):309–315

Onodera A, Nishiumi F, Kakiguchi K et al (2015) Short-term changes in intracellular ROS localization after the silver nanoparticles exposure depending on particle size. Toxicol Rep 2:574–579

Ortega FG, Fernandez-Baldo MA, Fernandez JG et al (2015) Study of antitumor activity in breast cell lines using silver nanoparticles produced by yeast. Int J Nanomedicine 10:2021–2031

Peterson MSM, Bouwman J, Chen A, Deutsch M (2007) Inorganic metallodielectric materials fabricated using two single-step methods based on the Tollens' process. J Colloid Interface Sci 306:41–49

Prabhu S, Poulose EK (2012) Silver nanoparticles: mechanism of antimicrobial action, synthesis, medical applications, and toxicity effects. Int Nano Lett 2(32)

Prasad R (2016) Advances and Applications through Fungal Nanobiotechnology. Springer, International Publishing Switzerland (ISBN: 978-3-319-42989-2)

Prasad R (2017) Fungal Nanotechnology: Applications in Agriculture, Industry, and Medicine. Springer Nature Singapore Pte Ltd. (ISBN 978-3-319-68423-9)

Prasad R, Pandey R, Barman I (2016) Engineering tailored nanoparticles with microbes: quo vadis. WIREs Nanomed Nanobiotechnol 8:316–330. doi: https://doi.org/10.1002/wnan.1363

Prasad R, Kumar V, Kumar M, Wang S (2018) Fungal Nanobionics: Principles and Applications. Springer Nature Singapore Pte Ltd. (ISBN 978-981-10-8666-3) https://www.springer.com/gb/book/9789811086656

Prasad R, Jha A, Prasad K (2018a) Exploring the Realms of Nature for Nanosynthesis. Springer International Publishing (ISBN 978-3-319-99570-0) https://www.springer.com/978-3-319-99570-0

Qian Y, Yu H, He D, Yang H, Wang W, Wan X, Wang L (2013) Biosynthesis of silver nanoparticles by the endophytic fungus *Epicoccum nigrum* and their activity against pathogenic fungi. Bioprocess Biosyst Eng 36:1613–1619

Raghupathi KR, Koodali RT, Manna AC (2011) Size-dependent bacterial growth inhibition and mechanism of antibacterial activity of zinc oxide nanoparticles. Langmuir 27:4020–4028

Raheman F, Deshmukh S, Ingle A et al (2011) Silver nanoparticles: novel antimicrobial agent synthesized from an endophytic fungus *Pestalotia* sp. isolated from leaves of *Syzygium cumini* (L). Nano Biomed Eng 3:174–178

Rahi DK, Parmar AS (2014) Mycosynthesis of silver nanoparticles by an endophytic *Penicillium* species of *Aloe vera* root, evaluation of their antibacterial and antibiotic enhancing activity. Int J Nanomater Biostruct 4:46–51

Rai M, Yadav A, Gade A (2008) Current trends in phytosynthesis of metal nanoparticles. Crit Rev Biotechnol 28:277–284

Rai M, Yadav A, Bridge P, Gade A (2009) Myconanotechnology: a new and emerging science. In: Rai MK, Bridge PD (eds) Applied mycology. CAB International, New York, pp 258–267

Rai V, Acharya S, Dey N (2012) Implications of nanobiosensors in agriculture. J Biomater Nanobiotechnol 3:315–324

Rai M, Ingle AP, Gade A, Duran N (2015) Synthesis of silver nanoparticles by *Phoma gardeniae* and *in vitro* evaluation of their efficacy against human disease causing bacteria and fungi. IET Nanobiotechnol 9:71–75

Rajan CS (2011) Nanotechnology in groundwater remediation. Int J Envir Sci Dev 2:182–187

Ramalingmam P, Muthukrishnan S, Thangaraj P (2015) Biosynthesis of silver nanoparticles using an endophytic fungus, *Curvularia lunata* and its antimicrobial potential. J Nanosci Nanoeng 1:241–247

Rathna GS, Elavarasi A, Peninal S et al (2013) Extracellular biosynthesis of silver nanoparticles by endophytic fungus *Aspergillus terreus* and its anti-dermatophytic activity. Int J Pharm Biol Arch 4:481–487

Ricco JB (2006) Inter Gard silver bifurcated graft: features and results of a multicenter clinical study. J Vasc Surg 44:39–46

Riddin TL, Gericke M, Whiteley CG (2006) Analysis of the inter- and extracellular formation of platinum nanoparticles by *Fusarium oxysporum* f. Sp. *lycopersici* using response surface methodology. Nanotechnology 17:3482–3489

Sanghi R, Verma P (2009) Biomimetic synthesis and characterization of protein capped silver nanoparticles. Bioresour Technol 100:502–504

Singh R, Shedbalkar UU, Wadhwani SA, Chopade BA (2015) Bacteriagenic silver nanoparticles: synthesis, mechanism, and applications. Appl Microbiol Biotechnol 99:4579–4593

Singh D, Rathod V, Ninganagouda S, et al (2013) Biosynthesis of silver nanoparticle by endophytic fungi Pencillium sp. isolated from Curcuma longa (turmeric) and its antibacterial activity against pathogenic Gram-negative bacteria. J Pharm Res. 7:448–453

Singh D, Rathod V, Ninganagouda S, Hiremath J, Singh AK, Mathew J (2014) Optimization and characterization of silver nanoparticle by endophytic Fungi sp. isolated from (Turmeric) and application studies against MDR *E. coli* and *S. aureus*. Bioinorg Chem Appl 1:1–8

Singh T, Jyoti K, Patnaik A, Singh A, Chauhan R, Chandel SS (2017) Biosynthesis, characterization and antibacterial activity of silver nanoparticles using an endophytic fungal supernatant of *Raphanus sativus*. J Gen Eng Biotechnol 15(1):31–39

Shanmukh S, Jones L, Zhao YP, Driskell JD, Tripp RA, Dluhy RA (2008) Identification and classification of respiratory syncytial virus (RSV) strains by surface-enhanced Raman spectroscopy and multivariate statistical techniques. Anal Bioanal Chem 390:1551–1555

Sharma VK, Yngard RA, Lin Y (2009) Silver nanoparticles: green synthesis and their antimicrobial activities. Adv Colloid Interf Sci 145:83–96

Sharma S, Kumar S, Bulchandani BD, Taneja S, Banyal S (2013) Green synthesis of silver nanoparticles and their antimicrobial activity against gram positive and gram negative bacteria. Int J Biotechnol Bioeng Res 7:711–714

Shivshankar SS, Ahmad A, Pasricha R et al (2003) Bioreduction of chloroaurate ions by *Geranium* leaves and its endophytic fungus yields gold nanoparticles of different shapes. J Mater Chem 13:1822–1826

Shrivastava S, Bera T, Singh SK (2009) Characterization of antiplatelet properties of silver nanoparticles. ACS Nano 3:1357–1364

Sibbald RJ, Contreras-Ruiz J, Coutts P, Fierheller M, Rothman A, Woo K (2007) Bacteriology, inflammation, and healing: a study of nanocrystalline silver dressings in chronic venous leg ulcers. Adv Skin Wound Care 20:549–558

Sondi I, Sondi B (2004) Silver nanoparticles as antimicrobial agent: a case study on *E. coli* as a model for gram-negative bacteria. J Colloid Interface Sci 275:177–182

Subbarao V, Kartheek V, Sirisha D (2013) Slow release ofpotash fertilizer through polymer coating. Int J Appl SciEng 11(1):25

Sun X, Luo Y (2005) Preparation and size control of silver nanoparticles by a thermal method. MaterLett 59:3847–3850

Sun RW, Chen R, Chung NP, Ho CM, Lin CL, Che CL (2005) Silver nanoparticles fabricated in Hepes buffer exhibit cytoprotective activities toward HIV-1 infected cells. Chem Commun 40:5059–5061

Sunkar S, Nachiyar CV (2012) Microbial synthesis and characterization of silver nanoparticles by using endophytic bacterium *Bacillus cereus*: a novel source in benign synthesis. Glob J Med Res 2(12):953–959

Syed A, Saraswati S, Kundu GC et al (2013) Biological synthesis of silver nanoparticles using the fungus *Humicola* sp. and evaluation of their cytoxicity using normal and cancer cell lines. Spectrochim Acta Part A Mol Biomol Spectrosc 14:144–147

Thakkar KN, Mhatre SS, Parikh RY (2010) Biological synthesis of metallic nanoparticles. Nanomedicine 6:257–262

Telling ND, Coker VS, Cutting RS, Laan G, Pearce E, Pattrick RAD, Arenholz E, Lloyd JR (2009) Remediation of Cr(VI) by biogenic magnetic nanoparticles: an X-ray magnetic circular dichroism study. Appl Phys Lett 95(16). https://doi.org/10.1063/1.3249578

Vago A, Szakacs G, Safran G, Horvath R, Pecz B, Lagzi I (2016) One-step green synthesis of gold nanoparticles by mesophilic filamentous fungi. Chem Phys Lett 645:1–4

Vahabi K, Mansoori GA, Karimi S (2011) Biosynthesis of silver nanoparticles by the fungus *Trichoderma reesei*. Insci J 1:65–79

Verma VC, Kharwar RN, Gange AC (2010) Biosynthesis of antimicrobial nanoparticles by the endophytic fungus *Aspergillus clavatus*. Nanomedicine 5(1):33–40

Verma SK,Gond SK (2016) Biofabrication of antibacterial and antioxidant silver nanoparticles (AgNps) by an endophytic fungus *Pestalotia* sp. isolated from *Madhuca longifolia*. J Nanomat Mol Nanotechnol 5(4): doi:10.4172/2324-8777.1000189

Vijayan S, Divya K, George TK, Jisha MS (2016) Biogenic synthesis of silver nanoparticles using endophytic fungi *Fusarium oxysporum* isolated from *Withania somnifera*, its antibacterial and cytotoxic activity. J Bionanosci 10:1–8

Vidhya S, Rajagopal K (2016) Isolation and identification of thermo tolerant endophytic fungi from Melia dubia and synthesis of Zinc nanoparticles. Int J NanoSci Nanotechnol 7: (2) 99–112

Wang L, Li Z, Zhang G et al (2007) Oil-in-water nanoemulsions for pesticide formulations. J Colloid Interface Sci 314:230–235

Wong KK, Cheung SO, Huang LM, Niu J, Tao C, Ho CM, Che CM, Tam PK (2009) Further evidence of the anti-inflammatory effects of silver nanoparticles. Chem Med Chem 4:1129–1135

Xiang DX, Chen Q, Pang L et al (2011) Inhibitory effects of silver nanoparticles on H1N1 influenza a virus in vitro. J Virol Methods 178:137–142

Yadav A, Kon K, Kratosova G, Duran N, Avinash P, Ingle RM (2015) Fungi as an efficient mycosystem for the synthesis of metal nanoparticles: progress and key aspects of research. Biotechnol Lett 37:2099–2120

Zhang WX (2003) Nanoscale iron particles for environmental remediation: an overview. J Nanopart Res 5:323–332

Zhang X, Yan S, Tyagi RD, Surampalli RY (2011) Synthesis of nanoparticles by microorganisms and their application in enhancing microbiological reaction rates. Chemosphere 82:489–494

Zhang T, Wang L, Chen Q et al (2014) Cytotoxic potential of silver nanoparticles. Yonsei Med J 55:283–291

Chapter 9
Applications for Nanotechnology in the Polyphagous Destructive Insect Pest Management of Agricultural Crops

Sunil Kumar Dwivedi and Ajay Tomer

Contents

9.1 Introduction

To reduce environmental pollution and harmful effects to the human body, new formulations have been developed. One of the most promising is the use of nanotechnology, which has a highly proficient grouping of an active compound in a medium that guards the core materials from undesirable reactions. Nanoformulations differ from conventional formulations principally in magnitude and surface characteristics (Prasad et al. 2019). Nanoparticles show promise for the well-organized and efficient control of pests; however, there is a lack of knowledge regarding their actions and practical use.

S. K. Dwivedi (✉)
Department of Entomology, School of Agriculture, Lovely Professional University, Phagwara, Punjab, India
e-mail: sunil.21186@lpu.co.in

A. Tomer
Department of Plant Pathology, School of Agriculture, Lovely Professional University, Phagwara, Punjab, India

D. Thangadurai et al. (eds.), *Nanotechnology for Food, Agriculture, and Environment*, Nanotechnology in the Life Sciences,
https://doi.org/10.1007/978-3-030-31938-0_9

There is a regulatory break before the discharge of nanoparticles into the environment, which is useful for farmers. Noteworthy differences can be found among nanotechnology-based formulations. Nanotechnology is the use of materials which measure between approximately 1 and 100 nanometers (nm), allowing for novel applications. A nanomaterial is one-billionth of a meter. Nanoscale materials have unique physical, chemical, and natural properties, which are totally distinct from their mass materials and individual atoms (Li et al. 2001). These extraordinary properties can be found in new applications in all fields. Nanoparticles have expansive surface-to-volume ratios, synthetically alterable physical characteristics, and solid attraction to targets, such as proteins (Kumar et al. 2010). Because of global environmental concerns such as climate change, urbanization, and ecological issues such as run-off and collection of agrochemicals, there is an increasing need to address the expected population growth from 6 billion to 9 billion by 2050 (Chen and Yada 2011). With constrained resources such as land, water, and soil, interest in sustainability has expanded greatly. The cost of synthetic manures, pesticides, and other necessities has also increased because of limited supplies of gasoline and oil (Ditta 2012). There is a need to overcome these obstacles with cultivation improvements and the successful use of nanotechnology in horticulture.

9.2 Role of Nanotechnology in Entomology and Agriculture

Nanoscale frameworks are used for agrosynthetic products, such as manures, insecticides, weedicides, plant growth regulators, and other dynamic substances by utilizing polymers, dendrimers, and surface ionic connections. Different systems may be used in the management of agricultural chemicals, which permit the moderate uptake of dynamic components and thusly diminish the measure of agrochemical applications by limiting waste. The significance of nanoscale conveyance frameworks in agribusiness is a direct result of their enhanced dissolvability and dependable debasement in natural components. Nanoscale conveyance vehicles restrict mobility to the plant canopy and diminish the effects of agricultural chemicals by avoiding discharge into the soil (Johnston 2010; Chen and Yada 2011). Nanoparticles are also used in advanced horticulture practices and provide the best nourishment all around (Gruère 2012). In developing nations, nanotechnology can potentially improve agrarian profitability, alongside other biotechnology applications including genetics, plant rearing, infection control, manure innovation, precision horticulture, and other partnered farm (Sastry et al. 2010; Jha et al. 2011; Chowdappa and Gowda 2013). Nanotechniques can treat plant disease by controlled conveyance of practical atoms or as analytic instruments to determine the illness source. Nanosensors and other field gadgets can be used to discover and estimate yield supplement status, insects, pathogens, weeds, dampness level, soil richness, temperature, and so forth. They allow the continuous observation of harvest development and provide basic information to accurately cultivate while working with limited rural data sources and augmenting resource yields (Scott and Chen 2013). Nanosensors can also provide data about ideal times to plant and reap crops, as well as when to use agrochemicals such as compost, pesticides, and herbicides.

Directed nanoparticles frequently have unique attributes such as additional common qualities, increased synthetic reactivity, and high electrical conductivity. Therefore, nanotechnology has turned out to be one of the most encouraging new advances in the past decade. Nanoparticles have particular physical, organic, and compound properties related to their nuclear composition (Leiderer and Dekorsy 2008). Nanoparticles (width of 1–100 nm) are agglomerated iotas by molecule; their size and sometimes shape may be maintained by specific exploratory methodology (Roy 2009). Nanomaterials are composed of or collected into the necessary layers (Ulrich et al. 2006). Such self-assembly occurs because of characteristics such as hydrogen holding, dipolar powers, hydrophilic or hydrophobic associations, surface strain, gravity, and others. Numerous ordinary organic structures such as films, vesicles, and deoxyribonucleic acid (DNA) are shaped without assistance.

"Nano" is from the Greek word signifying "predominate." In increasingly specialized terms, "nano" implies 10^{-9}, or 1 billionth of something. For instance, an infection is about 100 nm in size. The word nanotechnology was created to indicate the use of nanometer-sized particles (size range from 1 to 100 nm). Nanotechnology diminishes the extent of the smallest structures to the nanoscale, such as photonics applications in nanogadgets and nanobuilding. The base-up methodology includes singular iotas by particles and atoms into nanostructures, which are materials in a few natural procedures (Niemeyer and Doz 2001; Elibol et al. 2003; Wadhwa 2008). In this way, nanotechniques are focused on nanoparticles, with the particles displaying distinct physical qualities, reactivity, electrical conductance, and attractive properties (Nykypanchuk et al. 2008). As an emerging field of research, nanotechnology has expanded in the current decade to a wide variety of potential applications and specialized advancements in an assortment of modern areas. Specifically, in this chapter, our discussion is centered on nanoparticles for agricultural pest control.

9.3 Nanoscale Particles and Their Applications

Nanoscale particles and other attractive materials with comparable properties can be used in nanotechniques. In the environment, living beings from microorganisms to scarabs depend on nanometer-formed proteins, which play important roles ranging from the whipping of flagella to the utilization of muscles. Nanometer-sized particles have long been used to enhance the mechanical properties of tires (carbon dark), advance photographic film (silver), and serve as the basic impetus to the petrochemical industry (Hu et al. 2001; Fiorito 2008; Huck 2008).

Nanoparticles and other nanoorganized materials are frequently orchestrated using synthetic methods. It is extremely important to distinguish nanoparticles through a magnifying instrument like Scanning Tunneling Microscope or Atomic Force Microscope (STM or AFM). Nanoscale science most likely started with the pivotal development of the STM, for which Gerd Binnig and Heinrich Rohrer of International Business Machines (IBM) were granted the Nobel Prize for Physics in 1986. Nanotechnology has also shown potential for applications in ecological security (Nowack 2009; Ying 2009). Nano-organized materials are being used in gadgets for the detection, treatment, and control of contamination. The use of nanoparticles for contamination prediction with eco-friendly blends and assembly is also being investigated (Duebendorf 2008). Furthermore, nano-organized materials are being used in many fabricated items, particularly in composite materials and fillers. The capability of nanomaterials to change human services, materials, data, and innovation has been recorded. A few items powered by nanotechniques are currently in the market, including antibacterial dressings, sunscreens, strain-safe plants, and eco-friendly pesticides (Bhattacharyya 2009; Sukul et al. 2009; Barik et al. 2008; Gha-Young et al. 2008; Prasad et al. 2014, 2017).

The fruit borer *Helicoverpa armigera* (Hubner) (Lepidoptera: Noctuidae) is an exceedingly polyphagous pest, affecting tomatoes (Sharma et al. 2011), cotton (Malik et al. 2002), pigeon peas (Sreekanth and Seshamalakshmi 2012), gram, chickpea (Khalique and Ahmad 1989) and bhendi (Sarate et al. 2012). A few studies on the development of organic and microbial pesticides have been published. In the agricultural and horticultural crops, application of nanotechnology can suggest development of efficient and potential implications for overcoming of the management of polyphagous pests (Bhattacharyya et al. 2016a). Nanoparticles can be used in the production of new products such as pesticides, bug sprays, pheromones, and manure (Barik et al. 2008). The lifespan of the pod borer is between 35 and 40 days, depending on the yield from which it benefits.

The life cycle of this pest was explored by Singh (1972). In the months of July and August, the female moths created from the immature larvae stage deposited the following number of eggs: gram, 1125.4; maize, 1173.3; and chili, 481.5. From August to September, female moths laid 996.0, 847.9, 955.2, 771.7, 744.3, 673.7 and 540.0 eggs on bhendi, corn, tomatoes, tobacco, sunflower, Lucerne, and cabbage, respectively. The females converted from the immature stage benefited from tomatoes and deposited approximately 900 eggs in the midyear months. The brooding time of eggs was 2.7 days in July to August. The advancement of larvae and the quantity of instar rely upon the kind of sustenance given. In July to August, the female hatchlings on tomatoes matured in 18 days and weighed approximately 432 mg, whereas the male moths weighed approximately 444.5 mg. When they benefited from gram units, female hatchlings had the fastest formative period (9 days) and weighed 479.75 mg; the male hatchlings were created in 8.7 days and weighed 471.6 mg. The rate of advancement on other plants was somewhat slower, being the lowest for sunflowers (26 days).

9.4 Rearing of Polyphagous (*H. armigera*) on Artificial Diet

A lifecycle study of the pod borer on an artificial diet was completed by NBAII (Singh et al. 2002). The pod borer was raised on a semi-manufactured diet, as follows:

Part A: Kabuli chickpea (105 g), methyl para-hydroxyl benzoate (2 g), sorbic corrosive (1 g), yeast (2 containers)
Part B: Agar (12.75 g)
Part C: Ascorbic corrosive (3.25 g), multivitamin (2 cases), vitamin E (2 cases), streptomycin sulfate (0.25 g)

The ambient temperature was 26 ± 1 °C and relative humidity was kept at 60–70%. The elements of Part A were mixed in 390 mL of water for 2 min. Part B was mixed in 390 mL of water and added to Part A in a blender. Part C was added and mixed again for a moment. The eating regimen was then dispensed into clean glass vials or plastic containers of 2.5 cm in height, which were stopped with cotton fleece or a stick-holed cover. The hatchlings were exchanged at the rate of single vial or holder and the pupae were gathered over the following 20–25 days. The compartments in the internal multicellular plate isolated the eating regimen in the basal plate into pieces. The hatchlings were then separately put into the compartment through the opening. The top was fixed with wire to cover the openings of the internal plate. The caterpillar period lasted for 20–25 days, pupal period for 7–10 days, and the adult life span for 7–10 days.

9.5 Application of Nanotechnology in the Management of Polyphagous Pests

Nanotechniques have been used for vermin control through nanoparticle-based pesticides, bug sprays, and bio-jointed nanoparticles by adding supplements and water. Nanotechnology has promising applications to convey DNA and other desired synthetics into the internal parts of host plants to protect them from pests (Torney 2009).

9.5.1 Delivery of Silver Nanoparticles for Polyphagous Pest Management

The pediculocidal and larvicidal action of combined silver nanoparticles utilizing a fluid leaf concentrate of *Tinospora cordifolia* demonstrated extreme lethality against the head mite *Pediculus humanus* and fourth instar hatchlings of *Anopheles subpictus* and *Culex-quinque fasciatus* (Jayaseelan et al. 2011). A citronella oil nanoemulsion can be set up with a high-weight homogenization of 2.5% surfactant and 100% glycerol to make stable beads that expand the maintenance of the oil and moderate discharge. The discharge rate relies on the security time; therefore, a decline in the discharge rate can extend the mosquito protection time (Sakulk et al. 2009). Nanoembodiment is a procedure through which a compound is gradually yet productively discharged to a specific host for pest control. Discharge components incorporate disintegration, biodegradation, dissemination, and osmotic weight with explicit pH (Vidyalakshmi et al. 2009).

9.5.2 *Use of Nanopesticides for Pest Management*

Nanopesticides, nanofungicides, and nanoherbicides are being utilized effectively in horticulture (Owolade et al. 2008; Bhattacharyya et al. 2016b). Nanoparticles stacked with garlic essential oil are useful against stored grain pests like the red rust floor beetle *Tribolium castaneum* Herbst (Yang et al. 2009). Nanotubes loaded with aluminosilicate can stick to plant surfaces, whereas elements of the nanotube can adhere to the surface hair of the pest, eventually entering the body and affecting certain physical capacities (Vinutha et al. 2013).

9.6 Future Possibilities for Nanopesticides

Recent studies have demonstrated that nanopesticides have the ability to decrease the harmful effects of pesticides and give explicit control of the harvest targets. It can be useful to improve nanoframeworks to minimize problems in agribusiness such as natural unevenness, food security, and sustenance efficiency. These nanoframeworks have incredible potential capacity for the controlled discharge in dynamic fixing (AI), making them increasingly effective for long time periods. Their ease of use also can assist with eutrophication and lingering pesticide collection issues. In addition, nanopesticides have demonstrated enhanced dissolvability and sound qualities of dynamic elements for compelling control of irritation. In any case, there is potential to enhance the strategies for farming practices, as discussed by Fraceto et al. (2016).

9.7 Conclusion

The use of green science and sustainability principles in nanopesticide development can boost their productivity. However, much research is still required to understand the future of nanopesticides in farming, including the following topics: process improvements for upscaling nanopesticides for business levels, correlation of the nano-movement at the field level to determine commonsense utility, evaluation of the ecological effects to determine toxicity, and enhancement of nanomaterial applications in entomology. Improvements in nanopesticides will provide numerous solutions for the agro-compound industry, including the dissolvability, steadiness, controlled discharge, and dynamic fixing.

References

Barik TK, Sahu B, Swain V (2008) Nano silica-from medicine to pest control. Parasitol Res 103:253–258

Bhattacharyya A (2009) Nanoparticles-From Drug Delivery To Insect Pest Control. Akshar. 1(1): 1–7

Bhattacharyya A, Prasad R, Buhroo AA, Duraisamy P, Yousuf I, Umadevi M, Bindhu MR, Govindarajan M, Khanday AL (2016a) One-pot fabrication and characterization of silver nanoparticles using *Solanum lycopersicum*: An eco-friendly and potent control tool against Rose Aphid, *Macrosiphum rosae*. Journal of Nanoscience, Article ID 4679410, 7 pages, 2016. http://dx.doi.org/10.1155/2016/4679410

Bhattacharyya A, Duraisamy P, Govindarajan M, Buhroo AA, Prasad R (2016b) Nano-biofungicides: Emerging trend in insect pest control. In: Advances and Applications through Fungal Nanobiotechnology (ed. Prasad R), Springer International Publishing Switzerland 307–319

Chen H, Yada R (2011) Nanotechnologies in agriculture: new tools for sustainable development. Trends Food Sci Technol 22:585–594

Chowdappa P, Gowda S (2013) Nano technology in crop protection: status and scope. Pest Manag Hortic Ecosyst 19:131–151

Ditta A (2012) How helpful is nanotechnology in agriculture. Adv Nat Sci Nanosci Nanotechnol 3:033002. https://doi.org/10.1088/2043-6262/3/3/033002

Duebendorf St. GT (2008) How do nano particles behave in the environment? Nano Eco Empa organized international meeting on nano particles in the environment. Published by Bernd Nowack and Nicole Mueller. In Environ. Sci. Technol. http://pubs.acs.org/cgibin/ asap.cgi/est-hag/asap/html/es7029637.html

Elibol OH, Morisette DD, Denton JP, Bashir R (2003) Integrated nanoscale silicon sensors using top-down fabrication. Appl Phys Lett 83:4613–4615

Fraceto LF, Grillo R, de Medeiros GA, Scognamiglio V, Rea G, Bartolucci C (2016) Nanotechnology in agriculture: which innovation potential does it have? Front Environ Sci 4. https://doi.org/10.3389/fenvs.2016.00020

Gha-Young K, Joonmok S, Min-Su K, Seung-Hyeon M (2008) Preparation of a highly sensitive enzyme electrode using gold nanoparticles for measurement of pesticides at the ppt level. J Environ Monit 10:632–637

Hu J, Yang W, Alivisatos AP (2001) Band gap variation of size- and shape-controlled colloidal CdSe quantum rods. Nano Lett 1:349–351

Huck WTS (2008) Responsive polymers for nanoscale actuation. Mater Today 11:24–32

Jayaseelan C, Rahuman AA, Rajakumar G, Vishnu AK, Santhosh Kumar T, Marimuthu S, Bagavan A, Kamaraj C, Zahir AA, Elango G (2011) Synthesis of pediculocidal and larvicidal silver nanoparticles by leaf extract from heartleaf moonseed plant, *Tinospora cordifolia* Miers. Parasitol Res 109:185–194

Jha Z, Behar N, Sharma SN, Chandel G, Sharma D, Pandey M (2011) Nanotechnology: prospects of agricultural advancement. Nano Vision 1:88–100

Johnston CT (2010) Probing the nanoscale architecture of clay mineral. Clay Miner 45:245–279

Khalique S, Ahmad KF (1989) Effect of agronomic factors on the incidence of *Helicoverpa armigera* (hub.) and its parasite *Campoletis chloridae* in chickpea field. Pak J Sci Ind Res 32:694–667

Kumar R, Sharon M, Chaudhary AK (2010) Nanotechnology in agricultural diseases and food safety. J Phytology 2:83–92

Leiderer P, Dekorsy T (2008) Interactions of nano particles and surfaces Tag der mÄundlichen PrÄufung:25.April. URL: http://www.ub.unikonstanz.de/kops/volltexte/2008/5387/; URN: http://nbn-resolving.de/urn:nbn:de:bsz:352-opus-53877

Malik MF, Khan AG, Hussiany DU (2002) Scouting and control of *Helicoverpa armigera by* synthetic pheromones technology in apple. Asian J Plant Sci 3:663–664

Niemeyer CM, Doz P (2001) Nano particles, proteins, and nucleic acids: biotechnology meets materials science. J German Chem 40:4128–4158
Nowack B (2009) Is anything out there? What life cycle perspectives of nano-products can tell us about nanoparticles in the environment? Nano Today 4:11–12
Nykypanchuk D, Maye MM, Van Der Lelie D, Gang O (2008) DNA-guided crystallization of colloidal nanoparticles. Nature 451:549
Owolade OF, Ogunleti DO, Adenekan MO (2008) Titanium dioxide affects disease development and yield of edible cowpea. Elec J Env Agricult Food Chem Title 7:2942–2947
Prasad R, Kumar V, Prasad KS (2014) Nanotechnology in sustainable agriculture: present concerns and future aspects. Afr J Biotechnol 13(6):705–713
Prasad R, Bhattacharyya A, Nguyen QD (2017) Nanotechnology in sustainable agriculture: Recent developments, challenges, and perspectives. Front Microbiol 8:1014. doi: https://doi.org/10.3389/fmicb.2017.01014
Prasad R, Kumar V, Kumar M, Choudhary D (2019) Nanobiotechnology in Bioformulations. Springer International Publishing https://www.springer.com/gp/book/9783030170608
Roy SC (2009) There's plenty of holes at the bottom: the other side of Nano. Sci Cult 75:1–3
Sakulk U, Nuchuchua O, Uawongyart N, Puttipipatkhachorn S, Soottitantawat A, Ruktanonchai U (2009) Characterization and mosquito repellent activity of citronella oil nano emulsion. Int J Pharm 372:105–111
Sarate PJ, Tamhane VM, Kotkar HM, Ratnakaran N, Susan GVS, Giri AP (2012) Development and digestive flexibilities in the midgut of a polyphagous pest, the cotton bollworm, *Helicoverpa armigera*. J Insect Sci 12:12–42
Sastry RK, Rashmi H, Rao N, Ilyas S (2010) Integrating nanotechnology into agri-food systems research in India: a conceptual framework. Technol Forecast Soc Chang 77:639–648
Scott N, Chen H (2013) Nanoscale science and engineering for agriculture and food systems. Ind Biotechnol 9:17–18
Singh, SP (1972) Cotton bollworm in Kuban and basis for its biological control. (Ph.D, thesis Kuban Agri. Institute, Krasnodor) USSR. pp186
Singh, SP, Ballal, CR, and Poorani, J (2002) Old world bollworm, H. armigera, associated Heliothis and their natural enemies, NBAII, India. pp.135
Sharma KC, Bhardwaj SC, Sharma G (2011) Systematic studies, life history and infestation by *Helicoverpa armigera* on tomato in semiarid region of Rajasthan. Biol Forum 3:52–56
Sreekanth M, Seshamahalakshmi M (2012) Studies on regional toxicities of biopesticides to *Helicoverpa armigera* on pigeon pea (*Cajanus cajana*). J Biopest 5:191–195
Sukul NC, Singh RK, Sukul S, Sen P, Bhattacharyya A, Sukul A, Chakrabarty R (2009) Potentized drugs enhance growth of Pigeon pea. Environ Ecol 26:1115–1118
Torney F (2009) Nano particle mediated plant transformation. Emerging technologies in plant science research. Inter Departmental Plant Physiology Major Fall Seminar Physics, p 696
Ulrich C, Mewis I, Goswami A, Chatterjee SD, Banerjee SP, Adhikary S, Bhattacharyya A (2006) Biodiversity - macro and micro: to be nano or not to be! Everyman's. Science 11:433–443
Vinutha JS, Bhagat D, Bakthavatsalam N (2013) Nanotechnology in the management of polyphagous pest *Helicoverpa armigera*. J Acad Indust Res 1:606–608
Vidyalakshmi R, Bhakyaraj R, Subhasree RS (2009) Encapsulation the future of probiotics. Adv Biol Res 3:96–103
Wadhwa S (2008) Nanotechnology and its future applications. In Chilli breeze, 25th April. Indian Talent, Global Content. Composites-Part II: Silica-Chitin-Based Bio composites. J Nano Mater 1:1–8
Yang FL, Li XG, Zhu F, Lei CL (2009) Structural characterization of nanoparticles loaded with garlic essential oil and their insecticidal activity against *T. castaneum* (Herbst) (Coleoptera: Tenebrionidae). J Agr Food Chem 57:10156–10162
Ying JY (2009) Nano bio materials. Nano Today 4:1–2

Chapter 10
Myconanoparticles as Potential Pest Control Agents

Ajay Kumar Gautam and Shubhi Avasthi

Contents

A. K. Gautam (✉)
School of Agriculture, Abhilashi University,
Mandi, Himachal Pradesh, India

S. Avasthi
School of Studies in Botany, Jiwaji University,
Gwalior, Madhya Pradesh, India

D. Thangadurai et al. (eds.), *Nanotechnology for Food, Agriculture, and Environment*, Nanotechnology in the Life Sciences,
https://doi.org/10.1007/978-3-030-31938-0_10

10.1 Introduction

Nanotechnology is an innovative modern science which provides a platform for the synthesis and fabrication of materials or particles at the nanoscale. These particles are ranging in size between 1 and 100 nm, hence called nanoparticles and abbreviated as NPs (Mohanpuria et al. 2008; Singh et al. 2010; Liu et al. 2011). These nanoparticles are variable in shape, size, and composition and hence act as key player in multidisciplinary approaches. Nanoparticles have a wide range of applications in the field of agriculture, chemical industries, drug designing and drug delivery, electronics, fermentation technologies, and food processing industries (Prasad et al. 2016, 2017a, b). More specifically, nanoparticles are used in medical diagnostics, chemicals and materials industry, drug and gene delivery, tissue engineering and biochemical sensors, fluorescent biological labelling, cancer treatment, biodetection of pathogens, detection of proteins, and tissue engineering nowadays (Salata 2004). Nanoparticles can be organic (solid particles composed of organic compounds mainly lipids or polymeric) and inorganic (small particles with diameters in the range of 1–100 nm with enhanced physical and chemical properties depending on the particle size). For synthesis of nanoparticles, three basic methods are physical, chemical, and biological. Among these, biological methods are believed to be eco-friendly because of the use of microbial sources (e.g., fungi, yeasts or bacteria, actinomycetes), plant extracts, or enzymes or use of templates like DNA, membranes, and viruses (Prasad et al. 2016, 2018a, b). Besides the nature of NPs, their synthesis methods play significant role in generation of new nanoparticles with variable size, shape, and composition in nanometric scale. Two broad categories namely, top–down or bottom–up, are generally employed for the synthesis of nanoparticles, in which bottom–up approach is more advantageous.

The fungi are a relatively recent addition to the list of microorganisms used for nanoparticle synthesis. The fungal synthesized nanoparticles are known as myconanoparticles. It was found that fungi are more beneficial than other microorganisms as:

1. They grow in the form of mycelial mesh which helps them to bear flow pressure, agitation and other conditions to which microbes are subjected to in a bioreactor used for large scale production.

2. Though they require better precision and care to grow but they are easier to handle and manipulate.
3. They secrete enzymes in large amounts. The nanoparticles are generally produced extracellularly (sometimes intracellular production does occur) and are devoid of various impurities from the cell and can be used directly.

The multiple applications of myconanoparticles in the field of agriculture have made them as an emerging, important and integral part of nanotechnology. A number of bacteria, fungi, and yeasts are able to produce a wide variety of nanoparticles and can act as biofactories for nanoparticles (Prasad et al. 2016). The specificity of targeted delivery has made them a kingmaker in pesticide industry and emerging pesticides in agricultural fields (Bhattacharyya et al. 2016). A number of fungi have been studied extensively, and myconanoparticles have been synthesized using a variety of fungi, viz., *Aspergillus*, *Cladosporium*, *Fusarium*, *Phanerochaete*, *Phoma*, *Trichoderma*, *Verticillium*, and yeast. However, synthesis of myconanoparticles is not an easy job. It involves strategies for the synthesis of metal nanoparticles, mechanism, and factors affecting their mycosynthesis.

Fungus mediated synthesis of nanoparticles and their wide array of applications have recently attracted the attention of researchers towards myconanotechnology. To understand the applications and possible benefits of myconanoparticles in agriculture, it is important to analyse strategies, mechanisms and factors used in their mycosynthesis. Therefore, the current advances, challenges and potential applications in controlling pests in are discussed in this chapter.

10.2 General Background on Nanotechnology

Nanoscience is an interdisciplinary science which involves the study of phenomenon and manipulation of materials on ultra-small (nano)scale. The term nanotechnology (NT) was coined by Professor Norio Taniguchi of Tokyo Science University in 1974 to illustrate precision manufacturing of materials at the nanometer level (Taniguchi 1974). The nanoscale structures are too small and far away from general imagination and are popularly known as nanoparticles. One can draw the idea from the fact that a printed page is about 75,000 nanometers thick. The nanoparticles exhibit interesting and very useful properties with reduction of their size in the nanometer size range. The study of synthesis and applications of nanoscale materials is applied branch of nanoscience and known as nanotechnology.

Although nanotechnology is a relatively recent development in scientific research, but the earliest use of this concept is not so new. It can be traced back to era of carbon nanotubes and cementite nanowires found in the microstructure of wootz steel manufactured in ancient India from the time period of 600 BC and exported globally. However, invention of the scanning tunneling microscope in 1981 and the discovery of fullerenes in 1985 evidenced the emergence of nanotechnology in the 1980s. The application of nanotechnology commercially happened during the early years of twenty-first century, although these were limited to bulk applications of nanomaterials rather than the transformative applications (Gunter and Andreas 2010; Khan 2012; Luby et al. 2015). Nowadays, the nanomaterials (nanoparticles) have a wide range of applications in many areas including electronics, agriculture, defence, drug designing and drug delivery, fermentation technologies, food and food processing industries, and chemical industries.

10.2.1 What Are Nanoparticles (NPs)?

The word "nano" is derived from Latin language which means dwarf. A nanoparticle is an object that has all three dimensions within 1–100 nm. More precisely, at least one dimension of the object should be in the microscale. These nanoparticles (NPs) are not as new as stated in their laboratory studies; rather they existed in nature long before. Naturally occurring NPs can be exemplified with the fact that a single strand of DNA is 3 nanometers wide. Likewise, scales on a morpho butter-

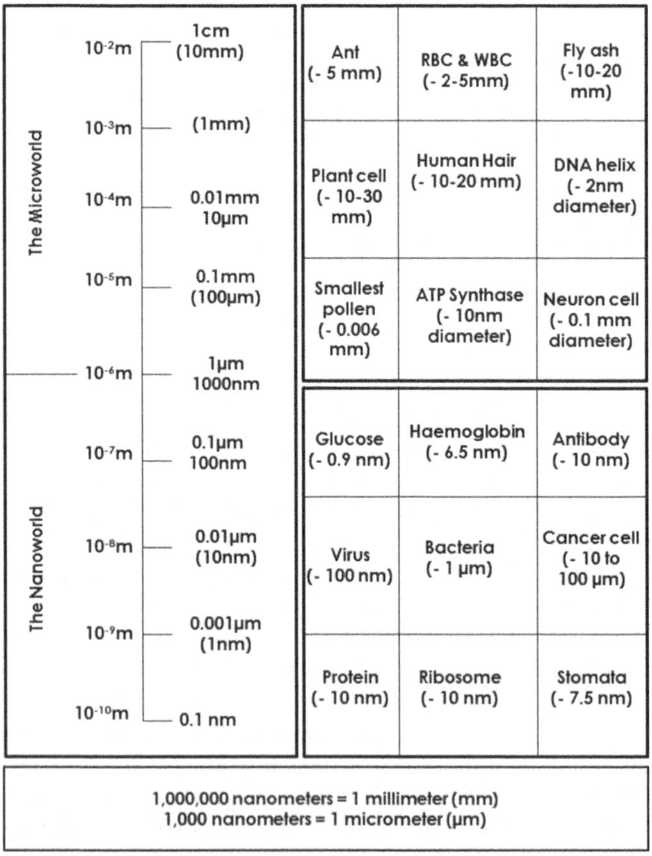

Fig. 10.1 Relative size of various naturally occurring nanoparticles

fly's wings contain nanostructures involved in colouring effects. Peacock feathers also get their iridescent coloration from light interacting with structures few tens of nanometers thick (Fig. 10.1). Even scientists have mimicked naturally occurring nanostructures to create NPs artificially in laboratory. Nanoparticles have now become an integral part of modern science. Nanoparticles can be classified primarily into natural and artificially fabricated types.

However, on the basis of their composition they can again be further categorized into two groups:

- *Organic nanoparticles*: Organic nanoparticles can be explained as solid particles composed of organic compounds (mainly lipids or polymeric) ranging in diameter from 10 nm to 1 μm. Thousands of organic chemicals are present in various pharmaceuticals to consumer products being used like inks, dyes, flavoring agents, and household cleaning products. These organic nanoparticles are environment friendly, economical and more suited for biological applications (Kumar and Lal 2014).
- *Inorganic nanoparticles*: Inorganic nanoparticles (NPs) are small particles with diameters in the range of 1–100 nm. Depending upon their size, they have shown special and enhanced physical and chemical properties. The four most common inorganic NPs are categorized as (1) noble metal, (2) magnetic, (3) fluorescence, and (4) multifunctional, e.g., luminescent magnetic.

On the basis of dimensions, nanomaterials can also be classified as:

- *Zero-dimensional nanomaterials (0-D)*: Materials wherein all the dimensions are measured within the nanoscale (no dimensions or 0-D are larger than 100 nm).
- *One-dimensional nanomaterials (1-D)*: One dimension that is outside the nanoscale. This leads to needle-like shaped nanomaterials, e.g., nanotubes, nanorods, and nanowires.
- *Two-dimensional nanomaterials (2-D)*: Two of the dimensions exhibit plate-like shapes and are not confined to the nanoscale, e.g., nanofilms, nanolayers, and nanocoatings.
- *Three-dimensional nanomaterials (3-D)*: It includes bulk nanomaterials having three arbitrary dimensions above 100 nm. Generally, these materials are not confined to the nanoscale in any dimension. Here, bulk nanomaterials can be composed of a multiple arrangement of nanosize crystals, most typically in different orientations.

What makes NPs so special? The most important reasons are size in nanoscale, fabrication atom by atom, and increased surface to volume ratio which make them so promising in material engineering and related sciences. The use of NPs now serves as a platform to harness the benefits of unique properties of atoms and molecules at the nanoscale (Singh et al. 2010). Besides this, the synthesis of nanomaterials with specific composition and size and of distinct properties has expanded the scope of their applications in myriad of industries including agriculture, textile, and food (Sharon et al. 2010; Sastry et al. 2011). Use of NPs in agriculture sector as nanopesticides, nanofertilizers, nanoherbicides, nanobiosensors, and nanobarcodes has further expanded their scope. They are also useful in maintaining overall health of crops and soil (Duhan et al. 2017).

10.2.2 Importance of Nanoparticles (NPs)

Nanoparticles are amazing materials that have been and are being explored and used in various scientific fields including biotechnological, pharmacological, pure technological, and agricultural. The increasing popularity and demand of nanoparticles is due to their variable physical and chemical properties with size over constant physical properties regardless of its size in bulk materials (Fig. 10.2). These properties which made nanoparticles, the today's key scientific tools are as:

- Nanoparticles are unique because of their large surface area and this dominates the contributions made by the small bulk of the material.
- Nanoparticles of yellow gold and grey silicon are red in colour as gold nanoparticles melt at much lower temperatures (~300 °C for 2.5 nm size) than the gold slabs (1064 °C).

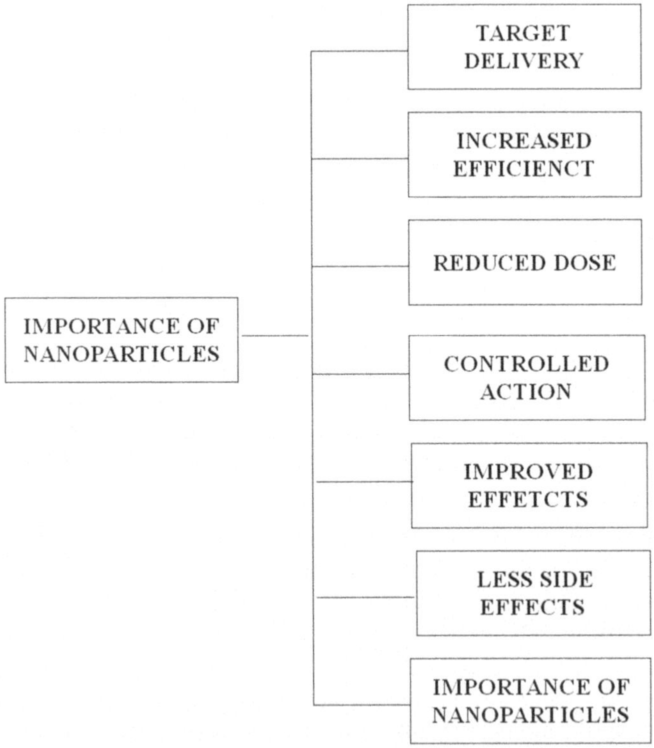

Fig. 10.2 Importance of nanoparticles (NPs)

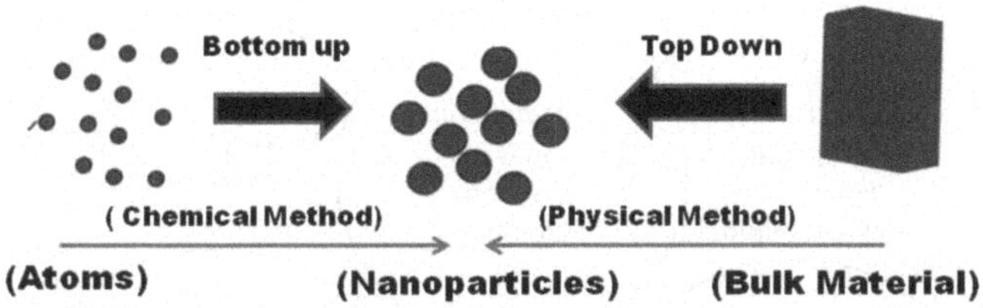

Fig. 10.3 Current applications of nanotechnology

- Absorption of solar radiation is higher in them.
- Nanoparticles also often possess unexpected optical properties as they are small enough to confine their electrons and produce quantum effects.
- They have unique ability to form suspensions and possess property of diffusion.

All these properties have made nanoparticles efficient nanocarriers in drug delivery and drug analysis in pharmaceutical and biomedical industry. This helped the clinicians in broadening the medical tools, knowledge, and therapies currently available and to produce precise solutions for disease prevention, diagnosis, and treatment. The nanosize of these NPs enables them to possess catalytic property to boost chemical reactions which result in the reduction of quantity of catalytic materials necessary to produce desired results, saving money and reducing pollutants. Use of nanomaterials in food and agriculture sector is also at the forefront of modern research. Nowadays, applications of nanoparticles provide better efficacy for the control of plant diseases (Fig. 10.3).

10.2.3 Synthesis of Nanoparticles

Nanoparticles are the fundamental components of nanotechnology made up of carbon, metal, metal oxides or organic matter. These practices are in nanometric scale and have improved properties such as high reactivity, strength, surface area, sensitivity, and stability. The small size of nanoparticles is mainly responsible for all such properties. However, the synthesis of nanoparticles plays an important role in their variable size, shape, and composition in nanometric scale. Synthesis refers to creating nanoparticles from larger molecules by various metabolic processes that are categorized into top–down or bottom–up method (Fig. 10.4). The bottom–up approach is more advantageous than the top–down approach because of less defects, homogeneity in chemical composition, and superior short- and long-range ordering.

A top–down synthesis method implies that the nanostructures are synthesized by top (larger) to bottom (smaller) approach. It includes etching out crystals planes (removing crystal planes) which are already present on the substrate. It found that building blocks are being removed from substrate in such a way to form the nanostructure.

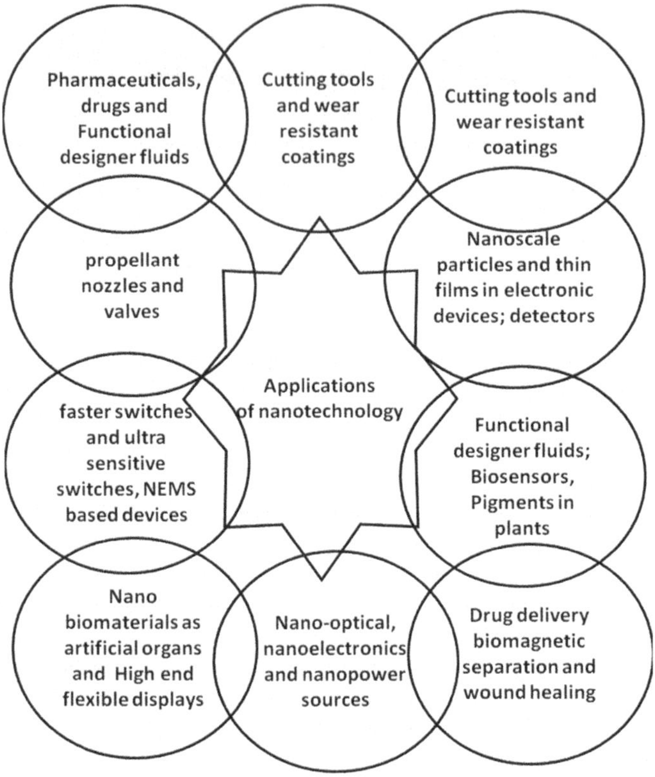

Fig. 10.4 Synthesis of nanoparticles

The process of bottom (smaller) to top or up (larger) is the basic approach behind bottom–up approach. In bottom–up synthesis, the nanostructures are synthesized onto the substrate by stacking atoms onto each other, which gives rise to crystal planes; crystal planes further stack onto each other, resulting in the synthesis of the nanostructures. A bottom–up approach can thus be viewed as a synthesis approach where the building blocks are added onto the substrate to form the nanostructures. Broadly nanoparticles synthesized by various methods are classified in three main types (Parvathi and Rajagopal 2017) which are as below: (1) physical methods, (2) chemical methods, and (3) biological methods.

10.2.3.1 Physical Method

The physical method of nanoparticle synthesis includes mechanical separation of metal aggregates. These methods lead to particles formation, the diameters of which have a large variability. The colloid particles prepared this way have usually diameters larger than 10 nm and they cannot be prepared and reproduced. The various physical methods involved in synthesis of nanoparticles include mechanical methods, ball milling, melt mixing, vapor deposition, physical vapor deposition, laser ablation, laser pyrolysis, sputter deposition, and electric arc deposition.

10.2.3.2 Chemical Method

Chemical methods are among the most widely used methods for preparation of nanoparticles. The methods involved in chemical synthesis of nanomaterials include chemical vapor deposition, epitaxial growth, colloidal dispersion, sol–gel, hydrothermal route, microemulsions, polymer route, and other precipitation processes. The synthesis of nanomaterials by all these methods is categorized into two categories, namely, colloidal synthesis and sol–gel method.

10.2.3.3 Biological Methods/Green Synthesis

These methods involve the use of biological (living) agents in the synthesis of nanoparticles. These are one of the most popular methods nowadays. Following are some biological sources used for the synthesis of nanoparticles:

- The use of microbial sources (fungi, yeasts, bacteria, actinomycetes, and even viruses also).
- Plant extracts or enzymes and templates like DNA.
- Membranes and viruses are some common substrates involved in biological methods.

The three main steps involved in synthesis of nanoparticles are the selection of solvent media, environmentally benign reducing agent, and a nontoxic material for stabilization of NPs. The advantages of biological methods over physical and chemical include the following: simple, eco-friendly, and time saving. Biological synthesis of NPs requires low-cost equipment, hence proving cost effective. Moreover, these methods would assist to remove ruthless processing conditions by allowing the synthesis at physiological pH, temperature, and pressure.

10.2.4 Substrates for Nanoparticle Synthesis

Several resources have been adopted for biosynthesis of NPs. These resources may be synthetic or natural. These NPs are inorganic and organic on the basis of source material used for synthesis. Inorganic nanoparticles are made up of inorganic materials. Examples are given as below:

Metal based: These types of NPs are synthesized from metals like aluminum (Al), cadmium (Cd), cobalt (Co), copper (Cu), gold (Au), iron (Fe), lead (Pb), silver (Ag), and zinc (Zn).

Metal oxide based: These are also metal based nanoparticles but synthesized to modify the properties of their respective metal based nanoparticles. Examples are as under:

- Iron oxide (Fe_2O_3), a nanoparticle of iron (Fe).
- Magnetite (Fe_3O_4), a nanoparticle of iron (Fe).
- Aluminum oxide (Al_2O_3), a nanoparticle of aluminum (Al).
- Cerium oxide (CeO_2), a nanoparticle of cerium (Ce).
- Silicon dioxide (SiO_2), a nanoparticle of silicon (Si).
- Titanium oxide (TiO_2), a nanoparticle of titanium (Ti).
- Zinc oxide (ZnO), a nanoparticle of zinc (Zn).

Carbon based: The nanoparticles are completely built of carbon and of hollow spheres, ellipsoids, or tubes. While spherical and ellipsoidal carbon nanomaterials are referred to as fullerenes, cylindrical ones as nanotubes (carbon nanotubes, CNTs).

Dendrimers: These are nanosized polymeric materials built from branched units. They are radially symmetrical molecules with numerous chain ends on their surface which provide them a catalytic property. These molecules have well-defined, homogeneous, monodisperse, three dimensional structures having a typically symmetric core, an inner shell and an outer shell into which other molecules could be placed. This property may be useful for drug delivery. These varied characteristics make dendrimers a good choice in the medical field (Abbasi et al. 2014).

Composites: It is the combination of components in which nanoparticles are combined with other nanoparticles or natural or artificial elements (with different physical or chemical properties). These combinations are stronger in comparison to individual players. The composites may be metal based, carbon based or polymer based nanomaterials with any form of metal, ceramic or polymer bulk materials. After combining, the component materials don't completely blend or lose their individual identity. They are designed with some specific characteristics like strength, efficacy or durability (Kaw 2005; Madbouly et al. 2015).

Organic nanoparticles: These include nanomaterials that are synthesized from materials that are biological in origin. The microbial sources including fungi, bacteria, viruses, actinomycetes, mushrooms, and plant extracts or enzymes are some commonly biological sources in synthesis of organic nanoparticles.

10.2.5 Microbes in Nanotechnology

Microorganisms play an important role in nanotechnology by involving in their synthesis. A variety of inorganic nanoparticles with well-defined chemical composition, size, and morphology has been synthesized. The main advantages of using microbes in nanotechnology are efficacy, economic, simplicity, and eco-friendly products as NPs (Prasad et al. 2018b). Contributing to this, increment in the use of microbes in nanoparticle (NP) synthesis is continuously taking place and attracting much attention. Overall we can say that it is the best solution for growing need to develop clean, nontoxic and environmentally friendly ("green chemistry") procedures for synthesis and assembly of nanoparticles. Presently, the potential of bioorganisms ranges from simple prokaryotic bacterial cells to eukaryotic fungus and even live plants are exploring for their possibilities in the synthesis of NPs (Bhattacharya and Gupta 2005).

There are numbers of nanoparticles synthesized with the involvement of microbes. Following is the categories of metallic nanoparticles including gold, silver, alloy, and other metal nanoparticles, oxide nanoparticles consisting of magnetic and nonmagnetic oxide nanoparticles, sulfide nanoparticles, and other miscellaneous nanoparticles. The most of microbial synthesized NPs are found to be metallic in nature (Xiangqian et al. 2011). Different microorganisms including algae, yeast, bacteria, viruses, and fungi are being used in synthesis of nanoparticles presently (Pantidos and Horsfall 2014; Li et al. 2011; Patel et al. 2015; Xue et al. 2016). The facts about growth in the form of mycelial mesh and secretion of various useful enzymes in large amount have popularized the fungi as microorganisms used for nanoparticle synthesis. Because of their mycelial mesh growth pattern, fungi can bear flow pressure, agitation, and other conditions. Their easy growth, maintenance, handling, and manipulation further support the usefulness of fungi. Enzyme production in larger amount also favors their suitability in nanotechnology. Moreover, the extracellular synthesis of nanoparticles also ensures their purity. Finally, microbial synthesized nanoparticles exhibit a wide range of applications. These are found to be very useful in targeted drug delivery. This enables them to be used in cancer treatment, gene therapy, and DNA analysis. Use of NPs in enhancing reaction rates, separation science, and magnetic resonance imaging (MRI) is another important application of these nanomolecules. Nowadays, these are used as antibacterial agents and biosensors also.

10.3 Myconanotechnology

The use of microorganisms in nanotechnology is one of the emerging scientific fields because of the fact that microorganisms aggregate inorganic material within or outside the cell to form nanoparticles (NPs). This microbial synthesis of NPs is one of the emerging eco-friendly approaches of green chemistry. It offers a connection between nanotechnology and microbial biotechnology. Use of fungi in nanotechnology for the synthesis of NPs is becoming popular day by day as they grow on readily available substrates which are inexpensive and produce wide range of commercially interesting metabolites.

10.3.1 General Introduction

Myconanotechnology is a branch of nanotechnology which involves the use of fungi for the synthesis of nanomaterials or nanostructures with desirable shape and size (Kashyap et al. 2013). Generally, it is defined as the fabrication of nanoparticles by fungi and their subsequent applications. It is an emerging, new, and pragmatic interdisciplinary science in the field of nanotechnology and leads to explore new commercial uses and benefits of fungi (Rai et al. 2009). Several fungal strains have been used as promising resources for nanoparticle fabrication and production of metal NPs both intra- and extracellularly. There are number of advantages of mycogenic synthesis of nanoparticles over the use of other microbes in nanotechnology (Sastry et al. 2003; Ingle et al. 2009; Hemath et al. 2010; Singh and Balaji 2011; Moharrer et al. 2012; Kashyap et al. 2013; Ortega et al. 2015; Xue et al. 2016; Rahimi et al. 2016; Ottoni et al. 2017). However, fungi in general show some distinctive advantages over bacteria, algae, and other microbes, which are as under:

- The use of filamentous fungi is a promising, reliable, and eco-friendly approach in the nanoparticle synthesis.
- These extracellular metabolites produced by fungi serve as agent for their own survival when exposed to environmental stresses like toxic materials (such as metallic ions), predators, and temperature variations. This production of extracellular enzymes eases downstream processing and handling of biomass.
- The use of fungi in the synthesis of NPs is potentially important since they produce large quantities of enzymes and are simpler to handle in the laboratory. Since produced extracellularly, they are easy to purify and can be directly used in various applications.
- Production of wide range of proteins by fungi aids higher yield of NPs. This range of protein secretion is lower in other microorganisms which encourages the use of fungi in nanotechnology.
- High metal tolerance and bioaccumulation ability is another advantage of use of fungi over other microbes.
- The potential of fungal mycelia to withstand in flow pressure, agitation and other conditions in bioreactors also promote their use. They offer better manipulation and control over crystal growth and their stabilization due to slower kinetics also supports their selection over other microbes in nanotechnology.
- Fungi can provide controlled synthesis of nanoparticles with well-defined shape and size and also helps in outside precipitation of nanoparticles, minimizes the unnecessary cellular components. It advocates another advantage on use of fungi in nanoscience.
- The nanoparticles synthesized using fungi present good polydispersity, dimensions and stability.
- Myconanoparticles are nontoxic for humans (in low concentrations), and safe inorganic antibacterial agents that have been shown to exhibit a strong toxicity to a wide range of microorganisms.

Apart from above advantages, the use of fungi, however, needs improved methods for the controlled synthesis of NPs of well-defined size, shape, and composition. Continuous research to develop advanced methods to synthesize myconanoparticles is regularly going on.

10.3.2 Biosynthesis of Nanoparticles Using Fungi

The use of fungi (eukaryotic organisms) for synthesis of nanoparticle was first demonstrated by use of *Verticillium* sp. Gold nanoparticles were synthesized in this experiment on the surface and cytoplasmic membrane of the fungal mycelia. Similarly, silver nanoparticles were also synthesized mycologically on exposure to silver nitrate solution. A number of myconanoparticles have been synthesized simultaneously, with the increased use of eukaryotic organisms in nanoscience. To understand the possible benefits of employing nanobiotechnology in biological sciences, it is necessary to analyze the exact mechanism of synthesis of nanoparticles by fungi, which is still not so clear. However, it is generally assumed that positively charged metal ions and negatively charged carboxylates on the enzymes present in fungal cell wall are interacted with each other (Fig. 10.5). The enzymes reduce these metal ions to elemental metal which serve as nucleation sites and further growth is carried out by subsequent reduction and accumulation. The general outline of biosynthesis of nanoparticles using fungi is presented as under.

Among the different microbes used for the synthesis of nanoparticles, fungi are proving excellent biological agents for both intra- and extracellular fabrication of metal nanoparticles. Several fungi have already been exploited for the synthesis of silver, gold, zirconium, silica, titanium, iron, and platinum nanoparticles (Table 10.1). Research is still going on at every corner of the world to find out new, noble and efficient fungal organisms which can be used in synthesis of nanoparticles (Krumov et al. 2009; Gade et al. 2010). In a survey carried out on nanoparticles (104 out of 502 nanoproducts surveyed), it was observed that silver nanoparticles (Ag NPs) have become one of the most commonly used nanomaterials in consumer products.

10.3.3 Strategies for Mycosynthesis of Nanoparticles

Nanotechnology emerges out from advancements in material science. The fabrication of nanoscale materials, in standardized and reliable manner at reasonable scale and cost, is usually carried out in nanoscience to manufacture crystals, particles, spheres, wires, and tubes. Two strategies are usually utilized in synthesis of metal nanoparticles: a bottom–up (self-assembly) and a top–down. While bottom–up strategy refers to the construction of a structure atom-by-atom, molecule-by-molecule, or by self-organization, a suitable starting material is reduced in size

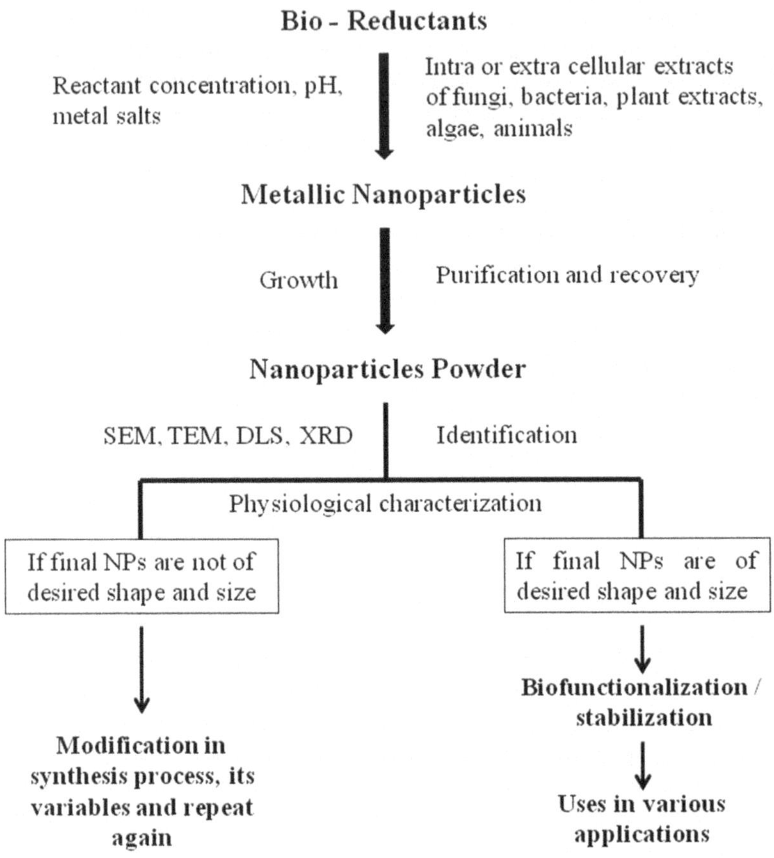

Fig. 10.5 Mechanism of nanoparticle synthesis

using physical or chemical means (cutting, grinding, and etching) in top–down approach (Singh et al. 2010; Marchiol 2012).

Mycosynthesis of nanoparticles is a kind of bottom–up approach; it give rise to colloidal structures as final products. It involves reduction/oxidation of substrates as main reaction to synthesize myconanoparticles (Moghaddam 2010; Singh et al. 2010). Fungi which are used in synthesis of NPs secrete certain metabolites or enzymes that exhibit antioxidant or reducing properties and can be used for reduction of metal compounds into their respective nanoparticles, which undergo further growth, leading to the generation of nanostructures. The reduction of Gibbs free energy strategies is mainly considered behind bottom–up approach, and synthesized nanostructures/nanomaterials are in a state closer to a thermodynamic equilibrium state. The myconanoparticles derived with bottom–up approach are more homogeneous in chemical compositions with minor defects (Behari 2010; Thakkar et al. 2010). However, research for finding more and better myconanoparticles, fungal components and efficient strategies is still a continuous process.

Table 10.1 Various types of nanoparticles and fungi involved in their synthesis

Nanoparticles types (products)	Fungi involved	References
Gold (au)	*Neurospora crassa*, *Alternaria alternata*, *Verticillium* sp., *Fusarium oxysporum*, *Aspergillus niger*, *Aspergillus clavatus*, *Colletotrichum* sp., *Trichothecium* sp., *Sclerotium rolfsii*, *Helminthosporium solani*, *Penicillium* spp.	Mukherjee et al. (2001), Ahmad et al. (2003a, b), Zhang et al. (2009), Verma et al. (2011), Soni and Prakash (2012), Shedbalkar et al. (2014), Geetanjali and Ashok (2015)
Silver (Ag)	*Arthroderma fulvum*, *Fusarium oxysporum*, *Aspergillus fumigatus*, *Aspergillus flavus*, *Aspergillus niger*, *Trichoderma viride*, *Trichoderma harzianum*, *Brevibacterium casei*, *Microporus xanthopus*, *Pleurotus florida*, *Inonotus obliquus*	Wang et al. (1998), Ahmad et al. (2003a, b), Singh and Balaji (2011), Dhandhukia et al. (2012), Balashanmugam et al. (2013), Gaikwad et al. (2013), Gurunathan et al. (2013), Ninganagouda and Rathod (2014), Xu et al. (2016), Ragunath et al. (2017)
Platinum (pt)	*F. oxysporum*	Riddin et al. (2006), Govender et al. (2009), Syed and Ahmad (2012)
Selenium (se)	*A. alternata*	Sarkar et al. (2011)
Ti	*F. oxysporum*	Bansal et al. (2005), Bansal et al. (2006)
Mn	*Fusarium oxysporum*	Chen et al. (2011)
Cd	*Fusarium oxysporum*, *Coriolus versicolor*, yeast	Ahmad et al. (2002), Bao et al. (2010), Abhilash et al. (2011)
Cadmium sulfide	*Coriolus versicolor*, *Fusarium oxysporum*	Ingle et al. (2008), Abhilash et al. (2011)
Zirconia	*Fusarium oxysporum*	Bansal et al. (2004), Abhilash et al. (2011)
Magnetite	*Fusarium oxysporum*, *Verticillium* spp.	Bharde et al. (2006)
Cobalt	*Neurospora crassa*	Rashmi et al. (2004)

10.3.4 Mechanism of Biosynthesis and Identification

The mycogenic synthesis of nanoparticles usually requires a biogenic source especially fungal agent, a reducing agent as well as a stabilizing agent in the system that would facilitate formation and stabilization of the metal nanoparticles. In general, fungi interact with the metals which come in contact with them through their cells and form nanoparticles. Several fungal strains have been used as promising resources for nanoparticle fabrication because of intra- or extracellular production of reducing enzyme, high wall-binding ability, and intracellular metal uptake capability (Ahmad et al. 2003a, b; Durán et al. 2005). Although a number of biogenic sources including fungi have been identified for synthesis of nanoparticles, the exact mechanism of synthesis is yet to be deciphered clearly. However, a number of hypothetical representations on synthesis of myconanoparticles have been suggested by researchers.

It has been observed that biogenic metal nanoparticles can be stable for 3 months and more. Their stability pointed out the fact that there should be a reducing agent as well as a stabilizing agent in the system that would facilitate formation and stabilization of the metal nanoparticles. It indicates that mycogenic synthesis of nanoparticles involves three main steps: choice of solvents, choice reducing agent, and choice of nontoxic materials required for stabilization (Moghaddam 2010; Singh et al. 2010). The identification of synthesized nanoparticles is enormously challenging. UV–vis absorption spectra (UV–Vis), transmission electron microscopy (TEM), energy-dispersive X-ray analysis (EDX), and X-ray diffraction analysis (XRD) can be generally employed in identification and size and distribution analysis of NPs (Singhal et al. 2011; Chan and Don 2012, 2013; Manzoor et al. 2014).

In general, peptides, proteins, polyols, and heterocyclic compounds have been suggested to have an important role in the synthesis of metal nanoparticles. It is believed that the fungus mycelium exposed to the metal salt solution stimulates the production of fungal enzymes and primary metabolites, which in turn switched the reduction of toxic metal ions to the nontoxic metallic nanoparticles through the catalytic effect of the extracellular enzymes and metabolites of the fungus. However, nature of the reducing agent and stabilizing agent also varies with the species employed for the synthesis, which can be a major limitation in the use of fungi in nanoparticle synthesis.

The trapping of metal ions over fungal cell surface is the initial step during intracellular mycogenic synthesis of nanoparticles. Electrostatic interaction of metal ions with the enzymes or proteins present on the fungal cell wall is believed to be behind this trapping. Thereafter, reduction of metal ions by enzymes occurs within the cell wall and then aggregation of metal ions and finally formation of nanoparticles takes place (Mukherjee et al. 2008). Some metal ions having the ability to diffuse through cell wall can be reduced intracellularly by enzymes present on the cytoplasmic membrane and cytoplasm (Sastry et al. 2003). Number of reductase enzymes like NADPH dependent reductase enzyme and hydroxyquinoline, phytochelatin, nitrate reductase (Hamedi et al. 2014; Duran et al. 2005) have been examined to explore their reductase activity both at intercellular and intracellular level. It is now believed to some extent that electrostatic interactions to trap metal ions with specific enzymes produced by fungi are major factors in the mycogenic synthesis of nanoparticles. Though the real mechanism of biosynthesis of nanoparticles is still unclear; however, the research to explore the real mechanism behind biosynthesis of nanoparticles is still going on.

10.3.5 Factors Affecting Mycosynthesis of Nanoparticles

Biosynthesis of myconanoparticles, their characterization and applications are affected by several factors. Change in nature of the synthesized nanoparticles has been reported with the type of the adsorbate and the activity of the catalysts used in the synthesis process. Broadly, optimum temperature, pH, exposure time to substrate, biomass, presence of specific enzyme, substrate concentration, and, above all, the protocols that are followed for the synthesis process are some major physical and chemical factors reported to affect size, shape, and monodispersity of nanoparticles (Fig. 10.6). Nature of the reducing agent and stabilizing agent varies with the fungal species employed for the synthesis, which can be a major limitation in the use of fungi in nanoparticle synthesis. The growth of fungi to produce biomass, production of specific enzyme by fungi involved, and substrate concentration can affect the physical and chemical nature of the synthesized myconanoparticles. Optimum temperature and pH can influence the growth and development of fungi and production of specific enzymes. In addition, agitation rate, incubation time and metal salt concentration equally distress formation process of nanoparticles. Some dominant factors that affect nanoparticle biosynthesis are described below.

10.3.5.1 Effect of pH

pH is an important factor that has prominent effect on the synthesis of nanoparticles by green technology methods. It is reported that pH of the solution medium influences the size and texture of the synthesized nanoparticles, therefore, alteration of solution pH can control nanoparticles size. However, there is no set rule to predict the best pH for preparation of nanomaterials. A number of research findings are available about the importance of pH in synthesis of nanoparticles. The increase in particle size from nanometers to micrometer was observed as the pH increases from 7. Simultaneously, keeping pH low but increasing reaction time, gives bigger size particles. Similarly, with high pH and low exposure, size almost remains the same. Apart from size; shape of NPs also gets affected by pH. Spherical shape of synthesized nanoparticles was observed at lower pH, while size became bigger with well-defined shapes such as hexagons, triangles, rods, and spheres up to pH 5.0. The increase in size is continuous with increase in pH; however, at pH 7 and 9, the particles formed were small spherical particles as well as bigger particles with irregular and undefined shapes (Gericke and Pinches 2006; Karbasian et al. 2008). Reduction reaction of metallic ions is also found sensitive to pH. Acidic pH increases the time taken for reduction reaction (Sanghi and Verma 2009; Zhang et al. 2010; Sanghi et al. 2011; Rac et al. 2014; Wagers et al. 2014; Phanjom and Ahmed 2017).

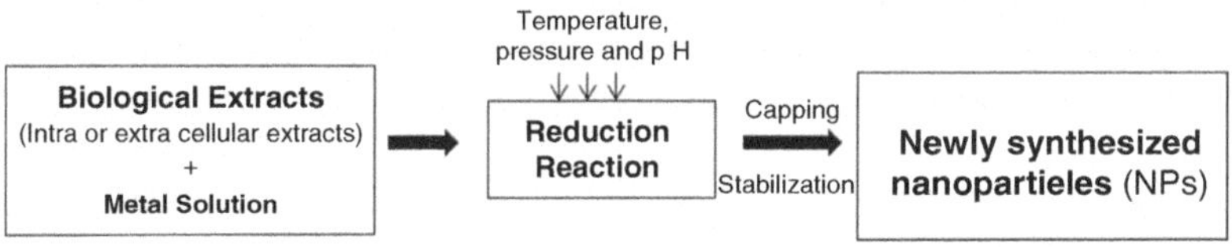

Fig. 10.6 Effects of pH, temperature and pressure on biosynthesis of nanoparticles

10.3.5.2 Effect of Temperature

Temperature plays a very prominent role on the morphology and size of nanoparticles. Generally elevated (ranging from 50 °C to 90 °C) temperatures offer small and high density uniform nanoparticles. The rate of reaction also increases. However, the best temperature for synthesis of nanoparticles is variable for different concentrations of precursor and reducing agent. Fungal metabolism and movement of metal ions are very much under the influence of temperature during nanoparticle synthesis. A promising effect of temperature on the growth of fungus as well as on metal uptake from the surrounding environment was observed which ultimately influenced the reaction rate and formation of the nanoparticles (Riddin et al. 2006; Dhillon et al. 2012). Fast or slow reaction rate during the synthesis process definitely could result in different sizes and morphology of the nanoparticles (Siva Kumar et al. 2011). Scientific studies showed that exposure to lower temperature for a time period of 1–24 hours resulted in the decrease in the number of smaller spherical particles; however, an increase in number of well-defined shaped larger particles was observed. Well-defined size and morphology of synthesized nanoparticles were observed to be unchanged upon an exposure up to 50 °C for 1–24 hours. The possible reason behind the synthesis of nanoparticles of well-defined size and shape at low temperature might be slower rate of particle formation (Sanghi et al. 2011; Pacioni et al. 2015; Reduan et al. 2016; Piñero et al. 2017; Phanjom and Ahmed 2017).

10.3.5.3 Effect of Enzymes

In synthesis process, use of specific enzymes is one of the crucial factors that influence the properties of the synthesized nanoparticles. Specificity of enzymes generates the need of use of specific fungal strains as all fungi are not able to produce required enzymes used in nanoparticle synthesis. The use of fungal enzymes is not only proving promising for the synthesis of predefined nanoparticles of different chemical composition, shape, and size, but its use also represents an important advancement in the biosynthesis of highly stable nanoparticles (Sanghi et al. 2011). Vigneshwaran et al. (2006) explored the lignolytic enzyme complex of *Phanerochaete chrysosporium* for the biosynthesis of gold NPs. Similarly, Sanghi et al. (2011) evaluated the same fungal strain (*P. chrysosporium*) to study the role of fungal enzymes in biosynthesis of gold NPs and observed that gold nanoparticles were formed within 90 minutes when challenged with gold ions, under normal pH in aqueous solution when temperature was increased to 37 °C. They further reported that the extracellular synthesis of nanoparticles was due to the presence of extracellular laccase enzyme and the intracellular nanoparticle synthesis on the mycelium surface was due to the presence of ligninase as evident by enzyme assays. Specificity of fungal enzymes also plays an important role in synthesis of nanoparticles. A study carried out by Ahmad et al. (2003a, b) revealed that synthesis of silver nanoparticles was not observed in the presence of *F. moniliforme*, but it was formed when *F. oxysporum* was present in the silver ion solution. The production of specific reductase enzyme (NADH-dependent reductase) by *F. oxysporum* was found to be the only reason behind this, although other reductases produced by the two fungi were the same. Similar findings were observed in mycosynthesis of zirconia and silica nanoparticles in the presence of fungal enzymes secreted from *F. oxysporum* (Bansal et al. 2006).

10.3.5.4 Effect of Metal Ion Concentration

The synthesis of nanoparticles of controlled size and morphology are some basic objectives as well as biggest challenges in nanotechnology. The concentration of metal ions is also one of the factors which play an important role in achieving these objectives and to overcome challenges. Many studies evaluated the metal ion concentration and nanoparticle synthesis and reported that nanoparticles with well-defined morphologies with an average standard size NPs were obtained at lower concentrations. Ahmad et al. (2014) studied the effect of gold ion concentration on size and properties of gold nanoparticles in Triton X-100 based inverse microemulsions. The results of their studies revealed that monodisperse and uniform gold nanoparticles with well-defined morphologies of an average grain size of 15 and 25 nm were obtained at lower concentrations (0.01 and 0.02 M). Similar observations were also proposed by Phanjom and Ahmed (2017) while studying the effect of different physicochemical conditions on the synthesis of silver nanoparticles using fungal cell filtrate of *Aspergillus oryzae*.

10.3.6 Applications of Myconanoparticles

Myconanotechnology is a new and emerging science deals with the synthesis of nanoparticles with the use of fungi. These myconanoparticles have great prospects for developing new products with industrial, agricultural, medicinal, and consumer applications in a wide range of sectors. Fungal products, with fungal nanomaterials are currently used in applications ranging from drug development to the food industry and agricultural biotechnology (Prasad 2017). Myconanoparticles have vast applications as they are reported to have antioxidant, anti-inflammatory, anticancer, antimicrobial, and catalytic activities and found very much effective in pathogen detection and control, wound healing, food preservation, textile fabrics, and many more (Saxena et al. 2014; Uzma et al. 2018; Aziz et al. 2016, 2019). The major applications of myconanoparticles in different sectors are as follows.

10.3.6.1 Agricultural Applications

The myconanoparticles have been reported to have inhibitory effects against plant pathogens. Number of silver and gold nanoparticles have been synthesized and evaluated against various phytopathogens (Kim et al. 2008, 2009; Gajbhiye et al.

2009; Min et al. 2009; Galdiero et al. 2011; Ray et al. 2011). The antibacterial activities of mycogenically synthesized silver nanoparticles (AgNPs) have also been evaluated (Matsumura et al. 2003; Bhainsa and D'Souza 2006; Afreen and Ranganath 2011; Singh et al. 2014) against a number of bacterial plant pathogens. Similarly, efficacy of nanoparticles against pathogenic fungi like *Sclerotia*, *Candida*, and *Phytophthora* (Gajbhiye et al. 2009; Shaligram et al. 2009; Ray et al. 2011) and viruses (Elechiguerra et al. 2005) has been well documented. The AgNPs synthesized from *T. viride* showed antibacterial activity and increased the shelf life of carrot and pear as compared to control when incorporated into thin film of sodium alginate in terms of weight loss and soluble protein content (Fayaz et al. 2009). The results of various studies revealed that these NPs can be considered as best alternate to chemical pesticides.

10.3.6.2 Healthcare Applications

Commercial applications have adapted gold nanoparticles as probes for the detection of targeted sequences of nucleic acids. This efficient approach increased the use of gold nanoparticles in clinical investigations and potential treatments for cancer and other diseases. However, the antimicrobial, healing, vector control properties of these nanomaterials has increased their publicity in healthcare sector. The antibacterial activities of mycogenically synthesized silver nanoparticles (AgNPs) were evaluated by Matsumura et al. (2003) who found that they have the ability to penetrate the bacterial cell wall and subsequently damage the cell membrane that leads to death of the cell. The silver nanoparticles synthesized with the help of *Rhizopus stolonifer* have been reported to have bactericidal against *P. aeruginosa* (Bhainsa and D'Souza 2006; Afreen and Ranganath 2011), *E. coli*, *P. aeruginosa*, and *S. aureus* (Birla et al. 2009; Singh et al. 2014). Several studies are available on anti-bactericidal effects of AgNPs against pathogenic bacteria (Shaligram et al. 2009; Ray et al. 2011). The antifungal effects against human pathogens like fungi (Khaydarov et al. 2011; Alghuthaymi et al. 2015; Prasad et al. 2017) and viruses (Elechiguerra et al. 2005; Ray et al. 2011) have been well studied. Recently, the effect of silver nanoparticles (AgNPs) synthesized from entomopathogenic fungus *Beauveria bassiana* is found effective against dengue vector *Aedes aegypti* (Banu and Balasubramanian 2014) and AgNPs synthesized from *Chrysosporium keratinophilum*, *Verticillium lecanii*, and *F. oxysporum* against filariasis vector *Culex quinquefasciatus* (Soni and Prakash 2012). Scientific studies also revealed that AgNPs have better wound healing properties (Sundaramoorthi et al. 2009; Thirumurugan et al. 2011).

The specificity in shape and size enables these nanoparticles to be more efficient and target specific which have made them popular leading to their use in drug industry. These NPs are capable of rapid detection of pathogenic fungi *Candida* sp. from low concentrated DNA (Bansod et al. 2013) which is helpful in early diagnosis of diseases and to produce precise solutions for disease prevention, diagnosis, and treatment. This technology has also broadened the medical tools, knowledge, and therapies currently available to clinicians.

10.3.6.3 Industrial Applications

Nanoparticles synthesized from fungi have remarkable biocatalytic properties. The metal nanoparticles obtained from fungi can be used in enzyme immobilization for improved enzymatic activity (Mishra et al. 2014a). This property is not only useful in industrial sector but found constructive in agriculture and heath sector also. The myconanoparticles are now used in cosmetic, fertilizer, and by-product industries and agro-industry also (Raj et al. 2012; Melo et al. 2015; Prasad 2016, Abdel-Aziz et al. 2018; Prasad et al. 2017).

10.4 Myconanotechnology in Agriculture

Due to the ever-growing global food demand due to changing climate, urbanization, and environmental issues such as run-off and accumulation of agrochemicals, there is an increasing need to feed an estimated population growth from 6 billion to 9 billion by 2050. Unfortunately, natural resources such as cultivable land, water, and soil fertility are decreasing continuously with the uncontrolled increase in population. The cost of chemical fertilizers, pesticides, and other production inputs has been drastically increased due to limited reserves of natural gas and petroleum. There is a need to overcome these constraints with the help of precision farming practices and effective applications of myconanotechnology to agriculture.

Mycogenic nanostructured materials provide intelligent and smart delivery systems because of their size and shape or functionality. These properties enable them to be used in nanodevices like biosensors and detectors to detect trace quantities of bacteria, airborne pathogens, biological hazards, and disease signatures well in advance (Chowdappa and Gowda 2013). Their smart functionality is expected to emerge as most important and powerful tool as alternate to conventional dosage form, and it is expected that the market will be flooded with nanotechnology devised pesticides and biocontrol practices in the next coming decades.

10.4.1 Myconanoparticles in Agriculture

It is now well established fact that fungal synthesized nanomaterials can be used for combating the plant diseases either by controlled delivery of functional molecules or as diagnostic tool for disease detection. These nanomolecules are not only limited to plant protection; they have a wide array of applications in other fields also. They are now used in detection and measurement of crop nutrient status, insects, pathogens, weeds, moisture level, soil fertility, and temperature, which helps in real-time monitoring of the crop growth and provides essential data for precision farming practices leading to minimize agricultural inputs and maximize resource

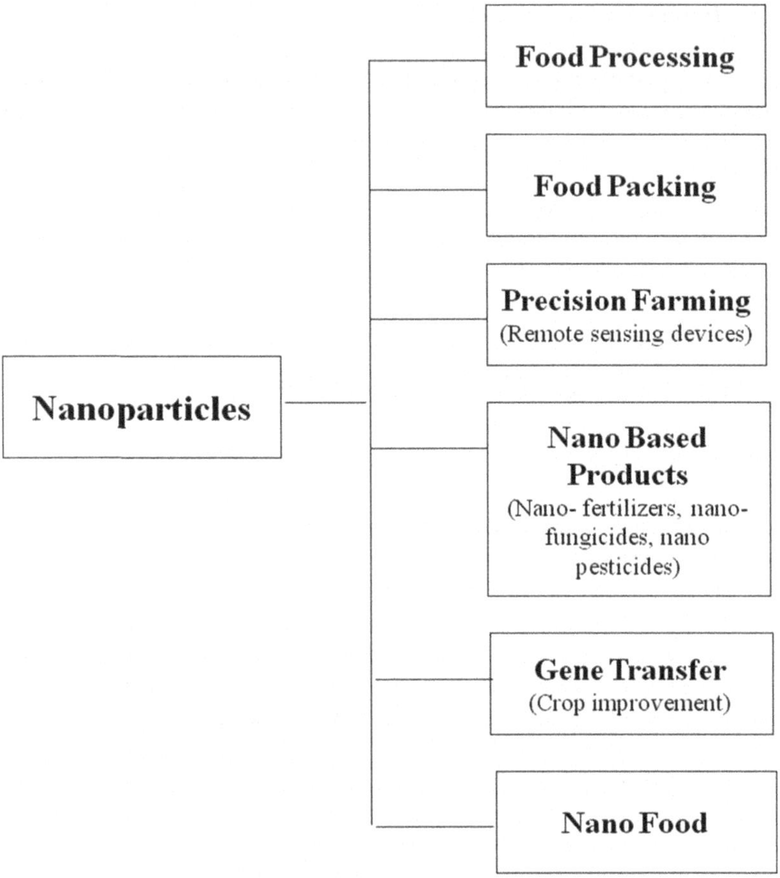

Fig. 10.7 Application of nanotechnology in different fields of agriculture

output and yield (Fig. 10.7). Important role of myconanoparticles in agriculture has been described in this section (Prasad et al. 2014).

10.4.1.1 Micronutrient Supply, Soil Health, and Plant Growth

Micronutrients are essential elements required for plant growth and play an important role in balanced crop nutrition. They include boron (B), copper (Cu), chloride (Cl), iron (Fe), manganese (Mn), molybdenum (Mo), nickel (Ni), and zinc (Zn). It is now well documented that green revolution has increased the crop yields, however, progressive decrease in micronutrients of soil was also accomplished. The external application of these micronutrients is the only way to balance soil micronutrients and make them available to growing plants. Use of fungi based nanomaterials is now a popular application of nanotechnology to enrich soil micronutrients. The micronutrients as nanoformulation can also be sprayed directly on plants. A number of studies have been conducted to test different nanoparticles as source of micronutrients in agricultural crops/plants (Alloway 2008; Graham 2008; Peteu et al. 2010; Tao et al. 2012) and as agents to improve soil health. It has been reported that nanoparticles like hydrogels, nanoclays, and nanozeolites enhance the water holding capacity of the soil and allow a slow release of water. This, in turn, reduces the hydric shortage, particularly during the crop season. Such practices are also helpful in the reforestation of degraded areas that have lost their soil fertility. The nanoparticles can be really helpful in remediating the plant growth by ensuring that the nutrients are utilized optimally by the plants. Moreover, these nanomaterials can be used to absorb the contaminants, thereby improving the soil remediation capacity and improving its health.

10.4.1.2 Role in Detecting Plant-Related Health Issues

The nanomaterials are now being used as preventive and early warning system. This approach is proving to be helpful in identifying and addressing plant health issues and offer remedial measures. The extreme harms caused by chemical pesticides can be neutralized by adopting integrated pest management system along with use of nanomaterials. These nanoparticles are useful in many ways such as efficient use of pesticides, management of resources, diffusion of precision agriculture, and reduction of waste. Role in management of phytopathogens is one of the most important applications of myconanoparticles in agriculture which have been described in this section.

10.4.1.3 Role in Agricultural Pest Management

Agriculture is one of the global leading sectors which grew and developed with progresses in civilizations and still on the path of continuous growth. It is the only way of producing food, feed, fiber, and many other desired products. The global increase in population leads to an ever-growing food demand and agriculture is the only solution to fulfill it. However, pests, including insects, mites, nematodes, and pathogens, are the major limiting factor in profitable crop production. Use of pesticides has resulted in development of pest and disease resistance, accumulating residues in produce and environmental pollution. Synthesis of various biological nanoparticles has emerged as an effective alternate to chemical pesticides as they are now causing harmful effects on users and consumers. Their applications not only advance the crop protection in management of insects and pathogens but also provide diagnostic tools for early detection. However, use in agriculture, especially for plant production and protection, is an under-explored area in the research community. The results of initial investigations on nanomaterials showed that these molecules have potential ability to improve seed germination and growth, plant protection, pathogen detection, and pesticide/herbicide residue detection (Ismail et al. 2017; Gupta et al. 2018; Abd-Elsalam and Prasad 2018, 2019).

Use of pesticides to control pest is facing a new challenge in form of insect pest resistance. Myconanotechnology represents a new model in the development of antimicrobial nanomaterial and it offers a new alternative to pesticides for the control of plant pathogens. Studies on testing efficacy of myconanoparticles against phytopathogens revealed that the size and shape of these particles play an important role in their effectiveness and accuracy in action.

10.4.1.4 As Fungicides

Nanotechnology offers an excellent alternate to the use of commercially available chemical fungicides as they cause detrimental effects to the health of soil and plants. Because of the target specificity and accuracy, these nanoparticles are playing an important role in solving this problem. A number of myconanoparticles have been investigated for their antifungal activity. Various nanoparticles of zinc oxide, silver, gold, and titanium dioxide have been synthesized and tested against plant pathogens like *Aspergillus flavus*, *Aspergillus fumigatus*, *Aspergillus niger* (Choudhury et al. 2011; Patra et al. 2012; Ramyadevi et al. 2012; Suriyaprabha et al. 2014; Jasim 2015), *Alternaria alternata*, *Alternaria solani*, *Alternaria citri* (Kim et al. 2012b; Patra et al. 2012; Abdelmalek and Salaheldin 2016), *Bipolaris sorokiniana* (Jo et al. 2009; Mishra et al. 2014b), *Botrytis cinerea* (He et al. 2011), *Candida albicans* (Jasim 2015), *Curvularia lunata* (Kanhed et al. 2014), *Fusarium equiseti* (Bramhanwade et al. 2016), *Fusarium oxysporum* (Choudhary et al. 2011; Choudhury et al. 2011; Patra et al. 2012; Kanhed et al. 2014; Suriyaprabha et al. 2014; Bramhanwade et al. 2016), *Magnaporthe grisea* (Jo et al. 2009), *Macrophomina phaseolina* (Saharan et al. 2013; Shyla et al. 2014), *Monosporascus cannonballus* (Kim et al. 2012b), *Penicillium digitatum* (Abdelmalek and Salaheldin 2016), *Penicillium expansum* (He et al. 2011), *Phoma destructiva* (Kanhed et al. 2014), *Pythium spinosum*, *Pythium aphanidermatum*, *Stemphylium lycopersici* (Kim et al. 2012a), and *Rhizoctonia solani* (Saharan et al. 2013). From the studies on antifungal effects on nanoparticles, it was revealed that silver has much higher antifungal activity at lower concentrations than that of other metals. It was reported that disruption of transmembrane, energy metabolism, and electron transport chain occurred because of the inactivation of thiol groups of fungal cell wall. A decrease in the efficacy of silver nanoparticles was observed with the increase in particle size. Shapes also affect the efficacy of NPs, truncated triangular shaped nanoparticles showed greater fungicidal effect than spherical and rod shaped particles (Panáček et al. 2006; Pal et al. 2007; Velmurugan et al. 2009).

Table 10.2 Insecticidal activities of various nanoparticles

Nanoparticles	Insecticidal effect against	References
Glycol-coated nanoparticles	*Tribolium castaneum* (red flour beetle)	Yang et al. (2009)
Silver nanoparticles, aluminum oxide, zinc oxide, and titanium dioxide	*Sitophilus oryzae*, *Bombyx mori*	Goswami et al. (2010)
Nanostructured alumina	*Sitophilus oryzae*, *Rhyzopertha dominica*	Teodoro et al. (2010)
Silica nanoparticles	*Sitophilus oryzae*	Debnath et al. (2011)
Nanoparticle	*Bombyx morinucleo*	Das et al. (2013)
Silver and zinc nanoparticles	*Aphis nerii*	Rouhani et al. (2012)
Amitraz with ZnO and ZnO–Al_2O_3 nanoparticles	*Agonoscena pistaciae*	Samih et al. (2011)
ZnO–TiO_2–Ag nanoparticles	*Frankliniella occidentalis*	Rouhani et al. (2011)
Nano-SDS/Ag/TiO_2–IMI	*Martianus dermestoides*	Guan et al. (2008)
Silica nanoparticles	*Tribolium castaneum*	Yang et al. (2009)
Silica nanoparticles	*Spodoptera littoralis*	Barik et al. (2008)

10.4.1.5 As Insecticides

Nanoparticles have a great potential to suppress insect pests affecting agricultural crops. The regular use of chemical agrochemicals has raised a number of problems like toxicity to plants, soil health, and even consumers. The development of insect pest resistance is another challenge developed by use of synthetic pesticides. Therefore, nanopesticides present an excellent alternative to synthetic agrochemicals, as they are target specific, nontoxic, and eco-friendly. They have a great promise for the management and control of insect pest of modern agriculture (Liu et al. 2006; Pérez-de-Luque and Rubiales 2009). The insecticidal activities of various nanoparticles have been investigated against a number of insect pests (Table 10.2).

10.4.1.6 As Herbicides

Another peril in agriculture is the weeds as they have the potential to reduce the yield up to a great extent. The management practices which are in fashion are either eradication or use of chemical herbicides. However, a number of herbicides are available in the market, their use posing a serious threat to delicate agro-ecosystems. Their design to act only upon aboveground parts of weeds is main disadvantage, while toxic residual effect contributes toward ill health of users, consumers, and even soil and environment. Nanotechnology has offered an excellent solution for the removal of weeds from the field and for the safety of users, consumers, soil, and environment. Here encapsulation of herbicides in polymeric nanoparticles can be carried out so that they can act on weeds in an eco-friendly way, without leaving any harmful residue in soil and environment. If the efficacy of herbicides improved through the use of nanotechnology, the crop production can be enhanced in an efficient and safe way. Research on formulation of nanoherbicides, testing their efficacy, and testing against weeds is still a continuous process. It has been reported that nanoparticles located in herbicides act smartly on weed roots, enter into their root system, translocate to cells, cause DNA damage, and inhibit metabolic pathway (Atha et al. 2012; Duhan et al. 2017) which ultimately leads to death of the weed plants. Herbicidal effects of nanoparticles were assessed in a number of studies against *Elodea densa* (Nekrasova et al. 2011), *Cucurbita pepo* (Hawthorne et al. 2012; Musante and White 2012), *Cucumis sativus* (Kim et al. 2012b), *Fagopyrum esculentum* (Lee et al. 2013), *Eichhornia crassipes* (Konotop et al. 2014; Namasivayam et al. 2014), and alga *Pseudokirchneriella subcapitata* and the microcrustacean *Daphnia similis* (Clemente et al. 2014). The biosynthesized nanoparticles also have potential application in the slow release and target delivery.

10.4.1.7 Role in Delivery of Chemical and Biofertilizers

Fertilizers are the off-screen players which played an important role in crop yields. This is one of the major reasons behind the success of green revolution. Although food production increased enormously by the use of fertilizers, however, there are many harmful impacts on beneficial soil microflora. Most of the fertilizers are not available to plants due to run-off and cause pollution. The use of biofertilizers like fungal mycorrhizae, *Rhizobium*, *Azotobacter*, *Azospirillum*, blue green algae, phosphate solubilizing bacteria has been emerged as best alternatives for synthetic fertilizers. They are considered safe and eco-friendly as they convert organic matter into simple compounds that provide essential nutrients to plants, improve soil fertility, maintain the natural habitat of the soil, and increase crop yield. There are, however, some drawbacks in their usage which include short shelf life, temperature sensitivity, and storage desiccation problems (Wu et al. 2005; Jha and Prasad 2006; Vandergheynst et al. 2006).

Nanotechnology provides an application that allow coating of fertilizers with polymeric nanoparticles, which enable them to be available to plants and avoid run-off. The coating also protects biofertilizers from desiccation and increases their shelf life (Santoso et al. 1995; Wilson et al. 2008). Moreover, target delivery is also enhanced which cause a positive impact on growth of plants. Improved root and stem growth, seed germination, development of healthy seedlings, and increased final yield are some positive effects on use of nanoparticles, whereas inhibition of cell growth and nitrogen fixation ability, transpiration, reduced germination, decreased shoot and root length and biomass, and decreased mitosis were reported as some adverse effects of nanoparticles on plants (Duhan et al. 2017).

10.4.1.8 Role in Food Industry

Apart from wide use of NPs in agriculture industry, they have been really integrated in a number of food and food packaging products during the last couple of years (Manimaran 2015). The conventional packing strategies are posing serious threats in terms of safety, performance, and economically. The materials used in conventional food packing are difficult to degrade and cause serious waste problems as solid waste material and threat to environment (Siracusa et al. 2008; Farris et al. 2009). The use of nanotechnology has solved this problem to a greater extent, but more work is still required to be carried out. It has been reported that incorporation of nanomaterials in packaging biopolymers has been proven to provide necessary mechanical strength, better reinforcement (Choudalakis and Gotsis 2009), increased thermal degradation (Bertini et al. 2006; Cava et al. 2006; Petersson and Oksman 2006), and barrier properties (Choudalakis and Gotsis 2009). The use of fungi-based silver nanoparticles and their incorporation into sodium alginate films for vegetable and fruit preservation have been studied by Fayaz et al. (2009). It has been reported that nanocapsules delivery systems play an important role in processing sector. They are also proved helpful in retaining the functional property of the processing material and keeping them healthy and fresh (Dingman 2008; Abbas et al. 2009).

After processing, nanoparticles play an important role in food packaging. The packing has a great significance in preserving the food, maintaining its quality, and shelf life. They are incorporated in the polymer matrix of the substances which develop a barrier for gases and help in maintaining an inert and low-oxygen atmosphere for inhibiting microbial growth and spoilage (Floros 2010). Another important commercial use of nanoparticles in food industry is its excellent role in antimicrobial packing. The biosynthesized nanoparticles have been reported to inhibit wide range of biological processes in microbes which are responsible for food spoilage in packing. Henriette and de Azeredo (2009) studied the antimicrobial activity of silver nanoparticles and reported that the AgNPs show needed action in bulk form, and its ions have the ability to inhibit wide range of biological processes in bacteria. Moreover, the incorporation of packaging material made of natural or synthetic nanoparticles has the properties like biodegradability which ensure their recycling. This approach protects soil from nondegradable materials and atmosphere from toxic gases (Momin et al. 2013). The use of nanosensors also reflects the shelf life of the packed products which is beneficial for consumers in terms of better quality identification and producers for rapid distribution and authentication of the food products (Singh et al. 2017).

10.4.2 Research Needs and Future Prospects

The boundaries of research on nanotechnology and their applications are increasing every year along with regular additions in published research. Enormous work has been carried out on synthesis of nanoparticles and their characterization and applications. However, investigations on NPs as nanocarriers for controlled release of active substances, their interaction (either positive or negative) with plants or microorganisms under controlled conditions, and development of nanodevices are still going on. Although use of different microbes in the synthesis of nanoparticles is carried out on large scale, however, a lot of research scope is still there to discover new, better, and efficient microbes. Despite tremendous advancements in nanotechnology, a number of questions need to be answered and can be areas of research for present and future researchers.

- What is the exact mechanism of nanoparticle synthesis using fungus?
- Evaluation of effects of varying parameters like fungus type, growth stage, cultivation medium, synthesis conditions, pH, substrate concentrations, source compound of target nanoparticle, temperature, reaction time, and addition of nontarget ions on shape and size of nanoparticles.
- Development of cost-effective methods of nanoparticle synthesis also requires more research so that nanodevices must be affordable with the current production cost of agricultural goods. Simultaneously, establishment of low-cost recovery techniques to make the synthesis process commercially feasible also needs to be undertaken.
- The pathways (apoplast vs. symplast) nanoparticles adapt to enter inside plants, factors affecting and final fate of interaction between nanoparticles (NPs) and plants still requires detailed investigations.

The nanotechnology is an important science which can be used in developing sustainable and environmentally friendly technologies for the future through green synthesis and other biological approaches. For this, it is necessary to frequently emphasize on the mechanisms and approaches for synthesis of myconanoparticles and their applications.

10.5 Conclusion

Nanoparticles have a great potential as 'magic bullets loaded with herbicides, fungicides, nutrients, fertilizers, or nucleic acids targeting specific sites to release their charge on the desired part of site to achieve the desired results. They are excellent in action and reaction and actually the basis for all systems in our living and mineral world. The smallest size of nanomaterials is among the major factors responsible for their ever increasing uses.

In this chapter we have tried to explore the potential of nanotechnology with respect to nanoparticles (NPs), their synthesis, mechanism of synthesis, and their properties. Special emphasis was given on the role of fungal-based nanoparticles and their application in the field of agriculture. Myconanoparticles are emerging as outstanding solutions for various problems in the field of agriculture, pharmacy, and other allied sciences. In agriculture, their role as pesticides, as biofertilizers, and in food industry is proving quite exceptional as they are eco-friendly, efficient and specific on target, easy to maintain, and cost effective. Fungal agents are environmentally friendly, clean, and nontoxic mediator for the synthesis of metal nanoparticles and employ both intracellular and extracellular methods. The simplicity of scaling up and downstream processing and the presence of fungal mycelia provide an increased surface area which added some key advantages in the synthesis of myconanoparticles and their importance in nanotechnology. Even though nanoscience is often perceived as science of future, certain issues are needed to be addressed. First, we need to be sure that a technology is safe before applying it to the environment, and second, consumers will distrust a technology unfamiliar for them in the same way that happened with transgenic crops. We need to broaden our horizons and start looking for real practical approaches, filling the main gaps that hamper our jump from laboratory research into field applications.

References

Abbas KA, Saleh AM, Mohamed A, Mohd Azhan N (2009) The recent advances in the nanotechnology and its applications in food processing: a review. J Food Agri Environ 7(3–4):14–17

Abbasi E, Aval SF, Akbarzadeh A, Milani M, Nasrabadi HT, Joo SW, Hanifehpour Y, Nejati-Koshki K, Pashaei-Asl R (2014) Dendrimers: synthesis, applications, and properties. Nanoscale Res Lett 9(1):247

Abdelmalek GMA, Salaheldin TA (2016) Silver nanoparticles as a potent fungicide for Citrus phytopathogenic fungi. J Nanomed Res 3:00065. https://doi.org/10.15406/jnmr.2016.03.00065

Abd-Elsalam KA, Prasad R (2018) Nanobiotechnology Applications in Plant Protection. Springer International Publishing (ISBN 978-3-319-91161-8) https://www.springer.com/us/book/9783319911601

Abdel-Aziz SM, Prasad R, Hamed AA, Abdelraof M (2018) Fungal nanoparticles: A novel tool for a green biotechnology? In: Fungal Nanobionics: Principles and Applications (eds. Prasad R, Kumar V, Kumar M, Wang S), Springer Singapore Pte Ltd. 61–87

Abd-Elsalam K, Prasad R (2019) Nanobiotechnology Applications in Plant Protection. Volume 2. Springer International Publishing (ISBN 978-3-030-13295-8) https://www.springer.com/gp/book/9783030132958

Abhilash, Revati K, Pandey BD (2011) Microbial synthesis of iron-based nanomaterials – a review. Bull Mater Sci 34(2):191–198

Afreen RV, Ranganath E (2011) Synthesis of monodispersed silver nanoparticles by *Rhizopus stolonifer* and its antibacterial activity against MDR strains of *Pseudomonas aeruginosa* from burnt patients. Int J Environ Sci 1(7):1582–1592

Ahmad A, Mukherjee P, Mandal D, Senapati S, Khan MI, Kumar R, Sastry M (2002) Enzyme mediated extracellular synthesis of CdS nanoparticles by the fungus *F. oxysporum*. J Am Chem Soc 124(141):12108–12109

Ahmad A, Mukharjee P, Senapati S, Mandal D, Khan MI, Kumar R (2003a) Extracellular biosynthesis of silver nanoparticles using the fungus *Fusarium oxysporium*. Colloid Surf B Biointerfaces 28(4):313–318

Ahmad A, Senapati S, Khan MI, Kumar R, Ramani R, Srinivas V, Sastry M (2003b) Intracellular synthesis of gold nanoparticles by a novel alkalotolerant actinomycete, *Rhodococcus* species. Nanotechnology 14:824–828

Ahmad T, Wani IA, Ahmed J, Al-Hartomy OA (2014) Effect of gold ion concentration on size and properties of gold nanoparticles in TritonX-100 based inverse microemulsions. Appl Nanosci 4:491–498

Alghuthaymi MA, Almoammar H, Rai M, Said-Galiev E, Abd-Elsalam KA (2015) Myconanoparticles: synthesis and their role in phytopathogens management. Biotechnol Biotechnol Equip 29(2):221–236

Alloway BJ (2008) Micronutrients and crop production: an introduction. In: Alloway BJ (ed) Micronutrient deficiencies in global crop production. Springer, Dordrecht, pp 1–39

Atha DH, Wang H, Petersen EJ, Cleveland D, Halbrook RD, Jaruga P, Dizdaroglu M, Xing B, Nelson BC (2012) Copper oxide nanoparticle mediated DNA damage in terrestrial plant models. Environ Sci Technol 46:1819–1827

Aziz N, Pandey R, Barman I, Prasad R (2016) Leveraging the attributes of *Mucor hiemalis*-derived silver nanoparticles for a synergistic broad-spectrum antimicrobial platform. Front Microbiol 7:1984. doi: https://doi.org/10.3389/fmicb.2016.01984

Aziz N, Faraz M, Sherwani MA, Fatma T, Prasad R (2019) Illuminating the anticancerous efficacy of a new fungal chassis for silver nanoparticle synthesis. Front Chem 7:65. doi: 10.3389/fchem.2019.00065

Balashanmugam P, Santhosh S, Giyaullah H, Balakumaran MD, Kalaichelvan PT (2013) Mycosynthesis, characterization and antibacterial activity of silver nanoparticles from *Microsporus xanthopus*: a macro mushroom. Int J Innov Res Sci Eng Technol 2(11):6262–6270

Bansal V, Rautaray D, Ahmad A, Sastry M (2004) Biosynthesis of zirconia nanoparticles using the fungus *Fusarium oxysporum*. J Mater Chem 14:3303–3305

Bansal V, Rautaray D, Bharde A, Ahire K, Sanyal A, Ahmad A, Sastry M (2005) Fungus-mediated biosynthesis of silica and titania particles. J Mater Chem 15(1):2583–2589

Bansal V, Poddar P, Ahmad A, Sastry M (2006) Room-temperature biosynthesis of ferroelectric barium titanate nanoparticles. J Am Chem Soc 128(36):11958–11963

Bansod S, Bonde S, Tiwari V, Bawaskar M, Deshmukh S, Gaikwad S, Gade A, Rai M (2013) Bioconjugation of gold and silver nanoparticles synthesized by *F. oxysporum* and their use in rapid identification of *Candida* species by using bioconjugate-nano-polymerase chain reaction. J Biomed Nanotechnol 9(12):1962–1971

Banu NA, Balasubramanian C (2014) Myco-synthesis of silver nanoparticles using *Beauveria bassiana* against dengue vector, *Aedes aegypti* (Diptera: Culicidae). Parasitol Res 113(8):2869–2877

Bao HF, Hao N, Yang YX, Zhao DY (2010) Biosynthesis of biocompatible cadmium telluride quantum dots using yeast cells. Nano Res 3(7):481–489

Barik TK, Sahu B, Swain V (2008) Nano-silica from medicine to pest control. Parasitol Res 103:253–258

Behari J (2010) Principles of nanoscience: an overview. Ind J Exp Biol 48:1008–1019

Bertini F, Canetti M, Audisio G, Costa G, Falqui L (2006) Characterization and thermal degradation of polypropylene–montmorillonite nanocomposites. Polymer Degrad Stab 91(3):600–605

Bhainsa KC, D'Souza SF (2006) Extracellular biosynthesis of silver nanoparticles using the fungus *Aspergillus fumigatus*. Colloids Surf B Biointerfaces 47(2):160–164

Bharde A, Rautaray D, Bansal V, Ahmad A, Sarkar I, Yusuf SM, Sanyal M, Sastry M (2006) Extracellular biosynthesis of magnetite using fungi. Small 2:135–141

Bhattacharya D, Gupta RK (2005) Nanotechnology and potential of microorganisms. Crit Rev Biotechnol 25(4):199–204

Bhattacharyya A, Duraisamy P, Govindarajan M, Buhroo AA, Prasad R (2016) Nano-biofungicides: Emerging trend in insect pest control. In: Advances and Applications through Fungal Nanobiotechnology (ed. Prasad R), Springer International Publishing Switzerland 307–319

Birla SS, Tiwari VV, Gade AK, Ingle AP, Yadav AP, Rai MK (2009) Fabrication of silver nanoparticles by *Phoma glomerata* and its combined effect against *Escherichia coli*, *Pseudomonas aeruginosa* and *Staphylococcus aureus*. Lett Appl Microbiol 48(2):173–179

Bramhanwade K, Shende S, Bonde S, Gade A, Rai M (2016) Fungicidal activity of cu nanoparticles against *Fusarium* causing crop diseases. Environ Chem Lett 14:229–235

Cava D, Giménez E, Gavara R, Lagaron J (2006) Comparative performance and barrier properties of biodegradable thermoplastics and nanobiocomposites versus PET for food packaging applications. J Plastic Film Sheeting 22(4):265–274

Chan YS, Don MM (2012) Characterization of Ag nanoparticles produced by white-rot fungi and its *in vitro* antimicrobial activities. Int Arab J Antimicrob Agents 2:1–8

Chan YS, Don MM (2013) Biosynthesis and structural characterization of Ag nanoparticles from white rot fungi. Mat Sci Eng C 33:282–288

Chen GQ, Zou ZJ, Zeng GM, Yan M, Fan JQ, Chen AW, Yang F, Zhang WJ, Wang L (2011) Coarsening of extracellularly biosynthesized cadmium crystal particles induced by thioacetamide in solution. Chemosphere 83(9):1201–1207

Choudalakis G, Gotsis A (2009) Permeability of polymer/clay nanocomposites: a review. Eur Pol J 45(4):967–984

Choudhury SR, Ghosh M, Mandal A, Chakravorty D, Pal M, Pradhan S, Goswami A (2011) Surface-modified sulfur nanoparticles: an effective antifungal agent against *Aspergillus niger* and *Fusarium oxysporum*. Appl Microbiol Biotechnol 90:733–743

Chowdappa P, Gowda S (2013) Nanotechnology in crop protection: status and scope. Pest Manag Hort Ecosys 19(2):131–151

Clemente Z, Grillo R, Jonsson M, Santos NZ, Feitosa LO, Lima R (2014) Ecotoxicological evaluation of poly (ε-caprolactone) nanocapsules containing triazine herbicides. J Nanosci Nanotechnol 14:4911–4917

Das S, Bhattacharya A, Debnath N, Datta A, Goswami A (2013) Nanoparticle-induced morphological transition of *Bombyx morinucleo* polyhedrovirus: a novel method to treat silkworm grasserie disease. Appl Microbiol Biotechnol 97:6019–6030

Debnath N, Das S, Seth D, Chandra R, Bhattacharya SC, Goswami A (2011) Entomotoxic effect of silica nanoparticles against *Sitophilus oryzae* (L.). J Pest Sci 84:99–105

Dhandhukia PC, Patel M, Thakker JN (2012) Biosynthesis of silver nanoparticles using a plant pathogenic fungus *Fusarium oxysporum* f. sp. *cubense*. J Pure Appl Sci 20:10–14

Dhillon GS, Brar SK, Kaur S, Verma M (2012) Green approach for nanoparticle biosynthesis by fungi: current trends and applications. Crit Rev Biotechnol 32(1):49–73

Dingman J (2008) Nanotechnology its impact on food safety. J Environ Health 70(6):47–50

Duhan JS, Kumar R, Kumar N, Kaur P, Nehra K, Duhan S (2017) Nanotechnology: the new perspective in precision agriculture. Biotechnol Rep 15:11–23

Durán N, Marcato PD, Alves OL, de Souza GIH, Esposito E (2005) Mechanistic aspects of biosynthesis of silver nanoparticles by several *Fusarium oxysporum* strains. J Nanobiotechnol 13(3):8. https://doi.org/10.1186/1477-3155-3-8

Elechiguerra JL, Burt JL, Morones JR, Camacho-Bragado A, Gao X, Lara HH (2005) Interaction of silver nanoparticles with HIV-1. J Nanobiotechnol 3:6–16

Farris S, Schaich KM, Liu L, Piergiovanni L, Yam KL (2009) Development of polyion-complex hydrogels as an alternative approach for the production of bio-based polymers for food packaging applications: a review. Trends Food Sc Tech 20(8):316–332

Fayaz MA, Balaji K, Girilal M, Kalaichelvan PT, Venkatesan R (2009) Mycobased synthesis of silver nanoparticles and their incorporation into sodium alginate films for vegetable and fruit preservation. J Agric Food Chem 57(14):6246–6252

Floros JD (2010) New and emerging applications of nanotechnology in our food supply. In: IFT international food nanoscience conference, Department of Food Science, Pennsylvania State University, 17 July, 2010, pp 25–28

Gade A, Ingle A, Whiteley C, Rai M (2010) Mycogenic metal nanoparticles: progress and applications. Biotechnol Lett 32:593–600

Gaikwad S, Birla SS, Ingle AP, Gade AK, Marcato PD, Rai MK, Duran D (2013) Screening of different *Fusarium* species to select potential species for the synthesis of silver nanoparticles. J Braz Chem Soc 24:1974–1982

Gajbhiye M, Kesharwani J, Ingle A, Gade A, Rai M (2009) Fungus mediated synthesis of silver nanoparticles and their activity against pathogenic fungi in combination with fluconazole. Nanomed Nanotechnol Biol Med 5:382–386

Galdiero S, Falanga A, Vitiello M, Cantisani M, Marra V, Galdiero M (2011) Silver nanoparticles as potential antiviral agents. Molecules 16(10):8894–8918

Geetanjali H, Ashok C (2015) Synthesis, characterization and stability of gold nanoparticles using the fungus *Fusarium oxysporum* and its impact on seed germination. Int J Recent Sci Res 6(3):3181–3185

Gericke M, Pinches A (2006) Microbial production of gold nanoparticles. Gold Bull 39:22–28

Goswami A, Roy I, Sengupta S, Debnath N (2010) Novel applications of solid and liquid formulations of nanoparticles against insect pests and pathogens. Thin Solid Films 519:1252–1257

Govender Y, Riddin T, Gericke M, Whiteley CG (2009) Bioreduction of platinum salts into nanoparticles: a mechanistic perspective. Biotechnol Lett 31(1):95–100

Graham RD (2008) Micronutrient deficiencies in crops and their global significance. In: Alloway BJ (ed) Micronutrient deficiencies in global crop production. Springer, Dordrecht, pp 41–61

Guan H, Chi D, Yu J, Li X (2008) A novel photodegradable insecticide: preparation, characterization and properties evaluation of nano-imidacloprid. Pest Biochem Physiol 92:83–91

Gunter R, Andreas H (2010) Magnetic nanoparticles. In: Sattler KD (ed) Handbook of nanophysics: nanoparticles and quantum dots. CRC Press, Boca Raton, pp 1–2

Gupta N, Upadhyaya CP, Singh A, Abd-Elsalam KA, Prasad R (2018) Applications of silver nanoparticles in plant protection. In: Nanobiotechnology Applications in Plant Protection (eds. Abd-Elsalam K, Prasad R), Springer International Publishing AG 247–266

Gurunathan S, Raman J, Abd Malek SN, John PA, Vikineswary S (2013) Green synthesis of silver nanoparticles using *Ganoderma neojaponicum* Imazeki: a potential cytotoxic agent against breast cancer cells. Int J Nanomedicine 8:4399–4413

Hamedi S, Shojaosadati SA, Shokrollahzadeh S, Hashemi-Najafabadi S (2014) Extracellular biosynthesis of silver nanoparticles using a novel and non-pathogenic fungus, *Neurospora intermedia*: controlled synthesis and antibacterial activity. World J Microbiol Biotechnol 30(2):693–704

Hawthorne J, Musante C, Sinha SK, White JC (2012) Accumulation and phytotoxicity of engineered nanoparticles to *Cucurbita pepo*. Int J Phytoremediation 14:429–442

He L, Liu Y, Mustapha A, Lin M (2011) Antifungal activity of zinc oxide nanoparticles against *Botrytis cinerea* and *Penicillium expansum*. Microbiol Res 166:207–215

Hemath NKS, Kumar G, Karthik L, Bhaskara Rao KV (2010) Extracellular biosynthesis of silver nanoparticles using the filamentous fungus *Penicillium* sp. Arch Appl Sci Res 2:161–167

Henriette MC, de Azeredo (2009) Nanocomposites for food packaging applications. Food Res Int 42:1240–1253

Ingle A, Gade A, Pierrat S, Sonnichsen C, Rai M (2008) Mycosynthesis of silver nanoparticles using the fungus *Fusarium acuminatum* and its activity against some human pathogenic bacteria. Curr Nanosci 4:141–144

Ingle A, Gade A, Bawaskar M, Rai M (2009) *Fusarium solani*: a novel biological agent for the extracellular synthesis of silver nanoparticles. J Nanopart Res 11:2079–2085

Ismail M, Prasad R, Ibrahim AIM, Ahmed ISA (2017) Modern prospects of nanotechnology in plant pathology. In: Nanotechnology (eds. Prasad R, Kumar M, Kumar V), Springer Nature Singapore Pte Ltd. 305–317

Jasim NO (2015) Antifungal activity of zinc oxide nanoparticles on *Aspergillus fumigatus* fungus and *Candida albicans* yeast. J Nat Sci Res 5:23–27

Jha MN, Prasad AN (2006) Efficacy of new inexpensive cyanobacterial biofertilizer including its shelf-life. World J Microbiol Biotechnol 22:73–79

Jo YK, Kim BH, Jung G (2009) Antifungal activity of silver ions and nanoparticles on phytopathogenic fungi. Plant Dis 93:1037–1043

Kanhed P, Birla S, Gaikwad S, Gade A, Seabra AB, Rubilar O, Durane N, Rai M (2014) In vitro antifungal efficacy of copper nanoparticles against selected crop pathogenic fungi. Mater Lett 115:13–17

Karbasian M, Atyabi SM, Siadat SD, Momem SB, Norouzian D (2008) Optimizing nano-silver formation by Fusarium oxysporum (PTCC 5115) employing response surface methodology. Am J AgricBiolSci 3:433–437

Kashyap PL, Kumar S, Srivastava AK, Sharma AK (2013) Myconanotechnology in agriculture: a perspective. World J Microbiol Biotechnol 29:191–207

Kaw AK (2005) Mechanics of composite materials, 2nd edn. CRC Press, Boca Raton

Khan AF (2012) Biotechnology fundamentals, 2nd edn. CRC Press, Boca Raton

Khaydarov RR, Khaydarov RA, Evgrafova S, Estrin Y (2011) Using silver nanoparticles as an antimicrobial agent. NATO Sci Peace Security Ser A 2011:169–177

Kim H, Kang H, Chu G, Byun G (2008) Antifungal effectiveness of nanosilver colloid against rose powdery mildew in greenhouses. Solid State Phenom 135:15–18

Kim SW, Kim KS, Lamsal K, Kim YJ, Kim SB, Jung M, Sim SJ, Kim HS, Chang SJ, Kim JK, Lee YS (2009) An in vitro study of the antifungal effect of silver nanoparticles on oak wilt pathogen *Raffaelea* sp. J Microbiol Biotechnol 19:760–764

Kim S, Lee S, Lee I (2012a) Alteration of phytotoxicity and oxidant stress potential by metal oxide nanoparticles in *Cucumis sativus*. Water Air Soil Pollut 223:2799–2806

Kim SW, Jung JH, Lamasal K, Kim YS, Min JS, Lee YS (2012b) Antifungal effects of silver nanoparticles (AgNPs) against various plant pathogenic fungi. Mycobiol 40:53–58

Konotop YO, Kovalenko MS, Ulynets VZ, Meleshko AO, Batsmanova LM, Taran NY (2014) Phytotoxicity of colloidal solutions of metal-containing nanoparticles. Cytol Genet 48:99–102

Krumov N, Perner-Nochta I, Oder S, Gotcheva V, Angelov A, Posten P (2009) Production of inorganic nanoparticles by microorganisms. Chem Eng Technol 32:1026–1035

Kumar R, Lal S (2014) Synthesis of organic nanoparticles and their applications in drug delivery and food nanotechnology: a review. J Nanomater Mol Nanotechnol 3(4). https://doi.org/10.4172/2324-8777.1000150

Lee S, Chung H, Kim S, Lee I (2013) The genotoxic effect of ZnO and CuO nanoparticles on early growth of buckwheat, *Fagopyrum esculentum*. Water Air Soil Poll 224:1668. https://doi.org/10.1007/s11270-013-1668-0

Li X, Xu H, Chen Z, Chen G (2011) Biosynthesis of nanoparticles by microorganisms and their applications. J Nanomaterials. https://doi.org/10.1155/2011/270974

Liu F, Wen L-X, Li Z-Z, Yu W, Sun HY, Chen JF (2006) Porous hollow silica nanoparticles as controlled delivery system for water-soluble pesticide. Mat Res Bull 41(12):2268–2275

Liu J, Qiao SZ, Hu QH (2011) Magnetic nanocomposites with mesoporous structures: synthesis and applications. Small 7:425–443

Luby Š, Lubyová M, Šiffalovič P, Jergel M, Majková E (2015) A brief history of nanoscience and foresight in nanotechnology. In: Bardosova M, Wagner T (eds) Nanomaterials and nanoarchitectures. NATO science for peace and security series C: environmental security. Springer, Dordrecht, pp 63–86

Madbouly S, Chaoqun Z, Kessler MR (2015) Bio-based plant oil polymers and composites. William Andrew Elsevier, Burlington

Manimaran M (2015) A review on nanotechnology and its implications in agriculture and food industry. Asian J Plant Sci Res 5(7):13–15

Manzoor-Ul-Haq RV, Patil S, Singh D, Krishnaveni R (2014) Isolation and screening of mushrooms for potent silver nanoparticles production from Bandipora district (Jammu and Kashmir) and their characterization. Int J Nat Prod Res 9:2399–2407

Marchiol L (2012) Synthesis of metal nanoparticles in living plants. Italian J Agron 7(37):274. https://doi.org/10.4081/ija.2012.e37

Matsumura Y, Yoshikata K, Kunisaki S, Tsuchido T (2003) Mode of bacterial action of silver zeolite and its comparison with that of silver nitrate. Appl Environ Microbiol 69(7):4278–4281

Melo A, Amadeu MS, Lancellotti M, Maria de Hollanda L, Machado D (2015) The role of nanomaterials in cosmetics: national and international legislative aspects. Quim Nova 38(4):599–603

Min JS, Kim KS, Kim SW, Jung JH, Lamsal K, Kim SB, Jung M, Lee YS (2009) Effects of colloidal silver nanoparticles on sclerotium-forming phytopathogenic fungi. Plant Pathol J 25:376–380

Mishra A, Kumari M, Pandey S, Chaudhry V, Gupta KC, Nautiyal CS (2014a) Biocatalytic and antimicrobial activities of gold nanoparticles synthesized by *Trichoderma* sp. Bioresour Technol 166:235–242

Mishra S, Singh BR, Singh A, Keswani C, Naqvi AH, Singh HB (2014b) Biofabricated silver nanoparticles act as a strong fungicide against *Bipolaris sorokiniana* causing spot blotch disease in wheat. PLoS One 9:e97881. https://doi.org/10.1371/journal.pone.0097881

Moghaddam KM (2010) An introduction to microbial metal nanoparticle preparation method. J Young Investig 19:1–7

Mohanpuria P, Rana NK, Yadav SK (2008) Biosynthesis of nanoparticles: technological concepts and future applications. J Nanopart Res 10:507–517

Moharrer S, Mohammad B, Gharamohammad RA, Yargol M (2012) Biological synthesis of silver nanoparticles by *Aspergillus flavus*, isolated from soil of Ahar copper mine. Indian J Sci Technol 5:2443–2447

Momin JK, Jayakumar C, Prajapati JB (2013) Nutrition in food and science, potential of nanotechnology in functional foods. Emir J Food Agric 25(1):10–19

Mukherjee P, Ahmad A, Mandal D, Senapati S, Sainkar SR, Khan MI, Ramani R, Parischa R, Ajayakumar PV, Alam M, Sastry M, Kumar R (2001) Bioreduction of $AuCl_4$ – ions by the fungus *Verticillium* sp. and surface trapping of the gold nanoparticles formed. Angew Chem Int Ed Engl 40:3585–3588

Mukherjee P, Roy M, Mandal BP, Dey GK, Mukherjee PK, Ghatak J, Tyagi AK, Kale SP (2008) Green synthesis of highly stabilized nanocrystalline silver particles by a nonpathogenic and agriculturally important fungus *Trichoderma asperellum*. Nano 19:75–103

Musante C, White JC (2012) Toxicity of silver and copper to *Cucurbita pepo*: differential effects of nano and bulk-size particles. Environ Toxicol 27:510–517

Namasivayam SKR, Aruna A, Gokila (2014) Evaluation of silver nanoparticles-chitosan encapsulated synthetic herbicide paraquate (AgNp-CS-PQ) preparation for the controlled release and improved herbicidal activity against *Eichhornia crassipes*. Res J Biotechnol 9:19–27

Nekrasova GF, Ushakova OS, Ermakov AE, Uimin MA, Byzov IV (2011) Effects of copper (II) ions and copper oxide nanoparticles on *Elodea densa* Planch. Russ J Ecol 42:458–463

Ninganagouda DS, Rathod V (2014) Characterization and biosynthesis of silver nanoparticles using a fungus *Aspergillus niger*. Int Lett Nat Sci 14-15:49–57

Ortega FG, Fernández-Baldo MA, Fernández JG, Serrano MJ, Sanz MI, Diaz- Mochón JJ, Lorente JA, Raba J (2015) Study of antitumor activity in breast cell lines using silver nanoparticles produced by yeast. Int J Nanomedicine 10:2021–2031

Ottoni CA, Simões MF, Fernandes S, Gomes dos Santos J, Sabino da Silva E, Brambilla de Souza EF, Maiorano AE (2017) Screening of filamentous fungi for antimicrobial silver nanoparticles synthesis. AMB Express 7:31. https://doi.org/10.1186/s13568-017-0332-2

Pacioni N, Borsarelli C, Rey V, Veglia A (2015) Synthetic routes for the preparation of silver nanoparticles, a mechanistic perspective. In: Alarcon EI, Griffith M, Udekwu KI (eds) Silver nanoparticles applications. Springer International Publishing, Cham, pp 13–46

Pal S, Tak YK, Song JM (2007) Does the antibacterial activity of silver nanoparticles depend on the shape of the nanoparticle? A study of the gram-negative bacterium *Escherichia coli*. Appl Environ Microbiol 73:1712–1720

Panáček A, Kvítek L, Prucek R, Kolář M, Večeřová R, Pizúrová N, Sharma VK, Nevecna T, Zboril R (2006) Silver colloid nanoparticles: synthesis, characterization, and their antibacterial activity. J Phys Chem B 110:16248–16253

Pantidos N, Horsfall LE (2014) Biological synthesis of metallic nanoparticles by bacteria, fungi and plants. J Nanomed Nanotechnol 5:233. https://doi.org/10.4172/2157-7439.1000233

Parvathi D, Rajagopal K (2017) A practical manual on synthesis of nanoparticles and its applications in biology. Digital Age Publishers, Chennai

Patel V, Berthold D, Puranik P, Gantar M (2015) Screening of cyanobacteria and microalgae for their ability to synthesize silver nanoparticles with antibacterial activity. Biotechnol Rep 5:112–119

Patra P, Mitra S, Debnath N, Goswami A (2012) Biochemical-, biophysical-, and microarray-based antifungal evaluation of the buffer-mediated synthesized nano zinc oxide: an in vivo and in vitro toxicity study. Langmuir 28:16966–16978

Pérez-de-Luque A, Rubiales D (2009) Nanotechnology for parasitic plant control. Pest Manag Sci 65(5):540–545

Petersson L, Oksman K (2006) Biopolymer based nanocomposites: comparing layered silicates and microcrystalline cellulose as nanoreinforcement. Composites Sci Tech 66(13):2187–2196

Peteu SF, Oancea F, Sicuia OA, Constantinescu F, Dinu S (2010) Responsive polymers for crop protection. Polymer 2:229–251

Phanjom P, Ahmed G (2017) Effect of different physicochemical conditions on the synthesis of silver nanoparticles using fungal cell filtrate of *Aspergillus oryzae* (MTCC no. 1846) and their antibacterial effect. Adv Nat Sci Nanosci Nanotechnol 8:045016

Piñero S, Camero S, Blanco S (2017) Silver nanoparticles: influence of the temperature synthesis on the particles' morphology. IOP Conf Series: J Physics 786:012020. https://doi.org/10.1088/1742-6596/786/1/012020

Prasad R (2016) Advances and Applications through Fungal Nanobiotechnology. Springer, International Publishing Switzerland (ISBN: 978-3-319-42989-2)

Prasad R (2017) Fungal nanotechnology: applications in agriculture, industry, and medicine. Springer, Cham

Prasad R, Kumar V, Prasad KS (2014) Nanotechnology in sustainable agriculture: present concerns and future aspects. Afr J Biotech 13(6):705–713

Prasad R, Pandey R, Barman I (2016) Engineering tailored nanoparticles with microbes: quo vadis. WIREs Nanomed Nanobiotechnol 8:316–330. doi: https://doi.org/10.1002/wnan.1363

Prasad R, Kumar M, Kumar V (2017a) Nanotechnology: An Agriculture paradigm. Springer Nature Singapore Pte Ltd. (ISBN: 978-981-10-4573-8)

Prasad R, Kumar V, Kumar M (2017b) Nanotechnology: Food and Environmental Paradigm. Springer Nature Singapore Pte Ltd. (ISBN 978-981-10-4678-0)

Prasad R, Kumar V, Kumar M, Wang S (2017) Fungal nanobionics: principles and applications. Springer, Cham

Prasad R, Jha A, Prasad K (2018a) Exploring the Realms of Nature for Nanosynthesis. Springer International Publishing (ISBN 978-3-319-99570-0) https://www.springer.com/978-3-319-99570-0

Prasad R, Kumar V, Kumar M, Wang S (2018b) Fungal Nanobionics: Principles and Applications. Springer Nature Singapore Pte Ltd. (ISBN 978-981-10-8666-3) https://www.springer.com/gb/book/9789811086656

Rac O, Suchorska-Woźniak P, Fiedot M, Teterycz H (2014) Influence of stabilising agents and pH on the size of SnO_2 nanoparticles. Beilstein J Nanotechnol 5:2192–2201

Ragunath A, Parveen Nisha A, Kumuthakalavalli R (2017) Mycosynthesis of silver nanoparticles: characterization, antioxidant and anti-inflammatory activity from *Pleurotus florida* (Mont) singer: macro fungi. Asian J Pharma Clin Res 10(6):186–191

Rahimi G, Alizadeh F, Khodavandi A (2016) Mycosynthesis of silver nanoparticles from *Candida albicans* and its antibacterial activity against *Escherichia coli* and *Staphylococcus aureus*. Trop J Pharm Res 15(2):371–375

Rai M, Yadav A, Bridge P, Gade A (2009) Myconanotechnology: a new and emerging science. In: Rai MK, Bridge PD (eds) Applied mycology. CAB International, New York, pp 258–267

Raj S, Jose S, Sumod US, Sabitha M (2012) Nanotechnology in cosmetics: opportunities and challenges. J Pharm Bioallied Sci 4(3):186–193

Ramyadevi J, Jeyasubramanian K, Marikani A, Rajakumar G, Rahuman AA (2012) Synthesis and antimicrobial activity of copper nanoparticles. Mater Lett 71:114–116

Rashmi K, Krishnaveni T, Ramanamurthy S, Mohan PM (2004) Characterization of cobalt nanoparticle from a cobalt resistant strain of *Neurospora crassa*. In: International symposium of research students on materials science and engineering, December 20–22, 2014, Chennai, India

Ray S, Sarkar S, Kundu S (2011) Extracellular biosynthesis of silver nanoparticles using the mycorrhizal mushroom *Tricholoma crassum* (Berk.) Sacc: its antimicrobial activity against pathogenic bacteria and fungus, including multidrug resistant plant and human bacteria. Dig J Nanomater Bios 6(3):1289–1299

Reduan NA, Ghazali S, Jamari SS (2016) Effect of temperature and NaCl concentration on synthesis of silver nanoparticles prepared in aqueous medium. ARPN J Eng Appl Sci 11(10):6399–6404

Riddin T, Gericke M, Whiteley C (2006) Analysis of inter and extracellular formation of platinum nanoparticles by *F. oxysporum* f. sp. *lycopersici* using response surface methodology. Nanotechnol 17(14):3482–3489

Rouhani M, Samih MA, Aslani A, Beiki KH (2011) Side effect of nano-ZnO-TiO_2-Ag mix-oxide nanoparticles on *Frankliniella occidentalis* Pergande (Thys.: Thripidae). In: Third international symposium on insect physiology, biochemistry and molecular biology, 2–5 July 2011, East China Normal University, Shanghai, pp 51

Rouhani M, Samih MA, Kalantari S (2012) Insecticide effect of silver and zinc nanoparticles against *Aphis nerii* Boyer de Fonscolombe (*Hemiptera*: *Aphididae*). Chilean J Agri Res 72(4):590–594

Saharan V, Mehrotra A, Khatik R, Rawal P, Sharma SS, Pal A (2013) Synthesis of chitosan based nanoparticles and their *in vitro* evaluation against phytopathogenic fungi. Int J Biol Macromol 62:677–683

Salata O (2004) Applications of nanoparticles in biology and medicine. J Nanobiotechnology 2(1):3. https://doi.org/10.1186/1477-3155-2-3

Samih MA, Rouhani M, Aslani A, Beiki KH (2011) Insecticidal properties of amitraz, nano-amitraz, nano-ZnO and nano-ZnO-Al_2O_3 nanoparticles on *Agonoscena pistaciae* (Hem.: Aphelaridae). In: Third international symposium on insect physiology, biochemistry and molecular biology, 2–5 July 2011, East China Normal University, Shanghai, pp 131

Sanghi R, Verma P (2009) A facile green extracellular biosynthesis of CdS nanoparticles by immobilized fungus. Chem Eng J 155(3): 886–891

Sanghi R, Verma P, Puri S (2011) Enzymatic formation of gold nanoparticles using *Phanerochaete chrysosporium*. Adv Chem Eng Sci 1:154–162

Santoso D, Lefroy RDB, Blair GJ (1995) Sulfur and phosphorus dynamics in an acid soil/crop system. Aust J Soil Res 33:113–124

Sarkar J, Dey P, Saha S, Acharya K (2011) Mycosynthesis of selenium nanoparticles. IET Micro Nano Lett 6(8):599–602

Sastry M, Ahmad A, Khan MI, Kumar R (2003) Biosynthesis of metal nanoparticles using fungi and actinomycete. Curr Sci 85:162–170

Sastry KR, Rashmi HB, Rao NH (2011) Nanotechnology for enhancing food security in India. Food Policy 36:391–400

Saxena J, Sharma MM, Gupta S, Singh A (2014) Emerging role of fungi in nanoparticle synthesis and their applications. World J Pharm Pharm Sci 3(9):1586–1613

Shaligram NS, Bule M, Bhambure R, Singhal RS, Singh SK, Szakac SG, Pandey A (2009) Biosynthesis of silver nanoparticles using aqueous extract from the compactin producing fungal strain. Process Biochem 44(8):939–943

Sharon M, Choudhary AK, Kumar R (2010) Nanotechnology in agricultural diseases and food safety. J Phytology 2:83–92

Shedbalkar U, Singh R, Wadhwani S, Gaidhani S, Chopade BA (2014) Microbial synthesis of gold nanoparticles: current status and future prospects. Adv Colloid Int Sci 209:40–48

Shyla KK, Natarajan N, Nakkeeran S (2014) Antifungal activity of zinc oxide, silver and titanium dioxide nanoparticles against *Macrophomina phaseolina*. J Mycol Pl Pathol 44:268–273

Singh P, Balaji R (2011) Biological synthesis and characterization of silver nanoparticles using the fungus *Trichoderma harzianum*. Asian J Exp Biol Sci 2:600–605

Singh M, Manikaran S, Kumaraguru AK (2010) Nanoparticles: a new technology with wide applications. Res J Nanosci Nanotech 1(1):1–11

Singh D, Rathod V, Ninganagouda S, Hiremath J, Singh AK, Mathew J (2014) Optimization and characterization of silver nanoparticle by endophytic fungi *Penicillium* sp. isolated from *Curcuma longa* (turmeric) and application studies against MDR *E. coli* and *S. aureus*. Bioinorg Chem. https://doi.org/10.1155/2014/408021

Singh T, Shukla S, Kumar P, Wahla V, Bajpai VK, Rather IA (2017) Application of nanotechnology in food science: perception and overview. Front Microbiol 8:1501. https://doi.org/10.3389/fmicb.2017.01501

Singhal G, Bhavesh R, Kasariya K, Sharma AR, Singh RP (2011) Biosynthesis of silver nanoparticles using *Ocimum sanctum* (Tulsi) leaf extract and screening its antimicrobial activity. J Nanopart Res 13:2981–2988

Siracusa V, Rocculi P, Romani S, Rosa MD (2008) Biodegradable polymers for food packaging: a review. Trends Food Sci Tech 19(12):634–643

Siva Kumar K, Divya A, Pavan Kumar S, Ajitha B, Chaithanya VS, Jaya Prasad K, Sreedhara Reddy P (2011) Influence of synthesis temperature on synthesis of PVA-stabilised silver nanoparticles. Micro Nano Letters 6(11):975–977

Soni N, Prakash S (2012) Fungal-mediated nano silver: an effective adulticide against mosquito. Parasitol Res 111(5):2091–2098

Sundaramoorthi C, Kalaivani M, Mathews DM, Palanisamy S, Kalaiselvan V, Rajasekaran A (2009) Biosynthesis of silver nanoparticles from *Aspergillus niger* and evaluation of its wound healing activity in experimental rat model. Int J Pharm Tech Res 1(4):1523–1529

Suriyaprabha R, Karunakaran G, Kavitha K, Yuvakkumar R, Rajendran V, Kannan N (2014) Application of silica nanoparticles in maize to enhance fungal resistance. IET Nanobiotechnol 8:133–137

Syed A, Ahmad A (2012) Extracellular biosynthesis of platinum nanoparticles using the fungus *F. oxysporum*. Colloids Surf B Biointerfaces 97:27–31

Taniguchi N (1974) On the basic concept of 'nano-technology'. In: Proceedings of the International Conference on Production Engineering, Tokyo, Part II. Japan Society for Precision Engineering, pp 1823

Tao S, Pang R, Chen C, Ren X, Hu S (2012) Synthesis, characterization and slow release properties of O-naphthylacetyl chitosan. Carbohydr Polym 88:1189–1194

Teodoro S, Micaela B, David K (2010) Novel use of nano-structured alumina as an insecticide. Pest Manag Sci 66:577–579

Thakkar KN, Mhatre SS, Parikh RY (2010) Biological synthesis of metallic nanoparticles. Nanomedicine 6(2):257–262

Thirumurugan G, Veni VS, Ramachandran S, Rao JV, Dhanaraju MD (2011) Superior wound healing effect of topically delivered silver nanoparticle formulation using eco-friendly potato plant pathogenic fungus: synthesis and characterization. J Biomed Nanotechnol 5:659–666

Uzma M, Raghavendra VB, Girisha ST (2018) Biogenesis of gold nanoparticles, role of fungal endophytes and evaluation of anticancer activity – a review. Eur J Biomed Pharm Sci 5(5):319–329

Vandergheynst JS, Scher H, Hong-Yun G (2006) Design of formulations for improved biological control agent viability and sequestration during storage. Ind Biotechnol 2:213–219

Velmurugan N, Gnana Kumar G, Sub Han S (2009) Synthesis and characterization of potential fungicidal silver nano-sized particles and chitosan membrane containing silver particles. Iran Polym J 18:383–392

Verma VC, Singh SK, Solanki R, Prakash S (2011) Biofabrication of anisotropic gold nanotriangles using extract of endophytic *Aspergillus clavatus* as a dual functional reductant and stabilizer. Nanoscale Res Lett 6(16). https://doi.org/10.1007/s11671-010-9743-6

Vigneshwaran N, Kathe AA, Varadarajan PV, Nachane RP, Balasubramanya RH (2006) Biomimetics of silver nanoparticles by white rot fungus, *Phaenerochaete Chrysosporium*. Colloids Surf B: Biointerfaces 53(1):55–59

Wagers K, Chui T, Adem S (2014) Effect of pH on the stability of gold nanoparticles and their application for melamine detection in infant formula. IOSR. J Appl Chem 7(8):15–20

Wang HX, Ng TB, Ooi VE (1998) Mycosynthesis: antibacterial, antioxidant and antiproliferative activities of silver nanoparticles synthesized from *Inonotus obliquus* (Chaga mushroom) extract. Biochem Biophys Res Commun 250:544–546

Wilson MA, Tran NH, Milev AS, Kannangara GSK, Volk H, Lu GQM (2008) Nanomaterials in soils. Geoderma 146:291–302

Wu SC, Cao ZH, Li ZG, Cheung KC, Wong MH (2005) Effects of biofertilizer containing N-fixer, P and K solubilizers and AM fungi on maize growth: a greenhouse trial. Geoderma 125:155–166

Xiangqian L, Huizhong X, Zhe-Sheng C, Guofang C (2011) Biosynthesis of Nanoparticles by Microorganisms and Their Applications. J Nanomaterials 21: https://doi.org/10.1155/2011/270974

Xue B, He D, Gao S, Wang D, Yokoyama K, Wang L (2016) Biosynthesis of silver nanoparticles by the fungus *Arthroderma fulvum* and its antifungal activity against genera of *Candida*, *Aspergillus* and *Fusarium*. Int J Nanomedicine 11:1899–1906

Yang FL, Li SG, Zhu F, Lei CL (2009) Structural characterization of nanoparticles loaded with garlic essential oil and their insecticidal activity against *Tribolium castaneum* (Herbst) (Coleoptera: Tenebrionidae). J Agric Food Chem 57:10156–10162

Zhang X, He X, Wang K, Wang Y, Li H, Tan W (2009) Biosynthesis of size-controlled gold nanoparticles using fungus, *Penicillium* sp. J Nanosci Nanotechnol 9:5738–5744

Zhang H, Chen B, Banfield JF (2010) Particle size and pH effects on nanoparticle dissolution. J Phys Chem C 114(35):14876–14884

Chapter 11
Nanoparticles and Their Role in Bioenergy Production

Amandeep Brar, Manish Kumar, Vivekanand Vivekanand, and Nidhi Pareek

Contents

A. Brar · M. Kumar · N. Pareek (✉)
Department of Microbiology, School of Life Sciences, Central University of Rajasthan, Bandarsindri, Kishangarh, Ajmer, Rajasthan, India
e-mail: nidhipareek@curaj.ac.in

V. Vivekanand
Centre for Energy and Environment, Malaviya National Institute of Technology, Jaipur, Rajasthan, India

D. Thangadurai et al. (eds.), *Nanotechnology for Food, Agriculture, and Environment*, Nanotechnology in the Life Sciences,
https://doi.org/10.1007/978-3-030-31938-0_11

11.1 Introduction

In recent years, energy consumption has reached its peak due to industrialization and globalization which pose serious concerns about environmental sustainability. It has been figured that the global oil consumption growth has averaged 1.6 million barrels per day (Mb/d) with China (400,000 b/d) and India (330,000 b/d) providing the largest increments and world's natural gas consumption has grown by 63 billion cubic meters (bcm) in 2017 (BP's Statistical Review of World Energy 2017). The global gas demand has been continuously rising since 2014, accelerated again in 2017, propelled by Asia, which has accounted for approximately one-third of the additional demand, and China was the largest contributor to the increase in gas consumption due to its coal-to-gas substitution policy. China, being the world's largest energy consumer since 2009, has doubled its energy consumption as compared to 2016 due to its strong industrial and economic demand. According to BP's Statistical Review of World Energy (2017), worldwide use of petroleum and other liquid fuels will grow from 95 million barrels per day in 2015 to 104 million barrels per day in 2030 and to 113 million barrels per day in 2040. Most of the growth in liquid fuels consumption is in the industrial and transportation sectors. According to Li (2017), although the nonfossil fuels are expected to grow at a faster rate, still the fossil fuels will account for 77% of the total energy use in 2040. It is estimated that 80% of the global energy demand has been fulfilled by fossil fuels alone and the remaining 20% is being met from the nonrenewable sources of energy, i.e., wind, solar, hydroenergy, nuclear power, and biomass processing. The major challenge of this decade is the greenhouse gases that are being emitted by the combustion of fossil fuels into the environment, leading to the global warming (Kumar et al. 2013). Hence, the focus has been shifted to sustainable and cost-effective alternative energy sources to meet the energy requirements of the future generation. These energy sources include biofuels in the form of biodiesel, bioethanol, biomethane biogas usually produced from the various organic biomasses. The first-generation energy biofuels (viz., biodiesel and bioethanol) that are being produced from vegetable oils and sugarcane having high sugar and starch composition are blessed with ease of abundance, cultivation, and processing. Nowadays, the first generation of biofuels is considered as unsustainable as it requires large cultivable land for biomass production and poses a serious threat to the food security as the population is ever increasing at a rapid pace. It directly impacts the economy as the intense competition will result in the price hike of the feedstocks. Technology was a boon as the focus from the first generation has been shifted to the second generation of biofuels, produced from the ligno-cellulosic waste materials such as agricultural and bioprocessing organic wastes (Sims et al. 2010). The lignocellulosic biomass is converted through biochemical processes into biodiesel (extraction and transesterification) and bioethanol (hydrolysis and fermentation). The pilot scale production of second generation biofuels by the utilization of low-cost ligno-cellulosic materials and organic waste aims for the establishment of energy security together with environmental and economic sustainability. However, their major constraints include low energy gain, high production costs, and variable conversion rate and yield (Damartzis and Zabaniotou 2011). There is another possibility for production of biofuels, i.e., the use of algal biomass as a raw material, commonly called as the third generation of biofuels (Milledge et al. 2014) which is devoid of farming and land inputs. This algal biomass is rich in oil/lipid, carbohydrate and protein content, and could be obtained in high yields in the bioreactors only.

11.2 Current Scenario of Bioenergy and Its Limitations

Bioenergy endeavors a possible route toward fulfilling the energy requirements using domestic renewable sources. Bioenergy is considered as one of the sustainable, alternative, and cheapest sources of energy. It is the energy stored in the materials that are composed of living things or biomass. Biomass refers to all the organic materials that originate from the plants, i.e., algae, crops, and trees. It is generated by the conversion of sunlight into plant materials by photosynthesis and the energy of sunlight is stored in the form of chemical energy. This chemical energy is released when the bonds present between carbon, hydrogen, and oxygen molecules are broken by combustion or digestion. There are many forms of bioenergy, viz., bioethanol, biodiesel, and biogas (Hussein 2015), and different routes could be employed for biomass conversion to bioenergy (Fig. 11.1).

The environmental gains of bioenergy include a reduced dependency on depleting fossil fuels, lower emission of greenhouse gases, reduced smog, and utilization of waste materials. It boosts the economy by using sub-utilized biomass resources as a renewable source of energy that is inexpensive and provides employment to the masses. One of the limitations of large-scale production of second generation biofuels is the innate cytotoxicity of short-chain aliphatic alcohols which results in the inhibition of biocatalyst productivity at relatively low concentrations (Ingram 1989; Bowles and Ellefson 1985). Due to this, conventional bioprocesses remain confined to dilute the fermentation conditions. This dilution makes a futile feed for the downstream separation and purification; thereby the bioprocess becomes economically unfeasible (Straathof 2003; Schugerl and Hubbuch 2005). It has been estimated that approximately 63% of the total energy requirement of the ethanol production is used in its distillation, constituting 6–10% of its total production cost (Alzate and Toro 2006; Galbe et al. 2007). Economic assessment has favored the use of lignocellulosic biomass as the potential source of biofuel production as their use has lowered the cost of production. However, the main problem associated with lignocellulosic biomass is their insolubility, crystallinity, and presence of lignin (Srivastava et al. 2015).

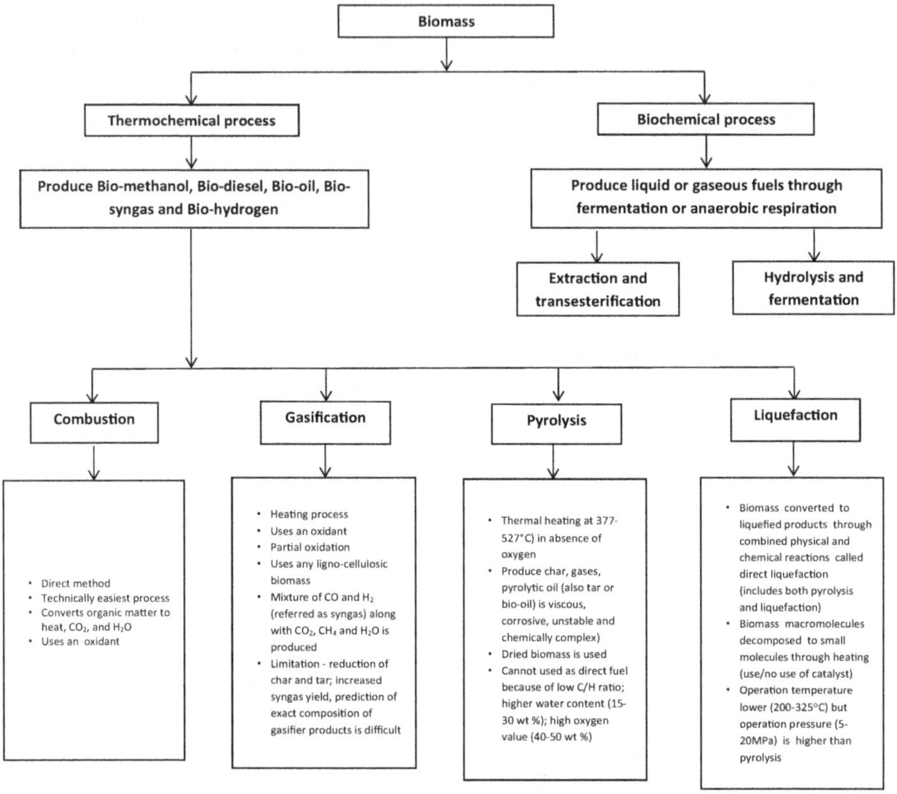

Fig. 11.1 Different routes of biomass conversion to bioenergy

11.3 Nanotechnology

Nanotechnology has gained tremendous attention in the recent years because of its vast technological applications. It is a branch of physical science that deals with the study and analysis of matter on an atomic or molecular scale to yield new structures, devices, materials, systems, and catalysts with unique or extraordinary properties. Nanoparticles are described as the materials having a structure in which at least one of its phases has one or more dimensions in the nanometer (1–100 nm) size range. In other words, it is also defined as building machines at the molecular scale involving the manipulation of materials on about two-tenths of a nanometer (atomic) scale. These materials include nanometer-sized crystallites, nanopore-separated materials, particles with sizes in the nanometer range, or nanometer-sized metallic clusters dispersed within a porous matrix (Budarin et al. 2013). Nanotechnology with biotechnology offers flashy advantages for improving the activity of enzymes, stability, potential, and overall performance in the bioprocessing applications. Present scenario demands highly specific, focused, and optimized biochemical pathways for bioenergy production, and nanotechnology is a blessing providing improved conventional bioprocess pathways due to its unique properties. These properties include improved materials for enzyme immobilization, materials with enhanced enzyme loading capacity, nanocatalysts, materials for separation and purification of liquid biofuels, and so on. It provides quicker, faster, and reliable methods for bioenergy generation from biological sources.

11.4 Nanoparticles in Bioenergy

Nanotechnology offers a meaningful solution to the conventional bioprocesses used for the biofuel production by changing the characteristics of the feed materials and biocatalysts. Nanoparticles depict electronic, magnetic, optical, physical, and chemical properties that completely differ from both the bulk and the constituent atoms or molecules. Nanoparticles manifest many unique characteristics like they are small enough to detain their electrons and can produce quantum effects, are chemically stable, uniform in size and well-dispersed in the liquid media (Rai et al. 2016). Nanoparticles act as potential catalysts for many processes, but their high cost and instability impart a major hindrance to their industrial applications. Different types of nanoparticles used in the bioenergy are mentioned in brief.

11.4.1 Magnetic Nanoparticles

Magnetic nanoparticles are of great interest as they have many applications in the field of biotechnology, biomedicine, material science, environmental science, and engineering (Lu et al. 2007; Abd-Elsalam et al. 2019). The magnetic properties are the result of both the intrinsic properties of a particle and the interaction between the particles. These particles perform best when their size is below a critical value (typically around 10–20 nm), depending upon the material used. At this size, each nanoparticle behaves as a single magnetic domain and above the blocking temperature, they array a superparamagnetic behavior. These individual particles have a large constant magnetic moment and in response to an applied magnetic field, behave like a giant paramagnetic atom. Their characteristics include a high surface area to volume ratio, quantum properties and ability to carry other compounds due to small size. Their nature shows the great potential for their use in the field of biofuel and bioenergy. These particles can be used as a catalyst in various reactions. The material used to build magnetic nanoparticle includes the alloys of cobalt, nickel, iron, and platinum (Misson et al. 2015; Rai et al. 2016). Magnetic nanoparticles are superior over their counter nanoparticles as they can be coated to use as a catalyst through immobilized particle and can be easily removed by the application of suitable magnetic field without applying any toxic effects (Nicolas et al. 2014).

In the absence of a magnetic field, they get dispersed and behave like any other nanoparticle.

The inevitable problem associated with magnetic nanoparticles is their intrinsic instability over a large period of time. These small particles tend to form clusters to reduce the energy associated with the high surface area to volume ratio. As the naked magnetic particles are chemically very active, they easily lose their magnetic properties by getting oxidized and get dispersed. Therefore, it is very important to develop protection strategies against chemical destruction during and after synthesis. The strategies include coating or modifications with organic species including surfactants or polymers, inorganic layers of silica or carbon. These functionalized nanoparticles have reposeful applications in catalysis and bio-separation as they are highly dispersible, are reactive, and are easily separable (Lu et al. 2007).

11.4.2 Gold Nanoparticles

One of the unique characteristics of gold nanoparticles is the presence of high relativistic contraction of 6s-orbitals resulting in their very small atomic radius. This means that their nucleus will have a large charge and this charge will result in the high velocity of an electron. Higher velocity means the electrons will remain nearer to the nucleus most of the time, thereby contracting the radius. As gold nanoparticles are nontoxic and biocompatible (both in vivo and in vitro environments), they are used as biosensors, antimicrobials, in drug-delivery and recently against cancer and HIV. They show high chemical stability as are resistant against oxidation (Rad et al. 2011).

11.4.3 Chitosan Nanoparticles

Chitosan is low cost, biocompatible, biodegradable, and renewable in nature, obtained from shells of crustaceans (mainly crabs, shrimps, lobsters) and wastes of the seafood processing industry. It is available in different geometric configurations, viz., powder, flakes, membranes, fibers, and others, together with the presence of hydroxyl and amino groups that enable it to interact with polymers and biomolecules. It is soluble in dilute acidic medium and forms a cationic polymer with an amino group that protonates to generate NH^{3+} ion. The nanochitosan has superior physical and chemical properties such as large surface area, high porosity, tensile strength, conductivity, and increased mechanical properties (Jia et al. 2011; Mendes et al. 2011). It has been extensively used for the enzyme immobilization.

11.4.4 Nanodiamond

These are nanoscale diamond particles having great potential for biological applications because of their chemical stability, biological compatibility, and nontoxicity (Cipolatti et al. 2014). Their surface provides a unique platform for the conjugation of biological molecules. There are many methods to produce nanoscale diamond particles, and the most commonly used at large scale is the transformation of graphitic material into diamond crystallites by the action of circular shock waves (Liu et al. 2008). They have gained attention due to their cost-effective large-scale synthesis based on the detonation of carbon-containing explosives, small particle size (approximately 4–5 nm), and effortless surface functionalization including bioconjugation with high biocompatibility (Schrand et al. 2009).

11.4.5 Heterogeneous Catalyst

The heterogeneous catalyst is nowadays used in the biofuel production and is noncorrosive, environment-friendly, easily separated from the products formed, and reusable with or without regeneration (Di Serio et al. 2007). The nanocatalytic system formed by the integration of heterogeneous catalyst and nanoparticles allows the rapid and selective chemical transformations with excellent product yield that is coupled with the effortless separation of catalysts and recovery. The recovery of a catalyst is the persuasive character of any industry depending upon the use of a catalyst (Yulianti et al. 2014). Due to their nanosize and large surface area, the contact between the reactants and catalyst is increased energetically forming a pseudo-homogeneous phase. The insolubility of the catalyst in the reaction solvent makes it heterogeneous and therefore, can easily be separated from the reaction mixture (Gurunathan and Ravi 2015). The nanosized catalysts are commonly prepared by sonochemical and hydrothermal methods. The sonochemical method promotes the reaction by means of cavitation that includes the formation of cavitation bubbles followed by growth and dramatic collapse and then the production of local high temperature and pressure at the time of the explosion. In the hydrothermal method, the reaction is completed in a closed container using water as a reaction medium. The high temperature and high-pressure reaction environment are created by heating the reaction vessel and the precursor is fully dissolved in the reaction system. The atom or molecule growth units are formed which is followed by the occurrence of nucleation or crystallization to yield nanomaterials (Zhou et al. 2015).

11.4.6 Nanoparticles Immobilized Enzymes

Immobilization of enzymes endeavors an encouraging path for improving the operational stability of enzymes in nonaqueous systems and allows their reusability. Solid carriers are required for the immobilization of enzymes and solid carrier influences the physio-chemical characteristics and biological activity of enzymes. These solid carriers allow the fast separation of products from the enzyme (Raita et al. 2015). Nanoparticle-immobilized enzymes can be easily recovered and reused. Recently, different types of nanostructure magnetic supports have been studied as solid matrices for enzyme immobilization. These supports allow the fast and smooth separation of products from enzymes (laccase, α-amylase, β-galactosidase, lipase) under the influence of magnetic field, by providing a large surface area and stability (Cipolatti et al. 2014).

11.4.7 Mesoporous Silica Nanoparticles

Mesoporous silica nanoparticles were first synthesized in 1990 by researchers in Japan (Kresge et al. 1992). Mesoporous silica materials have both hexagonal or cubic geometry and pore size between 2 and 50 nm (Reis et al. 2008). These particles have large surface area, adjustable pore size, and improved thermal stability. Silica nanoparticles have gained popularity in the biomedical field because of their well-defined chemical structure and controlled release that permits the modification by an amine, carboxyl and thiol groups, enzymes, proteins, and DNA (Souza et al. 2011). These particles are of high interest to many chemical applications, i.e., sorption, insulation, sensing, separation, and catalysis (Wang et al. 2012). Their synthesis can be carried out by various adjustments of reaction conditions, viz., variations in the pH and the nature of copolymer used.

11.4.8 Mesoporous Nanosized Zeolites

Zeolites are defined as alumino-silicate molecular sieves with pores of molecular dimensions having a large surface area, used for catalysis, ion exchange, and separation. Their identification is based on their specific chemical composition (silica/aluminum), pore size, and pore topology. When the zeolite crystals are scaled down from micrometer to nanometer, they get modified properties in the form of increased surface area and decreased diffusion path lengths. Due to the decrease in crystal size (below 100 nm), the external surface area of zeolite which is different from the internal pore surface, increases over 25% of the total surface area. The presence of active sites on the external surface of nanocrystalline zeolites results in high surface reactivity improving the catalytic properties. Their application potential for environmental catalysis, environmental remediation, and decontamination are credited to their improved properties for adsorption and intracrystalline diffusion (Larsen 2007).

11.5 Applications of Nanoparticles in Bioenergy

11.5.1 Biofuel

Biofuels, being the renewable sources of energy display exclusive characteristics, are biodegradable having low toxic effect, are diverse in nature, and are locally available. For the blazing future, biofuels are considered as a potential automotive fuel, produced through the present-day biological processes (anaerobic digestion) rather than geological processes (involved in the formation of fossil fuels, viz., coal and petrol). While the combustion process of fossil fuels produces the major part of CO_2 emissions in the atmosphere, the combustion of biofuels is considered carbon neutral. The largest constituent of the liquid biofuels within the conventional bioenergy industry is comprised of alcohol-based biofuel (mainly ethanol and n-butanol) (Zuliani et al. 2018). Currently, bioethanol is prepared by the hydrolysis of energy crops to starch and then subsequently converted into ethanol by fermentation. Among the enzymes used for the hydrolysis of starch, α-amylases are of commercial importance and they are usually considered for immobilization by nanoparticles. α-Amylases immobilized on the surface of nanoparticles are the preferred choice over the native enzymes for the industrial applications. These enzymes are stable against various types of physical, chemical denaturants, and poisons. Studies have reported the use of magnetic Fe_3O_4 and Fe_2O_3 nanoparticles for immobilization of α-amylases in the starch hydrolysis and their stability was proved by reusability experiments showing 83% residual activity after the eighth consecutive use. Experiments have shown that by increasing the temperature of the experiment, the rate of starch hydrolysis was also increased due to a decrease in the viscosity of the starting material and the product (Khan et al. 2012; Uygun et al. 2012).

Iron oxide nanoparticles were incorporated into single-walled carbon nanotubes to produce magnetic single-walled carbon nanotubes, used for the immobilization of amyloglucosidase enzyme. It was experimentally proved that enzyme retains up to 40% of its catalytic efficiency in starch prototype biomass hydrolysis when used repeatedly for 10 cycles and can be separated by a single magnet from the reaction mixture at the end of the experiment. It was also demonstrated that the enzyme following immobilization could be stored up to 1 month at 4 °C. This will further reduce the capital cost of enzymes used for the industrial biofuel production as the use of enzymes in biofuels is an important step in the overall production process (Goh et al. 2012).

The importance of cellulose hydrolysis comes in context to the second generation of biomass conversion to fuels. Cellulose is a polymer made up of D-glucose units, linked through β-1,4-glycosidic bonds. Intra/inter molecular hydrogen bonds and Van der Waals forces within and between cellulose strands make a stable crystalline structure showing great tensile strength. As cellulases are sensitive to temperature variations, their integrity for industrial use is questionable (Shuttleworth et al. 2014). Cellulase enzyme was immobilized onto a scaffold of mesoporous silica nanoparticles through two methods of immobilization, viz., physical and chemical binding. In the experiment for the production of bioethanol from glucose by fermentation, the reaction conditions, i.e., temperature, time, and the amount of cellulase required for hydrolysis were optimized along with a systematic investigation of the loading amount, catalytic activity, and stability of cellulase immobilized on mesoporous silica nanoparticles. It was concluded at the end of the experiment that the immobilized cellulase exhibiting carboxyl groups and large pore size is effective in achieving 80% glucose yield with excellent stability. Following storage for 23 days, the catalytic activity of free cellulase decreased significantly as compared to the immobilized cellulase. The glucose produced is converted into bioethanol through fermentation (Chang et al. 2011).

Together with enhanced thermal stability, an improvement in the production of cellulase and sugar productivity was reported in the presence of Fe_3O_4/alginate nanoparticles by using rice husk and rice straw as substrates in solid-state fermentation with thermotolerant *Aspergillus fumigatus* AA001. Enzyme production was increased by 35% and 40%, respectively, in the presence of only Fe_3O_4 nanoparticles and Fe_3O_4/alginate nanoparticles with respect to control. Cellulase treated with Fe_3O_4/alginate nanoparticles showed thermal stability for 8 h at 70 °C while retaining its 56% relative activity, whereas control was retaining only 19% activity (Srivastava et al. 2015). The thermostable cellulase from *Aspergillus fumigatus* AA001 on treatment with ZnO nanoparticles showed the considerable stability (10 h) at 65 °C at pH 10.5. The above-mentioned features portray cellulase as an asset to the bioethanol production from the cellulosic waste in conjunction with nanoparticles (Srivastava et al. 2016).

The above-mentioned studies support the fact that nanoparticles play a pivotal role in amending the bioconversion processes to produce enhanced levels of biofuels (Table 11.1). Moreover, the enhanced thermal stability of the enzyme cellulase may be attributed to the nanoparticles. There are many studies that have reported the improved cellulase production, thermal stability, as well as hydrolysis efficiency in the presence of different nanoparticles, viz., iron oxide and zinc oxide. It has been postulated that nanoparticles act as a carrier and not only helps to provide thermal stability by covalent bonding but also protect enzymes against pH variations and inhibitors. The exact mechanism of action of nanoparticles in the processes is still unknown (Ansari and Husain 2012; Verma et al. 2013).

Palladium (Pd) nanoparticles were used as a catalyst supported on mesoporous nitrogen-doped carbon in the biofuel production. During the catalytic process, deposition of noble metals like Pt, Pd, and Ru on carbon scaffold gets easily leached out due to the weak interaction between the metal nanoparticles and the surface of the carbon. Hence, modification of carbon scaffold is necessary by oxidizing it with

Table 11.1 Nanoparticles used in the production of biofuel

Nanoparticle	Target and source	Conclusion	Reference
Cu–Co	Cu–Co bimetallic nanoparticles coated with carbon layers were developed for the hydrogenolysis of 5-hydroxymethylfurfural to biofuel 2,5-dimethylfuran	Cu–Co–C shows the highest yield of 99.4% for dimethylfuran Catalytic activity and selectivity was constant in six-run recycling test The synergistic effect between bimetallic nanoparticles enhanced the catalytic performance	Chen et al. (2017)
$Mg(NO_3)_2.6H_2O$; $Al(NO_3)_3.9H_2O$	Hydrotalcite nanoparticles were synthesized by coprecipitation of aqueous $Mg(NO_3)_2.6H_2O$ and $Al(NO_3)_3.9H_2O$ and were used to hydrolyze to cellulose for glucose production	47.4% hydrolysis yield with high glucose selectivity of 85.8% was obtained at 423K The nanocatalyst was stable, can be easily separated from the mixture and was reused at least four times	Fang et al. (2011)
Mesoporous zeolite (HZSM-5) and metal nanoparticles	Combination of metal nanoparticles and acidic sites is apt for catalytic hydrodeoxygenation of pyrolysis oil	Mesoporous zeolites showed improved catalytic activity, better selectivity, longer lives for upgrading of oil pyrolysis and transformation of lipids into biofuels	Zhang et al. (2018)
HZSM-5, zeolite beta and zeolite Y	Catalytic cracking of waste cooking palm oil to biofuels by nanocrystalline (0.54–0.80 nm) zeolites in a fixed bed reactor	86.4 wt% oil conversion and 46.5 wt% organic liquid product yield was obtained with nanocrystalline zeolite Y having pore size 0.67 nm, optimum reaction temperature of 458 °C with oil/catalyst ratio of 6	Taufiqurrahmi et al. (2011)
Magnetic iron oxide	Thermostable β-glucosidase immobilized for biofuel production	Thermal stability of immobilized enzyme enhanced at 70 °C Immobilized nanoparticle enzyme conjugate retained more than 50% activity up to 16th cycle Maximum cellobiose hydrolysis was achieved at 16 h	Verma et al. (2013)

H_2O_2 or HNO_3 to impart strength to metallic nanoparticles and surface carbon. The enhancement of nitrogen to carbon scaffold (homogenous incorporation of nitrogen to carbon) is preferred as it improves the chemical, electrical, and functional properties of the scaffold. The modified Pd nanoparticles, used as a catalyst can be easily separated from the reaction solution by simple filtration. The Pd nanoparticles are highly stable and can be reused for several cycles without losing its activity. The study concluded that Pd nanoparticles heterogeneously supporting mesoporous N-doped carbon possess special base properties and electronic behavior and exhibit high activity for hydrodeoxygenation of lignin-derived biofuels, under mild conditions and water as clean solvent (Xu et al. 2012).

Lin et al. (2012) synthesized ZrO_2 nanoparticles through the facile hydrothermal process having a diameter in the range of 5–10 nm and surface area of $102m^2/g$. ZrO_2 nanoparticles were used for the removal of phospholipids from *Jatropha* oil solutions for the first time as Zr atoms had partial positive charge in a Lewis acid ZrO_2 nanoparticle and phospholipid was having a negative charge. Removal of phospholipids from *Jatropha* oil makes it potential oil for biofuel application. The capacity of phospholipid adsorption was higher with ZrO_2 nanoparticles as compared to the commercial ZrO_2 powder.

11.5.2 Bioethanol

Bioethanol is the most common substitute fuel after petrol for road transport. The enhanced development of bioethanol may be attributed to its use as a favorable and near carbon-neutral renewable fuel that reduces the CO_2 emissions and associated climate change. The use of bioethanol blend of 10% reduces green house gases emission by 12–19% in comparison with the conventional fossil fuels (Saini et al. 2015). The second generation of bioethanol is produced by saccharification of ligno-cellulosic biomass (agricultural residues, wood, paper, municipal waste, and dedicated energy crops) followed by microbial fermentation. Enhancements in ethanol yield were investigated in simultaneous saccharification and fermentation (SSF) reactions following adsorption of cellulase on silica nanoparticles with reference to the free enzyme (Lupoi and Smith 2011). The study reported that the ethanol yields were 2.1–2.3 times higher in enzyme immobilized system along with higher conversion rates of insoluble cellulose and cellulose hydrolysis products to the soluble oligomers. Cellulose is hydrolyzed into glucose by the enzyme and simultaneously converted into ethanol by yeast in the SSF reactions. The temperature (35 °C) and presence of ethanol (32 mg/mL) are not favorable for the enzyme but its immobilization on to the solid support stabilized it and increased the overall yield of the reaction along with recycling of the enzyme (Lupoi and Smith 2011). The potential of *Saccharomyces* sp. was investigated for the ethanol production by entrapping it in the matrix of alginate and magnetic nanoparticles. It was covalently immobilized on to the magnetite containing chitosan and cellulose coated α-amylase magnetic nanoparticles. The immobilized cells and enzymes were applied in the column reactors and the ethanol yield reached up to 91% with immobilized glucoamylase and 86% with immobilized α-amylase. These magnetic nanoparticles were stable for more than 1 month at 4 °C in saline (Ivanova et al. 2011). Kim et al. (2014) studied the role played by nanoparticles in bioethanol production by syngas fermentation by the use of *Clostridium ljungdahlii.* The primary method used for bioenergy production is a saccharification–fermentation process, but due to the complexity of the process, gasification–fermentation has emerged as an alternative where lignocellulosic biomass is gasified to produce synthetic gas (syngas) followed by fermentation to produce bioethanol. Among the nanoparticles of palladium on carbon, palladium on alumina, silica, hydroxyl-functionalized single-walled nanotubes, alumina, and iron (III) oxide that was employed in the study, silica nanoparticles at a concentration of 0.3 weight % appeared the best in enhancing gas–liquid mass transfer. Application of nanoparticles has resulted in improvements in ethanol production levels by 166.1% during syngas fermentation.

An effective method for the encapsulation of gold nanoparticles in zeolite silicate was experimentally demonstrated by Mielby et al. (2014). The study reported the high activity and selectivity of gold nanoparticles for the catalytic gas phase oxidation of ethanol. Modification of zeolites was done by recrystallization process in which intraparticle voids and mesopores are formed that help in the formation of small and dispersed nanoparticles by simple impregnation. A single zeolite crystal has a broad range of mesopores and can accumulate several hundreds of gold nanoparticles (diameter of 2–3 nm). These encapsulated gold nanoparticles displayed good stability and resulted in a 50% conversion of ethanol with 98% selectivity towards acetaldehyde at 200 °C (Table 11.2).

11.5.3 Biogas

Biogas is produced by anaerobic digestion of biomass and sludge and is a flammable mixture of gases mainly methane and carbon-dioxide. Its major source includes sewage treatment plants, landfills, organic industrial streams, and anaerobic digestion of organic wastes. The nature and texture of the source of biogas are biological; it is referred to as a type of biofuel. The production of biogas has four steps, viz., hydrolysis, acidogenesis, acetogenesis, and methanogenesis (Madsen et al. 2011). Through the use of nanoparticles, nanotechnology paves the way for the amelioration of biogas production by ensuring more efficient bioconversion strategies, better breakdown of substrates, and more optimized output delivery (Table 11.3). Nanoparticles provide a significant boost to achieve powerful and more controlled ways for the feedstock processing, genetic engineering, and biofuel harvesting. Inner surroundings of the anaerobic digester play a significant role in the biogas generation and problems including corrosion, wear and tear resistance limit the energy output from the biogas plant. The use of nanoparticles as the construction agents of the plant design can prove to be a hefty resource for the harness of biogas (Malik and Sangwan 2012).

Table 11.2 Nanoparticles used in the production of bioethanol

Nanoparticle	Target	Conclusion	Reference
Mesoporous silica	Cellulase enzyme immobilized on scaffold of mesoporous silica nanoparticles	Immobilized cellulase effective in achieving 80% yield of cellulose to glucose Have excellent stability and can be stored for 23 days	Chang et al. (2011)
MnO_2	Cellulase produced by *Aspergillus fumigatus* JCF was immobilized onto MnO_2 nanoparticles for bioethanol production from agricultural waste	Cellulase binding efficiency was 75% The properties of immobilized cellulase like activity, operational pH, temperature, thermal stability, and reusability were more stable than free enzyme This enzyme–nanoparticles combination with yeast for SSF, yielded a bioethanol concentration of about 21.96 g/l	Cherian et al. (2015)
Iron oxide – Fe_3O_4 and Fe_3O_4/alginate	Cellulase from *Aspergillus fumigatus* AA001	Cellulase showed thermal stability for 8 h at 70 °C	Srivastava et al. (2015)
Zinc oxide (ZnO)	Cellulase from *Aspergillus fumigatus* AA001	Cellulase showed stability at 65 °C for 10 h at pH of 10.5	Srivastava et al. (2015)
Cu	Electrochemical sensor was modified with graphene oxide containing Cu nanoparticles for bioethanol production	The study showed that the method has good accuracy and repeatability, with recovery of 92.8 ± 0.5% This sensor has good long-term stability, sensitivity, and is free from interference	Santos et al. (2016)
$FeCl_3$	Cellulase bound magnetic nanoparticles was used as nanocatalyst for the hydrolysis of *Sesbania aculeata* biomass for bioethanol production	The optimized conditions – 1.5% (w/v) nanocatalyst concentration, 4% (w/v) biomass concentration at 30 °C temperature 5.31 g/l bioethanol yield was obtained Nanocatalyst was reused for successive cycles and was efficient	Baskar et al. (2016)

Bio-stimulation of methanogenic bacteria by nanoparticles has been observed to increase their activity. Trace elements in the form of Fe^{2+}, Ni^{2+} or Co^{2+} (in μM concentrations) are required for anaerobic digestion as they improve the digestion process by exciting the bacterial action (Rao and Seenayya 1994; Donoso-Bravo and Mairet 2012). They work as an electron donor in an anaerobic process and release ions contributing to enzyme production. Trace metals can optimize the microbial

Table 11.3 Nanoparticles used in the production of biogas

Nanoparticle	Target	Conclusion	Reference
Ferrous oxide	Biostimulation of methanogenic bacteria	Enhance the methane production	Donso-Bravo and Mairet (2012)
CuO; ZnO	Effect of particle-size on methane and biogas production during anaerobic digestion of cattle manure were studied for 14 days	CuO nanoparticles (30 nm), ZnO nanoparticles (50–70 nm), bulk ZnO and CuO were most toxic to anaerobic bacteria Biogas production was inhibited at 15 mg/l and 120 mg/l concentrations of CuO and ZnO, respectively	Lunadel-Risco et al. (2011)
CeO_2, TiO_2; Ag, Au	Inhibitory effect of metal oxide and zero-valent nanoparticles was studied on the activity of microbial communities involved in the biogas production in wastewater treatment	CeO_2 nanoparticles have inhibited 100% biogas production Ag nanoparticles caused 33–50% biogas production inhibition Au and TiO_2 were not having any inhibitory effect	Garcia et al. (2012)
Iron oxide	Response of anaerobic granular sludge in the beet sugar industrial wastewater	Increased biogas and methane production Higher COD removal	Ambuchi et al. (2017)
Zero-valent iron	Benefits of zero-valent iron (nZVI) on sludge stabilization for the biogas improvement was studied	Addition of 0.10 wt% of nZVI in anaerobic digestion for 17 days enhanced the methane concentration in biogas by 5.1–13.2% It improved methane and biogas production by 30.4 and 40.4%, respectively The amendment of iron nanoparticles during anaerobic digestion effectively reduced the H_2S production in biogas	Su et al. (2013)

population and change the hydrolysis fermentation type. It has been reported that the addition of Fe^{2+} (up to 20 mM) in the form of chloride at 37 °C in a reactor resulted in a modest increase in biogas production for 24–48 h and then gradually decreased afterward (Ram et al. 2000). In another study, iron sulfate was used at room temperature and 40% increase was detected in methane production but after some time, the high concentration of iron has an inhibitory effect on bacterial population and the biogas production decreased dramatically (Rao and Seenayya 1994). This proves that no doubt iron is an essential component playing a crucial role in biogas production, but its high concentration is toxic to the microbial community (Rao and Seenayya 1994).

Zaidi et al. (2018) evaluated the influence of metal (1 mg/l) and metal-oxide (10 mg/l) nanoparticles on biogas production from microalgae *Enteromorpha.* In a batch experiment, anaerobic digestion was carried with 500 ml working volume at 37 °C (150 rpm, 170 h). After 60 h of retention time, the influence of nanoparticles was significantly improved. The maximum biogas yield of 624 ml was recorded by Fe_3O_4 nanoparticles. The increase in biogas production was 28% for Fe_3O_4, 26% with Ni, 9% with Ni and 8% with MgO. This increase in biogas production can be attributed to the stimulatory effect of nanoparticles responsible for the disintegration of carbohydrates, celluloses, and proteins. Nanoparticles also dissolve the organic compounds of microalgae and cause their lysis. Nanoparticles clobber the external layer of proteins and carbohydrates by hydrolyzing the glycosidic bond into simpler sugars followed by solubilizing the internal layer of cellulose and hemicellulose into oligosaccharides like cellobiose and cellodextrin. The use of nickel nanoparticles (60 mg/l) in the anaerobic digestion of industrial wastewater containing mono-ethylene glycol has resulted in the 23% increase in the biogas production (Elreedy et al. 2017).

Small iron oxide nanoparticles at nonsaturation conditions dissolve slowly, act as a carrier, and release the ion as an active compound. These particles have a unique feature of first dispersion and then dissolve progressively over a period of time supplying iron to microorganisms. In a closed digester, the addition of iron under the anaerobic methanogenic conditions gives rise to insoluble precipitates of ferric hydroxide that are subsequently reduced to Fe^{2+} (soluble form) and becomes available to the microorganisms (Casals et al. 2014). The study included the addition of 100 ppm of 7 nm iron oxide nanoparticles into the closed reactor and the biogas production per gram of organic matter was increased by 180%, recorded as the largest improvement till date. The results showed that the biogas was having a higher concentration of methane (8% more than control) and methane percentage in the biogas was 56% with Fe_3O_4 nanoparticles and was 48% in control. Iron oxide nanoparticles are the exclusive Fe^{2+} source due to their liability, denseness, and large size specifically in the anaerobic methanogenic conditions of the reactor. The response of anaerobic granular sludge to the use of iron oxide nanoparticles in the beet sugar industrial wastewater was studied by Ambuchi et al. (2017). Iron oxide nanoparticles significantly increased the biogas and methane production rate in comparison to the control samples. The possible reason might be the use of nanoparticles as a channel for electron transfer towards the methanogens. The use of nanoparticles also showed higher chemical oxygen demand removal efficiency after 12 h hydraulic retention time as compared to the control. The influence of nanoparticles on biogas production was also revealed by volatile fatty acid content as their generation is peculiar to more cell wall disruption and breakage of glycoproteins, carbohydrates, and celluloses. High concentration of acetic acid and butyric acid caused more biogas yield. Inhibitory effect of 10 mg/kg of silver nanoparticles released from industry was observed on anaerobic digestion in a bench scale landfill bioreactor. This might be due to the accumulation of volatile fatty acids and low pH causing deterioration of the methanogenic population. High concentration of silver nanoparticles has an inhibitory effect on methanogenesis as it is sensitive to various toxic substances (Yang et al. 2012).

The effect of methyl-functionalized silica and methyl-functionalized cobalt ferrite–silica nanoparticles on the enhancement of bioethanol production in syngas fermentation and to improve syngas–water mass transfer by *Clostridium ljungdahlii* was examined (Kim and Lee 2016). These nanoparticles were recovered using a magnet, recycled and reused for five times. Methyl-functionalized cobalt ferrite–silica nanoparticles were efficient in syngas fermentation due to improved syngas mass transfer. In comparison with control, methyl-functionalized cobalt ferrite–silica nanoparticles showed 227.6% increased biomass, 213.5% increased ethanol, and 59.6% increased acetic acid levels (Kim and Lee 2016).

11.5.4 Biodiesel

Biodiesel is captivating as biofuels as it does not contain particulate matter in comparison with nonrenewable diesel. Biodiesel is the mono-alkyl ester of long-chain fatty acids derived from the edible oils, nonedible oils, waste cooking oil, and animal fat or algal oils. It has a clean burning and is biodegradable fuel containing almost no sulfur and no aromatic compounds. It burns completely due to its higher cetane number and 10% built-in oxygen (also called oxygenated fuel as it contains a reduced amount of carbon and higher hydrogen to oxygen content than fossil diesel). Due to these properties, its combustion is improved further on its blending with petroleum diesel (Pathak et al. 2017). It can be blended up to 20% petroleum diesel with no engine modifications.

Pyrolysis and transesterification are the methodologies that are used for the production of biodiesel. Pyrolysis is the thermal treatment requiring high temperature and consumes high energy. The commercial biodiesel is synthesized through liquid base-catalyzed transesterification of C_{14}–C_{20} components of triacylglyceride (TAG) lipids into fatty acid methyl esters (FAMEs) along with glycerol as by-product using methanol and an alkaline catalyst (NaOH, NaOMe) (Lee et al. 2014). These are the homogeneous catalysts that are being used popularly as they have high efficiency. Homogeneous catalyst also carries some limitations like it is corrosive to the equipment, generates wastewater, and is always unrecoverable at the end. In comparison, heterogeneous catalyst is insoluble, can be separated at the end simply by filtration and again be reused. The requirement for high reaction temperature (>170 °C), long reaction time (>24 hours) and low catalytic stability add to the limitation of heterogenous catalyst. Hence, this paves the way for the new trends in the preparation of biodiesel, i.e., green method based on heterogeneous catalysts and nanoparticles (Dehkordi and Ghasemi 2012; Gurunathan and Ravi 2015). The use of nanocatalyst provides them higher catalytic activity due to nanodimensions and morphology. Till date, various catalysts have been explored for biodiesel production (Table 11.4).

Table 11.4 Nanoparticles used in the production of biodiesel

Nanoparticle	Source	Conclusion	Reference
Fe_3O_4	Lipase was immobilized by covalently bound to the amino-functionalized magnetic nanoparticles by using glutaraldehyde as a coupling reagent for biodiesel production	With 60% immobilized lipase, 90% yield of biodiesel was obtained from soybean oil by 3-step addition of methanol Immobilized enzyme was reused four times without decrease in its activity	Xie and Ma (2009)
Fe_3O_4 (magnet)	Magnetic nanoparticles was used to immobilize lipase for biodiesel production	Porcine pancreas lipase, *Candida rugosa* lipase and *Pseudomonas cepacia* lipase were immobilized onto the amino-functionalized Fe_3O_4 nanoparticles More than 30% activity of enzyme was tested even after 10 cycles	Wang et al. (2009)
Copper doped zinc oxide (CZO)	Transesterification of neem oil using CZO as catalyst	97.18% biodiesel yield was obtained CZO nanoparticles form multilayered nanostructures with nonuniform surface having pores 73.95% yield was obtained in sixth cycle of nanocatalyst reuse	Gurunathan and Ravi (2015)
CaO	Nanoparticles were synthesized from calcium nitrate (CaO/CaN) and Snail shell (CaO/SS) and tested for transesterification	Biodiesel yield increased from 93% to 96% CaO/SS exhibited excellent catalytic activity, stability, and reusability for 5 cycles	Gupta and Agarwal (2016)
Fe_3O_4	Lipase–Fe_3O_4 nanoparticle biocomposite catalyst developed for biodiesel production from soybean oil methanolysis	This catalyst showed high activity and stability in the single packed bed reactor After 192 h of reaction conversion rate was 88% and 240 h of reaction, conversion rate was 75%	Wang et al. (2011)

Meher et al. (2006) have developed a favorable route for the production of biodiesel through transesterification of vegetable oil and animal fat with the use of methanol under very mild conditions using nanoparticles as a catalyst. They grafted lipase onto the magnetic Fe_3O_4 nanoparticles by the 1-ethyl-3-(3-dimethylaminopropyl) carbodiimide activation. The study showed that modified nanoparticles were having an average diameter of 12.7 nm and maintained their activity through a wide range of pH and temperature as compared to the free lipase. They have shown the maximum conversion rate of 94% for methyl esters by using bound lipase as a catalyst. Nanoparticles of CaO were used as a catalyst for the transesterification of soybean oil and poultry fat with methanol at room temperature and 99% conversion was obtained quantitatively for biodiesel (Xie and Ma 2010). Methyl ester yield up to 97.7% was obtained in the presence of KF-impregnated nanoparticles of gamma-Al_2O_3. It was concluded by the experimental proceedings that relatively high conversion of vegetable oil to biodiesel was due to the high basicity of the catalyst and high surface to volume ratio of the gamma-Al_2O_3 (Hutchings 2013). KF/CaO nanoparticles were used as nanocatalysts in the production of biodiesel from Chinese tallow seed oil (Wen et al. 2010). The biodiesel yield was reached up to 96.8% in the presence of nanocatalyst and could be used to convert the oil with higher acid value into biodiesel. Magnetic solid nanoparticles of KF/CaO–`Fe_3O_4 were used as a base catalyst for biodiesel production, and the results had indicated that when the reaction was performed at 65 °C with 12:1 molar ratio of methanol/oil and the catalyst ratio of 4% by weight, the yield exceeded up to 95% at 3 h of reaction time. It was concluded that the nanomagnetic particles when used as a solid base catalyst for the biodiesel production gave a good prospect for the development and application (Hu et al. 2011). An experiment was carried out to probe into the possibility of using well-dispersed sulfated zirconia nanoparticles synthesized using poly(*N*-vinylpyrrolidone) (PVP) as a surfactant for the production of methane and biodiesel. The presence of PVP improves the uniformity and has promoted the formation of sulfated zirconia nanoparticles. These nanoparticles were directly used as a catalyst for the synthesis of biodiesel through the esterification of long-chain fatty acids. Varying concentration of catalyst was used and at 15% by weight, the active sites were close to saturation. The conversion of palmitic acid at this concentration was up to 97% and satisfactory esterification ratio was obtained over the conventional method (Chen et al. 2013).

Cationic nanoparticles in the form of amino-clay (containing cationic metals like Al^{3+}, Ca^{2+}, Mg^{+2} and organofunctional materials such as APTES) have been used nowadays for the extraction of oil from *Chlorella* sp. These cationic nanoparticles played a role by weakening the algal cells and enhancing the permeability of cells as their amino group contribute to the destabilization of plasma membrane and thereby, facilitating the contact to the hydrophobic solvent for the release of microalgal oil (Lee et al. 2013). The potential of lipase from *Pseudomonas cepacia* immobilized on $FeCl_3$ magnetic nanoparticles used as a biocatalyst for the synthesis of FAME using waste cooking oil as substrate was investigated (Yu et al. 2013). The proposed process can lower the production cost of biodiesel and help to dispose of the waste cooking oil. The optimum dosage of lipase-bound magnetic nanoparticle was 40% (w/w of oil), and the optimal reaction conditions were 44.2 °C, substrate molar ratio of 5.2, and 12.5% water content. Nearly 80% FAME conversion was retained after three cycles of the use of the immobilized magnetic nanoparticles as a biocatalyst. This biocatalyst showed good storage ability at 4 °C and can be easily recovered by applying the magnetic field (Yu et al. 2013).

11.5.5 Microbial Fuel Cells

The major issue that the world is facing is the generation of electricity and nowadays, research is being focused on to the new energy resource in the form of fuel cells to generate electricity by employing various catalysts. The fuel cells are pollution free, have high efficiency and are without any mechanical motion but its high cost acts as a barrier to its popularity (Hussain et al. 2017). Microbial fuel cells were considered in the early 1990s for the bioenergy production using sewage wastewater (Shah et al. 2019). They are capable of converting 50–90% of energy obtained by oxidation of organic matter into electricity. It works under specific conditions and uses biodegradable substrates as a fuel. The substrate is catabolized by the microorganisms to produce electricity or oxidative metabolism of electrochemically active bacterial species is the energy source as it catalyzes the generation of electrons from inorganic sources (like acetate, glucose, starch or even water). These cells generate low power energy, so they are used in combinations to generate power at a comparatively higher level (Malik and Sangwan 2012). However, the power density of microbial fuel cells at present is significantly lower than the theoretically calculated values due to technical limitations in the form of low bacteria loading capacity and difficulty between the transfer of electrons from bacteria to the electrode. To overcome this problem, 3-D (three-dimensional) graphene aerogel having platinum nanoparticles was used as an efficient freestanding anode in microbial fuel cells. This anode has continuous 3-D macroporous structure favorable for the immobilization of microorganisms and electrolyte transfer. This graphene aerogel scaffold is uniformly loaded with Pt nanoparticles, enhancing the transfer of charge between the bacteria and anode. This construction of microbial fuel cell generated a remarkable power density of 1460 mW/m^2, which is 5.3 times higher than that based on carbon cloth (273 mW/m^2) (Zhao et al. 2015).

Nafion membranes were modified with TiO_2/SnO_2 nanoparticles to enhance the proton exchange between the membranes and it was observed that the performance of the modified Nafion membranes was increased at high temperature when compared to the traditional Nafion membranes (Abbaraju et al. 2008). Microorganisms are used as biocatalysts that produce electrons as anode chamber and transfer electrons through an external circuit to the cathode chamber. Oxygen is mostly the electron acceptor in the cathode chamber due to its high oxidation potential and infiniteness. In microbial fuel cells, the cathode is the limiting factor affecting their efficiency because of the poor kinetics of oxygen reduction in the medium. Till today, platinum (Pt) is a common catalyst for oxygen reduction reaction. Due to the high cost of Pt (approximately its cost covers 50% of the total capital cost), it is not preferred commercially (Fan et al. 2008). Ghasemi et al. (2013a, b) studied different catalysts (viz., carbon black, nickel nanoparticles, phthalocyanine, and copper–phthalocyanine) in a two-chamber microbial fuel cell and compared their performance with platinum (Pt is the most commonly used cathode catalyst). The results proved nickel nanoparticles as the potential alternative catalyst for Pt as by decreasing the size of the metal; their catalytic activity can be increased. The Coulombic efficiency of nickel nanoparticles was about 20.7% which suggests that by further improving the surface area, higher power output can be obtained.

A reliable method was developed to synthesize nanohybrids of anatase TiO_2 nanoparticles and carbon nanotubes (Wen et al. 2013). This hybrid consists of carbon nanotubes encapsulated inside and anatase of TiO_2 nanoparticles coating on the surface of carbon nanotubes. The characterization revealed that the hybrid exhibits unique properties of carbon nanotubes and TiO_2 nanoparticles with one-dimensional structure, high surface area, good compatibility, and excellent electrical conductivity. Their electrochemical properties revealed that they have exceptional properties to be used as anode materials in microbial fuel cells. When hybrid was used as anode material, it exhibited higher output current, power density, and Coulombic efficiency in comparison with the carbon nanotubes and TiO_2 nanoparticles alone. This might be due to the synergistic effect of both carbon nanotubes and TiO_2 nanoparticles (Wen et al. 2013). An et al. (2011) examined the use of silver nanoparticles (AgNPs) as cathode for microbial fuel cells that are suffering from organic contamination and oxygen depletion. They prepared four treated cathodes (AgNPs-coated, Pt/C-coated, Pt/C+AgNPs-coated, and plain graphite cathodes) and tested them under high levels of organic load. When 50 mM acetate was used as the organic load, the system having AgNPs-coated cathode showed the highest dissolved oxygen concentration (0.8 mg/l) as well as the highest current ranging from 0.04 to 0.12 mA. Their finding concluded that the AgNPs could act as inhibitors for the growth of oxygen-consuming heterotrophic microbes leading to the increased concentration of dissolved oxygen in the cathode chamber in comparison to other cathodes. Hence, AgNPs can be used as a cathode catalyst for the reduction of oxygen. Graphene is the first isolated two-dimensional material having good electronic conductivity, large surface area, high mechanical strength, high thermal stability, and durability that makes it a potential catalyst support material for fuel cell applications. Reduced graphene oxide was investigated as a support material and ethylene glycol was used as reducing agent to fabricate Pt–Co alloy nanoparticles in order to develop low-cost, stable, and active electro-catalyst that was able to replace Pt. Pt–Co/graphene alloy nanoparticles contain reduced content of Pt and were investigated as a replacement to Pt/C cathode catalyst in the microbial fuel cells. These alloy nanoparticles were having a size of 4.1 nm and produced maximum power density of 1378 mW/m^2, which is close to Pt/C cathode (1406 mW/m^2) but the Coulombic efficiency (71.6%) was better than of Pt/C (52.0%). The price of Pt–Co/graphene alloy nanoparticles was just one-third of Pt/C cathode which means it has the potential to be used as cost-effective cathode in air microbial fuel cells (Yan et al. 2013) (Table 11.5).

Table 11.5 Nanoparticles used in the production of microbial fuel cell

Nanoparticle	Source	Conclusion	Reference
Fe_3O_4	*Saccharomyces cerevisiae* used as an active biocatalyst and neutral red with low concentration (200 mol l^{-1}) selected as electron shuttle in anode chamber	15% Fe_3O_4 nanoparticles generated maximum power (20 mWm^{-2}) and current (148 mA m^{-2}) In comparison to Nafion membranes the power was 29% more Nanoparticles were prepared by phase inversion method	Rahimnejad et al. (2012)
Pt	Composite electrode of carbon nanotube and Pt was prepared as a cathode catalyst	This composite electrode has reduced the amount of microbial fuel cell up to 25% Power output increased by 8.7–32.2% with respect to pristine Pt catalyst	Ghasemi et al. (2013a, b)
Zn, Fe, Co	Carbonized nanoparticles having nanoscale metal organic framework was prepared by using metal ions for their use as electrocatalyst	The onset potential and the half-wave potential for the ORR at the carbonized nanoparticles are up to 1.03 and 0.92 V in 0.1 M KOH solution, respectively, which is the best ORR activity of all the nonnoble metal catalysts reported Carbonized nanoparticles when used as the cathode of the alkaline direct fuel cell, the power density obtained is 22.7 mW/cm^2, 1.7 times higher than the commercial Pt/C catalysts	Zhao et al. (2014)
Tin oxide	SnO_2 nanoparticles dispersed on or encapsulated in reduced graphene oxide (SnO_2–rGO) hybrids	SnO_2–rGO hybrids exhibit remarkable lithium storage capacity and cycling stability Their capacity is 1222 $mAhg^{-1}$in the first cycle and maintains at 700 $mAhg^{-1}$ after 100 cycles	Tan et al. (2015)

11.6 Conclusion

Nanoparticles have been observed to play an important role in bioenergy production via increasing the efficiency by interlinking biological materials with nanoparticles. They use biochemical and thermochemical reactions for the production of bioenergy. In biochemical conversions, nanoparticles can be applied to fermentation, microbial fuel cells, biomass pretreatment, hydrolysis, separation of products, and recovery of catalysts. For thermochemical conversions, they are involved in the development of different catalysts, improving the mass transfer, product separation, and recovery. Their use can increase the biofuel production, separation, storage, and energy outputs. Nanoparticles can be used to chemically catalyze the depolymerization of readily available biomass for bioenergy production.

The expected use of nanoparticles in bioenergy production is not without the concern with regard to their potential toxic effects on humans or environmental impacts (Colvin 2003). It is still untimely to infer the chronic effects of nanoparticles; however, the studies on mice have shown that acute exposure to these particles can result in various inflammatory diseases. Since nanoparticles cause oxidation reactions and are small enough to enter human cells, it might cause cytotoxic effects by interacting with the biomolecules. The nanoparticles are usually made up of metals that are extremely toxic to organs and can be bioaccumulated in the body. The subacute and intraperitoneal exposure of TiO_2 nanoparticles was studied in rat and it was founded that their exposure has resulted in the pathological changes in the liver, has increased the platelet count and caused mild inflammation. It was further recorded that the bioaccumulation of titanium occurred in the liver, lungs, and brain of rats resulting in neurobehavioral changes (Younes et al. 2015). The exposure of silver nanoparticles to the bacterial community might result in the loss of DNA replication ability as reported by Raffi et al. (2008). Inhalation of mesoporous silica nanoparticles has serious toxic effects similar to aerosols. To add to the worry is the processing cost of these particles for the large-scale applications in the production of bioenergy (Pugh et al. 2011). Still, in spite of these threats, the potential utility of these particles makes them an accomplished platform for the recent research and studies have been stimulated to determine the safety measures required to limit the exposure of nanoparticles to human and environment.

References

Abbaraju RR, Dasgupta N, Virkar AV (2008) Composite Nafion membranes containing nanosize TiO_2/ SnO_2 for proton exchange membrane fuel cells. J Electrochem Soc 155(12):B1307–B1313

Abd-Elsalam K, Mohamed AA, Prasad R (2019) Magnetic Nanostructures: Environmental and Agricultural Applications. Springer International Publishing (ISBN 978-3-030-16438-6) https://www.springer.com/gp/book/9783030164386

Alzate CC, Toro OS (2006) Energy consumption analysis of integrated flowsheets for production of fuel ethanol from lignocellulosic biomass. Energy 31(13):2447–2459

Ambuchi JJ, Zhang Z, Shan L, Liang D, Zhang P, Feng Y (2017) Response of anaerobic granular sludge to iron oxide nanoparticles and multi-wall carbon nanotubes during beet sugar industrial wastewater treatment. Water Res 117:87–94

An J, Jeon H, Lee J, Chang IS (2011) Bifunctional silver nanoparticle cathode in microbial fuel cells for microbial growth inhibition with comparable oxygen reduction reaction activity. Environ Sci Technol 45(12):5441–5446

Ansari SA, Husain Q (2012) Potential applications of enzymes immobilized on/in nano materials: a review. Biotechnol Adv 30(3):512–523

Baskar G, Kumar RN, Melvin XH, Aiswarya R, Soumya S (2016) *Sesbania aculeate* biomass hydrolysis using magnetic nanobiocomposite of cellulase for bioethanol production. Renew Energ 98:23–28

Bowles LK, Ellefson WL (1985) Effects of butanol on *Clostridium acetobutylicum*. Appl Environ Microbiol 50(5):1165–1170

BP's Statistical Review of World Energy (2017) https://www.bp.com/content/dam/bp-country/de_ch/PDF/bp-statistical-review-of-world-energy-2017-full-report.pdf

Budarin V, Shuttleworth PS, Lanigan B, Clark JH (2013) Nanocatalysts for biofuels. In: Polshettiwar V, Asefa T (eds) Nanocatalysis synthesis and applications. Wiley, Hoboken, pp 595–614

Casals E, Barrena R, García A, González E, Delgado L, Busquets-Fité M, Font X, Arbiol J, Glatzel P, Kvashnina K, Sánchez A (2014) Programmed iron oxide nanoparticles disintegration in anaerobic digesters boosts biogas production. Small 10(14):2801–2808

Chang RH, Jang J, Wu KC (2011) Cellulase immobilized mesoporous silica nanocatalysts for efficient cellulose-to-glucose conversion. Green Chem 13(10):2844–2850

Chen G, Guo CY, Qiao H, Ye M, Qiu X, Yue C (2013) Well-dispersed sulfated zirconia nanoparticles as high-efficiency catalysts for the synthesis of bis (indolyl) methanes and biodiesel. Catal Commun 41:70–74

Chen B, Li F, Huang Z, Yuan G (2017) Carbon-coated Cu-Co bimetallic nanoparticles as selective and recyclable catalysts for production of biofuel 2,5-dimethylfuran. Appl Catal B 200:192–199

Cherian E, Dharmendirakumar M, Baskar G (2015) Immobilization of cellulase onto MnO_2 nanoparticles for bioethanol production by enhanced hydrolysis of agricultural waste. Chinese J Catal 36(8):1223–1229

Cipolatti EP, Silva MJ, Klein M, Feddern V, Feltes MM, Oliveira JV, Ninow JL, de Oliveira D (2014) Current status and trends in enzymatic nanoimmobilization. J Mol Catal B Enzym 99:56–67

Colvin VL (2003) The potential environmental impact of engineered nanomaterials. Nat Biotechnol 21(10):1166–1170

Damartzis T, Zabaniotou A (2011) Thermochemical conversion of biomass to second generation biofuels through integrated process design – a review. Renew Sust Energ Rev 15(1):366–378

Dehkordi AM, Ghasemi M (2012) Transesterification of waste cooking oil to biodiesel using Ca and Zr mixed oxides as heterogeneous base catalysts. Fuel Process Technol 97:45–51

Di Serio M, Tesser R, Pengmei L, Santacesaria E (2007) Heterogeneous catalysts for biodiesel production. Energy Fuel 22(1):207–217

Donoso-Bravo A, Mairet F (2012) Determining the limiting reaction in anaerobic digestion processes. How has this been tackled? J Chem Technol Biotechnol 87(10):1375–1378

Elreedy A, Ibrahim E, Hassan N, El-Dissouky A, Fujii M, Yoshimura C, Tawfik A (2017) Nickel-graphene nanocomposite as a novel supplement for enhancement of biohydrogen production from industrial wastewater containing mono-ethylene glycol. Energy Convers Manage 140:133–144

Fan Y, Sharbrough E, Liu H (2008) Quantification of the internal resistance distribution of microbial fuel cells. Environ Sci Technol 42(21):8101–8107

Fang Z, Zhang F, Zeng HY, Guo F (2011) Production of glucose by hydrolysis of cellulose at 423 K in the presence of activated hydrotalcite nanoparticles. Bioresour Technol 102(17):8017–8021

Galbe M, Sassner P, Wingren A, Zacchi G (2007) Process engineering economics of bioethanol production. Adv Biochem Eng Biotechnol 108:303–327

García A, Delgado L, Torà JA, Casals E, González E, Puntes V, Font X, Carrera J, Sánchez A (2012) Effect of cerium dioxide, titanium dioxide, silver, and gold nanoparticles on the activity of microbial communities intended in wastewater treatment. J Hazard Mater 199:64–72

Ghasemi M, Daud WR, Rahimnejad M, Rezayi M, Fatemi A, Jafari Y, Somalu MR, Manzour A (2013a) Copper-phthalocyanine and nickel nanoparticles as novel cathode catalysts in microbial fuel cells. Int J Hydrogen Energy 38(22):9533–9540

Ghasemi M, Ismail M, Kamarudin SK, Saeedfar K, Daud WR, Hassan SH, Heng LY, Alam J, Oh SE (2013b) Carbon nanotube as an alternative cathode support and catalyst for microbial fuel cells. Appl Energy 102:1050–1056

Goh WJ, Makam VS, Hu J, Kang L, Zheng M, Yoong SL, Udalagama CN, Pastorin G (2012) Iron oxide filled magnetic carbon nanotube–enzyme conjugates for recycling of amyloglucosidase: toward useful applications in biofuel production process. Langmuir 28(49):16864–16873

Gupta J, Agarwal M (2016) Preparation and characterization of CaO nanoparticle for biodiesel production. AIP Conf Proc. https://doi.org/10.1063/1.4945186

Gurunathan B, Ravi A (2015) Process optimization and kinetics of biodiesel production from neem oil using copper doped zinc oxide heterogeneous nanocatalyst. Bioresour Technol 190:424–428

Hu S, Guan Y, Wang Y, Han H (2011) Nano-magnetic catalyst KF/CaO–Fe_3O_4 for biodiesel production. Appl Energy 88(8):2685–2690

Hussain M, Ahmad R, Liu Y, Liu B, He M, He N (2017) Applications of nanomaterials and biological materials in bioenergy. J Nanosci Nanotechnol 17(12):8654–8666

Hussein AK (2015) Applications of nanotechnology in renewable energies – a comprehensive overview and understanding. Renew Sust Energ Rev 42:460–476

Hutchings G (2013) Nanocatalysis: synthesis and applications. John Wiley and Sons, Weinheim

Ingram LO (1989) Ethanol tolerance in bacteria. Crit Rev Biotechnol 9(4):305–319

Ivanova V, Petrova P, Hristov J (2011) Application in the ethanol fermentation of immobilized yeast cells in matrix of alginate/magnetic nanoparticles, on chitosan-magnetite microparticles and cellulose-coated magnetic nanoparticles. Int Rev Chem Eng 3:289–299

Jia Y, Hu Y, Zhu Y, Che L, Shen Q, Zhang J, Li X (2011) Oligoamines conjugated chitosan derivatives: synthesis, characterization, in vitro and in vivo biocompatibility evaluations. Carbohydr Polym 83(3):1153–1161

Khan MJ, Husain Q, Azam A (2012) Immobilization of porcine pancreatic α-amylase on magnetic Fe_2O_3 nanoparticles: applications to the hydrolysis of starch. Biotechnol Bioprocess Eng 17(2):377–384

Kim YK, Lee H (2016) Use of magnetic nanoparticles to enhance bioethanol production in syngas fermentation. Bioresour Technol 204:139–144

Kim YK, Park SE, Lee H, Yun JY (2014) Enhancement of bioethanol production in syngas fermentation with *Clostridium ljungdahlii* using nanoparticles. Bioresour Technol 159:446–450

Kresge CT, Leonowicz ME, Roth WJ, Vartuli JC, Beck JS (1992) Ordered mesoporous molecular sieves synthesized by a liquid-crystal template mechanism. Nature 359(6397):710

Kumar Gupta S, Kumari S, Reddy K, Bux F (2013) Trends in biohydrogen production: major challenges and state-of-the-art developments. Environ Technol 34(13–14):1653–1670

Larsen SC (2007) Nanocrystalline zeolites and zeolite structures: synthesis, characterization, and applications. J Phys Chem C 111(50):18464–18474

Lee YC, Huh YS, Farooq W, Chung J, Han JI, Shin HJ, Jeong SH, Lee JS, Oh YK, Park JY (2013) Lipid extractions from docosahexaenoic acid (DHA)-rich and oleaginous *Chlorella* sp. biomasses by organic-nanoclays. Bioresour Technol 137:74–81

Lee AF, Bennett JA, Manayil JC, Wilson K (2014) Heterogeneous catalysis for sustainable biodiesel production via esterification and transesterification. Chem Soc Rev 43(22):7887–7916

Li M (2017) World Energy 2017-2050: Annual report. https://content.csbs.utah.edu/~mli/2017/World%20Energy%202017-2050.pdf

Lin YF, Chen JH, Hsu SH, Hsiao HC, Chung TW, Tung KL (2012) The synthesis of Lewis acid ZrO_2 nanoparticles and their applications in phospholipid adsorption from *Jatropha* oil used for biofuel. J Colloid Interface Sci 368(1):660–662

Liu KK, Chen MF, Chen PY, Lee TJ, Cheng CL, Chang CC, Ho YP, Chao JI (2008) Alpha-bungarotoxin binding to target cell in a developing visual system by carboxylated nanodiamond. Nanotechnology 19(20):205102

Lu AH, Salabas EE, Schüth F (2007) Magnetic nanoparticles: synthesis, protection, functionalization, and application. Angew Chem Int Ed 46(8):1222–1244

Luna-del Risco M, Orupõld K, Dubourguier HC (2011) Particle-size effect of CuO and ZnO on biogas and methane production during anaerobic digestion. J Hazard Mater 189(1–2):603–608

Lupoi JS, Smith EA (2011) Evaluation of nanoparticle-immobilized cellulase for improved ethanol yield in simultaneous saccharification and fermentation reactions. Biotechnol Bioeng 108(12):2835–2843

Madsen M, Holm-Nielsen JB, Esbensen KH (2011) Monitoring of anaerobic digestion processes: a review perspective. Renew Sust Ener Rev 15(6):3141–3155

Malik P, Sangwan A (2012) Nanotechnology: a tool for improving efficiency of bio-energy. J Eng Appl Sci 1:37–49

Meher LC, Sagar DV, Naik SN (2006) Technical aspects of biodiesel production by transesterification – a review. Renew Sust Energ Rev 10(3):248–268
Mendes AA, Oliveira PC, Castro HF, Giordano RD (2011) Application of chitosan as support for immobilization of enzymes of industrial interest. Quím Nova 34(5):831–840
Mielby J, Abildstrøm JO, Wang F, Kasama T, Weidenthaler C, Kegnæs S (2014) Oxidation of bioethanol using Zeolite-encapsulated gold nanoparticles. Angew Chem Int Ed 126(46):12721–12724
Milledge JJ, Smith B, Dyer PW, Harvey P (2014) Macroalgae-derived biofuel: a review of methods of energy extraction from seaweed biomass. Energies 7(11):7194–7222
Misson M, Zhang H, Jin B (2015) Nanobiocatalyst advancements and bioprocessing applications. J R Soc Interface 12(102):20140891. https://doi.org/10.1098/rsif.2014.0891
Nicolas P, Lassalle V, Ferreira ML (2014) Development of a magnetic biocatalyst useful for the synthesis of ethyloleate. Bioprocess Biosyst Eng 37(3):585–591
Pathak PK, Raj J, Saxena G, Sharma US (2017) A review on production of biodiesel by transesterification using heterogeneous nanocatalyst. Int J Sci Res Dev 5(2):631–636
Pugh S, McKenna R, Moolick R, Nielsen DR (2011) Advances and opportunities at the interface between microbial bioenergy and nanotechnology. Can J Chem Eng 89(1):2–12
Rad AG, Abbasi H, Afzali MH (2011) Gold nanoparticles: synthesizing, characterizing and reviewing novel application in recent years. Phys Procedia 22:203–208
Raffi M, Hussain F, Bhatti TM, Akhter JI, Hameed A, Hasan MM (2008) Antibacterial characterization of silver nanoparticles against *E. coli* ATCC-15224. J Mater Sci Technol 24(2):192–196
Rahimnejad M, Ghasemi M, Najafpour GD, Ismail M, Mohammad AW, Ghoreyshi AA, Hassan SH (2012) Synthesis, characterization and application studies of self-made Fe_3O_4/PES nanocomposite membranes in microbial fuel cell. Electrochim Acta 85:700–706
Rai M, dos Santos JC, Soler MF, Marcelino PR, Brumano LP, Ingle AP, Gaikwad S, Gade A, da Silva SS (2016) Strategic role of nanotechnology for production of bioethanol and biodiesel. Nanotechnol Rev 5(2):231–250
Raita M, Arnthong J, Champreda V, Laosiripojana N (2015) Modification of magnetic nanoparticle lipase designs for biodiesel production from palm oil. Fuel Process Technol 134:189–197
Ram MS, Singh L, Suryanarayana MV, Alam SI (2000) Effect of iron, nickel and cobalt on bacterial activity and dynamics during anaerobic oxidation of organic matter. Water Air Soil Poll 117(1–4):305–312
Rao PP, Seenayya G (1994) Improvement of methanogenesis from cow dung and poultry litter waste digesters by addition of iron. World J Microbiol Biotechnol 10(2):211–214
Reis P, Witula T, Holmberg K (2008) Mesoporous materials as host for an entrapped enzyme. Micropor Mesopor Mat 110(2–3):355–362
Saini JK, Saini R, Tewari L (2015) Lignocellulosic agriculture wastes as biomass feedstocks for second-generation bioethanol production: concepts and recent developments. 3 Biotech 5(4):337–353
Santos FC, Paim LL, da Silva JL, Stradiotto NR (2016) Electrochemical determination of total reducing sugars from bioethanol production using glassy carbon electrode modified with graphene oxide containing copper nanoparticles. Fuel 163:112–121
Schrand AM, Hens SA, Shenderova OA (2009) Nanodiamond particles: properties and perspectives for bioapplications. Crit Rev Solid State Mater Sci 34(1–2):18–74
Schügerl K, Hubbuch J (2005) Integrated bioprocesses. Curr Opin Microbiol 8(3):294–300
Shah S, Venkatramanan V, Prasad R (2019) Microbial fuel cell: Sustainable green technology for bioelectricity generation and wastewater treatment. In: Sustainable Green Technologies for Environmental Management (eds. Shah S, Venkatramanan V, Prasad R), Springer Springer Nature Singapore Pte Ltd. 199–218
Shuttleworth PS, Parker HL, Hunt AJ, Budarin VL, Matharu AS, Clark JH (2014) Applications of nanoparticles in biomass conversion to chemicals and fuels. Green Chem 16(2):573–584
Sims RE, Mabee W, Saddler JN, Taylor M (2010) An overview of second generation biofuel technologies. Bioresour Technol 101(6):1570–1580
Souza KC, Mohallem ND, Sousa EM (2011) Magnetic nanocomposites: potential for applications in Biomedicine. Quím Nova 34(10):1692–1703

Srivastava N, Srivastava M, Mishra PK, Singh P, Ramteke PW (2015) Application of cellulases in biofuels industries: an overview. J Biofuel Bioenerg 1(1):55–63
Srivastava N, Srivastava M, Mishra PK, Ramteke PW (2016) Application of ZnO nanoparticles for improving the thermal and pH stability of crude cellulase obtained from *Aspergillus fumigatus* AA001. Front Microbiol 7:514. https://doi.org/10.3389/fmicb.2016.00514
Straathof AJ (2003) Auxiliary phase guidelines for microbial biotransformations of toxic substrate into toxic product. Biotechnol Prog 19(3):755–762
Su L, Shi X, Guo G, Zhao A, Zhao Y (2013) Stabilization of sewage sludge in the presence of nanoscale zero-valent iron (nZVI): abatement of odor and improvement of biogas production. J Mater Cycle Waste Manage 15(4):461–468
Tan C, Zhao S, Yang G, Hu S, Qin X (2015) Facile and surfactant-free synthesis of SnO_2-graphene hybrids as high performance anode for lithium-ion batteries. Ionics 21(4):987–994
Taufiqurrahmi N, Mohamed AR, Bhatia S (2011) Production of biofuel from waste cooking palm oil using nanocrystalline zeolite as catalyst: process optimization studies. Bioresour Technol 102(22):10686–10694
Uygun DA, Öztürk N, Akgöl S, Denizli A (2012) Novel magnetic nanoparticles for the hydrolysis of starch with *Bacillus licheniformis* α-amylase. J Appl Polym Sci 123(5):2574–2581
Verma ML, Chaudhary R, Tsuzuki T, Barrow CJ, Puri M (2013) Immobilization of β-glucosidase on a magnetic nanoparticle improves thermostability: application in cellobiose hydrolysis. Bioresour Technol 135:2–6
Wang X, Dou P, Zhao P, Zhao C, Ding Y, Xu P (2009) Immobilization of lipases onto magnetic Fe_3O_4 nanoparticles for application in biodiesel production. Chem Sus Chem 2(10):947–950
Wang X, Liu X, Zhao C, Ding Y, Xu P (2011) Biodiesel production in packed-bed reactors using lipase–nanoparticle biocomposite. Bioresour Technol 102(10):6352–6355
Wang W, Martin JC, Fan X, Han A, Luo Z, Sun L (2012) Silica nanoparticles and frameworks from rice husk biomass. ACS Appl Mater Interfaces 4(2):977–981
Wen L, Wang Y, Lu D, Hu S, Han H (2010) Preparation of KF/CaO nanocatalyst and its application in biodiesel production from Chinese tallow seed oil. Fuel 89(9):2267–2271
Wen Z, Ci S, Mao S, Cui S, Lu G, Yu K, Luo S, He Z, Chen J (2013) TiO_2 nanoparticles-decorated carbon nanotubes for significantly improved bioelectricity generation in microbial fuel cells. J Power Sources 234:100–106
Xie W, Ma N (2009) Immobilized lipase on Fe_3O_4 nanoparticles as biocatalyst for biodiesel production. Energ Fuel 23(3):1347–1353
Xie W, Ma N (2010) Enzymatic transesterification of soybean oil by using immobilized lipase on magnetic nano-particles. Biomass Bioenergy 34(6):890–896
Xu X, Li Y, Gong Y, Zhang P, Li H, Wang Y (2012) Synthesis of palladium nanoparticles supported on mesoporous N-doped carbon and their catalytic ability for biofuel upgrade. J Am Chem Soc 134(41):16987–16990
Yan Z, Wang M, Huang B, Liu R, Zhao J (2013) Graphene supported Pt-Co alloy nanoparticles as cathode catalyst for microbial fuel cells. Int J Electrochem Sci 8:149–158
Yang Y, Xu M, Wall JD, Hu Z (2012) Nanosilver impact on methanogenesis and biogas production from municipal solid waste. Waste Manag 32(5):816–825
Younes NR, Amara S, Mrad I, Ben-Slama I, Jeljeli M, Omri K, El Ghoul J, El Mir L, Rhouma KB, Abdelmelek H, Sakly M (2015) Subacute toxicity of titanium dioxide (TiO_2) nanoparticles in male rats: emotional behavior and pathophysiological examination. Environ Sci Pollut Res 22(11):8728–8737
Yu CY, Huang LY, Kuan I, Lee SL (2013) Optimized production of biodiesel from waste cooking oil by lipase immobilized on magnetic nanoparticles. Int J Mol Sci 14(12):24074–24086
Yulianti CH, Ediati R, Hartanto D, Purbaningtias TE, Chisaki Y, Jalil AA, Ku CK, Prasetyoko D (2014) Synthesis of CaO-ZnO nanoparticles catalyst and its application in transesterification of refined palm oil. Bull Chem Reac Eng Cat 9(2):100–110
Zaidi AA, RuiZhe F, Shi Y, Khan SZ, Mushtaq K (2018) Nanoparticles augmentation on biogas yield from microalgal biomass anaerobic digestion. Int J Hydrogen Energy 18:1–12

Zhang J, Wang L, Ji Y, Chen F, Xiao FS (2018) Mesoporous zeolites for biofuel upgrading and glycerol conversion. Front Chem Sci Eng 26(1):132–144

Zhao S, Yin H, Du L, He L, Zhao K, Chang L, Yin G, Zhao H, Liu S, Tang Z (2014) Carbonized nanoscale metal–organic frameworks as high performance electrocatalyst for oxygen reduction reaction. ACS Nano 8(12):12660–12668

Zhao S, Li Y, Yin H, Liu Z, Luan E, Zhao F, Tang Z, Liu S (2015) Three-dimensional graphene/Pt nanoparticle composites as freestanding anode for enhancing performance of microbial fuel cells. Sci Adv 1(10):e1500372. https://doi.org/10.1126/sciadv.1500372

Zhou Q, Zhang H, Chang F, Li H, Pan H, Xue W, Hu DY, Yang S (2015) Nano La_2O_3 as a heterogeneous catalyst for biodiesel synthesis by transesterification of *Jatropha curcas* L. oil. J Ind Eng Chem 31:385–392

Zuliani A, Ivars F, Luque R (2018) Advances in nanocatalyst design for biofuel production. Chem Cat Chem 10(9):1968–1981

Chapter 12
Remediation of Heavy Metal Ions Using Nanomaterials Sourced from Wastewaters

Paul Thomas, Nelson Pynadathu Rumjit, Pynadathu Jacob George, Chin Wei Lai, Preeti Tyagi, Mohd Rafie Bin Johan, and Manickam Puratchiveeran Saravanakumar

Contents

P. Thomas · C. W. Lai (✉) · P. Tyagi · M. R. B. Johan
Nanotechnology and Catalysis Research Centre, Institute for Advanced Studies, University of Malaya, Kuala Lumpur, Malaysia
e-mail: cwlai@um.edu.my

P. J. George
Centre for Human Resource Development, Sam Higginbottom University of Agriculture, Technology and Sciences, Allahabad, Uttar Pradesh, India

N. P. Rumjit · M. P. Saravanakumar
Department of Environmental and Water Resources Engineering, School of Civil Engineering, Vellore Institute of Technology, Vellore, Tamil Nadu, India

D. Thangadurai et al. (eds.), *Nanotechnology for Food, Agriculture, and Environment*, Nanotechnology in the Life Sciences,
https://doi.org/10.1007/978-3-030-31938-0_12

12.1 Introduction

Preserving water is one of the most valuable resources required for the survival and future betterment of human life and other life forms on earth. With the proliferation of population, urbanisation, technology and industrial explosion, the demand for fresh water is increasing day by day (Yang et al. 2019). Heavy metals are one of the main contributors of water contamination which are sourced from mining activities, metallurgy, nuclear power plants, chemical treatment plants, electroplating and home-based and agriculture effluents (Li et al. 2018; Vilardi et al. 2018; Yin et al. 2018; Yu et al. 2018b). Examples of toxic heavy metals include Cu, Pb, Zn and Hg which pose severe long-range dangers to human and other life forms due to their accumulation in the food cycle (Wang et al. 2018b; Zhang et al. 2018; Jaishankar et al. 2014). Studies have shown that heavy metal causes severe impairments to lungs, kidney function, central nervous system and mental health and other organs (Lentini et al. 2017).

Heavy metals cannot be decomposed by microbial activity due to its toxic and carcinogenic nature exerting harmful effects on the ecosystem and environment and further resulting in the accumulation in the food cycle (Cocarta et al. 2016). Hence, remediation of heavy metals from waste effluents is of much critical importance and acquired much concern tremendously. In order to tackle this issue, commonly used techniques for the removal of heavy metals from various wastewater sources include ion-exchange process (Borklu Budak 2013), chemical precipitation methods (Chen et al. 2018), membrane filtration (Khulbe and Matsuura 2018), extraction by solvent (Li et al. 2017), electrodialysis (Barakat 2011), photocatalytic degeneration, oxidation and reduction methods (Sheng et al. 2016), adsorption (Renu et al. 2017) and coagulation process (Zou et al. 2016).

Compared to other prevailing techniques, adsorption methods are found to be feasible due to their cost-effectiveness, efficient practicability, simple process and eco-friendly and simple redevelopment of sorbents towards the removal of heavy metal contaminants (Hu et al. 2015; Wang et al. 2018c; Zhao et al. 2018). Recently, poriferous materials prepared from metal-organic frameworks (MOFs) exhibited superior applicability in removing hazardous contaminants from the environment owing to their highly tunable porosities and functional properties of various poriferous structures (Li et al. 2018).

Ricco and his colleagues synthesised aluminium-based magnetic nanohybrid material MOF which showed high adsorption capacity of Pb^{2+} ions at 492.4 mg/g (Ricco et al. 2015). Within the past few years, various nanomaterials have shown their supreme dominance in multiple fields such as energy and health sector (Pumera 2011; Kumar et al. 2015), electronic devices (Fatikow et al. 2012) and environmental remediation (Guerra et al. 2018).

Nanomaterials are particles of nanolevel dimensions mostly in the range 1–100 nm (Buzea and Pacheco 2017). Since nanomaterials are of nanolevel dimensions, such particles usually display unique properties such as the effect of small size, high surface area/volume ratio and quantum and macro-quantum tunnel effect (Ansari et al. 2010; Prasad et al. 2016; Chen et al. 2017; Hossein Beyki et al. 2017; Singh et al. 2017). These unique properties promote excellent sorption capacity, selectivity and reactiveness favouring effective adsorption of heavy metal ions from wastewater effluents (Vunain et al. 2016; Abdullah et al. 2018b).

This chapter focused on reviewing recent advancements of various nanomaterials such as graphene oxide (GO) (Nujic and Habuda-Stanic 2019), carbon nanotubes (CNTs) (Ihsanullah et al. 2016), metal oxides (MOs) (Hua et al. 2012), carbon-based nanoparticles (NPs), zero-valent metal (ZVM)-based NPs, polymer-based adsorbents and other nano-combinations (Wang et al. 2012), which were used for remediation of heavy metals from wastewater, as shown in Fig. 12.1. The effect of various operating parameters influencing the sorption behaviours has been investigated in this chapter. Recent challenges and outlooks of CNTs were also discussed.

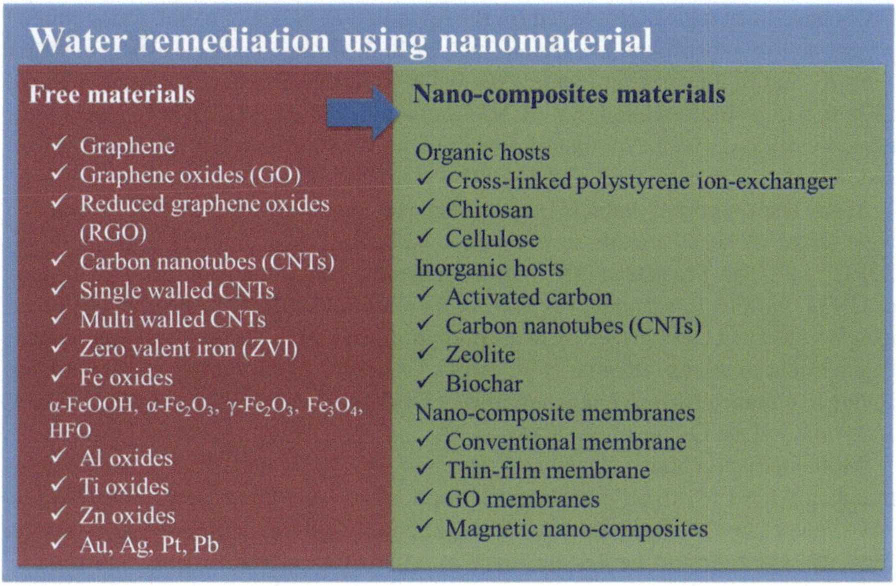

Fig. 12.1 Various nanomaterials used for heavy metal remediation from contaminated wastewater. (Copyright© Elsevier 2016, reprinted with permission from Zhang et al. (2016))

Different preparation and characterisation approaches of nanomaterials were presented in this chapter. Furthermore, this chapter examines future insights for the enhancement of nanomaterials in the remediation of heavy metal-contaminated water resources.

12.2 Various Nanomaterials Used for Removal of Heavy Metals

12.2.1 Carbon-Derived Nanomaterials

Carbon-derived nanomaterials have been widely implemented for the treatment of contaminated wastewaters (Al-Anzi and Siang 2017). Their extraordinary properties consist of enormous surface area characteristics, easy modification by physical or chemical activation process and capability in treating both organic and inorganic contaminants proving carbon-derived nanomaterials as promising adsorbents for heavy metal remediation source from wastewater (Smith and Rodrigues 2015). Carbon-derived nanomaterials are classified into two types, namely, carbon nanotubes (CNTs) and graphene-derived nanomaterials.

12.2.1.1 Carbon Nanotubes (CNTs)

CNTs have been widely employed for the remediation of heavy metal-contaminated waters due to exceptional properties such as high surface characteristics, sorption capacity and rapid sorption kinetics (Gupta et al. 2016; Yu et al. 2018a). CNTs are classified into single-walled (SW) CNTs and multi-walled (MW) CNTs which are a type of carbon-derived nanomaterial whose length varies between 100 and 1000 nm and diameter size between 1 and 3 nm (Lu et al. 2016).

It has been reported that CNTs exhibited excellent sorption capacities towards a wide range of heavy metals such as Cu^{2+}, Cr^{6+}, Pb^{2+}, Mn^{7+} and TI ions (Pu et al. 2013; Yadav and Srivastava 2017). To enhance the heavy metal sorption behaviour of CNTs, various functional groups (COOH (carboxylic acid), –OH and $–NH_2$) are incorporated on CNT surface by heat irradiation, endohedral filling and amendment by chemical treatment (Kumar et al. 2014). Examples of commonly used oxidising agents to enhance surface characteristics of CNTs consist of HNO_3, H_2SO_4, $KMnO_4$ and NaOCl (Duc Quyen et al. 2018).

Mohamed and colleagues studied the removal of Hg^{2+} ions from water by functionalising CNTs with allyl triphenyl phosphonium (ATPB) and glycerol ($C_3H_8O_3$) (Al Omar et al. 2017). From batch experimental results, maximum Hg^{2+} sorption capacity was found to be 186.98 mg/g at optimised conditions (pH, 5.5; time of contact, 28 min). The adsorption results were well correlated with the Freundlich isotherm and followed pseudo-second-order kinetics. In another study, Zhan and his colleagues synthesised magnetic NH_2-functionalised Fe_3O_4/COOH MWCNT hybrid composite by one-pot solvothermal technique and investigated the applicability towards the removal of Cu^{2+} ions from contaminated waters (Zhan et al. 2016). The newly functionalised CNTs exhibited high Cu^{2+} sorption efficacy of 30.48 mg/g, and adsorption data were well suited with the Langmuir isotherm model. The nanocomposite (NC) was readily isolated from contaminated waters with the help of a strong magnetic field.

Magnetic hydroxypropyl (HP) chitosan (CS)/oxidised MWCNT hybrid composites were synthesised effectively and investigated as a sorbent for the remediation of Pb^{2+} from aqueous media, as illustrated in Fig. 12.2 (Wang et al. 2015). Initially, magnetic Fe_3O_4 NPs were developed from $FeCl_3{\cdot}6H_2O$ (ferric chloride hexahydrate), and MWCNTs have undergone oxidation by a mixture of conc. H_2SO_4 and HNO_3 to form O-MWCNTs. Furthermore, magnetic Fe_3O_4 NPs and O-MWCNTs were added back to back to the HP CS solution, and nanohybrid composites were fabricated. The novel hybrid NC exhibited excellent sorption behaviour with optimum sorption capacity of 116.4 mg/g. Additionally, rapid easy separation of NC was achieved within 3 min after the sorption process. The adsorption results were correlated with the Sips isotherm model.

Even though CNTs displayed many advantages towards the removal of heavy metals from contaminated waters, there are significant downsides. CNTs are of high cost which inhibited their commercial utilisation; hence, more research has to be carried out in order to synthesise both cost-effective and productive CNTs. The

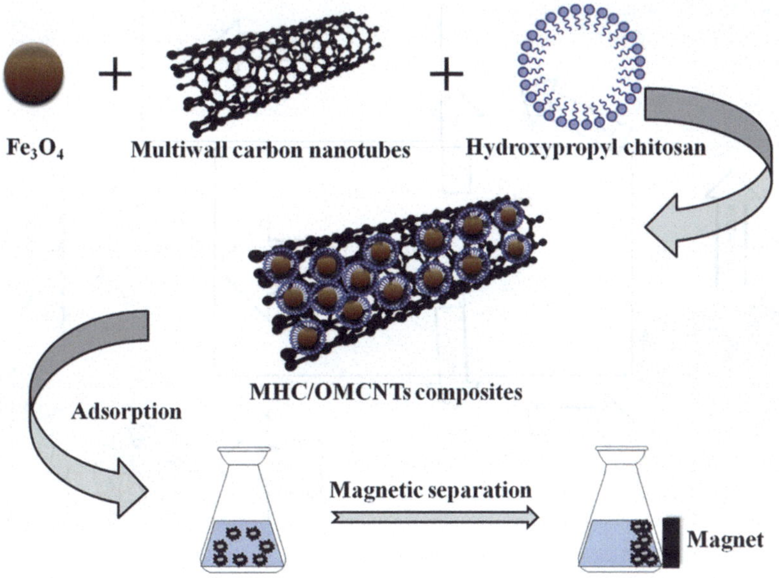

Fig. 12.2 Synthesis of MHC/O-MWCNT composites and their applicability towards Pb^{2+} metal ion removal. (Copyright© Elsevier 2015, reprinted with permission from Wang et al. (2015))

segregation of CNTs from contaminated waters is often difficult which escalates the cost of treatment and the possibility of secondary contamination. Ultimately, the toxic nature of CNTs has to be carefully reviewed.

12.2.2 Graphene-Based NPs

Graphene is another vital category of carbon-derived nanomaterial which is widely exploited for the treatment of heavy metals from contaminated waters. The 2D graphene structure possesses exceptional properties such as rigidity and plasticity, mechanical strength and thermal and electrical conductivity (Novoselov et al. 2012). Graphene-derived nanomaterials are classified into two types, namely, reduced graphene oxide (RGO) and GO, which can be employed for the remediation of heavy metals from contaminated waters. Graphene is oxidised to form GO which comprises of O_2 functional groups including –OH, –COOH, carbonyl (C=O) and epoxide which played a significant role in the remediation of heavy metals (Gao et al. 2011). RGO which is formed by the reduction of GO can be easily amended by –OH and –COOH functional groups (Avouris and Dimitrakopoulos 2012).

Various materials procured from graphene structure are displayed in Fig. 12.3 (Suárez-Iglesias et al. 2017). The graphene-derived nanomaterials exhibited excel-

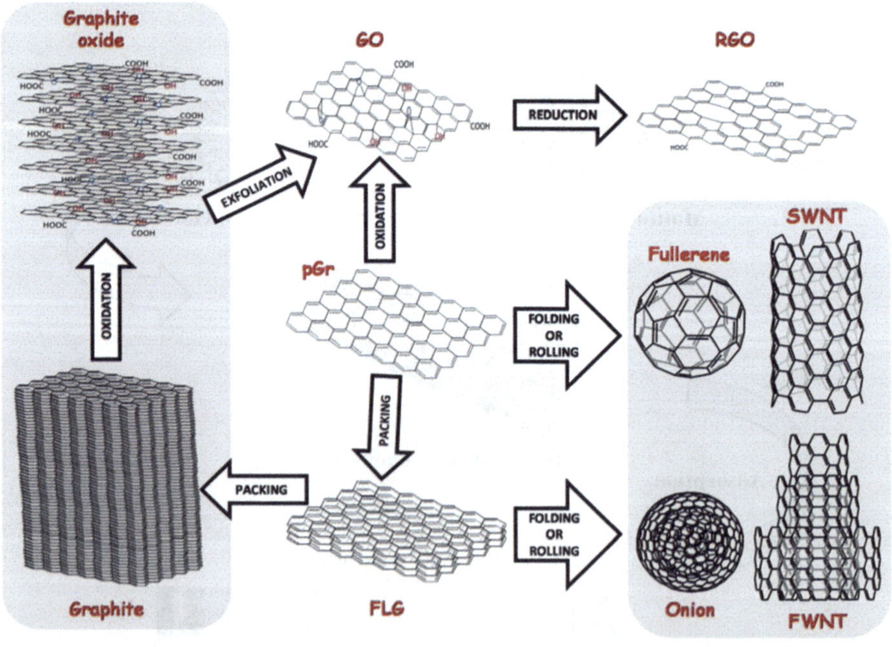

Fig. 12.3 Various materials procured from the 2D graphene structure. (Copyright© Elsevier 2017, reprinted with permission from Suárez-Iglesias et al. (2017))

lent properties such as high surface characteristics, abundant functional groups, hydrophilic nature and immense negative charge density towards the treatment of heavy metals (Xu and Wang 2017).

Wang and colleagues studied the sorption behaviour of GO towards Zn^{2+} heavy metals (Ho 2014). From the batch experimental results, optimum sorption capacity of Zn^{2+} was about 246.1 mg/g. The adsorption data were well correlated with the Langmuir adsorption model and followed pseudo-second-order kinetics. Zhao and his colleagues synthesised single-layered GO nanosheets by Hummer's process and employed to study the sorption of Cd^{2+} and Co^{2+} heavy metal ions from water by batch technique (Zhao et al. 2011).

The maximum removal capacities were shown to be 106.4 mg/g for Co^{2+} and 68.3 mg/g for Cd^{2+} ions. The thermodynamic variables for this sorption process were spontaneous and endothermic. Recent advancement of graphene-derived NCs is summarised in Tables 12.1 and 12.2. Presently, Arshad and his colleagues developed a newly amended graphene-modified sorbent by incorporation of calcium alginate (CA) beads that have further undergone reduction with polyethyleneimine

Table 12.1 Recent studies on graphene-derived nanocomposites (NCs) used for the remediation of heavy metals

Adsorbents	Heavy metals	Optimum sorption capacity (mg/g)	References
Functionalised GOCA (calcium alginate) beads	Pb^{2+} Hg^{2+} Cd^{2+}	602.1 374.02 181.2	Arshad et al. (2019)
RGO/NiO (nickel oxide)	Cr^{6+}	198.2	Zhang et al. (2018)
PAS (siloxane)-GO	U^{6+} Eu^{3+}	310.65 243.80	Zhao et al. (2017a)
GO-glycol	Pb^{2+}	146.2	Fang et al. (2017)
PS (polystyrene)/MGO	As^{3+} As^{5+}	104.2 68.3	Kang et al. (2017)
GO	Cd^{2+}	1792.65	Zhang et al. (2017)
Cellulose membrane-GO	Cd^{2+}	26.78	Sitko et al. (2016)
HPEI (polyethyleneimines)-GO	Pb^{2+}	438.65	Liu et al. (2016b)
TiO_2 (titanium dioxide)/GO	Cu^{2+}	45.30	Yu et al. (2016)
RGO/NiO	Cr^{6+}	198.1	Zhang et al. (2018)
IT(2-imino-4-thiobiuret)/PRGO	Cr^{6+}	63.01	Awad et al. (2017)
Ozonised GO	Co^{2+}	371.95	Liu et al. (2016a)
EDTA-modified GO	U^{6+}	277.45	Zhao et al. (2017b)

Table 12.2 Recent studies on nanomaterials used for heavy metal remediation

Adsorbents	Heavy metals	Sorption capacity (mg/g)	References
DES-functionalised CNTs	Hg^{2+}	186.93	Al Omar et al. (2017)
Amino-functionalised Fe_3O_4 (iron oxide)/MWCNTs	Cu^{2+}	30.5	Zhan et al. (2016)
Functionalised GO (graphene oxide) CA beads	Pb^{2+}	601	Arshad et al. (2019)
Si-nanospheres	Cu^{2+}	139.7	Kotsyuda et al. (2017)
Si-phy-(nano-polyaniline) NPANI	Pb^{2+}	186.1	Mahmoud et al. (2016b)
Hematite magnetic nanomaterials	Cr^{6+}, Cu^{2+} and Pb^{2+}	201, 34.2, 68.8	Rajput et al. (2017), and Kefeni et al. (2018)
Amino-functionalised Fe_3O_4 nanomaterials	Ni^{2+}, Cr^{6+}	222.15, 232.50	Norouzian Baghani et al. (2016)
PTMT (organodisulphide polymer)/ Fe_3O_4 NC	Hg^{2+}, Cd^{2+} and Pb^{2+}	603.2, 216.6, 533.2	Huang et al. (2018b)
HFO-P(TAA(trans-aconitic acid) / HEA(2-hydroxyethyl acrylate)	Pb^{2+}, Cu^{2+} and Ni^{2+}	303.7, 107.6, 85.88	Zhang and Li (2017)
Hydrous ferric oxide (HFO)-CMC-NC	As^{5+}	355.1	Huo et al. (2017)
MnO_2/gelatin	Cd^{2+}, Pb^{2+}	105.2, 318.8	Wang et al. (2018a)
Casein-coated ZnO nanomaterials	Cd^{2+}, Co^{2+} and Pd^{2+}	156.8, 67.83, 194.97	Somu and Paul (2018)
TiO_2-chitosan nanomaterials	Cd^{2+}	1800 μmol/g	Mahmoud et al. (2018)
Al_2O_3 (aluminium oxide) nanomaterials	Pb^{2+}, Cd^{2+}	47.1, 17.3	Tabesh et al. (2018)
MgO (magnesium oxide) nanomaterials	Cd^{2+}, Pb^{2+}	2293, 2615	Xiong et al. (2015)
CeO_2 (cerium oxide) nanomaterials	Pb^{2+}, As^{3+}	23.2, 71.8	Mishra et al. (2018), and Meepho et al. (2018)
CNTs/PAMAM (polyamidoamine dendrimer) NPs	As^{3+}, Zn^{2+}, Co^{2+}	433, 471, 495	Hayati et al. (2018)
HAp(hydroxyapatite)/zeolite (NaP) nanomaterials	Cd^{2+}	40.18	Zendehdel et al. (2016)
Polypyrrole(PPy)-polyaniline (PAN)/ Fe_3O_4 NPs	Pb^{2+}	243.8	Afshar et al. (2016)
ZnO/chitosan nanomaterials	Cd^{2+}	135.2	Saad et al. (2018)
Alginate/chitosan nanomaterials	Cr^{6+}	108.6	Gokila et al. (2017)
Fe_3O_4 (iron oxide)-SiO_2/Zr-MOFs (metal organic frameworks)	Pb^{2+}	102.1	Huang et al. (2018a)
Fe/MgO NPs	Pb^{2+}	1476.5	Ge et al. (2018)

(PEI) to enhance the removal capacity of heavy metals (Arshad et al. 2019). The maximum sorption capacities of Pb^{2+}, Hg^{2+} and Cd^{2+} heavy metal ions were 602.1, 374.2 and 181.3 mg/g. The adsorption results were well suited with the Langmuir isotherm and followed pseudo-second-order kinetics. Furthermore, GO NC found applicability for reuse and achieved about 75–80% removal rate for Pb^{2+} ions even after five cycles. These NCs facilitated easy separations of GO from aqueous solution. Vilela and colleagues developed GO-based microbots (GOx-microbots) which performed as a self-boosted system for capture, transfer and removal of heavy metals (Vilela et al. 2016). The microbot framework was composed of Ni, GO and Pt.

Optimum results showed that GOx-microbots achieved a reduction of Pb^{2+} heavy metal ion concentration from 1000 ppb to less than 50 ppb after a 60-min treatment. The GOx-microbots can be reprocessed after separating Pb from microbot surface. Since real-time wastewater contains several contaminants, practical applicability of graphene-derived nanomaterials is still limited, and research is in the initial phase. Hence, the recycling and reuse of carbon-derived nanomaterials have to be investigated in a cost-effective manner (Xu and Wang 2017).

12.3 Silica-Derived NPs

Silica-derived NPs are another vital category of nanomaterials employed for the removal of heavy metals owing to excellent properties such as high surface area and non-toxic nature (Mahmoud et al. 2016b). Nano-based silica can be amended by $-NH_2$ and –SH groups which serve as a supporting agent for NCs. Kotsyuda and colleagues developed silica nanostructures functionalised by phenyl (C_6H_5) and 3-aminopropyl (3-AP) groups and researched their sorption behaviour towards Cu^{2+} ions (Kotsyuda et al. 2017). From the batch experimental results, it was depicted that functionalised silica nanostructures intensified Cu^{2+} sorption and also showcased antibacterial action. Najafi and his colleagues synthesised NH_2-functionalised nano-silica hollow spheres (NH_2-SNHS) and studied adsorption effects towards Pb^{2+}, Ni^{2+} and Cd^{2+} heavy metal ions by the batch process (Najafi et al. 2012).

The optimum removal capacities for Pb^{2+}, Cd^{2+} and Ni^{2+} metal ions by NH_2–SNHS were 96.78, 31.2 and 40.79 mg/g. The sorption results were well suited with the Langmuir-Freundlich and the Sips isotherm and followed pseudo-second-order kinetics. It has been reported that silica has been widely used for the development of NCs for industrial applications (Pogorilyi et al. 2014).

In another study, Mahmoud and colleagues generated NCs by immobilising nano-polyaniline (NPANI) and cross-linked NPANI into nano-based silica and investigated sorption effects towards Cu^{2+}, Cd^{2+}, Hg^{2+} and Pb^{2+} heavy metals by the batch process (Mahmoud et al. 2016b). The maximum sorption capacities of silica NPANI NCs for Cu^{2+}, Cd^{2+}, Hg^{2+} and Pb^{2+} were 1701, 800, 602 and 901 mol/g, while sorption capacities of silica-cross-linked NPANI nanomaterial for these heavy metals were 1650, 1051, 1352 and 1451 mol/g, and adsorption results were well correlated with the Langmuir adsorption isotherm.

12.4 Zero-Valent Metal (ZVM)-Based NPs

In recent years, ZVM-derived NPs have shown many potentialities in remediating wastewater contaminants (Lu et al. 2016). Gold NPs are used to decontaminate wastewater due to its antimicrobic activity (Amin et al. 2014). It was reported that nanoscopic zero-valent Zn exhibited exceptional dioxin degeneration ability (Bokare et al. 2013). Various zero-valent iron (ZVI) NPs employed for heavy metal removal are reviewed in this chapter.

12.4.1 Silver (Ag)-Based NPs

Ag nanomaterials were found to be efficient in the removal of Hg^{2+} ions from aqueous medium (Vélez et al. 2018). Ag nanomaterials exhibited higher interaction with Hg^{2+} ions due to the reduced reduction potential of Ag and diminishment of particle size (Pradhan et al. 2002). Sumesh et al. (2011) synthesised nanomaterial by blending with mercaptosuccinic acid (MSA). A ratio of 1:6 (Ag/MSA) showed high adsorption capacity of Hg^{2+} ions (800 mg/g). Hence, this NC proved to be a promising adsorbent for the remediation of Hg^{2+} heavy metal. Recently, zeolite-derived coal fly ash (CFA) doped with Ag nanomaterials (particle size 5–40 nm) is used for the enhanced removal of Hg^{2+} ions from the aqueous medium, as shown in Fig. 12.4. A high Hg^{2+} removal rate of 99% was obtained with less leaching of NC (Tauanov et al. 2018).

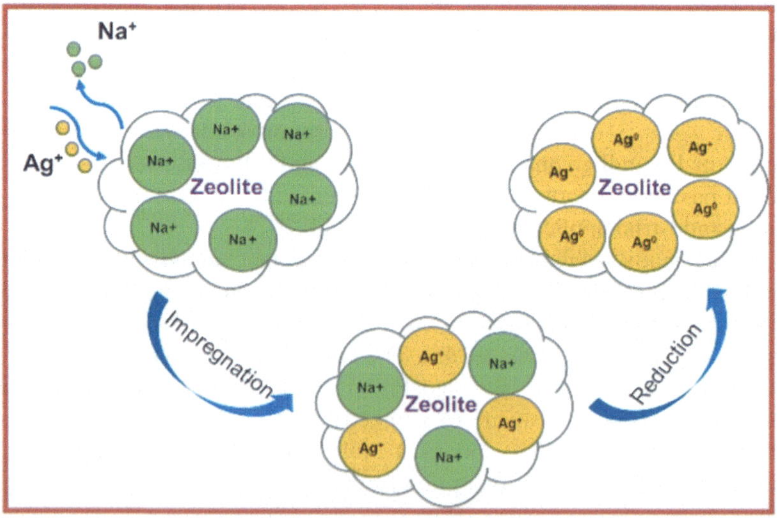

Fig. 12.4 Ag nanomaterial coated with CFA for the removal of Hg^{2+} from the aqueous medium. (Copyright© Elsevier 2018, reprinted with permission from Tauanov et al. (2018))

12.4.2 Gold (Au)-Based Nanomaterials

Au-based nanomaterials exhibit strong affinity towards Hg ions by forming complexes such as AuHg, Au_3Hg and $AuHg_3$ (Zhang et al. 2016). Lisha and her colleagues synthesised aluminium (Al)-supported Au NPs for the removal of Hg^{2+} ions from potable water (Lisha et al. 2009). $NaBH_4$ (sodium borohydride) was employed as a reducing agent to reduce the oxidation state of Hg^{2+} to Hg, and experimental results showed enhanced adsorption capacity of about 4.065 g/g. Al-supported Au NPs were shown to be cost-effective and were found to be feasible for real-time wastewater remediation applications. Jimenez and his colleagues synthesised citrate-supported Au NPs for the removal of Hg^{2+} from contaminated waters (Ojea-Jiménez et al. 2012). The presence of citrate ions facilitated as a scavenging agent for the reduction of Hg^{2+} to Hg, avoiding the use of $NaBH_4$. From the batch experimental results, the toxicity of Hg^{2+} was reduced from 65 to 5 ppm, and Au can be recovered by treating Au_3Hg at elevated temperature or pressure conditions.

Recently, Biao and his colleagues synthesised green Au NPs through functionalising with pomegranate skin waste and studied the suitability in treating heavy metals from aqueous medium (Biao et al. 2018), as shown in Fig. 12.5. The water extracted from pomegranate skin wastes is a high source of proanthocyanidins and polyphenol contents which exhibit high antioxidant properties and acts as a coating and reducing agent during the synthesis of Au NPs by the hydrothermal process (Ambigaipalan et al. 2016). From the batch experiments, removal efficiency of various heavy metals was found to be 98.8% (Pb^{2+}), 96.7% (Ni^{2+}), 95.5% (Cu^{2+}) and 96.4% (Cd^{2+}) at optimised conditions (pH, 8; sorption time, 3–6 h). The interactivity between the heavy metal ions and –OH groups of proanthocyanidin on Au surface played a major role in the separation of heavy metal ions. Au NPs can be easily separated by centrifugation and can be further reused.

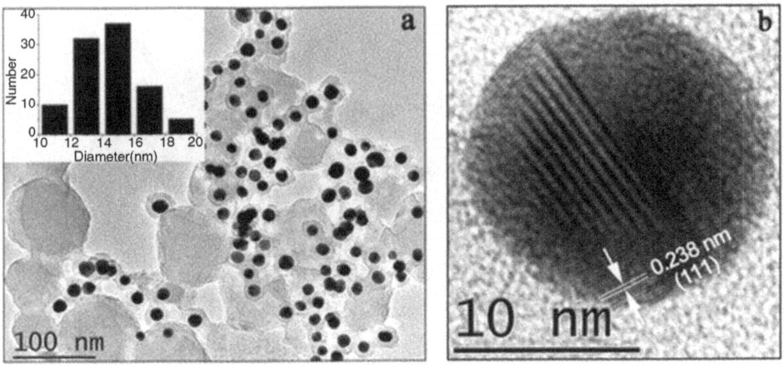

Fig. 12.5 (**a**) TEM pictures of proanthocyanidin-functionalised Au NPs, (**b**) high-resolution TEM pictures of Au NPs (Biao et al. 2018)

12.4.3 Zero-Valent Iron (ZVI)

Nanolevel zero-valent iron (nZVI) is a hybrid comprising of ferrous oxide (FeO) and ferric oxide (Fe_2O_3) coating, as shown in Fig. 12.6 (O'Carroll et al. 2013). Nano-ZVI has gained much research interest towards the treatment of heavy metals, namely, Cu^{2+}, Cd^{2+}, Cr^{6+}, Ni^{2+} and Hg^{2+} ions (Liu et al. 2015; Seyedi et al. 2017). FeO is a good reducing agent; at the same time, Fe_2O_3 provides abundant active sites and enhances interactions towards heavy metals (Hashim et al. 2011). Nano-ZVI exhibited high surface characteristics and reducing capacity proving it as a promising candidate for the treatment of heavy metals sourced from contaminated waters (Huang et al. 2013).

The removal process of nZVI for various heavy metals differs depending upon standard potential E^0 (Huang et al. 2013). As an illustration, for Pb^{2+} ions, E^0 value is −0.13 which is partly greater than Fe^{2+} ions (−0.41); hence, the removal process encompassed both sorption and reduction mechanisms signified by the following reactions (Wang et al. 2016):

$$\text{Reduction}: Fe^0 + \mathrm{M}^{2+} \rightarrow Fe^{2+} + \mathrm{M}^0 \tag{12.1}$$

$$\text{Sorption}: \mathrm{FeOOH} + \mathrm{M}^{2+} \rightarrow \mathrm{FeOOM}^{+} + \mathrm{H}^{+} \tag{12.2}$$

For Cr^{6+} ions, E^0 value (1.36) is considerably higher compared to Fe^{2+} ions; hence, the removal process encompasses both precipitation and reduction mechanisms. Despite several advantages, nZVI possesses certain drawbacks which cannot be avoided. It was reported that nZVI undergoes easy oxidation with O_2 and H_2O in aqueous media which further inhibits the reduction mechanism of heavy metals (Zhang et al. 2013b). The segregation of nZVI from contaminated waters is a tedious

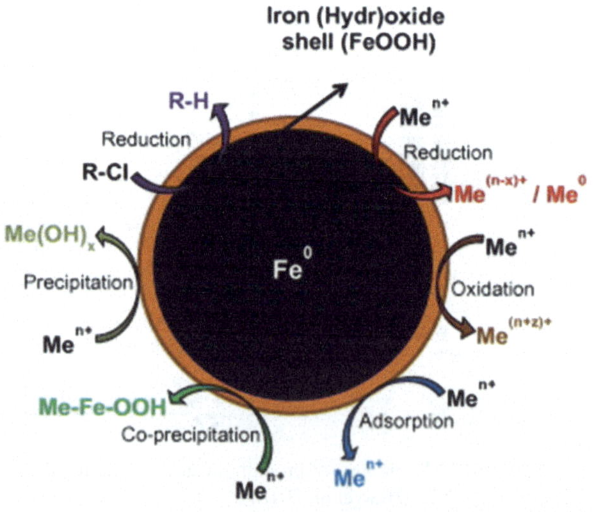

Fig. 12.6 Core-shell framework of nanolevel zero-valent iron (nZVI) portraying various processes for the removal of heavy metals and chlorinated combinations. (Copyright© Elsevier 2013, reprinted with permission from O'Carroll et al. (2013))

process due to the aggregation of nZVI particles which reduces the motility and reacting surface area (O'Carroll et al. 2013). Either amendment with chemical agents or doping with certain heavy metals (Cu, Ni, Pd and Pt) overcame the shortcomings of nZVI (Fu et al. 2014).

Huang and colleagues developed innovative nZVI NPs by modifying with sodium dodecyl sulphate (SDS) and investigated sorption behaviour towards Cr^{6+} from contaminated waters. SDS exhibited exceptional migration capabilities and diffusion (Huang et al. 2015). From batch experimental results, optimum Cr^{6+} sorption capacity of nZVI NPs was found to be 253.65 mg/g and achieved 98.92% removal rate. This novel adsorbent enhanced sorption capacity and reduced the problem of aggregation. The sorption results were well correlated with the Freundlich isotherm model and followed pseudo-second-order kinetics.

In another study, Au-spiked nZVI NPs was synthesised and found useful in simultaneous removal of nitrate and Cd^{2+} heavy metal ions from water by the batch process (Su et al. 2014). The fabrication of novel nZVI with inexpensive bentonite for the treatment of a wide range of heavy metals such as Cu^{2+}, Cd^{2+}, Co^{2+}, Ni^{2+} and Pb^{2+} ions from water was investigated (Zarime et al. 2018). The incorporation of bentonite onto nZVI NPs reduced the problem of aggregation that could decrease the aggregation of nZVI particles and contribute more active sites for effective sorption of heavy metals.

12.5 Metal Oxide (MO)-Derived NPs

Nanoscopic MO possesses many remarkable properties, such as high sorption capacity and selectiveness proving as a high potential adsorbent for remediation of heavy metals. Metal oxide-derived nanomaterials are classified into nanoscopic iron oxides, manganese oxides (MnO), titanium oxides (TiO_2), zinc oxides (ZnO), magnesium oxides (MgO), aluminium oxides (Al_2O_3), zirconium oxides (ZrO_2) and cerium oxides (CeO_2).

12.5.1 Iron Oxide-Derived NPs

Over the past few years, iron oxide-derived nanomaterials have gained much research attention in remediating heavy metals from contaminated waters due to their abundant availability and easy synthesis process (Saharan et al. 2014). The most commonly studied iron oxide nanomaterials are goethite (α-FeOOH), maghemite (γ-Fe_2O_3), hematite (α-Fe_2O_3), hydrous iron oxides (HFO) and magnetite (Fe_3O_4) (Hua et al. 2012), which are reviewed in this chapter. For goethite, maghemite, hematite and HFO, the iron valences are trivalent, while for magnetite iron, both bivalent and trivalent existed.

12.5.1.1 Goethite (α-FeOOH)-Based NPs

Goethite previously existed in nature as inorganic form, proved as a competent sorbent for heavy metals, due to excellent sorption efficacy, cost-effectiveness and environmental security (Massalimov et al. 2014). Sun and his colleagues synthesised nanolevel α-FeOOH utilising various ferric and ferrous salts for the removal of uranium from aqueous medium (Sun et al. 2011). The results exhibited that nanolevel α-FeOOH showed the high capacity of uranium removal in the pH range 5.5–7.5.

Chen and his colleagues synthesised nanolevel α-FeOOH which performed both as photocatalyst for methylene blue (MB) dye solution under the influence of UV light and as sorbent with high sorption capacity of 149.25 mg/g for the removal of Cu^{2+} heavy metal by co-precipitation methods (Chen and Li 2010). The adsorption values were well suited with the Langmuir adsorption isotherm and followed pseudo-second-order kinetics. α-FeOOH/CS nanohybrid composites (diameter range, 10–60 nm) were found useful in the removal of Pb^{2+} from water at optimised conditions such as initial Pb^{2+} ion concentration 74.4 mg/l, pH 6 and adsorbent dosage and achieved over 98.26% removal capacity (Rahimi et al. 2015), as shown in Fig. 12.7.

In another study, Khezami and his colleagues synthesised nanoscale crystalline α-FeOOH powders by energy ball milling process for the efficient removal of Cd^{2+} heavy metal (Khezami et al. 2016). Maximum sorption was found to be 167 mg/g under optimised condition pH 7 and temperature 328 K. The adsorption values were well correlated with the Langmuir-Freundlich adsorption and followed the pseudo-second-order kinetic model. From the thermodynamic results, the sorption process was found to be endothermic and spontaneous throughout the process. α-FeOOH-based NPs found applicability in remediating other heavy metals which include Mn^{2+}, Co^{2+}, V^{5+}, Th^{2+}, Zn^{2+} and Ni^{2+} (Yan et al. 2011; Leiviska et al. 2017).

12.5.1.2 Maghemite (γ-Fe_2O_3)

It has been reported that maghemite NPs were abundantly used for the treatment of heavy metals from contaminated waters (Etale et al. 2016; Ahmadi et al. 2017). γ-Fe_2O_3 NPs exhibited exceptional properties for remediating heavy metals such as high surface characteristics contributing to excellent sorption capacity, simple synthesis and easy separation of NPs from contaminated waters by incorporating magnetic field and were environmentally safe without generating secondary contamination (Cheng et al. 2012).

Akhbarizadeh and colleagues developed γ-Fe_2O_3 NPs (14 nm size) and have undergone batch studies for the treatment of heavy metals such Cu^{2+}, Cd^{2+}, Cr^{6+}, Ni^{2+} and Mn^{2+} ions sourced from contaminated wastewaters (Akhbarizadeh et al. 2014). The optimum removal rates for heavy metals by γ-Fe_2O_3 NPs were shown to be 88.1% for Cu^{2+}, 8.5% for Cd^{2+}, 84.3% for Cr^{6+}, 15.8% for Ni^{2+} and 18.4% for

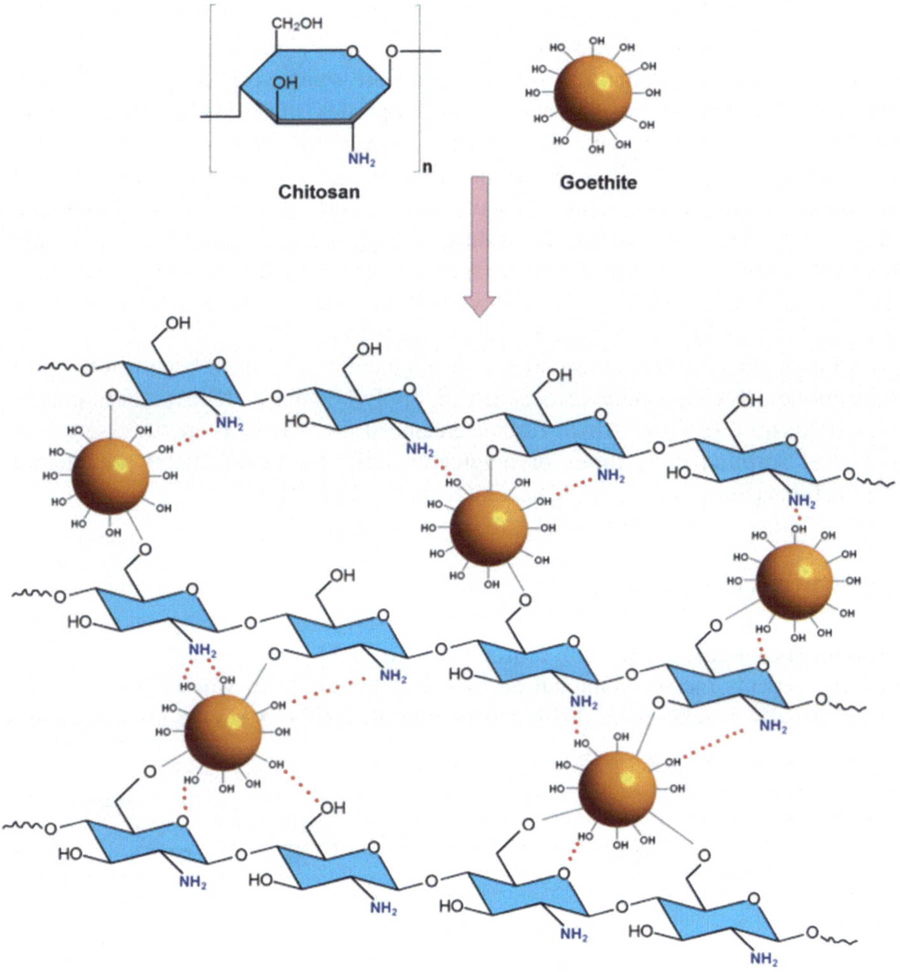

Fig. 12.7 Suggested structure of goethite/chitosan nanohybrid material. (Copyright© Elsevier 2015, reprinted with permission from Rahimi et al. (2015))

Mn^{2+} heavy metal ions. Rajput and his colleagues synthesised γ-Fe_2O_3 NPs exhibiting superparamagnetism with tuneful arrangement by implementing flame spray pyrolysis (Rajput et al. 2017). The newly developed γ-Fe_2O_3 NPs showed a high surface area of about 79.35 m^2/g and were applied for the removal of Cu^{2+} and Pb^{2+} heavy metal ions from contaminated waters.

The adsorption data showed a strong correlation with the Langmuir adsorption, and maximum sorption capacities were found to be 34.1 mg/g for Cu^{2+} and 68.8 mg/g for Pb^{2+} heavy metal ions. Electrostatic interactiveness facilitated sorption of heavy metal ions. In water, γ-Fe_2O_3 exterior was embedded with FeOH groups, resulting in the formation of $Fe(OH)_2$ or FeO^- groups during alteration of pH. As the pH was raised, there was an increase in the creation of Fe^{3+} O^- or $Fe^{3+}OH$ sites which further enhanced the sorption capacities of Cu^{2+} and Pb^{2+} heavy metal ions.

Recently, polymer-amended γ-Fe_2O_3 NPs were synthesised and researched for sorption of heavy metals by combining properties of γ-Fe_2O_3 and polymers. Madrakian and his colleagues developed mercaptoethylamino monomer-modified maghemite NPs (MAMNPs) by synthesis mechanism, as illustrated in Fig. 12.8, and studied sorption behaviour of heavy metals from aqueous media (Madrakian et al. 2015). The novel MAMNPs exhibited high surface area (92.42 m^2/g), and optimum sorption capacities for heavy metals Cd^{2+}, Hg^{2+}, Ag^+ and Pb^{2+} were 91.56, 237.61, 260.6 and 237.58 mg/g. The adsorption results were well suited with the Sips isotherm and followed pseudo-second-order kinetics.

Additionally, it has been reported that maghemite NPs have been reformed by polyaniline (PANI), polypyrrole (PPy), polyrhodanine (PRd) and poly(1-vinylimidazole) (PVIm), which further enhanced the overall properties, selectiveness and sorption capabilities of maghemite NPs for heavy metal remediation applications (Song et al. 2011; Chávez-Guajardo et al. 2015).

12.5.1.3 Hematite (α-Fe_2O_3)

Hematite is a highly stable form of iron oxide (α-Fe_2O_3). NPs were demonstrated as superb sorbents for the remediation of heavy metals from wastewaters (Ahmed et al. 2013; Adegoke et al. 2014; Dickson et al. 2017). Adegoke and colleagues

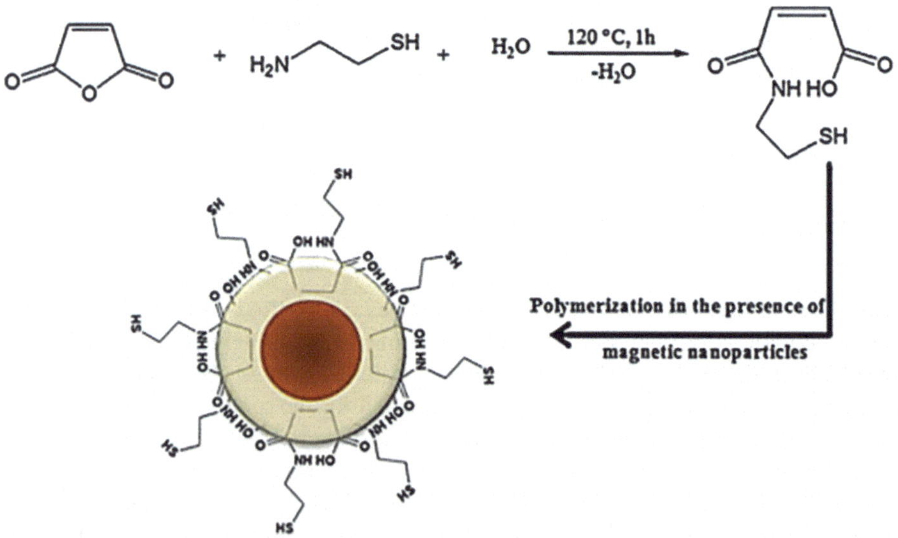

Fig. 12.8 Schematic illustration of mercaptoethylamino monomer-modified maghemite NP (MAMNP) formation. (Copyright© Elsevier 2015, reprinted with permission from Madrakian et al. (2015))

prepared various arrangements of α-Fe_2O_3 NPs, such as round, sub-round, dish-like, spherical and nano-cubical shapes, and studied the effective removal of Cr^{6+} heavy metal ions (Adegoke et al. 2014). Various morphologies of α-Fe_2O_3 NPs influenced Cr^{6+} removal, and maximum sorption capacity was found to be 200 mg/g.

In another study, α-Fe_2O_3 NPs showed effectiveness in treating a wide range of heavy metals (Cu^{2+}, Cd^{2+}, Pb^{2+} and Zn^{2+}) simultaneously with satisfactory sorption capacities (Shipley et al. 2013). The sorption results were well fitted and followed pseudo-second-order kinetics, and thermodynamic values were exothermic. Presently, magnetic α-Fe_2O_3 NPs were developed and employed for treating heavy metals (Zn^{2+}, Mg^{2+}, Mn^{2+}, Al^{3+} and Ni^{2+}) sourced from AMD (acid mine drainage) (Kefeni et al. 2018). The results exhibited that about 100% removal was obtained for heavy metals (Mn^{2+}, Mg^{2+} and Al^{3+}), while for Zn^{2+} and Ni^{2+}, 80% removal rate was achieved. Hence, α-Fe_2O_3 NPs displayed various plusses such as high sorption capacity and steadiness proving as assuring sorbent for heavy metal remediation applications.

12.5.1.4 Hydrous Iron Oxide (HFO)-Based NPs

HFO-based NPs have exhibited remarkable potential application in treating heavy metals from contaminated waters owing to their high affinity, high surface characteristics and cost-effectiveness (Qiu et al. 2012). HFO-based NPs cannot be directly used for the treatment of heavy metals owing to feeble mechanical strength, excess in pressure drop and less hydraulic conductivity (Qiu et al. 2013). Instead, HFO has undergone incorporation with poriferous nanomaterials which aided in the formation of various NCs.

Hydrogel-assisted HFO-P [TAA (trans-aconitic acid)/HEA (2-hydroxyethyl acrylate)] NPs were synthesised, and adsorption effect towards heavy metals such as Cu^{2+}, Cd^{2+}, Ni^{2+} and Pb^{2+} from contaminated waters was studied (Zhang and Li 2017). The batch experimental results revealed that maximum sorption capacities of these four heavy metals were found to be 0.232, 0.0933, 0.1617 and 0.431 mmol/g. Also, HFO-based NPs (1–60 nm) have shown applicability towards As ions with optimum sorption capacity of 74.1 mg/g, and sorption data were well correlated with the Langmuir adsorption model (Zhang et al. 2017). In another study, Huo and colleagues synthesised HFO amended with CMC (carboxymethyl cellulose) for the treatment of As^{5+} ions from water (Huo et al. 2017). Optimum As^{5+} sorption capacity of 355.1 mg/g was achieved and well suited with pseudo-second-order kinetics. The treatment process consisted of surface complexing and precipitation method. Also, HFO-CMC NPs have shown feasibility towards the treatment of As^{5+} from real-time contaminated water derived from regular mine sites and achieved a removal rate of about 90.5%.

12.5.1.5 Magnetite

Magnetite (Fe_3O_4)-derived nanomaterials are extensively employed as nanoadsorbents owing to their low price, usage simplicity, easy accessibility and eco-friendly nature (Hao et al. 2010). Fe_3O_4 NPs can be easily segregated from aqueous media under the influence of the external magnetic field, as shown in Fig. 12.9. Various studies on heavy metal remediation by magnetite NPs have been reported (Shan et al. 2015; Watts et al. 2015; Mahmoud et al. 2016a). Giraldo and his colleagues prepared Fe_3O_4 NPs by co-precipitation process, which are employed for the treatment of Cu^{2+}, Mn^{2+}, Pb^{2+} and Zn^{2+} heavy metal ions in a batch system (Giraldo et al. 2013). The batch experimental results indicated magnetite NPs exhibited high sorption effect towards Pb^{2+} ions (0.181 mmol/g) and minimal sorption capacity for Mn^{2+} ions (0.142 mmol/g).

The adsorption results were well correlated with the Langmuir isotherm model and followed pseudo-second-order kinetics. It was inferred that the size of hydrated ionic radius influences the interactivity towards negatively charged sorption site. As the hydrated ionic radius escalated, the distance to the sorbent surface would increase, and further the adsorption mechanism will be inefficient. Pb^{2+} ions displayed minimal hydrated ionic radii and exhibited excellent ability for proton compete, which was the main reason for excellent sorption capacity compared to other heavy metal ions.

Bare Fe_3O_4 NPs are easily oxidised by O_2 due to the presence of Fe^{2+} in their frameworks and easily tarnished by acid or bases. To overcome this, Fe_3O_4 NPs are surface-amended by functional groups such as –SH, –COOH and $–NH_2$ groups (Pan et al. 2012; Tan et al. 2012; Shi et al. 2015). Baghani and his colleagues developed $–NH_2$-functionalised Fe_3O_4 NPs by one-pot synthesis and investigated sorption behaviour towards Cr^{6+} and Ni^{2+} heavy metal ions from aqueous media (Norouzian Baghani et al. 2016).

From the batch experimental results, optimum removal capacities were 232.52 mg/g for Cr^{6+} and 222.13 mg/g for Ni^{2+} metal ions. The adsorption results were well suited with the Langmuir adsorption model and followed pseudo-second-order kinetics. The novel sorbent can be easily segregated from contaminated waters

Fig. 12.9 (**a**) Pb^{2+} solution with distributed Fe_3O_4@ SiO_2composite and (**b**) Pb^{2+} solution after employing magnetic separator. (Copyright© Elsevier 2010, reprinted with permission from Hu et al. (2010))

within 30 s with the aid of an external magnetic field. It has been reported that SiO_2, SDS, polyethylene glycol (PEG), tannic acid and CS are employed as coating agents for Fe_3O_4 NPs for the treatment of heavy metals from contaminated waters.

In a study, Huang and his colleagues fabricated innovative sorbent by coating organodisulphide polymer (PTMT) onto $-NH_2$-functionalised Fe_3O_4 NPs (Huang et al. 2018b). During batch studies, the novel sorbent exhibited significant removal of heavy metals and achieved sorption capacities of 603.2, 533.18 and 216.6 mg/g for Hg^{2+}, Pb^{2+} and Cd^{2+} ions. Additionally, this novel sorbent displayed regeneration capacity up to five cycles with stable removal capacities.

12.5.1.6 Manganese Oxide-Derived NPs

It has been reported that manganese oxide (MnO), manganese dioxide (MnO_2) and hydrous manganese oxide (HMO) NPs showed potentials in the treatment of heavy metals from contaminated waters (Su et al. 2010). MgO exhibited high surface characteristics, and the presence of $M\text{-}O^{\delta-}$ and $M\text{-}O^{\delta+}$ units enhanced the performance of heavy metal sorption (Mukherjee et al. 2013). Wang and his colleagues developed dumbbell-shaped MgO/gelatin NC and studied sorption behaviour towards Cd^{2+} and Pb^{2+} ions (Wang et al. 2018a). From the batch experimental results, maximum removal capacities were found to be 105.2 mg/g for Cd^{2+} and 318.8 mg/g for Pb^{2+} ions. The adsorption data were well suited with the Langmuir adsorption isotherm and followed pseudo-second-order kinetics. NH_2-functionalised poly(methyl methacrylate plate) (PMMA)-assisted NC exhibited high sorption and steadiness for the treatment of heavy metals in real-time wastewater. Moreover, MnO_2 NPs were employed for the oxidation and removal of TI (I) from contaminated waters (Huangfu et al. 2015). The batch sorption process was finished within 15 min and exhibited optimum sorption capacity of 672.2 mg/g, followed by Langmuir adsorption isotherm. Additionally, it has been reported that MnO-based nanomaterials have shown significant potential in treating a wide range of heavy metals such as Cu^{2+}, Cd^{2+}, Hg^{2+}, Pb^{2+} and Zn^{2+} (Kim et al. 2013; Abdullah et al. 2018a).

Hydrous manganese oxide (HMO) is the type of MnO which has displayed its dominance towards heavy metal remediation due to high surface characteristics, poriferous frameworks and abundant availability of active sites for sorption (Wan et al. 2010). The presence of OH^- groups on HMO surface facilitated more coordination with heavy metals and enhanced the sorption process (Yang et al. 2019). The sorption process of heavy metal ions onto the HMO surface is carried out by two steps. Firstly, rapid sorption of heavy metals takes on HMO surface, and further gradual intraparticle diffusion takes place through micro-poriferous HMO walls.

Presently, Wan and colleagues synthesised HMO-BC (biochar) NC and studied sorption behaviour towards Cd^{2+} and Pb^{2+} heavy metal ions (Wan et al. 2018). The results indicated that a combination of HMO along with BC enhanced removal capacity of heavy metals four to six times compared to BC by overcoming disadvantages such as selectivity issues and inadequate sorption capacity.

12.5.1.7 Titanium Oxide (TiO_2)-Derived NPs

TiO_2 is commonly used for the photodegradation of natural pollutants using productive photocatalytic activity and further has shown applicability in the removal of heavy metals (Anandan et al. 2009; George et al. 2016). Gulaim and his colleagues synthesised nano-based titania and studied the adsorption behaviour of dichromate ($Cr_2O_7^{2-}$) under various concentrations ranging from 20 to 300 mg/l (Seisenbaeva et al. 2012). TiO_2 exhibited maximum adsorption capacity of about 12.6 mg/g for Cr^{6+} and 26.1 mg/g for $Cr_2O_7^{2}$. Yoseff and his colleagues synthesised titanium oxide nanowires with diameter range 30–50 nm found effective in removing Fe^{3+} and Pb^{2+} heavy metals from drinking water (Youssef and Malhat 2014).

Also, TiO_2 nanomaterials doped with iron were found productive in treating arsenic contaminants (Nabi et al. 2009). TiO_2 nanomaterials coated with starch were synthesised and were found efficient in removing heavy metals such as Co^{2+}, Cd^{2+}, Ni^{2+} and Cu^{2+} from contaminated tap water achieving a removal rate of more than 90% (Baysal et al. 2018).

Currently, the microwave-assisted sorption process was used to synthesise TiO_2-CS nanomaterials for the effective removal of Cd^{2+} and Cu^{2+} heavy metals from water (Mahmoud et al. 2018), as shown in Fig. 12.10. The highest sorption capacity was found to be 1800 μmol/g for Cd^{2+}. About 72.56% of Cd^{2+} and 88% of Cu^{2+} removal from the water was achieved during microwave heating-assisted process for 60–70 s. Nevertheless, the significant obstacles of TiO_2 NPs lie in the complexity of the synthesis process and separation complicacy during semi-liquid suspension (Du et al. 2008; Lu et al. 2016).

12.5.1.8 Zinc Oxide (ZnO)-Based NPs

ZnO-based NPs have attained much notability as heavy metal sorbents have gained their popularity as adsorbents for heavy metals due to immense surface area characteristics, low production cost and excellent sorption capacities (Kumar et al. 2013). A wide range of heavy metals (Cu^{2+}, Cr^{6+}, Ni^{2+}) has been revealed to be treated by ZnO NPs (Rafiq et al. 2014; Hadadian et al. 2018). Sheela and her colleagues synthesised ZnO NPs and studied the feasibility towards the treatment of heavy metals (Cd^{2+}, Hg^{2+} and Zn^{2+}) from aqueous medium (Sheela et al. 2012). The batch experimental results demonstrated that optimum sorption capacities were found to be 387.2 mg/g for Cd^{2+}, 714.3 mg/g for Hg^{2+} and 357.4 mg/g for Zn^{2+}. The sorption data were well fitted by the Langmuir adsorption isotherm, and the removal of heavy metals from wastewater by ZnO NPs was enhanced in the pH range 4–8. The result indicated that pH had a significant influence on the adsorption behaviours which could be explained by the surface charge of ZnO and the degree of speciation of sorbents.

Ghiloufi and his colleagues modified ZnO NPs by incorporating calcium (Ca) and studied sorption behaviour towards Cd^{2+}, Ni^{2+} and Cr^{6+} from aqueous medium

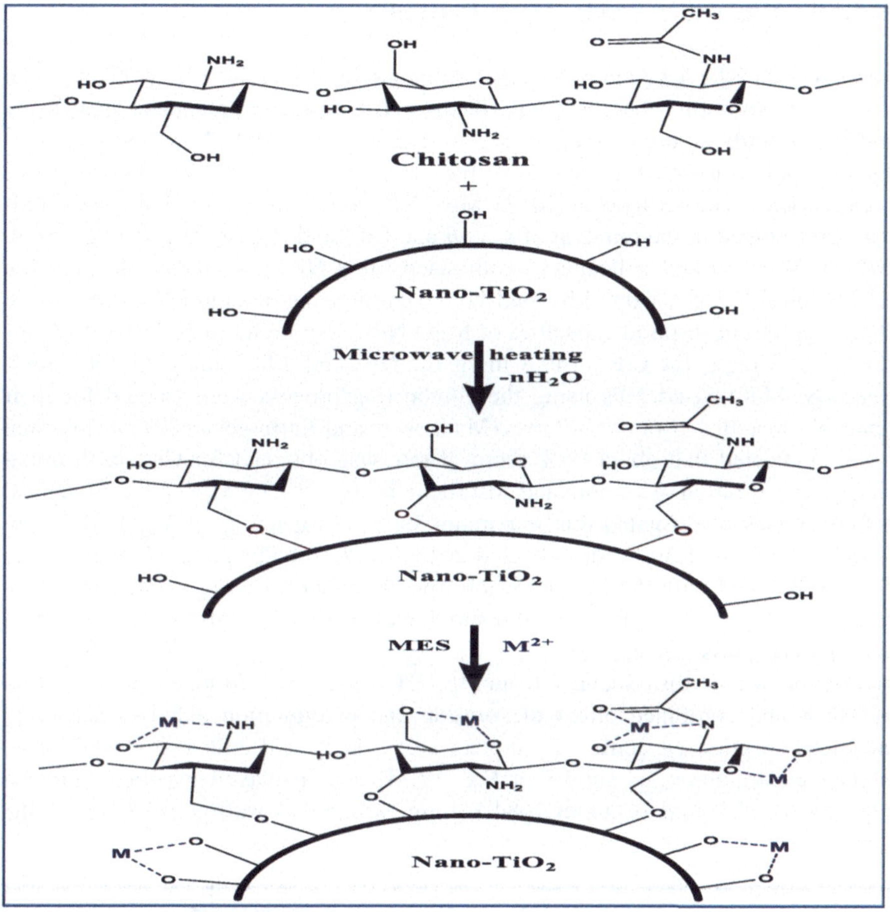

Fig. 12.10 Microwave-assisted synthetic pathway of TiO_2-chitosan NPs. (Copyright© Elsevier 2018, reprinted with permission from Mahmoud et al. (2018))

(Ghiloufi et al. 2015). The experimental results displayed that Ca-doped ZnO NPs optimised the heavy metal uptake. In a similar study, Somu and colleagues developed amended ZnO NPs by employing casein as a capping and reducing agent (Somu and Paul 2018). The casein-doped ZnO NPs (10 nm) were implemented for treating two dyes and three heavy metals from wastewater.

Sorption results were well correlated with the Langmuir adsorption isotherm, and optimum removal capacities were found to be 156.8 mg/g for Cd^{2+}, 67.9 mg/g for Co^{2+} and 194.91 mg/g for Pd^{2+}, and sorption capacities were 115.42 mg/g for MB and 62.2 mg/g for Congo red (CR). Apart from heavy metal removal, casein-doped ZnO NPs also displayed unusual antimicrobic activity, proving as an assuring sorbent towards real-time wastewater.

12.5.1.9 Magnesium Oxide (MgO)-Derived NPs

MgO-derived NPs have many benefits as potential sorbents for heavy metals, such as excellent sorption capacity, low production cost, non-toxic, abundant availability and eco-friendly nature (Cai et al. 2017b). Furthermore, MgO NPs exhibited anti-bacterial action towards the spores of bacteria, Gram-positive and Gram-negative bacteria (Krishnamoorthy et al. 2012). MgO NPs prepared by sol-gel process simultaneously helped in the removal of *E. coli* and Cd^{2+} and Pb^{2+} from water (Cai et al. 2017b). Mahdavi and colleagues synthesised MgO NPs and studied the sorption effects towards Cu^{2+}, Cd^{2+}, Pb^{2+} and Ni^{2+} from aqueous medium (Mahdavi et al. 2013). Optimum sorption capacities of MgO NPs were found to be 135.1 mg/g for Cu^{2+}, 149.2 mg/g for Cd^{2+}, 149.8 mg/g for Ni^{2+} and 148.7 mg/g for Pb^{2+} ions. Recently, MgO-based NPs using the combustion process were studied for their removal capacities towards Cu^{2+} ions (Madzokere and Karthigeyan 2017). Optimum results indicated that about 96% removal rate was obtained for Cu^{2+} with initial $CuCl_2$ concentration of 10 ppm and adsorbent dosage of 0.2 g MgO NPs. Xiong and his colleagues investigated that maximum sorption capacities of MgO NPs were found to be 2294.2 mg/g for Cd^{2+} and 2614.3 mg/g for Pb^{2+}, and adsorption data were well suited with the Langmuir adsorption isotherm (Xiong et al. 2015). The sorption process was inspired by the simultaneous effect of intraparticle diffusion and external mass transfer reactions. The high sorption capacities were due to the production of OH^- disassociated from $Mg(OH)_2$, which was formed from hydration of MgO, and combined effect of sorption and precipitation aided in achieving enhanced sorption capacities. In another study, Feng and colleagues synthesised MgO nanostructures, as shown in Fig. 12.11, and displayed excellent removal capacity for Ni^{2+} ions of about 1684.3 mg/g, which was well correlated with the

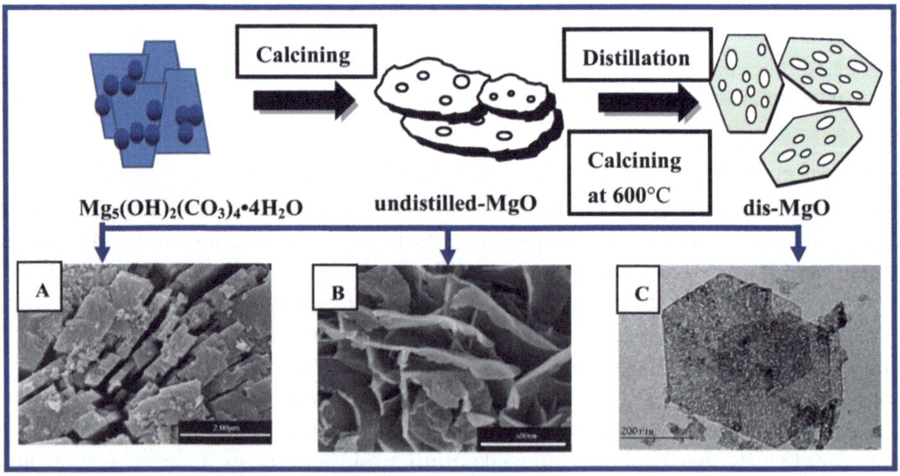

Fig. 12.11 Diagrammatic representation of MgO nanosheet formation. (Copyright© Elsevier 2015, reprinted with permission from Feng et al. (2015))

Langmuir isotherm model and followed pseudo-second-order kinetics (Feng et al. 2015). Further treatment by distillation process improved MgO nanosheet BET surface area to 181.69 m^2/g.

12.5.1.10 Aluminium Oxide (Al_2O_3)-Based NPs

NPs are another category of heavy metal sorbent with benefits such as low production cost and high detoxification efficiency (Giles et al. 2011; Prabhakar and Samadder 2018). Al_2O_3 has various crystalline forms such as α, θ, γ and η, and γ-Al_2O_3 is commonly used (Xie et al. 2016). γ-Al_2O_3 NPs have immense potential as promising sorbents due to their exceptional sorption capacity, high surface characteristics, excellent adsorption capacity, low-temperature modifications and mechanical strength (Saadi et al. 2013). Presently, Tabesh and his colleagues synthesised Al_2O_3 NPs (6–13 mm size) using the sol-gel method and studied the sorption effect for the removal of Cd^{2+} and Pb^{2+} heavy metal ions (Tabesh et al. 2018). The experimental results exhibited excellent removal rates of 87.2% (Cd^{2+}) and 97% (Pb^{2+}) with sorption capacities of 17.23 mg/g and 47.1 mg/g, and adsorption data are well fitted by the Freundlich isotherm.

In another study, Al_2O_3 NPs were modified by humic acid (HA), citrate and phosphate (PO_4) which further enhanced the sorption capacity of towards Cd^{2+} and Zn^{2+} remediation (Stietiya and Wang 2014). Also reports have shown that Al_2O_3 NPs are effectively used to remediate As^{3+}, Cr^{6+}, Cu^{2+}, Hg^{2+} and Ni^{2+} and displayed excellent sorption capacities (Patra et al. 2012; Mahdavi et al. 2015; Shokati Poursani et al. 2015; Wang et al. 2015).

12.5.1.11 Zirconium Oxide (ZrO_2)-Derived NPs

ZrO_2-derived NPs are another encouraging metallic oxide (MO) sorbent employed for the remediation of heavy metals from contaminated waters that are used to remove heavy metals in wastewater. Their benefits include high surface characteristics and abundant availability of hydroxyl groups on ZrO_2 surface.

The ZrO_2 NPs exhibited excellent chemical stability and sorption capacities towards Cd^{2+}, Pb^{2+} and Zn^{2+} heavy metal ions (Jiang and Xiao 2014). ZrO_2-based nanomaterials are categorised into nanoscopic zirconia and hydrous zirconium oxide (HZO). Gulaim and his colleagues developed ZrO_2 NPs and studied sorption behaviour towards Cr^{6+} heavy metals from aqueous media. Maximum sorption capacity of about 73.1 mg/g was attained proving ZrO_2 NPs as promising sorbent for Cr^{6+} ion removal (Seisenbaeva et al. 2014).

Yalçinkaya and colleagues developed ZrO_2/B_2O_3 (boric oxide) NC and investigated the remediation of heavy metals such as Cd^{2+}, Cu^{2+} and Co^{2+} ions by column sorption studies (Yalçinkaya et al. 2011). The adsorption results were well correlated with the Langmuir adsorption with optimum removal capacities of 109.8 mg/g for Cd^{2+}, 46.6 mg/g for Cu^{2+} and 32.3 mg/g for Co^{2+}. The nanohybrid composite was found suitable for regeneration and reuse.

Zhang and colleagues prepared polystyrene (PS)-assisted nanoscopic HZO-PS NPs and researched its sorption behaviour towards Cd^{2+} heavy metal ions (Zhang et al. 2015). The batch studies showed that Cd^{2+} ion removal was enhanced by HZO-PS NPs in the pH range 2.5–7. Column adsorption studies indicated that this nanohybrid composite showed good suitability towards Cd^{2+} heavy metal ions with a treatment capacity of 750-bed volume/run.

Hua and his colleagues synthesised HZO/D-001 (cation-exchange resin) NC and studied sorption efficacy towards Cd^{2+} and Pb^{2+} from water by column sorption method (Hua et al. 2013). Optimum removal capacities were found to be 319.5 mg/g for Pb^{2+} and 214.8 mg/g for Cd^{2+} and followed pseudo-first-order kinetics. HZO/D-001 NCs showed promising potential in remediating both non-natural and real-time acid mine-contaminated waters without losing sorption capacity by employing HNO_3-$Ca(NO_3)_2$ as a regeneration agent.

12.5.1.12 Cerium Oxide (CeO_2)-Derived Nanomaterials

Nano-range CeO_2, a non-toxic and very rare MO, was found to be applicable in various areas such as UV blocking (Umar et al. 2015), sensor and photocatalyst (Anupriya et al. 2014) and remediation of wastewater (Recillas et al. 2011). The surface pore characteristics, crystalline structure and bulk density of CeO_2 NPs have a significant effect on functioning, steadiness, specificity and distribution behaviour towards remediation of heavy metals. Recillas and his colleagues synthesised CeO_2 NPs (12 nm size) and studied the sorption influence of Cr^{6+} from fresh water (Recillas et al. 2010).

Optimum sorption capacity for Cr^{6+} was found to be 121.96 mg/g at initial Cr^{6+} concentration of 80 mg/l. In a similar study, Mishra and his colleagues synthesised CeO_2 NPs (3–5 nm size) for the effective removal of As^{3+} and As^{5+} from contaminated water by the batch process (Mishra et al. 2018). The sorption capacities were found to be 71.8 mg/g for As^{3+} and 36.7 mg/g for As^{5+}. The adsorption data were well fitted by the Langmuir adsorption isotherm, and adsorption reached equilibrium after 10 min.

Furthermore, CeO_2 NPs were incorporated with other metal oxides (MOs) for the treatment of heavy metals (Ayawanna et al. 2015; Ayawanna and Sato 2019). Recently, various frameworks of CeO_2 NPs were synthesised, and removal influence of samaria-doped ceria (SDC) nanopowders was studied towards Zn^{2+}, Cu^{2+} and Pb^{2+} heavy metal ions (Meepho et al. 2018). The experimental results indicated that spherical SDC nanopowders (SDC-F) exhibited high sorption capacity in comparison with cluster plate-like SDC nanopowders (SDC-I).

12.6 Nanohybrid Nanomaterials

Even though nanomaterials possess several advantages, some disadvantages cannot be disregarded. Most of the CNTs found complexity for the uniform suspension in various solvents, whereas nZVIs are susceptible to oxidation (Hayati et al. 2016). Additionally, nanoparticles often exhibit issue of accumulation, inadequate segregation and high-pressure drop when applied in fixed-bed and continuous flow process (Hotze et al. 2010; Hua et al. 2012). A general approach to overcome these challenges is to synthesise nanohybrid composite with the help of various nanomaterials (Zhang et al. 2016). Multiple NCs such as organic and inorganic nanohybrids along with magnetic-based NCs were reviewed intensively.

12.6.1 Organic Polymer Nanohybrids

Organic polymers possess superior properties such as tuneful functional groups, mechanical strength, practicable regeneration, eco-friendliness and easy degradability which implement natural polymers as supporting agents for nano-based composites (Lu and Astruc 2018). Polymer-assisted nanomaterials are classified into two types: artificial organic polymer-assisted NC and biobased polymer-assisted NC (Lu and Astruc 2018).

The synthesis of nanohybrids can be attained in two ways, in situ and direct synthesising process, as shown in Fig. 12.12 (Zhang et al. 2016). Examples for artificial organic polymer include PANI and PS which are used for the synthesis of nanohybrid composites for the treatment of heavy metals (Musico et al. 2013; Rajakumar et al. 2014). Afshar and his colleagues synthesised polypyrrole (PPy)-PANI/Fe_3O_4 magnetic nanohybrid composite material and studied sorption capacity of Pb^{2+} in an aqueous medium (Afshar et al. 2016). NC exhibited about 100% removal rate of Pb^{2+} at optimum conditions [initial Pb^{2+} concentration, 2omg/L; pH, 8–10)].

Alginate, cellulose and CS were under the category of biobased polymer and were used as a supporting agent for NC. Celluloses are commonly used biobased polymers containing hydroxyl groups (OH) on its glucose loop providing ample coordinative sites for heavy metals (Cai et al. 2017a). Suman and his colleagues synthesised nanocellulose (NC)-AgNP-supported pebble hybrid composite material that was implemented for the removal of heavy metals, dyes and microbes in water using the sorption column method (Suman et al. 2015). From the batch experimental results, about 98.4% of Cr^{3+} and 99.5% of Pb^{2+} removal rate was achieved from contaminant water with microbial load detoxification efficiency of about 99%.

CS is another type of green biobased polymer which exhibits great sorption ability of heavy metals and is an eco-friendly and biodegradable material which has excellent potential for the sorption of heavy metals due to the existence of –OH (hydroxyl) and –NH_2 (amine) groups in the framework. Saad and his colleagues developed ZnO/CS core-shell NC (ZOCS) which was found to be productive in the

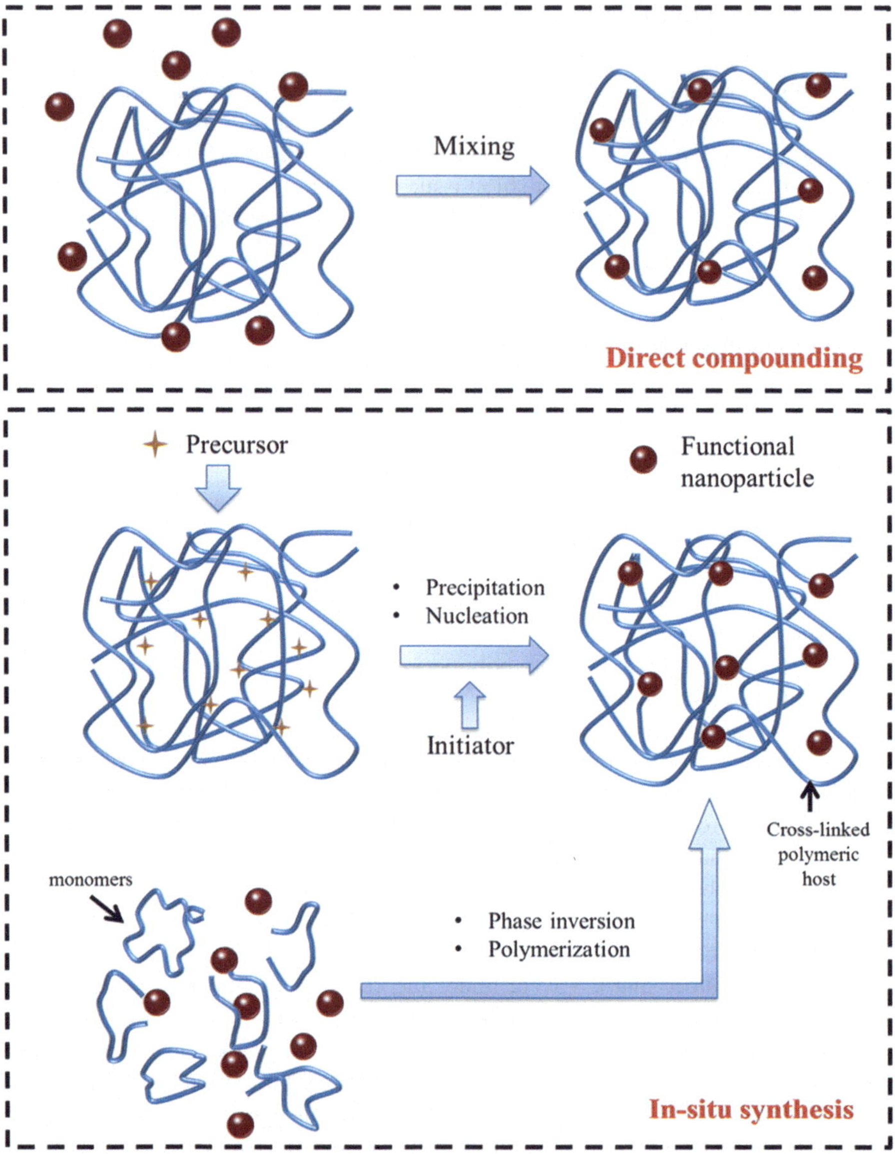

Fig. 12.12 Synthesis process of polymer nanocomposites (NC). (Copyright© Elsevier 2016, reprinted with permission from Zhang et al. (2016))

remediation of heavy metal such as Cu^{2+}, Cd^{2+} and Pb^{2+} from contaminated waters. CS-derived biopolymers exhibited low biologic noxiousness and viability (Saad et al. 2018). The batch experimental results indicated that optimum sorption capacities were found to be 117.5 mg/g for Cu^{2+}, 135.2 mg/g for Cd^{2+} and 476.2 mg/g for

Pb^{2+} and were well correlated with the Langmuir adsorption isotherm. Furthermore, this NC can be regenerated and reused without diminishing the sorption capacity.

Alginate is a green polysaccharide derived from brown algae which is a non-hazardous and biodegradable biobased polymer (Esmat et al. 2017). Gokila and her colleagues created CS/alginate NC for the removal of Cr^{6+} from contaminated wastewater (Gokila et al. 2017). The maximum sorption capacity of Cr^{6+} ions was found to be 108.7 mg/g and followed the multilayer sorption mechanism.

12.6.2 Inorganic Anchored Nanohybrids

NCs are supported with inorganic polymers to enhance the functional properties of NPs towards heavy metal treatment from wastewaters. Examples for inorganic materials include CNTs, AC (activated carbon) and some nature-derived materials such as zeolite and bentonite (Tounsadi et al. 2016). AC-strengthened NCs presented immense potential in the treatment of heavy metals (Cd^{2+}, Cr^{6+}, Pb^{2+}) from water due to cost-effectiveness, economic value and simplicity of AC sorbent (Fernando et al. 2015; Parlayici et al. 2015; Jafari Kang et al. 2016; Jayaweera et al. 2018). Salam and his colleagues developed MWCNT/CS NCs by the process of sonication between CS and CNT suspensions that have further undergone cross-linking with glutaraldehyde, as shown in Fig. 12.13 (Salam et al. 2011). NCs were packaged into glass column for the removal of Cu^{2+}, Cd^{2+}, Ni^{2+} and Zn^{2+} ions, as shown in Fig. 12.14, and showcased high sorption capacities towards chosen heavy metal ions.

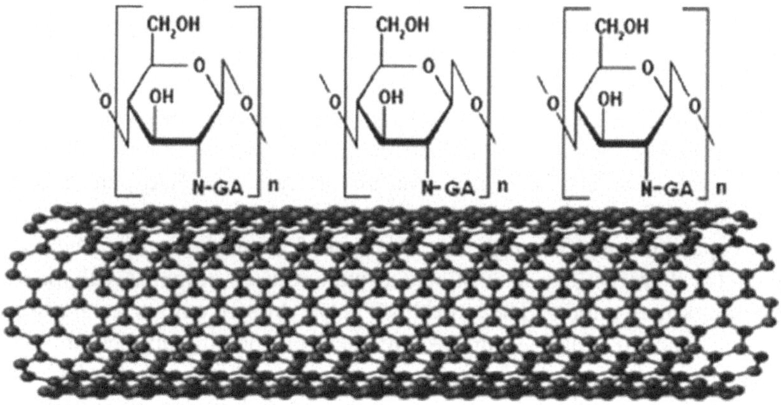

Fig. 12.13 Physical arrangement of multi-walled CNT/chitosan (CS) nanocomposite. (Copyright© Elsevier 2011, reprinted with permission from Salam et al. (2011))

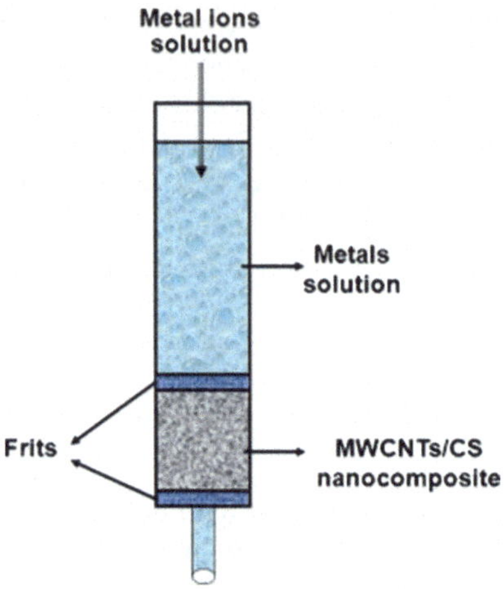

Fig. 12.14 Multi-walled CNT/chitosan (CS) nanocomposite packaged in a glass column. (Copyright© Elsevier 2011, reprinted with permission from Salam et al. (2011))

Recently, Hayati and his colleagues synthesised CNT overlaid poly(amidoamine) (PAMAM) dendrimer NC and examined its sorption efficacy towards Co^{2+}, As^{3+} and Zn^{2+} heavy metal ions in a continuous packed-bed system (Hayati et al. 2018). From the column sorption studies, optimum removal capacities were found to be 432.1 mg/g for As^{3+}, 494.3 mg/g for Co^{2+} and 470.5 mg/g for Zn^{2+} heavy metal ions.

Bentonite possesses impressive properties such as high surface characteristics, sorptive affinity and cation-exchange capacity which proved to be a promising sorbent for the treatment of heavy metals with high concentration (Ntwampe and Moothi 2018).

It has been reported that ethylenediaminetetraacetic acid (EDTA), Fe_3O_4, cetyltrimethylammonium bromide (CTMAB), 2-mercaptobenzothiazole (MBT), cellulose and NZVI were compounded with bentonite for the effective removal of heavy metals from water (Ma et al. 2016; Putro et al. 2017; Eskandari et al. 2018).

Zeolite is another category of assuring candidate and acts as a stabilising agent for NPs owing to its high surface characteristics, exceptional ion-exchange capacity, water-loving and eco-friendly nature and chemical properties that can be easily controlled (Alswata et al. 2017). Zeolite (NaP)/hydroxyapatite (HAp) NC was synthesised for the treatment of Cd^{2+} and Pb^{2+} from water by batch process and achieved sorption capacities of 40.2 mg/g for Cd^{2+} and 55.6 mg/g for Pb^{2+} heavy metal (Zendehdel et al. 2016). The adsorption data were well-coordinated with the pseudo-second-order kinetic model. The hybrid NC also exhibited antibacterial action concerning Gram-negative and Gram-positive bacteria, showing applicability towards wastewater treatment. Furthermore, inorganic supporting agents for NCs such as clay, sand and GO established their potential applicability in remediating heavy metals from contaminated waters (Nujić and Habuda-Stanić 2019).

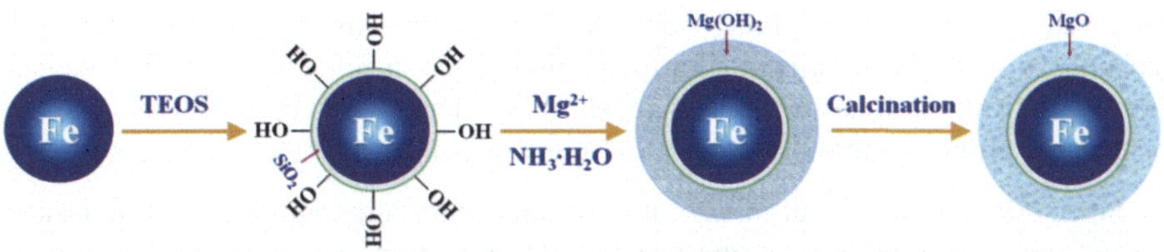

Fig. 12.15 Synthesis process of Fe/MgO nanohybrid. (Copyright© Elsevier 2018, reprinted with permission from Ge et al. (2018))

12.6.3 Magnetic Nanohybrids

Magnetic nanohybrids are the typical type of nanomaterial which has gained much attention due to the capability of easy separation. Magnetic nanohybrids consist of iron oxide (Fe_3O_4) and magnetic iron. The synthesis of the magnetic nano-based composite is carried out in three steps: (1) Surface improvement of magnetic iron/Fe_3O_4 NPs is facilitated by means of –SH and –NH_2 functional groups. (2) Encapsulation of iron/Fe_3O_4 NPs is carried out by HA, PPy, PRd and PEI in order to construct a core cell framework (Song et al. 2011; Kim et al. 2013; Mirrezaie et al. 2013; Lü et al. 2018). (3) Coating of iron/Fe_3O_4 NPs is carried out by poriferous materials such as CNTs and GO (Zhang et al. 2013a; Elmi et al. 2017).

Currently, Huang and his colleagues synthesised Fe_3O_4/SiO_2 NPs as core and amino-enriched Zr-MOFs as a shell. The amino-enhanced Zr-MOFs showed adsorption for MB dye and Pb^{2+} from aqueous medium (Huang et al. 2018a). Ge et al. synthesised Fe/MgO nanohybrid composite, as shown in Fig. 12.15, which integrated the advantageousness of magnesium oxide's sorption capacity and active magnetic property of zero-valent irons (Ge et al. 2018).

Optimum sorption capacity for this nanohybrid material was found to be 1476.5 mg/g for Pb^{2+} and 6947.8 mg/g for methyl orange (MO). The adsorption data were well suited with the Langmuir adsorption isotherm and followed pseudo-second-order kinetics. Hence, magnetic nanohybrids were proved to be a potential adsorbent in remediating heavy metal from contaminated waters due to the ease of separation.

12.7 Conclusion and Future Directions

Nanomaterials have been abundantly utilised for the remediation of heavy metal ions from wastewater due to their remarkable characteristics such as high surface area, abundance of active sites, good applicability and excellent rate of removal and adsorption capacity of heavy metals. In this chapter, various nanomaterials, such as carbon-based NPs, metal-based NPs and other nanohybrid composites, were reviewed in detail. Even though advanced nanomaterials possess a number of advantages, still they confront significant challenges in remediating heavy metals from wastewater contaminants. Hence, some hurdles have to be overcome to have better utilisation of nanomaterials in the future as follows:

1. Since NPs are nanolevel in size, it may cause easy aggregations, fouling issues and clogging problems during functional applications and ultimately diminish the adsorption capacity which results in the formation of secondary contaminations.
2. Surface characteristics of specific NPs undergo easy oxidation and are inconsistent.
3. Majority of the nanomaterials are synthesised by a chemical process, which gives rise to toxic contaminants.
4. Global production of nanomaterials for wastewater remediation is much limited due to biologic unsuitability and high value.

Green synthesis of innovative nanomaterials has to be widely explored. Various agri-based wastes (e.g. bagasse waste), peanut and coconut shell, corn cobs, mushrooms (*Bacillus subtilis* and *Dictyophora indusiata*), clay components and mineral slags (e.g. coal fly ash) can be effectively used for the synthesis of nanoadsorbents due to its cost-effectiveness, eco-friendly nature and non-hazardous nature. Hence, wide-scale utilisation of nature-derived nanomaterials should be promoted in the future. To enhance the overall stability, sorption capacity and selectivity for heavy metal remediation can be improved by optimising the synthesis process, surface characteristics and consistent alignment of NPs. To be cost-effective, recycling technology for advanced nanomaterials has to be encouraged.

Since nanomaterials are widely used for wastewater treatment, their environmental impacts and toxicity towards humans and other life forms have to be well addressed and assured for safe applicability. The removal capacities of NPs are commonly analysed for spiked water with simple constituents. However, there are no enough data or reports of nanomaterials in realistic wastewater. Hence, more research has to be devoted towards practical applications. By overcoming the abovementioned drawbacks and undergoing more research in this area, advanced eco-friendly nanomaterials can be commercially produced globally in wastewater treatment applications for the betterment of the future.

Acknowledgments This research work was financially supported by the University of Malaya Research Grant (No. RP045B-17AET), Impact-Oriented Interdisciplinary Research Grant (No. IIRG018A-2019) and Global Collaborative Programme - SATU Joint Research Scheme (No.ST012-2019).

References

Abdullah J, Al Lafi AG, Amin Y, Alnama T (2018a) A Styrofoam-nano manganese oxide based composite: preparation and application for the treatment of wastewater. Appl Radiat Isot 136:73–81. https://doi.org/10.1016/j.apradiso.2018.02.013

Abdullah EC, Mubarak NM, Inamuddin MK et al (2018b) Recent trends in the synthesis of graphene and graphene oxide based nanomaterials for removal of heavy metals – a review. J Ind Eng Chem 66:29–44. https://doi.org/10.1016/j.jiec.2018.05.028

Adegoke HI, Amoo Adekola F, Fatoki OS, Ximba BJ (2014) Adsorption of Cr (VI) on synthetic hematite (α-Fe_2O_3) nanoparticles of different morphologies. Korean J Chem Eng 31:142–154. https://doi.org/10.1007/s11814-013-0204-7

Afshar A, Sadjadi SAS, Mollahosseini A, Eskandarian MR (2016) Polypyrrole-polyaniline/Fe_3O_4 magnetic nanocomposite for the removal of Pb(II) from aqueous solution. Korean J Chem Eng 33:669–677. https://doi.org/10.1007/s11814-015-0156-1

Ahmadi M, Hazrati Niari M, Kakavandi B (2017) Development of maghemite nanoparticles supported on cross-linked chitosan (γ-Fe_2O_3@CS) as a recoverable mesoporous magnetic composite for effective heavy metals removal. J Mol Liq 248:184–196. https://doi.org/10.1016/J.MOLLIQ.2017.10.014

Ahmed MA, Ali SM, El-Dek SI, Galal A (2013) Magnetite–hematite nanoparticles prepared by green methods for heavy metal ions removal from water. Mater Sci Eng B 178:744–751. https://doi.org/10.1016/J.MSEB.2013.03.011

Akhbarizadeh R, Shayestefar MR, Darezereshki E (2014) Competitive removal of metals from wastewater by maghemite nanoparticles: a comparison between simulated wastewater and AMD. Mine Water Environ 33:89–96. https://doi.org/10.1007/s10230-013-0255-3

Al Omar MK, Alsaadi MA, Hayyan M et al (2017) Allyl triphenyl phosphonium bromide based DES-functionalized carbon nanotubes for the removal of mercury from water. Chemosphere 167:44–52. https://doi.org/10.1016/J.CHEMOSPHERE.2016.09.133

Al-Anzi BS, Siang OC (2017) Recent developments of carbon based nanomaterials and membranes for oily wastewater treatment. RSC Adv 7:20981–20994. https://doi.org/10.1039/C7RA02501G

Alswata AA, Ahmad MB, Al-Hada NM et al (2017) Preparation of Zeolite/Zinc Oxide nanocomposites for toxic metals removal from water. Results Phys 7:723–731. https://doi.org/10.1016/J.RINP.2017.01.036

Ambigaipalan P, de Camargo AC, Shahidi F (2016) Phenolic compounds of pomegranate byproducts (outer skin, mesocarp, divider membrane) and their antioxidant activities. J Agric Food Chem 64:6584–6604. https://doi.org/10.1021/acs.jafc.6b02950

Amin MT, Alazba AA, Manzoor U (2014) A review of removal of pollutants from water/wastewater using different types of nanomaterials. Adv Mater Sci Eng 2014:1–24. https://doi.org/10.1155/2014/825910

Anandan S, Kathiravan K, Murugesan V, Ikuma Y (2009) Anionic (IO3-) non-metal doped TiO_2 nanoparticles for the photocatalytic degradation of hazardous pollutant in water. Catal Commun 10:1014–1019. https://doi.org/10.1016/J.CATCOM.2008.12.054

Ansari AA, Alhoshan M, Alsalhi MS, Aldwayyan AS (2010) Prospects of nanotechnology in clinical immunodiagnostics. Sensors (Basel) 10:6535–6581. https://doi.org/10.3390/s100706535

Anupriya K, Vivek E, Subramanian B (2014) Facile synthesis of ceria nanoparticles by precipitation route for UV blockers. J Alloys Compd 590:406–410. https://doi.org/10.1016/J.JALLCOM.2013.12.121

Arshad F, Selvaraj M, Zain J et al (2019) Polyethylenimine modified graphene oxide hydrogel composite as an efficient adsorbent for heavy metal ions. Sep Purif Technol 209:870–880. https://doi.org/10.1016/J.SEPPUR.2018.06.035

Avouris P, Dimitrakopoulos C (2012) Graphene: synthesis and applications. Mater Today 15:86–97. https://doi.org/10.1016/S1369-7021(12)70044-5

Awad FS, AbouZeid KM, El-Maaty WMA, et al (2017) Efficient Removal of Heavy Metals from Polluted Water with High Selectivity for Mercury(II) by 2-Imino-4-thiobiuret–Partially Reduced Graphene Oxide (IT-PRGO). ACS Appl Mater Interfaces 9:34230–34242. https://doi.org/10.1021/acsami.7b10021

Ayawanna J, Sato K (2019) Photoelectrodeposition effect of lanthanum oxide-modified ceria particles on the removal of lead (II) ions from water. Catal Today 321–322:128–134. https://doi.org/10.1016/J.CATTOD.2017.11.010

Ayawanna J, Teoh W, Niratisairak S, Sato K (2015) Gadolinia-modified ceria photocatalyst for removal of lead (II) ions from aqueous solutions. Mater Sci Semicond Process 40:136–139. https://doi.org/10.1016/J.MSSP.2015.06.058

Barakat MA (2011) New trends in removing heavy metals from industrial wastewater. Arab J Chem 4:361–377. https://doi.org/10.1016/j.arabjc.2010.07.019

Baysal A, Kuznek C, Ozcan M (2018) Starch coated titanium dioxide nanoparticles as a challenging sorbent to separate and preconcentrate some heavy metals using graphite furnace atomic absorption spectrometry. Int J Environ Anal Chem 98:45–55. https://doi.org/10.1080/03067319.2018.1427741

Biao L, Tan S, Meng Q et al (2018) Green synthesis, characterization and application of proanthocyanidins-functionalized gold nanoparticles. Nanomater (Basel, Switzerland) 8. https://doi.org/10.3390/nano8010053

Bokare V, Jung J, Chang Y-Y, Chang Y-S (2013) Reductive dechlorination of octachlorodibenzo-p-dioxin by nanosized zero-valent zinc: modeling of rate kinetics and congener profile. J Hazard Mater 250–251:397–402. https://doi.org/10.1016/J.JHAZMAT.2013.02.020

Borklu Budak T (2013) Removal of heavy metals from wastewater using synthetic ion exchange resin. Asian J Chem 25:4207–4210. https://doi.org/10.14233/ajchem.2013.13902

Buzea C, Pacheco I (2017) Nanomaterial and nanoparticle: origin and activity. Springer, Cham, pp 71–112

Cai J, Lei M, Zhang Q et al (2017a) Electrospun composite nanofiber mats of Cellulose@ Organically modified montmorillonite for heavy metal ion removal: design, characterization, evaluation of absorption performance. Compos Part A Appl Sci Manuf 92:10–16. https://doi.org/10.1016/j.compositesa.2016.10.034

Cai Y, Li C, Wu D et al (2017b) Highly active MgO nanoparticles for simultaneous bacterial inactivation and heavy metal removal from aqueous solution. Chem Eng J 312:158–166. https://doi.org/10.1016/J.CEJ.2016.11.134

Chávez-Guajardo AE, Medina-Llamas JC, Maqueira L et al (2015) Efficient removal of Cr (VI) and Cu (II) ions from aqueous media by use of polypyrrole/maghemite and polyaniline/maghemite magnetic nanocomposites. Chem Eng J 281:826–836. https://doi.org/10.1016/J.CEJ.2015.07.008

Chen Y-H, Li F-A (2010) Kinetic study on removal of copper(II) using goethite and hematite nano-photocatalysts. J Colloid Interface Sci 347:277–281. https://doi.org/10.1016/J.JCIS.2010.03.050

Chen K, He J, Li Y et al (2017) Removal of cadmium and lead ions from water by sulfonated magnetic nanoparticle adsorbents. J Colloid Interface Sci 494:307–316. https://doi.org/10.1016/J.JCIS.2017.01.082

Chen Q, Yao Y, Li X et al (2018) Comparison of heavy metal removals from aqueous solutions by chemical precipitation and characteristics of precipitates. J Water Process Eng 26:289–300. https://doi.org/10.1016/J.JWPE.2018.11.003

Cheng Z, Tan ALK, Tao Y et al (2012) Synthesis and characterization of iron oxide nanoparticles and applications in the removal of heavy metals from industrial wastewater. Int J Photoenergy 2012:1–5. https://doi.org/10.1155/2012/608298

Cocarta DM, Neamtu S, Resetar Deac AM (2016) Carcinogenic risk evaluation for human health risk assessment from soils contaminated with heavy metals. Int J Environ Sci Technol 13:2025–2036. https://doi.org/10.1007/s13762-016-1031-2

Dickson D, Liu G, Cai Y (2017) Adsorption kinetics and isotherms of arsenite and arsenate on hematite nanoparticles and aggregates. J Environ Manag 186:261–267. https://doi.org/10.1016/J.JENVMAN.2016.07.068

Du P, Carneiro JT, Moulijn JA, Mul G (2008) A novel photocatalytic monolith reactor for multiphase heterogeneous photocatalysis. Appl Catal A Gen 334:119–128. https://doi.org/10.1016/J.APCATA.2007.09.045

Duc Quyen NV, Ngoc Tuyen T, Quang Khieu D et al (2018) Lead ions removal from aqueous solution using modified carbon nanotubes. Bull Mater Sci 41:6. https://doi.org/10.1007/s12034-017-1541-7

Elmi F, Hosseini T, Taleshi MS, Taleshi F (2017) Kinetic and thermodynamic investigation into the lead adsorption process from wastewater through magnetic nanocomposite Fe_3O_4/CNT. Nanotechnol Environ Eng 2:13. https://doi.org/10.1007/s41204-017-0023-x

Eskandari M, Khatir MZ, Darban AK, Meshkini M (2018) Decreasing Ni, Cu, Cd, and Zn heavy metal using magnetite-bentonite nanocomposites and adsorption isotherm study. Mater Res Express 5:045030. https://doi.org/10.1088/2053-1591/aabb1d

Esmat M, Farghali AA, Khedr MH, El-Sherbiny IM (2017) Alginate-based nanocomposites for efficient removal of heavy metal ions. Int J Biol Macromol 102:272–283. https://doi.org/10.1016/J.IJBIOMAC.2017.04.021

Etale A, Tutu H, Drake DC (2016) The effect of silica and maghemite nanoparticles on remediation of Cu(II)-, Mn(II)- and U(VI)-contaminated water by *Acutodesmus* sp. J Appl Phycol 28:251–260. https://doi.org/10.1007/s10811-015-0555-z

Fang Q, Zhou X, Deng W, Liu Z (2017) Hydroxyl-containing organic molecule induced self-assembly of porous graphene monoliths with high structural stability and recycle performance for heavy metal removal. Chem Eng J 308:1001–1009. https://doi.org/10.1016/J.CEJ.2016.09.139

Fatikow S, Eichhorn V, Bartenwerfer M (2012) Nanomaterials enter the silicon-based CMOS era: nanorobotic technologies for nanoelectronic devices. IEEE Nanotechnol Mag 6:14–18. https://doi.org/10.1109/MNANO.2011.2181735

Feng J, Zou L, Wang Y et al (2015) Synthesis of high surface area, mesoporous MgO nanosheets with excellent adsorption capability for Ni(II) via a distillation treating. J Colloid Interface Sci 438:259–267. https://doi.org/10.1016/J.JCIS.2014.10.004

Fernando MS, de Silva RM, de Silva KMN (2015) Synthesis, characterization, and application of nano hydroxyapatite and nanocomposite of hydroxyapatite with granular activated carbon for the removal of Pb^{2+} from aqueous solutions. Appl Surf Sci 351:95–103. https://doi.org/10.1016/J.APSUSC.2015.05.092

Fu F, Dionysiou DD, Liu H (2014) The use of zero-valent iron for groundwater remediation and wastewater treatment: a review. J Hazard Mater 267:194–205. https://doi.org/10.1016/J.JHAZMAT.2013.12.062

Gao W, Majumder M, Alemany LB et al (2011) Engineered graphite oxide materials for application in water purification. ACS Appl Mater Interfaces 3:1821–1826. https://doi.org/10.1021/am200300u

Ge L, Wang W, Peng Z et al (2018) Facile fabrication of Fe@MgO magnetic nanocomposites for efficient removal of heavy metal ion and dye from water. Powder Technol 326:393–401. https://doi.org/10.1016/J.POWTEC.2017.12.003

George R, Bahadur N, Singh N et al (2016) Environmentally benign TiO_2 nanomaterials for removal of heavy metal ions with interfering ions present in tap water. In: Materials Today: Proceedings. Elsevier, pp 162–166

Ghiloufi I, Khezami L, El Mir L (2015) Preparation and characterization of nanoporous resin for heavy metal removal from aqueous solution. J Water Supply Res Technol 64:316–325. https://doi.org/10.2166/aqua.2014.086

Giles DE, Mohapatra M, Issa TB et al (2011) Iron and aluminium based adsorption strategies for removing arsenic from water. J Environ Manag 92:3011–3022. https://doi.org/10.1016/J.JENVMAN.2011.07.018

Giraldo L, Erto A, Moreno-Piraján JC (2013) Magnetite nanoparticles for removal of heavy metals from aqueous solutions: synthesis and characterization. Adsorption 19:465–474. https://doi.org/10.1007/s10450-012-9468-1

Gokila S, Gomathi T, Sudha PN, Anil S (2017) Removal of the heavy metal ion chromium(VI) using Chitosan and Alginate nanocomposites. Int J Biol Macromol 104:1459–1468. https://doi.org/10.1016/J.IJBIOMAC.2017.05.117

Guerra FD, Attia MF, Whitehead DC, Alexis F (2018) Nanotechnology for environmental remediation: materials and applications. Molecules 23:1–23. https://doi.org/10.3390/molecules23071760

Gupta VK, Moradi O, Tyagi I et al (2016) Study on the removal of heavy metal ions from industry waste by carbon nanotubes: effect of the surface modification: a review. Crit Rev Environ Sci Technol 46:93–118. https://doi.org/10.1080/10643389.2015.1061874

Hadadian M, Goharshadi EK, Fard MM, Ahmadzadeh H (2018) Synergistic effect of graphene nanosheets and zinc oxide nanoparticles for effective adsorption of Ni (II) ions from aqueous solutions. Appl Phys A Mater Sci Process 124:239. https://doi.org/10.1007/s00339-018-1664-8

Hao Y-M, Man C, Hu Z-B (2010) Effective removal of Cu (II) ions from aqueous solution by amino-functionalized magnetic nanoparticles. J Hazard Mater 184:392–399. https://doi.org/10.1016/J.JHAZMAT.2010.08.048

Hashim MA, Mukhopadhyay S, Narayan Sahu J, Sengupta B (2011) Remediation technologies for heavy metal contaminated groundwater. J Environ Manag 92:2355–2388. https://doi.org/10.1016/j.jenvman.2011.06.009

Hayati B, Maleki A, Najafi F et al (2016) Synthesis and characterization of PAMAM/CNT nanocomposite as a super-capacity adsorbent for heavy metal (Ni^{2+}, Zn^{2+}, As^{3+}, Co^{2+}) removal from wastewater. J Mol Liq 224:1032–1040. https://doi.org/10.1016/J.MOLLIQ.2016.10.053

Hayati B, Maleki A, Najafi F et al (2018) Heavy metal adsorption using PAMAM/CNT nanocomposite from aqueous solution in batch and continuous fixed bed systems. Chem Eng J 346:258–270. https://doi.org/10.1016/J.CEJ.2018.03.172

Ho Y-S (2014) Comments on adsorption characteristics and behaviors of graphene oxide for Zn(II) removal from aqueous solution. Appl Surf Sci 301:584. https://doi.org/10.1016/J.APSUSC.2014.02.040

Hossein Beyki M, Ghasemi MH, Jamali A, Shemirani F (2017) A novel polylysine–resorcinol base γ-alumina nanotube hybrid material for effective adsorption/preconcentration of cadmium from various matrices. J Ind Eng Chem 46:165–174. https://doi.org/10.1016/J.JIEC.2016.10.027

Hotze EM, Phenrat T, Lowry GV (2010) Nanoparticle aggregation: challenges to understanding transport and reactivity in the environment. J Environ Qual 39:1909. https://doi.org/10.2134/jeq2009.0462

Hu H, Wang Z, Pan L (2010) Synthesis of monodisperse Fe3O4@silica core–shell microspheres and their application for removal of heavy metal ions from water. J Alloys Compd 492:656–661. https://doi.org/10.1016/J.JALLCOM.2009.11.204

Hu R, Wang X, Dai S et al (2015) Application of graphitic carbon nitride for the removal of Pb(II) and aniline from aqueous solutions. Chem Eng J 260:469–477. https://doi.org/10.1016/J.CEJ.2014.09.013

Hua M, Zhang S, Pan B et al (2012) Heavy metal removal from water/wastewater by nanosized metal oxides: a review. J Hazard Mater 211–212:317–331. https://doi.org/10.1016/J.JHAZMAT.2011.10.016

Hua M, Jiang Y, Wu B et al (2013) Fabrication of a new hydrous Zr(IV) oxide-based nanocomposite for enhanced Pb(II) and Cd(II) removal from waters. ACS Appl Mater Interfaces 5:12135–12142. https://doi.org/10.1021/am404031q

Huang P, Ye Z, Xie W et al (2013) Rapid magnetic removal of aqueous heavy metals and their relevant mechanisms using nanoscale zero valent iron (nZVI) particles. Water Res 47:4050–4058. https://doi.org/10.1016/J.WATRES.2013.01.054

Huang D-L, Chen G-M, Zeng G-M et al (2015) Synthesis and application of modified zero-valent iron nanoparticles for removal of hexavalent chromium from wastewater. Water Air Soil Pollut 226:375. https://doi.org/10.1007/s11270-015-2583-3

Huang L, He M, Chen B, Hu B (2018a) Magnetic Zr-MOFs nanocomposites for rapid removal of heavy metal ions and dyes from water. Chemosphere 199:435–444. https://doi.org/10.1016/J.CHEMOSPHERE.2018.02.019

Huang X, Yang J, Wang J et al (2018b) Design and synthesis of core–shell Fe_3O_4@PTMT composite magnetic microspheres for adsorption of heavy metals from high salinity wastewater. Chemosphere 206:513–521. https://doi.org/10.1016/J.CHEMOSPHERE.2018.04.184

Huangfu X, Jiang J, Lu X et al (2015) Adsorption and oxidation of Thallium(I) by a nanosized Manganese Dioxide. Water Air Soil Pollut 226:2272. https://doi.org/10.1007/s11270-014-2272-7

Huo L, Zeng X, Su S et al (2017) Enhanced removal of As (V) from aqueous solution using modified hydrous ferric oxide nanoparticles. Sci Rep 7:40765. https://doi.org/10.1038/srep40765

Ihsanullah AA, Al-Amer AM et al (2016) Heavy metal removal from aqueous solution by advanced carbon nanotubes: critical review of adsorption applications. Sep Purif Technol 157:141–161. https://doi.org/10.1016/J.SEPPUR.2015.11.039

Jafari Kang A, Baghdadi M, Pardakhti A (2016) Removal of cadmium and lead from aqueous solutions by magnetic acid-treated activated carbon nanocomposite. Desalin Water Treat 57:18782–18798. https://doi.org/10.1080/19443994.2015.1095123

Jaishankar M, Tseten T, Anbalagan N et al (2014) Toxicity, mechanism and health effects of some heavy metals. Interdiscip Toxicol 7:60–72. https://doi.org/10.2478/intox-2014-0009

Jayaweera HDAC, Siriwardane I, de Silva KMN, de Silva RM (2018) Synthesis of multifunctional activated carbon nanocomposite comprising biocompatible flake nano hydroxyapatite and natural turmeric extract for the removal of bacteria and lead ions from aqueous solution. Chem Cent J 12:18. https://doi.org/10.1186/s13065-018-0384-7

Jiang C, Xiao DA (2014) Nanosized zirconium dioxide particles as an efficient sorbent for lead removal in waters. Adv Mater Res 926–930:166–169. https://doi.org/10.4028/www.scientific.net/AMR.926-930.166

Kang BK, Lim BS, Yoon Y, et al (2017) Efficient removal of arsenic by strategically designed and layer-by-layer assembled PS@+rGO@GO@Fe3O4 composites. J Environ Manage 201:286–293. https://doi.org/10.1016/J.JENVMAN.2017.05.066

Kefeni KK, Msagati TAM, Nkambule TTI, Mamba BB (2018) Synthesis and application of hematite nanoparticles for acid mine drainage treatment. J Environ Chem Eng 6:1865–1874. https://doi.org/10.1016/J.JECE.2018.02.037

Khezami L, Ould M'hamed M, Lemine OM et al (2016) Milled goethite nanocrystalline for selective and fast uptake of cadmium ions from aqueous solution. Desalin Water Treat 57:6531–6539. https://doi.org/10.1080/19443994.2015.1010231

Khulbe KC, Matsuura T (2018) Removal of heavy metals and pollutants by membrane adsorption techniques. Appl Water Sci 8:19. https://doi.org/10.1007/s13201-018-0661-6

Kim E-J, Lee C-S, Chang Y-Y, Chang Y-S (2013) Hierarchically structured manganese oxide-coated magnetic nanocomposites for the efficient removal of heavy metal ions from aqueous systems. ACS Appl Mater Interfaces 5:9628–9634. https://doi.org/10.1021/am402615m

Kotsyuda SS, Tomina VV, Zub YL et al (2017) Bifunctional silica nanospheres with 3-aminopropyl and phenyl groups. Synthesis approach and prospects of their applications. Appl Surf Sci 420:782–791. https://doi.org/10.1016/J.APSUSC.2017.05.150

Krishnamoorthy K, Manivannan G, Kim SJ et al (2012) Antibacterial activity of MgO nanoparticles based on lipid peroxidation by oxygen vacancy. J Nanopart Res 14:1063. https://doi.org/10.1007/s11051-012-1063-6

Kumar KY, Muralidhara HB, Nayaka YA et al (2013) Low-cost synthesis of metal oxide nanoparticles and their application in adsorption of commercial dye and heavy metal ion in aqueous solution. Powder Technol 246:125–136. https://doi.org/10.1016/J.POWTEC.2013.05.017

Kumar R, Khan MA, Haq N (2014) Application of carbon nanotubes in heavy metals remediation. Crit Rev Environ Sci Technol 44:1000–1035. https://doi.org/10.1080/10643389.2012.741314
Kumar S, Ahlawat W, Kumar R, Dilbaghi N (2015) Graphene, carbon nanotubes, zinc oxide and gold as elite nanomaterials for fabrication of biosensors for healthcare. Biosens Bioelectron 70:498–503. https://doi.org/10.1016/J.BIOS.2015.03.062
Kumar PS, Sivaranjanee R, Rajan PS, Saravanan A (2018) Carbon sphere: synthesis, characterization and elimination of toxic Cr(VI) ions from aquatic system. J Ind Eng Chem 60:307–320. https://doi.org/10.1016/J.JIEC.2017.11.017
Leiviska T, Khalid MK, Sarpola A, Tanskanen J (2017) Removal of vanadium from industrial wastewater using iron sorbents in batch and continuous flow pilot systems. J Environ Manag 190:231–242. https://doi.org/10.1016/J.JENVMAN.2016.12.063
Lentini P, Zanoli L, Granata A et al (2017) Kidney and heavy metals – the role of environmental exposure (Review). Mol Med Rep 15:3413–3419. https://doi.org/10.3892/mmr.2017.6389
Li Y, Yang L, Xu Z, Sun Q (2017) Separation and recovery of heavy metals from waste water using synergistic solvent extraction. IOP Conf Ser Mater Sci Eng 167:012005. https://doi.org/10.1088/1757-899X/167/1/012005
Li J, Wang X, Zhao G et al (2018) Metal–organic framework-based materials: superior adsorbents for the capture of toxic and radioactive metal ions. Chem Soc Rev 47:2322–2356. https://doi.org/10.1039/C7CS00543A
Lisha KP, Anshup, Pradeep T (2009) Towards a practical solution for removing inorganic mercury from drinking water using gold nanoparticles. Gold Bull 42:144–152. https://doi.org/10.1007/BF03214924
Liu T, Wang Z-L, Sun Y (2015) Manipulating the morphology of nanoscale zero-valent iron on pumice for removal of heavy metals from wastewater. Chem Eng J 263:55–61. https://doi.org/10.1016/J.CEJ.2014.11.046
Liu X, Huang Y, Duan S, et al (2016a) Graphene oxides with different oxidation degrees for Co(II) ion pollution management. Chem Eng J 302:763–772. https://doi.org/10.1016/J.CEJ.2016.05.107
Liu Y, Xu L, Liu J, et al (2016b) Graphene oxides cross-linked with hyperbranched polyethylenimines: Preparation, characterization and their potential as recyclable and highly efficient adsorption materials for lead(II) ions. Chem Eng J 285:698–708. https://doi.org/10.1016/J.CEJ.2015.10.047
Lu F, Astruc D (2018) Nanomaterials for removal of toxic elements from water. Coord Chem Rev 356:147–164. https://doi.org/10.1016/j.ccr.2017.11.003
Lu H, Wang J, Stoller M et al (2016) An overview of nanomaterials for water and wastewater treatment. Adv Mater Sci Eng 2016:1–10. https://doi.org/10.1155/2016/4964828
Lü T, Qi D, Zhang D et al (2018) A facile method for emulsified oil-water separation by using polyethylenimine-coated magnetic nanoparticles. J Nanopart Res 20:88. https://doi.org/10.1007/s11051-018-4193-7
Ma J, Su G, Zhang X, Huang W (2016) Adsorption of heavy metal ions from aqueous solutions by bentonite nanocomposites. Water Environ Res 88:741–746. https://doi.org/10.2175/106143016X14609975747081
Madrakian T, Afkhami A, Zadpour B, Ahmadi M (2015) New synthetic mercaptoethylamino homopolymer-modified maghemite nanoparticles for effective removal of some heavy metal ions from aqueous solution. J Ind Eng Chem 21:1160–1166. https://doi.org/10.1016/J.JIEC.2014.05.029
Madzokere TC, Karthigeyan A (2017) Heavy metal ion effluent discharge containment using magnesium oxide (MgO) nanoparticles. Mater Today Proc 4:9–18. https://doi.org/10.1016/J.MATPR.2017.01.187
Mahdavi S, Jalali M, Afkhami A (2013) Heavy metals removal from aqueous solutions using TiO_2, MgO, and Al_2O_3 nanoparticles. Chem Eng Commun 200:448–470. https://doi.org/10.1080/00986445.2012.686939

Mahdavi S, Jalali M, Afkhami A (2015) Heavy metals removal from aqueous solutions by Al_2O_3 nanoparticles modified with natural and chemical modifiers. Clean Techn Environ Policy 17:85–102. https://doi.org/10.1007/s10098-014-0764-1

Mahmoud ME, Abdelwahab MS, Abdou AEH (2016a) Enhanced removal of lead and cadmium from water by Fe_3O_4-cross linked-O-phenylenediamine nano-composite. Sep Sci Technol 51:237–247. https://doi.org/10.1080/01496395.2015.1093505

Mahmoud ME, Fekry NA, El-Latif MMA (2016b) Nanocomposites of nanosilica-immobilized-nanopolyaniline and crosslinked nanopolyaniline for removal of heavy metals. Chem Eng J 304:679–691. https://doi.org/10.1016/J.CEJ.2016.06.110

Mahmoud ME, Abou Ali SAA, Elweshahy SMT (2018) Microwave functionalization of titanium oxide nanoparticles with chitosan nanolayer for instantaneous microwave sorption of Cu(II) and Cd(II) from water. Int J Biol Macromol 111:393–399. https://doi.org/10.1016/J.IJBIOMAC.2018.01.014

Massalimov IA, Il'yasova RR, Musavirova LR et al (2014) Use of micrometer hematite particles and nanodispersed goethite as sorbent for heavy metals. Russ J Appl Chem 87:1456–1463. https://doi.org/10.1134/S1070427214100115

Meepho M, Sirimongkol W, Ayawanna J (2018) Samaria-doped ceria nanopowders for heavy metal removal from aqueous solution. Mater Chem Phys 214:56–65. https://doi.org/10.1016/J.MATCHEMPHYS.2018.04.083

Mirrezaie N, Nikazar M, Hasan Zadeh M (2013) Synthesis of magnetic nanocomposite Fe_3O_4 coated polypyrrole (PPy) for Chromium(VI) removal. Adv Mater Res 829:649–653. https://doi.org/10.4028/www.scientific.net/AMR.829.649

Mishra PK, Saxena A, Rawat AS et al (2018) Surfactant-free one-pot synthesis of low-density cerium oxide nanoparticles for adsorptive removal of arsenic species. Environ Prog Sustain Energy 37:221–231. https://doi.org/10.1002/ep.12660

Mukherjee J, Ramkumar J, Chandramouleeswaran S et al (2013) Sorption characteristics of nano manganese oxide: efficient sorbent for removal of metal ions from aqueous streams. J Radioanal Nucl Chem 297:49–57. https://doi.org/10.1007/s10967-012-2393-7

Musico YLF, Santos CM, Dalida MLP, Rodrigues DF (2013) Improved removal of Lead(II) from water using a polymer-based graphene oxide nanocomposite. J Mater Chem A 1:3789. https://doi.org/10.1039/c3ta01616a

Nabi D, Aslam I, Qazi IA (2009) Evaluation of the adsorption potential of titanium dioxide nanoparticles for arsenic removal. J Environ Sci 21:402–408. https://doi.org/10.1016/S1001-0742(08)62283-4

Najafi M, Yousefi Y, Rafati AA (2012) Synthesis, characterization and adsorption studies of several heavy metal ions on amino-functionalized silica nano hollow sphere and silica gel. Sep Purif Technol 85:193–205. https://doi.org/10.1016/J.SEPPUR.2011.10.011

Norouzian Baghani A, Mahvi AH, Gholami M et al (2016) One-Pot synthesis, characterization and adsorption studies of amine-functionalized magnetite nanoparticles for removal of Cr (VI) and Ni (II) ions from aqueous solution: kinetic, isotherm and thermodynamic studies. J Environ Heal Sci Eng 14:11. https://doi.org/10.1186/s40201-016-0252-0

Novoselov KS, Fal′ko VI, Colombo L et al (2012) A roadmap for graphene. Nature 490:192–200. https://doi.org/10.1038/nature11458

Ntwampe OI, Moothi K (2018) Removal of heavy metals using bentonite clay and inorganic coagulants. In: Heavy metals. InTech, Croatia

Nujic M, Habuda-Stanic M (2019) Toxic metal ions in drinking water and effective removal using graphene oxide nanocomposite. In: A new generation material graphene: applications in water technology. Springer International Publishing, Cham, pp 373–395

O'Carroll D, Sleep B, Krol M et al (2013) Nanoscale zero valent iron and bimetallic particles for contaminated site remediation. Adv Water Resour 51:104–122. https://doi.org/10.1016/J.ADVWATRES.2012.02.005

Ojea-Jiménez I, López X, Arbiol J, Puntes V (2012) Citrate-coated gold nanoparticles as smart scavengers for Mercury(II) removal from polluted waters. ACS Nano 6:2253–2260. https://doi.org/10.1021/nn204313a

Pan S, Shen H, Xu Q et al (2012) Surface mercapto engineered magnetic Fe_3O_4 nanoadsorbent for the removal of mercury from aqueous solutions. J Colloid Interface Sci 365:204–212. https://doi.org/10.1016/J.JCIS.2011.09.002

Parlayici S, Eskizeybek V, Avcı A, Pehlivan E (2015) Removal of chromium (VI) using activated carbon-supported-functionalized carbon nanotubes. J Nanostructure Chem 5:255–263. https://doi.org/10.1007/s40097-015-0156-z

Patra AK, Dutta A, Bhaumik A (2012) Self-assembled mesoporous γ-Al_2O_3 spherical nanoparticles and their efficiency for the removal of arsenic from water. J Hazard Mater 201–202:170–177. https://doi.org/10.1016/J.JHAZMAT.2011.11.056

Pogorilyi RP, Melnyk IV, Zub YL et al (2014) New product from old reaction: uniform magnetite nanoparticles from iron-mediated synthesis of alkali iodides and their protection from leaching in acidic media. RSC Adv 4:22606–22612. https://doi.org/10.1039/C4RA02217C

Prabhakar R, Samadder SR (2018) Low cost and easy synthesis of aluminium oxide nanoparticles for arsenite removal from groundwater: a complete batch study. J Mol Liq 250:192–201. https://doi.org/10.1016/J.MOLLIQ.2017.11.173

Pradhan N, Pal A, Pal T (2002) Silver nanoparticle catalyzed reduction of aromatic nitro compounds. Colloids Surfaces A Physicochem Eng Asp 196:247–257. https://doi.org/10.1016/S0927-7757(01)01040-8

Prasad R, Pandey R, Barman I (2016) Engineering tailored nanoparticles with microbes: quo vadis. WIREs Nanomed Nanobiotechnol 8:316–330. doi: https://doi.org10.1002/wnan.1363

Pu Y, Yang X, Zheng H et al (2013) Adsorption and desorption of thallium(I) on multiwalled carbon nanotubes. Chem Eng J 219:403–410. https://doi.org/10.1016/J.CEJ.2013.01.025

Pumera M (2011) Graphene-based nanomaterials for energy storage. Energy Environ Sci 4:668–674. https://doi.org/10.1039/C0EE00295J

Putro JN, Santoso SP, Ismadji S, Ju Y-H (2017) Investigation of heavy metal adsorption in binary system by nanocrystalline cellulose – Bentonite nanocomposite: improvement on extended Langmuir isotherm model. Microporous Mesoporous Mater 246:166–177. https://doi.org/10.1016/J.MICROMESO.2017.03.032

Qiu H, Zhang S, Pan B et al (2012) Effect of sulfate on Cu(II) sorption to polymer-supported nano-iron oxides: behavior and XPS study. J Colloid Interface Sci 366:37–43. https://doi.org/10.1016/J.JCIS.2011.09.070

Qiu H, Zhang S, Pan B et al (2013) Oxalate-promoted dissolution of hydrous ferric oxide immobilized within nanoporous polymers: effect of ionic strength and visible light irradiation. Chem Eng J 232:167–173. https://doi.org/10.1016/J.CEJ.2013.07.092

Rafiq Z, Nazir R, Durr-e-Shahwar N et al (2014) Utilization of magnesium and zinc oxide nano-adsorbents as potential materials for treatment of copper electroplating industry wastewater. J Environ Chem Eng 2:642–651. https://doi.org/10.1016/J.JECE.2013.11.004

Rahimi S, Moattari RM, Rajabi L, Derakhshan AA (2015) Optimization of lead removal from aqueous solution using goethite/chitosan nanocomposite by response surface methodology. Colloids Surf A Physicochem Eng Asp 484:216–225. https://doi.org/10.1016/J.COLSURFA.2015.07.063

Rajakumar K, Kirupha SD, Sivanesan S, Sai RL (2014) Effective removal of heavy metal ions using Mn_2O_3 doped polyaniline nanocomposite. J Nanosci Nanotechnol 14:2937–2946. https://doi.org/10.1166/jnn.2014.8628

Rajput S, Singh LP, Pittman CU, Mohan D (2017) Lead (Pb^{2+}) and copper (Cu^{2+}) remediation from water using superparamagnetic maghemite (γ-Fe_2O_3) nanoparticles synthesized by Flame Spray Pyrolysis (FSP). J Colloid Interface Sci 492:176–190. https://doi.org/10.1016/J.JCIS.2016.11.095

Recillas S, Colón J, Casals E et al (2010) Chromium VI adsorption on cerium oxide nanoparticles and morphology changes during the process. J Hazard Mater 184:425–431. https://doi.org/10.1016/J.JHAZMAT.2010.08.052

Recillas S, García A, González E et al (2011) Use of CeO_2, TiO_2 and Fe_3O_4 nanoparticles for the removal of lead from water: toxicity of nanoparticles and derived compounds. Desalination 277:213–220. https://doi.org/10.1016/J.DESAL.2011.04.036

Renu, Agarwal M, Singh K (2017) Heavy metal removal from wastewater using various adsorbents: a review. J Water Reuse Desalin 7:387–419. https://doi.org/10.2166/wrd.2016.104

Ricco R, Konstas K, Styles MJ et al (2015) Lead(II) uptake by aluminium based magnetic framework composites (MFCs) in water. J Mater Chem A 3:19822–19831. https://doi.org/10.1039/C5TA04154F

Saad AHA, Azzam AM, El-Wakeel ST et al (2018) Removal of toxic metal ions from wastewater using ZnO@Chitosan core-shell nanocomposite. Environ Nanotech Monit Manag 9:67–75. https://doi.org/10.1016/J.ENMM.2017.12.004

Saadi Z, Saadi R, Fazaeli R (2013) Fixed-bed adsorption dynamics of Pb (II) adsorption from aqueous solution using nanostructured γ-alumina. J Nanostructure Chem 3:48. https://doi.org/10.1186/2193-8865-3-48

Saharan P, Chaudhary GR, Mehta SK, Umar A (2014) Removal of water contaminants by iron oxide nanomaterials. J Nanosci Nanotechnol 14:627–643. https://doi.org/10.1166/jnn.2014.9053

Salam MA, Makki MSI, Abdelaal MYA (2011) Preparation and characterization of multi-walled carbon nanotubes/chitosan nanocomposite and its application for the removal of heavy metals from aqueous solution. J Alloys Compd 509:2582–2587. https://doi.org/10.1016/J.JALLCOM.2010.11.094

Seisenbaeva GA, Daniel G, Nedelec J-M et al (2012) High surface area ordered mesoporous nanotitania by a rapid surfactant-free approach. J Mater Chem 22:20374. https://doi.org/10.1039/c2jm33977c

Seisenbaeva GA, Daniel G, Kessler VG, Nedelec J-M (2014) General facile approach to transition-metal oxides with highly uniform mesoporosity and their application as adsorbents for heavy-metal-ion sequestration. Chem - A Eur J 20:10732–10736. https://doi.org/10.1002/chem.201402691

Seyedi SM, Rabiee H, Shahabadi SMS, Borghei SM (2017) Synthesis of zero-valent iron nanoparticles via electrical wire explosion for efficient removal of heavy metals. Clean (Weinh) 45:1600139. https://doi.org/10.1002/clen.201600139

Shan C, Ma Z, Tong M, Ni J (2015) Removal of Hg(II) by poly(1-vinylimidazole)-grafted Fe_3O_4@SiO_2 magnetic nanoparticles. Water Res 69:252–260. https://doi.org/10.1016/J.WATRES.2014.11.030

Sheela T, Nayaka YA, Viswanatha R et al (2012) Kinetics and thermodynamics studies on the adsorption of Zn(II), Cd(II) and Hg(II) from aqueous solution using zinc oxide nanoparticles. Powder Technol 217:163–170. https://doi.org/10.1016/J.POWTEC.2011.10.023

Sheng G, Yang P, Tang Y et al (2016) New insights into the primary roles of diatomite in the enhanced sequestration of UO^{22+} by zerovalent iron nanoparticles: an advanced approach utilizing XPS and EXAFS. Appl Catal B Environ 193:189–197. https://doi.org/10.1016/J.APCATB.2016.04.035

Shi J, Li H, Lu H, Zhao X (2015) Use of carboxyl functional magnetite nanoparticles as potential sorbents for the removal of heavy metal ions from aqueous solution. J Chem Eng Data 60:2035–2041. https://doi.org/10.1021/je5011196

Shipley HJ, Engates KE, Grover VA (2013) Removal of Pb(II), Cd(II), Cu(II), and Zn(II) by hematite nanoparticles: effect of sorbent concentration, pH, temperature, and exhaustion. Environ Sci Pollut Res 20:1727–1736. https://doi.org/10.1007/s11356-012-0984-z

Shokati Poursani A, Nilchi A, Hassani AH et al (2015) A novel method for synthesis of nano-γ-Al_2O_3: study of adsorption behavior of chromium, nickel, cadmium and lead ions. Int J Environ Sci Technol 12:2003–2014. https://doi.org/10.1007/s13762-014-0740-7

Singh DK, Verma DK, Singh Y, Hasan SH (2017) Preparation of CuO nanoparticles using *Tamarindus indica* pulp extract for removal of As(III): optimization of adsorption process by ANN-GA. J Environ Chem Eng 5:1302–1318. https://doi.org/10.1016/J.JECE.2017.01.046

Sitko R, Musielak M, Zawisza B, et al (2016) Graphene oxide/cellulose membranes in adsorption of divalent metal ions. RSC Adv 6:96595–96605. https://doi.org/10.1039/C6RA21432K

Smith SC, Rodrigues DF (2015) Carbon-based nanomaterials for removal of chemical and biological contaminants from water: a review of mechanisms and applications. Carbon NY 91:122–143. https://doi.org/10.1016/J.CARBON.2015.04.043

Somu P, Paul S (2018) Casein based biogenic-synthesized zinc oxide nanoparticles simultaneously decontaminate heavy metals, dyes, and pathogenic microbes: a rational strategy for wastewater treatment. J Chem Technol Biotechnol 93:2962–2976. https://doi.org/10.1002/jctb.5655

Song J, Kong H, Jang J (2011) Adsorption of heavy metal ions from aqueous solution by polyrhodanine-encapsulated magnetic nanoparticles. J Colloid Interface Sci 359:505–511. https://doi.org/10.1016/J.JCIS.2011.04.034

Stietiya MH, Wang JJ (2014) Zinc and cadmium adsorption to aluminum oxide nanoparticles affected by naturally occurring ligands. J Environ Qual 43:498. https://doi.org/10.2134/jeq2013.07.0263

Su Q, Pan B, Wan S et al (2010) Use of hydrous manganese dioxide as a potential sorbent for selective removal of lead, cadmium, and zinc ions from water. J Colloid Interface Sci 349:607–612. https://doi.org/10.1016/J.JCIS.2010.05.052

Su Y, Adeleye AS, Huang Y et al (2014) Simultaneous removal of cadmium and nitrate in aqueous media by nanoscale zerovalent iron (nZVI) and Au doped nZVI particles. Water Res 63:102–111. https://doi.org/10.1016/J.WATRES.2014.06.008

Suárez-Iglesias O, Collado S, Oulego P, Díaz M (2017) Graphene-family nanomaterials in wastewater treatment plants. Chem Eng J. https://doi.org/10.1016/j.cej.2016.12.022

Suman KA, Gera M, Jain VK (2015) A novel reusable nanocomposite for complete removal of dyes, heavy metals and microbial load from water based on nanocellulose and silver nano-embedded pebbles. Environ Technol 36:706–714. https://doi.org/10.1080/09593330.2014.959066

Sumesh E, Bootharaju MS, Anshup PT (2011) A practical silver nanoparticle-based adsorbent for the removal of Hg^{2+} from water. J Hazard Mater 189:450–457. https://doi.org/10.1016/J.JHAZMAT.2011.02.061

Sun YB, Wang Q, Yang ST et al (2011) Characterization of nano-iron oxyhydroxides and their application in UO^{22+} removal from aqueous solutions. J Radioanal Nucl Chem 290:643–648. https://doi.org/10.1007/s10967-011-1325-2

Tabesh S, Davar F, Loghman-Estarki MR (2018) Preparation of γ-Al_2O_3 nanoparticles using modified sol-gel method and its use for the adsorption of lead and cadmium ions. J Alloys Compd 730:441–449. https://doi.org/10.1016/J.JALLCOM.2017.09.246

Tan Y, Chen M, Hao Y (2012) High efficient removal of Pb (II) by amino-functionalized Fe_3O_4 magnetic nano-particles. Chem Eng J 191:104–111. https://doi.org/10.1016/J.CEJ.2012.02.075

Tauanov Z, Tsakiridis PE, Mikhalovsky SV, Inglezakis VJ (2018) Synthetic coal fly ash-derived zeolites doped with silver nanoparticles for mercury (II) removal from water. J Environ Manag 224:164–171. https://doi.org/10.1016/J.JENVMAN.2018.07.049

Tounsadi H, Khalidi A, Abdennouri M, Barka N (2016) Activated carbon from *Diplotaxis harra* biomass: optimization of preparation conditions and heavy metal removal. J Taiwan Inst Chem Eng 59:348–358. https://doi.org/10.1016/J.JTICE.2015.08.014

Umar A, Kumar R, Akhtar MS et al (2015) Growth and properties of well-crystalline cerium oxide (CeO_2) nanoflakes for environmental and sensor applications. J Colloid Interface Sci 454:61–68. https://doi.org/10.1016/J.JCIS.2015.04.055

Vélez E, Campillo G, Morales G et al (2018) Silver nanoparticles obtained by aqueous or ethanolic *Aloe vera* extracts: an assessment of the antibacterial activity and mercury removal capability. J Nanomater 2018:1–7. https://doi.org/10.1155/2018/7215210

Vilardi G, Ochando-Pulido JM, Verdone N et al (2018) On the removal of hexavalent chromium by olive stones coated by iron-based nanoparticles: equilibrium study and chromium recovery. J Clean Prod 190:200–210. https://doi.org/10.1016/J.JCLEPRO.2018.04.151

Vilela D, Parmar J, Zeng Y et al (2016) Graphene-based microbots for toxic heavy metal removal and recovery from water. Nano Lett 16:2860–2866. https://doi.org/10.1021/acs.nanolett.6b00768

Vunain E, Mishra AK, Mamba BB (2016) Dendrimers, mesoporous silicas and chitosan-based nanosorbents for the removal of heavy-metal ions: a review. Int J Biol Macromol 86:570–586. https://doi.org/10.1016/j.ijbiomac.2016.02.005

Wan S, Zhao X, Lv L et al (2010) Selective adsorption of Cd(II) and Zn(II) ions by nano-hydrous manganese dioxide (HMO)-encapsulated cation exchanger. Ind Eng Chem Res 49:7574–7579. https://doi.org/10.1021/ie101003y

Wan S, Wu J, Zhou S et al (2018) Enhanced lead and cadmium removal using biochar-supported hydrated manganese oxide (HMO) nanoparticles: behavior and mechanism. Sci Total Environ 616–617:1298–1306. https://doi.org/10.1016/J.SCITOTENV.2017.10.188

Wang X, Guo Y, Yang L et al (2012) Nanomaterials as sorbents to remove heavy metal ions in wastewater treatment. J Environ Anal Toxicol 2:1–5. https://doi.org/10.4172/2161-0525.1000154

Wang X, Zhan C, Kong B et al (2015) Self-curled coral-like γ-Al_2O_3 nanoplates for use as an adsorbent. J Colloid Interface Sci 453:244–251. https://doi.org/10.1016/J.JCIS.2015.03.065

Wang W, Hua Y, Li S et al (2016) Removal of Pb(II) and Zn(II) using lime and nanoscale zero-valent iron (nZVI): a comparative study. Chem Eng J 304:79–88. https://doi.org/10.1016/j.cej.2016.06.069

Wang X, Huang K, Chen Y et al (2018a) Preparation of dumbbell manganese dioxide/gelatin composites and their application in the removal of lead and cadmium ions. J Hazard Mater 350:46–54. https://doi.org/10.1016/J.JHAZMAT.2018.02.020

Wang X, Liu Y, Pang H et al (2018b) Effect of graphene oxide surface modification on the elimination of Co(II) from aqueous solutions. Chem Eng J 344:380–390. https://doi.org/10.1016/J.CEJ.2018.03.107

Wang X, Yu S, Wu Y et al (2018c) The synergistic elimination of uranium (VI) species from aqueous solution using bi-functional nanocomposite of carbon sphere and layered double hydroxide. Chem Eng J 342:321–330. https://doi.org/10.1016/J.CEJ.2018.02.102

Watts MP, Coker VS, Parry SA et al (2015) Biogenic nano-magnetite and nano-zero valent iron treatment of alkaline Cr(VI) leachate and chromite ore processing residue. Appl Geochem 54:27–42. https://doi.org/10.1016/J.APGEOCHEM.2014.12.001

Xie Y, Kocaefe D, Kocaefe Y et al (2016) The effect of novel synthetic methods and parameters control on morphology of nano-alumina particles. Nanoscale Res Lett 11:259. https://doi.org/10.1186/s11671-016-1472-z

Xiong C, Wang W, Tan F et al (2015) Investigation on the efficiency and mechanism of Cd(II) and Pb(II) removal from aqueous solutions using MgO nanoparticles. J Hazard Mater 299:664–674. https://doi.org/10.1016/J.JHAZMAT.2015.08.008

Xu L, Wang J (2017) The application of graphene-based materials for the removal of heavy metals and radionuclides from water and wastewater. Crit Rev Environ Sci Technol 47:1042–1105. https://doi.org/10.1080/10643389.2017.1342514

Yadav DK, Srivastava S (2017) Carbon nanotubes as adsorbent to remove heavy metal ion (Mn+7) in wastewater treatment. Mater Today Proc 4:4089–4094. https://doi.org/10.1016/J.MATPR.2017.02.312

Yalçınkaya Ö, Kalfa OM, Türker AR (2011) Chelating agent free-solid phase extraction (CAF-SPE) of Co(II), Cu(II) and Cd(II) by new nano hybrid material (ZrO_2/B_2O_3). J Hazard Mater 195:332–339. https://doi.org/10.1016/j.jhazmat.2011.08.048

Yan L, Qiaohui F, Wangsuo W (2011) Sorption of Th(IV) on goethite: effects of pH, ionic strength, FA and phosphate. J Radioanal Nucl Chem 289:865–871. https://doi.org/10.1007/s10967-011-1166-z

Yang J, Hou B, Wang J et al (2019) Nanomaterials for the removal of heavy metals from wastewater. Nano 9:424. https://doi.org/10.3390/nano9030424

Yin L, Song S, Wang X et al (2018) Rationally designed core-shell and yolk-shell magnetic titanate nanosheets for efficient U(VI) adsorption performance. Environ Pollut 238:725–738. https://doi.org/10.1016/J.ENVPOL.2018.03.092

Youssef AM, Malhat FM (2014) Selective removal of heavy metals from drinking water using titanium dioxide nanowire. Macromol Symp 337:96–101. https://doi.org/10.1002/masy.201450311
Yu S, Wang X, Ai Y, et al (2016) Spectroscopic and theoretical studies on the counterion effect of Cu(ii) ion and graphene oxide interaction with titanium dioxide. Environ Sci Nano 3:1361–1368. https://doi.org/10.1039/C6EN00297H
Yu G, Lu Y, Guo J et al (2018a) Carbon nanotubes, graphene, and their derivatives for heavy metal removal. Adv Compos Hybrid Mater 1:56–78. https://doi.org/10.1007/s42114-017-0004-3
Yu S, Liu Y, Ai Y et al (2018b) Rational design of carbonaceous nanofiber/Ni-Al layered double hydroxide nanocomposites for high-efficiency removal of heavy metals from aqueous solutions. Environ Pollut 242:1–11. https://doi.org/10.1016/J.ENVPOL.2018.06.031
Zarime NA, Yaacob WZW, Jamil H (2018) Removal of heavy metals using bentonite supported nano-zero valent iron particles. In: AIP Conference Proceedings. AIP Publishing LLC, p 20029
Zendehdel M, Shoshtari-Yeganeh B, Cruciani G (2016) Removal of heavy metals and bacteria from aqueous solution by novel hydroxyapatite/zeolite nanocomposite, preparation, and characterization. J Iran Chem Soc 13:1915–1930. https://doi.org/10.1007/s13738-016-0908-9
Zhan Y, Hu H, He Y et al (2016) Novel amino-functionalized Fe_3O_4/carboxylic multi-walled carbon nanotubes: one-pot synthesis, characterization and removal for Cu(II). Russ J Appl Chem 89:1894–1902. https://doi.org/10.1134/S1070427216110227
Zhang Y, Li Z (2017) Heavy metals removal using hydrogel-supported nanosized hydrous ferric oxide: synthesis, characterization, and mechanism. Sci Total Environ 580:776–786. https://doi.org/10.1016/J.SCITOTENV.2016.12.024
Zhang W, Shi X, Zhang Y et al (2013a) Synthesis of water-soluble magnetic graphene nanocomposites for recyclable removal of heavy metal ions. J Mater Chem A 1:1745–1753. https://doi.org/10.1039/C2TA00294A
Zhang Y, Su Y, Zhou X et al (2013b) A new insight on the core–shell structure of zerovalent iron nanoparticles and its application for Pb(II) sequestration. J Hazard Mater 263:685–693. https://doi.org/10.1016/j.jhazmat.2013.10.031
Zhang Q, Teng J, Zhang Z et al (2015) Unique and outstanding cadmium sequestration by polystyrene-supported nanosized zirconium hydroxides: a case study. RSC Adv 5:55445–55452. https://doi.org/10.1039/C5RA09628F
Zhang Y, Wu B, Xu H et al (2016) Nanomaterials-enabled water and wastewater treatment. Nano Impact 3–4:22–39. https://doi.org/10.1016/j.impact.2016.09.004
Zhang X, Wang Y, Chang X et al (2017) Iron oxide nanoparticles confined in mesoporous silicates for arsenic sequestration: effect of the host pore structure. Environ Sci Nano 4:679–688. https://doi.org/10.1039/C6EN00514D
Zhang C, Li X, Chen Z et al (2018) Synthesis of ordered mesoporous carbonaceous materials and their highly efficient capture of uranium from solutions. Sci China Chem 61:281–293. https://doi.org/10.1007/s11426-017-9132-7
Zhao G, Li J, Ren X et al (2011) Few-layered graphene oxide nanosheets as superior sorbents for heavy metal ion pollution management. Environ Sci Technol 45:10454–10462. https://doi.org/10.1021/es203439v
Zhao D, Chen L, Xu M, et al (2017a) Amino Siloxane Oligomer Modified Graphene Oxide Composite for the Efficient Capture of U(VI) and Eu(III) from Aqueous Solution. ACS Sustain Chem Eng 5:10290–10297. https://doi.org/10.1021/acssuschemeng.7b02316
Zhao D, Zhang Q, Xuan H, et al (2017b) EDTA functionalized Fe3O4/graphene oxide for efficient removal of U(VI) from aqueous solutions. J Colloid Interface Sci 506:300–307. https://doi.org/10.1016/J.JCIS.2017.07.057
Zhao G, Huang X, Tang Z et al (2018) Polymer-based nanocomposites for heavy metal ions removal from aqueous solution: a review. Polym Chem 9:3562–3582. https://doi.org/10.1039/C8PY00484F
Zou Y, Wang X, Ai Y et al (2016) Coagulation behavior of graphene oxide on nanocrystallined Mg/Al layered double hydroxides: batch experimental and theoretical calculation study. Environ Sci Technol 50:3658–3667. https://doi.org/10.1021/acs.est.6b00255

Chapter 13
Application of Nanotechnology in the Bioremediation of Heavy Metals and Wastewater Management

Shanthala Mallikarjunaiah, Mahesh Pattabhiramaiah, and Basavaraja Metikurki

Contents

13.1 Introduction

Environmental pollution is an important issue these days, mounting every day with population growth and rapid industrialization, which is a great challenge posed by pollutants (Chong et al. 2010). The sustainability of agriculture and the civilization

S. Mallikarjunaiah (✉) · M. Pattabhiramaiah
Centre for Applied Genetics, Department of Zoology, Bangalore University, Jnana Bharathi, Bengaluru, Karnataka, India

B. Metikurki
Department of Pharmaceutical Chemistry, Vivekananda College of Pharmacy, Bengaluru, Karnataka, India

D. Thangadurai et al. (eds.), *Nanotechnology for Food, Agriculture, and Environment*, Nanotechnology in the Life Sciences,
https://doi.org/10.1007/978-3-030-31938-0_13

of mankind rely on precious natural resources like land and water which have been subjected to maximum exploitation and severely degraded or polluted due to anthropogenic activities. The problem of environmental pollution, importantly water pollution, has become a global threat to mankind.

Water pollution remains as an enduring environmental problem along with a worldwide increase in human population and economic growth. Water pollution includes industrial effluents, solid discharges, vehicle exhaustion, metals from mining and smelting, use of insecticides/pesticides, disposal of municipal wastes in agriculture, and excessive use of fertilizers (Eduardo and Ines 1988; McGrath 1999). Human health is affected by numerous biological infectious agents (bacteria, viruses, and parasites) that contaminate the water through sewage, human waste, and animal excreta. The chemical pollutants might be either organic (pesticides, fertilizers, oil, detergents, and plastic discharged from domestic, industrial/agricultural waste) or inorganic (metals, acids, salts of domestic and industrial effluents). The health problems caused by potable contaminated water may range from simple toxication and stomach ache to deadly diseases or sudden death.

13.2 Wastewater

Wastewater is a contaminated water due to human exploitation and is a by-product of domestic, commercial, industrial, or agricultural activities, and its distinctiveness varies depending on the source containing physical, chemical, and biological pollutants (Tilley et al. 2016). Wastewater without suitable treatment causes water pollution.

Discharging of wastewater with different kinds of pollutants contaminates the water bodies posing a serious risk to the environment and living organisms (Schwarzenbach et al. 2010). The major wastewater contaminants include inorganic compounds, organic pollutants, and many other complex compounds (Li et al. 2011). Among various types of aquatic pollutants, heavy metals are the most significant ones, because they are very toxic even at very low concentrations and persist in the environment, thereby menacing the environment and biota (Seiler et al. 1988; Yadanaparthi et al. 2009).

13.3 Heavy Metal Pollution

All metals have the potential to parade harmful effects at higher concentrations, and the toxicity of each metal depends on the quantity accessible to organisms, the absorbed dose, the route, and the duration of exposure (Mani and Kumar 2014). Heavy metal pollution is currently a major environmental hazard because of their persistence and nondegradable nature. The toxicity and bioaccumulation tendency of heavy metals in the water is a serious threat to the living organisms. In contrast to organic contaminants, heavy metals cannot be broken down easily by chemical or biological processes. Toxic effects of heavy metals on different organisms have been investigated by several researchers (Chehregani et al. 2004; Mohsenzadeh et al. 2011; Yousefi et al. 2011). The majority of heavy metals are noxious at low concentrations and are proficient in entering the food chain, where they gather and inflict damage to living organisms. Due to the noxious effects of these heavy metals, there is a need to increase awareness in order to remediate the heavy metal-polluted water.

Many heavy metals are essential trace elements acting as micronutrients for humans, animals, and plants in small amounts but in larger amounts cause acute and chronic toxicity which is linked to learning disabilities, cancers, and even death. All the heavy metals at high concentrations have strong toxic effects and are therefore regarded as environmental pollutants (McGrath 1999). A toxic heavy metal with impending toxicity is a major public concern.

Industrial sources, namely, printed board manufacturing, metal finishing, electroplating industry, tanneries, electronics manufacturing industry, coal-fired power plants and mining operation, semiconductor manufacturing, textile dyes, smelting, mining, energy, fuel production, melting operations, and power transmission, are key sources of heavy metals, which are released into the environment as a consequence of human activities. Therefore, it is essential to treat the industrial effluents to prevent heavy metal pollution of water before their discharge (Wang and Chen 2009; Yin et al. 2010; Pang et al. 2011).

In wastewater, the most recurrently occurring toxic heavy metals are cadmium, mercury, lead, and arsenic, whereas less common are chromium, copper, nickel, zinc, manganese, cobalt, selenium, silver, antimony, and thallium. The highly toxic and nondegradable being cadmium, lead, and mercury can be harmful to human health even at very low concentrations. They accumulate in the organisms, which occupy the highest levels in the food pyramid. The toxic environmental pollutants include other heavy metals with their potentially hazardous nature: manganese (central nervous system damage); cobalt and nickel (carcinogens); zinc, copper, selenium, and silver (endocrine disruption, congenital disorders, or general lethal effects in aquatic organisms, plants, and birds); tin as organotin (damage to central nervous system); antimony (an alleged carcinogen); and thallium (central nervous system impairment).

Heavy metals are found naturally on Earth. The occurrence of heavy metals in the environment is due to natural processes like erosion, volcanic eruptions, and the weathering of minerals. According to D'Amore et al. (2005), the geochemical cycle of heavy metals results in the buildup of heavy metals in the environment; this could cause risk to all life forms when they are above-permitted levels. Contamination of the environment with heavy metals has increased beyond the recommended limit and is detrimental to all life forms (Tak et al. 2013; Gaur et al. 2014; Dixit et al. 2015). The determined acceptable concentration of some heavy metals in water is 0.01, 0.01, 0.002, 0.05, 0.015, and 0.05 mg/L for Ar, Cr, Hg, Cd, Pb, and Ag, respectively, as specified by the Comprehensive Environmental Response Compensation and Liability Act (CERCLA), USA (Chaturvedi et al. 2015).

The heavy metals have negatively obstructed the environment, causing alteration and destruction of the ecosystem, including accretion of pollutants in the environment and a loss of biodiversity. Heavy metals become concentrated as a result of anthropogenic activities with high bioavailability due to their soluble and mobile reactive forms. Abandoned mines contaminate water bodies through chemical runoff and particulates that accumulate in water sources (Adler et al. 2007).

The anthropogenic causes comprise of atmospheric deposition, battery production, alloy production, explosive manufacturing, improper stacking of industrial solid waste, biosolids, coating, mining, leather tanning, pesticides, phosphate fertilizer, printing pigments, photographic materials, smelting, steel and electroplating industries, sewage irrigation, textiles, dyes, and wood preservation (Fulekar et al. 2009; Dixit et al. 2015).

Advancement in technology and industrialization has put an escalating burden on the environment by releasing large quantities of perilous waste like heavy metals (cadmium, lead, and chromium), metalloids (elements having intermediate properties between those of typical metals and nonmetals, such as arsenic and antimony), and organic contaminants that have inflicted serious grievance on the ecosystem. The buildup of heavy metals in water continues to create serious global health concerns, as they cannot be degraded into nontoxic forms, but persist in the ecosystem. Speciation of metal and its bioavailability determines the physiological and toxic effects of metal on living organisms (Olaniran et al. 2013).

There are numerous methods to exterminate these heavy metals, including chemical precipitation, filtration, oxidation or reduction, reverse osmosis, ion exchange, evaporation, membrane technology, and electrochemical treatment. But most of these techniques become ineffective when the concentrations of heavy metals are less than 100 mg/L (Ahluwalia and Goyal 2007). Most heavy metal salts are water-soluble and get dissolved in wastewater, which means they cannot be separated by physical separation methods. Physicochemical methods are ineffective and expensive for the very low concentration of heavy metals. Bioaccumulation and biosorption are the biological methods used for thr elimination of heavy metals may be an attractive method and an alternative to physicochemical methods (Kapoor and Viraraghvan 1995).

13.4 Route of Entry

Heavy metal pollution of water sources like lakes, streams, rivers, and groundwater is by leaching from industrial and consumer waste. Acid rain can exacerbate the process of pollution by releasing intent heavy metals of soils. Motor vehicle emissions include arsenic, cobalt, nickel, cadmium, lead, vanadium, zinc, platinum, antimony, palladium, and rhodium, which are the main sources of airborne pollutants (Balasubramanian et al. 2009). Heavy metals enter the food chain via air inhalation and diet and lead to their biomagnifications and bioaccumulate in living organisms as they are hard to metabolize in their body. Heavy metals get incorporated in plants through the uptake of water, and the animals and humans feeding on these plants accumulate heavy metals (Kamal et al. 2010).

13.5 Toxicity of Heavy Metals to Living Organisms

The toxicity of metals is the capability of a metal to cause deleterious effects on organisms, which depends on the bioavailability of heavy metal and their absorbed dose (Rasmussen et al. 2000). The threat posed by heavy metals to the well-being of living organisms is worsened by their persistent nature in the environment. Some heavy metals play significant roles in the physiological, biochemical, and metabolic processes of living organisms. They act as cofactors for some enzymes and micronutrients and regulate osmotic pressure and consistency of molecules, and the majority of them have no known biological function in living organisms and become toxic when generated in excess (Fashola et al. 2016).

Heavy metals bind to vital cellular components, namely, structural proteins, enzymes, and nucleic acids, and interfere with their functioning (Landis et al. 2000). Symptoms and effects of metal or metal compound can vary according to the dose involved. Broadly, long-term exposure to toxic heavy metals can have an effect on nervous and circulatory systems and also shows a carcinogenic effect on humans. For humans, effects associated with exposure to the toxic heavy metals are listed in Table 13.1 (Afal and Weiner 2014).

Microbial population size, diversity, and activity are affected by heavy metal toxicity and also affects the morphology, metabolism, and growth of microorganisms by altering the nucleic acid structure, disrupting the cell membranes, causing a functional disturbance, inhibiting enzyme activity of oxidative phosphorylation and lipid peroxidation, and altering the osmotic balance and protein denaturation (Fashola et al. 2016; Xie et al. 2016).

Toxic heavy metals like lead, cadmium, mercury, chromium, and arsenic have the maximum potential to cause harm on account of their extensive use, toxicity in elemental or combined forms, and widespread distribution in the environment (Baird and Cann 2012). These five elements have a strong affinity for sulfur in the human body, and usually they bind via thiol groups (–SH) to enzymes responsible for controlling the speed of metabolic reactions. The resulting sulfur-metal bonds inhibit the proper functioning of the enzymes, which deteriorates human health and sometimes leads to death. Mercury and lead damage the central nervous system, and cadmium causes degenerative bone disease, whereas chromium (hexavalent form) and arsenic are carcinogens that may induce cancer.

Exposure to lead and mercury can cause the development of autoimmunity, which can result in joint diseases (rheumatoid arthritis), kidney diseases, circulatory and nervous system disorders, and fatal brain damage in humans. In children, exposure to lead and mercury causes reduced intelligence, impaired development, and an increased risk of cardiovascular disease. Cadmium can disrupt the endocrine system, damage fragile bones, and affect the regulation of calcium in biological

Table 13.1 Effect of toxic heavy metals on humans

Element	Acute exposure	Chronic exposure
Lead	Nausea Vomiting Encephalopathy (brain dysfunction)	Anemia Encephalopathy Foot drop/wrist drop (palsy) Nephropathy (kidney disease)
Cadmium	Pneumonitis (lung inflammation)	Osteomalacia (softening of bones) Proteinuria (excess protein in urine; possible kidney damage) Lung cancer
Mercury	Vomiting Diarrhea Fever	Stomatitis (inflammation of gums and mouth) Nausea Nephrotic syndrome (nonspecific kidney disorder) Neurasthenia (neurotic disorder) Parageusia (metallic taste) Pink disease (pain and pink discoloration of hands and feet) Tremor
Arsenic	Nausea Vomiting Diarrhea Encephalopathy Multi-organ effects Arrhythmia Painful neuropathy	Diabetes Hypopigmentation/hyperkeratosis Cancer
Chromium	Gastrointestinal hemorrhage (bleeding) Hemolysis (red blood cell destruction) Acute renal failure	Pulmonary fibrosis (lung scarring) Lung cancer

systems and is known to be a mutagen and carcinogen. Hair loss, headaches, diarrhea, nausea, and vomiting in humans are caused by chromium.

The presence of lead in water may be due to the application of lead and PVC pipes in addition to a spill of sewage from industries such as battery making, metal plating, electrical equipment, chemicals, steel, iron, and copper (Hakim and Philippe 2006). Lead compounds are generally toxic pollutants which have bioaccumulation property in tissues of the human body (Khayat and Sarkar 2013).

Human intestine absorbs lead, which may cause colics, skin pigmentation, and paralysis due to overexposure. Exposure to high levels of Pb (II) could damage the central nervous system and may lead to death (Sanders et al. 2009). Chromium (VI), another toxic heavy metal pollutant, might lead to gastrointestinal disorders; liver, kidney, and lung cancer; cardiovascular shocks; and other health-related problems.

At acidic pH levels, heavy metals tend to form free ionic species, with more protons available to saturate metal binding sites. This means that at higher hydrogen ion concentrations, the adsorbent surface is further positively charged, thus reducing the attraction between an adsorbent and metal cation. Hence, heavy metals become more available, thereby increasing their toxicity to microorganisms and plants. At basic conditions, metal ions replace protons to form other species, such as hydroxo-metal complexes that are soluble as in the case of Cd, Ni, and Zn, while those of Cr and Fe are insoluble. A small change in the pH level can influence the solubility and bioavailability of heavy metals. Owing to large changes on the Earth's resources, environmental sustainability finds ways to reduce the harvesting of non-renewable resources, as well as the effects of the activities associated with them on the Earth's biosphere.

13.6 Environmental Sustainability

Environmental sustainability is nothing but the responsible interaction with the environment to avoid depletion or degradation of natural resources and for enhanced environmental quality. The advancement in science and technology contributes directly or indirectly to the increase in waste and toxic materials in the environment. Environmental sustainability programs include protection and restoration of the natural environment in which one of the restoration strategies used currently is bio-remediation which makes use of microorganisms.

Exploitation of microorganisms and plants for remediation purposes is a possible solution for heavy metal pollution since it includes sustainable remediation technologies. Therefore, it is very important to eliminate or diminish heavy metal contamination in order to prevent/reduce contaminating the environment and the possibility of uptake in the food web. To achieve this, bioremediation is employed in order to increase mental stability (speciation), which in turn reduces the bioavailability of metal (Abbas et al. 2014; Akcil et al. 2015; Ndeddy and Babalola 2016).

13.7 Bioremediation

Remediation is the solution to a problem, and "bioremediation" encompasses the process by which various biological agents, such as bacteria, microalgae, fungi, protists, plants, or their enzymes, are used to degrade the environmental contaminants into less toxic forms. These microbes are indigenous to the contaminated area and are nonpathogenic. The most commonly used microbes are aerobic microbes in view of the fact that they are very effective and easier to isolate and control during the process of biodegradation.

Bioremediation is an environmentally friendly and cost-effective technique that utilizes inherent biological mechanisms of microorganisms and plants or their products, to eradicate hazardous contaminants including heavy metals to restore polluted environments to their original condition (Mani and Kumar 2014; Akcil et al. 2015; Dixit et al. 2015), whereas conventional chemical and physical techniques are more often expensive and unproductive and generate significant amounts of toxic sludge (Kratochvil and Volesky 1998).

If the bioremediation occurs on its own, then it is known as natural attenuation or intrinsic bioremediation, and when it is encouraged to occur with the addition of fertilizers for the improvement of bioavailability within the medium, then it is known as stimulated bioremediation. Most common bioremediation technologies include bioreactor, bioaugmentation, bioventing, bioleaching, composting, biostimulation, land farming, rhizofiltration, and phytoremediation (Li and Li 2011).

In the contaminated site, bioremediation naturally works in one of two ways. In the first case, temperature, nutrients, and amount of oxygen are used to improve the growth of indigenous microorganisms, which might already be living at the contaminated site and feed on pollutants. In the second case, less common, specialized microbes or exogenous microorganisms are added to mortify the contaminants. But in both cases, the microbes die, once the harmful chemicals are cleaned up by the microbes fed on available pollutants as their food.

Biological wastewater treatment by microorganisms has a long history. This worldwide approach has been proven as an effective environmentally friendly strategy. The metabolic diversity of microorganisms ensures a variety of substrates to be consumed. It has been investigated that bacteria like *Pseudomonas aeruginosa*, *Aspergillus niger*, and *Rhodopseudomonas sphaeroides* (Liu et al. 2015) are used in wastewater treatment. These well-known microbes can degrade toxic pollutants in aqueous media during their metabolisms. The slow biodegradation processes and hard-to-recover cells, which are significantly inhibited by substrates, are the limitations of biological wastewater treatment.

13.8 Mechanisms of Bioremediation

Bioremediation utilizes microorganisms for converting the organic pollutants to metabolic intermediate or culmination products such as carbon dioxide and water. These metabolic end or intermediate products are utilized by microorganisms as primary substrates for their growth. For bioremediation, microorganisms such as bacteria, fungi, and algae are the frequently used bioagents (Al-Rub et al. 2004; Abdel Hameed 2006).

Microorganisms show two-way defense to pollutants: (i) production of degradative enzymes for the target pollutants and (ii) resistance to relevant heavy metals. The diverse mechanisms of bioremediation include metal-microbe interactions, biosorption, biomineralization, bioaccumulation, biotransformation, and bioleaching. Removal of heavy metals by microorganisms is performed using chemicals, which are available in wastewater for their growth and development. They are proficient in dissolving metals and reducing/oxidizing transition metals. Different methods are followed by microbes to restore the polluted environment through the process of binding, immobilizing, volatizing, oxidizing, and transforming the heavy metals. For understanding the mechanism of precise growth and activity of microorganisms in the contaminated sites of a specific location, the bioremediation process is conceivable through the approach of a designer microbe.

Microorganisms, specifically the phytoplankton/microalgae, could bind to heavy metal ions utilized in the bioremoval of metallic ions present in the polluted environments (Bitton 2011). They are greatly recognized for their capacity and efficiency in the process of bioaccumulation and biosorption of noxious heavy metals. The phytoplankton has shown a great heavy metal affinity, efficiency, sequestration, and consistently good performance when compared to another form of bioremediation.

Microalgae are known for their competence in the production of several important biological materials either at a viable or nonviable state or when immobilized, and the ease at which algae can be grown has made them dominant useful biomass when treating environmental problems. The ability of microalgae in the uptake of heavy metals from wastewater has been recognized, and this has been the major spotlight on the exploit of microalgae in bioremediation (Abdel Hameed and Ebrahim 2007). Few vital microalgal strains which have demonstrated heavy metal removal ability and efficiency from wastewater include *Botryococcus*, *Chlorella*, *Phormidium*, *Scenedesmus*, *Chlamydomonas*, *Desmodesmus*, and *Spirulina* (Rawat et al. 2011; Kshirsagar 2013).

There are two steps involved in the assimilation of heavy metals. First, the metals are adsorbed over the cell very quickly by physical adsorption. Next, these metals are assimilated slowly into the cytoplasm in a process of chemisorptions. However, absorption of heavy metal depends on the other parameter, viz., pH. As reported by Dwivedi (2012), surface charge studies showed that the accessibility of free sites depends on pH. With increasing pH, the surface-charged sites of calcium alginate became more negative, and then the uptake of metal increased with increasing pH.

The uptake of metal depends on the cellular surface of different microorganisms. Metals are attached to cell surfaces, which displace essential metals from their normal binding sites. Once the metals are bound, microbial cells can transform them from one oxidation state to another, thus reducing their toxicity (Chaturvedi et al. 2015). The cell walls of bacteria are polyelectrolyte, which interacts with metal ions to maintain electroneutrality by mechanisms of extracellular precipitations, redox interactions, covalent bonding, and van der Waals forces (Gavrilescu 2004). The fungi are made up of a rigid cell wall which is composed of chitin, lipids, inorganic ions, polyphosphates, polysaccharide, and proteins. Fungi endure and detoxify metal ions by dynamic uptake of heavy metals into their mycelium and spores by means of extracellular and intracellular precipitation. The surface of their cell wall acts as a ligand for binding metal ions, resulting in the removal of metals (Gupta et al. 2015a, b).

The excreted substances such as organic acids and/or proteins act as the first barrier with a capability to immobilize heavy metals. The second barrier comprises the (unspecific) binding of heavy metals by the cell wall and melanins located in the cell wall. Toxic heavy metals that could not be detained outside the cell must be detoxified inside the cell (Mishra and Malik 2013). The cell wall of brown and red algae is made of cellulose with a sulfonated polysaccharide present in it. Polysaccharides such as alginic acid, glycan, mannan, proteins, and xylans act as other binding sites in algae. The cell wall of cyanobacteria is chiefly made up of peptidoglycan, which produces sheaths and extracellular polymeric substances for sorption. Characteristics of the biomass, chemical and physical properties of the metal of interest, and pH of the solution influence the sorption capacity of algae (Lesmana et al. 2009).

Microorganisms are used extensively in the process of remediation until now, but nowadays the application of nanotechnology and nanoparticles has become a boon to solve several problems of pollution because nanoparticles have more advancement than microorganisms. More effectual and powerful technologies for the treatment of municipal and industrial wastewaters can be achieved by developing an entirely new approach or by refining the prevailing methods (Ferroudj et al. 2013). Among the diverse emerging technologies, the progression in nanotechnology has proved an unbelievable latent for the remediation of wastewater and various other environmental problems (Zare et al. 2013; Sadegh et al. 2014; Gupta et al. 2015a, b).

Nanotechnology first emerged billions of years ago in nature, at the point where molecules began to arrange in composite forms and structures that launched life on Earth. Plants owing to adaptations, mutation, and evolution can convert carbon dioxide and oxygen by means of photosynthesis which occurs in "chloroplasts" encompassing "thylakoid disks" with a green pigment (chlorophyll). Another example of natural nanotechnology is "biochemical catalysts," also called as "enzymes" that are biomolecules catalyzing chemical reactions.

The ability to reduce toxic substances to safe levels efficiently and at a reasonable cost is therefore very important. In this respect, nanotechnology that is the engineering and art of manipulating matter at the nanoscale (1–100 nm) can play a vital role (USEPA 2009). Nanoscience and nanotechnology are the emerging fields of science, and their applications have changed the face of science and technology.

Nanoscience deals with the study of atoms, molecules, and objects having size on the nanometer scale, which are not visible on the macroscale, and is gaining much importance nowadays. Nanotechnology is the exploitation of a matter for the use in particular applications through certain chemical/physical processes to construct materials having a nanosized magnitude in the range of 1–100 nm with the definite properties. "Nano" means dwarf, derived from the Greek word. Owing to the unique active surface area of nanomaterials, it offers a wide range of applications like bioactive nanoparticles, nanosorbents, and metal nanoparticles and catalytic nanomembranes. In the past two decades, nanoscale materials have been used as a substitute for existing treatment materials due to their efficiency, cost-effectiveness, and eco-friendly nature (Dastjerdi and Montazer 2010).

The nanotechnology manages the water problem by solving the technical challenges by removing water pollutants like pathogenic viruses, bacteria, and harmful chemicals like pesticides, insecticides, and others. The occurrence of large quantities of toxic heavy metals such as lead, mercury, cadmium, arsenic, chromium, and others in the environment has major health risks to humans, and this hazard pressurizes the scientists to develop innovative approaches to observe and remove toxic pollutants from wastewaters in economically viable ways. Utilizing nanotechnology for wastewater treatment would certainly help our environment and living organisms (Pandey et al. 2011; Siegert et al. 2019).

The combination of bioremediation and nanotechnology is suggested as an impending efficient and low-cost technique. The use of nanomaterials, as an innovative method to the contaminated site for remediation, has received greater awareness recently (Prasad and Aranda 2018). A wide range of applications of nanoparticles in the field of biomedical, electrical, and environmental engineering fields is due to their unique optical, thermal, electrical, chemical, and physical properties.

The nanoparticle is a small object, behaves as a whole unit with respect to its transport and properties, and is the smallest structure that humans have developed, having the size of a few nanometers. The term "nanoparticle" refers to inorganic materials. The size of nanoparticles varies between 1 and 100 nanometers (nm) with a surrounding interfacial layer as structural component typically consists of ions and inorganic and organic molecules. NPs have greater surface area-to-volume ratios than larger particles. Nanomaterials are developed in a variety of forms such as nanowires, nanotubes, particles, quantum dot films, and colloids (Edelstein and Cammaratra 1998; Lubick and Betts 2008).

In the present scenario, effective treatment of wastewater is a major prerequisite, due to the scarcity of water resources for a growing economy. It is vital to develop and implement complex wastewater treatment technologies with high efficiency and low capital requirement. Recent sophisticated processes in nanomaterial sciences have been attracting the interest of scientists nowadays.

NPs that have been developed for the treatment of wastewater are efficient, eco-friendly, and cost-effective with exclusive functionalities for latent decontamination of industrial effluents, surface water, drinking water, and groundwater (Brumfiel 2003; Theron et al. 2008; Gupta et al. 2015a, b; Prasad and Thirugnanasanbandham 2019). Nanotechnology has been investigated as one of the most advanced processes for wastewater treatment.

The utilization of various classes of nanomaterials for wastewater treatment processes includes four main classes:

(i) Nanoadsorbents like activated carbon, carbon nanotubes, titanium oxide, magnesium oxide, manganese oxide, zinc oxide, graphene, and ferric oxides that are applied for the removal of heavy metals from the wastewater
(ii) Nanocatalysts like electrocatalyst, photocatalyst, Fenton-based catalyst, and chemical oxidant that have shown the potential for removing both organic and inorganic contaminants
(iii) Nanomembranes that are used for the effective removal of dyes, foulants, and heavy metals using carbon nanotube membranes, electrospun nanofibers, and hybrid nanomembranes
(iv) The combination of nanotechnology with biological processes such as anaerobic digestion, algal membrane bioreactor, and microbial fuel cell that is used with respect to its potential for wastewater purification

13.9 Synthesis of Nanoparticles

Various methods are employed for the synthesis of NPs broadly classified into two main classes, i.e., (1) bottom-up approach and (2) top-down approach (Iravani 2011). These approaches were further divided into various subclasses based on the operation, reaction condition, and adopted protocols.

The buildup of material from the bottom (atom-by-atom, molecule-by-molecule, or cluster-by-cluster) is referred to as the bottom-up approach. Here, the miniaturization of material components starts with atoms or molecules with the process of self-assembly, leading to the formation of nanostructures. At the time of self-assembly, the physical forces operating at the nanoscale are used to combine basic units into larger stable structures. This process has the ability to generate a uniform size, shape, and circulation during the formation of NPs from colloidal dispersion. It effectively covers chemical synthesis and precisely controls the reaction to inhibit further particle growth. Even though the bottom-up approach is not new, it plays an important role in the fabrication and processing of nanostructures and nanomaterials.

In the top-down approach, larger (macroscopic) initial structures are used, which can be externally controlled in the processing of nanostructures. By nature, this approach is not cheap and quick to manufacture and also not suitable for large-scale production. Top-down routes are enclosed in the distinctive solid-state processing of the materials. This route is based on the bulk material and makes it smaller, thus breaking up larger particles by the use of physical processes like crushing, milling, or grinding. Typically, this route is not appropriate for formulating consistently shaped materials. The biggest problem of the top-down approach is the limitation of the surface structure, which would have a significant impact on physical properties and surface chemistry of nanostructures and nanomaterials. The conventional top-down technique can cause important crystallographic impairment to the processed patterns.

Many adverse effects have been associated with synthesis methods, leading to the development of eco-friendly alternative biological way of nanoparticle synthesis using microorganisms (Klaus et al. 1999; Konishi and Uruga 2007; Prasad et al. 2016), enzymes (Willner et al. 2006), fungi (Vigneshwaran et al. 2007; Prasad 2016, 2017; Prasad et al. 2018a), and plant extracts or plants (Shankar et al. 2004; Ahmad et al. 2011; Prasad 2014). This method is evolving into an important branch of nanotechnology with many applications (Kyriacou et al. 2004; Kim et al. 2010).

Nanoparticles were produced traditionally only by physical and chemical methods. Due to the high cost of physical and chemical processes, the biosynthesis of nanoparticles is required. In the search of cheaper pathways for nanoparticle synthesis, microorganisms and then plant extracts were used for synthesis. The antioxidant or reducing properties of the microbial enzymes or the plant phytochemicals are usually responsible for the reduction of metal compounds into their respective nanoparticles (Prasad et al. 2018b).

Microorganisms such as bacteria, algae, actinomycetes, and fungi can be used for biosynthesis of nanoparticles (Prasad et al. 2016). Nanoparticle biosynthesis by microorganisms is a green and eco-friendly technology. Different microorganisms, both prokaryotes and eukaryotes, are used for the synthesis of metallic nanoparticles, namely, iron, cadmium, gold, silver, platinum, zirconium, palladium, and metal oxides such as titanium oxide and zinc oxide. The synthesis of nanoparticles may be intracellular or extracellular according to the location of nanoparticles (Mann 2001; Hulakoti and Taranath 2014).

Microorganisms for production of nanoparticles that comprise of both unicellular and multicellular organisms produce inorganic materials either intra- or extracellularly (Shankara et al. 2004). Intracellular synthesis of nanoparticles involves the transport of ions into microbial cells to form nanoparticles in the presence of enzymes. In comparison with the size of extracellularly reduced nanoparticles, the nanoparticles formed inside the organism are smaller. The size limit may be related to the particles nucleating inside the organisms (Narayanan and Sakthivel 2010). Extracellular synthesis of nanoparticles by fungi has more applications as compared to intracellular synthesis. Nanoparticles produced extracellularly by fungi are due to their massive secretory components, which are involved in the lessening and capping of nanoparticles (Narayanan and Sakthivel 2010).

A large number of plants are currently being explored for their role in the synthesis of nanoparticles. Fungi and bacteria require a comparatively longer incubation time for the reduction of metal ions, whereas water-soluble phytochemicals do this in a much lesser time. Hence, plants are better applicants for the synthesis of nanoparticles when compared to bacteria and fungi. With the aid of plant tissue culture techniques and downstream processing procedures, it is possible to synthesize metallic as well as oxide nanoparticles on an industrial scale. It is apparent from congregated information that effect of nanoparticles varies from plant to plant and depends on their mode of application, size, and concentrations (Manzer et al. 2015); further work is needed to explore the mode of action of NPs, their interaction with biomolecules, and their impact on the regulation of gene expressions in plants.

13.10 Classification of Nanoparticles

Nanoparticles (NPs) may be either metallic or nonmetallic and are differently shaped. A variety of metallic and nonmetallic NPs of different shapes and sizes can be used for environmental cleanup, for example, single metal NPs, bimetallic NPs, and carbon-based NPs. The nanoparticles are broadly classified into two types: (1) natural NPs and (2) artificial/engineered NPs. Natural NPs are naturally occurring molecules available in either organic or inorganic forms. Natural organic NPs include natural and functional nanomaterials attributed to biological organizations. The protein/capsid of viruses, silk of mite/spider, natural colloids (milk, blood), horn materials (skin, claws, beaks, feathers, horns, hair), paper, cotton, nacre, corals, and even our own bone matrix are all natural organic nanomaterials.

Natural inorganic nanomaterials crop up through crystal growth in the diverse chemical conditions of the Earth's crust. For example, clays show evidence of complex nanostructures due to the anisotropy of their underlying crystal structure, and volcanic activity can give rise to an opal hydrated amorphous form of silica of nanoscale structure. Natural sources of nanoparticles contain combustion products, forest fires, volcanic ash, ocean spray, and the radioactive decay of radon gas. Natural nanomaterials can be created through weathering processes of metal- or anion-containing rocks, as well as at acid mine drainage sites (Letterman and Mitsch 1978).

Artificial/engineered NPs are intentionally and artificially produced by man. Furthermore, artificial/engineered NPs are broadly classified into two: (i) organic and (ii) inorganic nanoparticles. Organic NPs are carbon-based dendrimers and single-enzyme NPs. Inorganic NPs are metal based and metal oxides.

Carbon-based NPs are nonmetallic nanoparticles. These nanomaterials are composed of carbon, commonly taking the form of a hollow sphere, ellipsoids, or tubes. Ellipsoidal and spherical carbon nanomaterials are referred to as fullerenes, while cylindrical ones are called nanotubes. Two major classes of carbon-based NPs are represented by fullerenes and carbon nanotubes (CNTs). Fullerenes contain nanomaterial that is made of globular hollow cage such as allotropic forms of carbon of commercial interest due to their electrical conductivity, electron affinity, high strength, structure, and versatility (Astefanei and Nunez 2015). These materials possess arranged pentagonal and hexagonal sp^2-hybridized carbon units. Carbon nanotubes can remove pollutants from industrial wastewater due to their electrostatic interactions (Wang et al. 2012).

Dendrimers are nanosized polymers built from branched units. The dendrimer surface has numerous chain ends, which are tailored to achieve precise chemical functions. This property could also be useful for catalysis. Dendrimer-NP composite can be used in water treatment and dye treatment industries due to more reactivity and more surface area, and less toxicity enhances catalytic activity.

Single-enzyme nanoparticles are attracting great interest for their unique properties and potential for application in diverse areas. For the preparation of single-enzyme nanoparticles (SENs), each enzyme molecule is surrounded by a hybrid organic/inorganic polymer network. The artificial technique followed is two orthogonal polymerization steps and its variation. This SEN approach has been successful in stabilizing several enzymes in a vivid way. The enzyme surrounding nanoscale structure is adequately thin and does not enforce a significant mass transfer restraint on the substrate. Since these nanoparticles can be processed into a diverse form, they persist to be soluble or suspended in solutions. SEN approach is promising for numerous applications including biosensors and bioconversion, since they are enzymatically specific, have highly diversified reactions, and have the flexibility in the use of single enzyme nanoparticles.

13.11 Metal-Based Materials

These nanomaterials comprise of quantum dots, nanogold, nanosilver, and metal oxides, such as titanium dioxide. The size of quantum dot is of few nanometers to a few hundred nanometers and is meticulously packed with semiconductor crystal comprised of hundreds or thousands of atoms. Altering the size of quantum dots changes their optical properties.

The uses of nanoparticles are as follows: (i) they can diffuse or penetrate into a contamination zone, wherein microparticles cannot reach, and (ii) they have higher reactivity to redox-amenable contaminants. By applying an external magnetic field to the microbial cells, magnetic NP-coated cells are concentrated at a specific location on the reactor wall, separated from the bulk solution, and recycled for the treatment of the same substrate in a bioreactor with a high biomass concentration.

13.12 Nanobioremediation

Nanobioremediation is the removal of environmental pollutants from contaminated sites using nanoparticles formed by bacteria, algae, actinomycetes, fungi, and plants, by the help of nanotechnology. It is the emerging technologies for the removal of pollutants from the environmental cleanup. Through current advances, bioremediation proposes an environmentally friendly and economically feasible choice to eradicate contaminants from the environment (Singh and Walker 2006; Prasad and Aranda 2018). Nanoparticles can be used for remediation of water contaminated with heavy metals and organic and inorganic pollutants.

It is broadly known that bacteria possess the capacity to degrade organic and transform inorganic contaminants to less hazardous products, either through direct enzymatic processes or mediated by the production of reactive biogenic nanoparticles. The harnessing of this ability for the treatment of "real-world" contaminant issues is a key area of research for the effective application of these techniques in wastewater management.

From the past few years, nanomaterials are being explored in water treatment applications owing to their beneficial properties which include higher surface area, adsorption capacity, enhanced reactivity, and an increased surface/volume ratio (Hristovski et al. 2007). Among these methods, adsorption has attracted the attention of many researchers because it is simple, low cost, and effective for the removal of heavy metal ions in low and medium concentrations.

The mechanism for the removal of hazardous heavy metal ions from wastewater includes physical adsorption, surface complexation, ion exchange (Di Natale et al. 2008), electrostatic interaction, acid-base interaction, redox reactions, photocatalytic transformation, and size exclusion, which can be enhanced or initiated by nanoparticles. Adsorption coefficient and recitation partitioning of a contaminant under equilibrium conditions explain the adsorption of nanoparticles (Hu and Wang,

2010). The additional method is a technique in which the heavy metal ions pass across the cell membrane into the cytoplasm, through the cell metabolic cycle, which is referred to as active uptake or bioaccumulation. Bioaccumulation is a method of a living cell that is dependent on a variety of physical, chemical, and biological mechanisms. These factors include intracellular and extracellular processes, where biosorption plays a limited and well-defined role (Fomina and Gadd 2014). The organism that will hoard heavy metals should have a tolerance to higher concentrations of one or more metals and must display greater transformational abilities, changing toxic chemicals to harmless forms that allow the organism to decrease the toxic effect of the metal and, at the same time, keep the metal contained (Mosa et al. 2016).

Through the studies, adsorption mechanisms were elucidated at the molecular level. Various functional groups like -COOH, -C=O, and -OH were found to be introduced onto nanoparticle surfaces by acid or air oxidation. These functional groups made nanoparticle more hydrophilic and suitable for the adsorption of relatively low molecular weight and polar contaminants (Iijima 1991). Mechanisms of contaminant adsorption from wastewater by modified iron oxide nanoparticles show surface site binding (Li et al. 2002), magnetic selective adsorption (Li et al. 2003), electrostatic interaction (Li et al. 2003), and modified ligand combination (Gao et al. 2008). The adsorption process, tailed by magnetic separation, leads to the quick and inexpensive removal of metal ions.

For the rapid adsorption of heavy metals from wastewater, the most widely studied nanomaterials include activated carbon, carbon nanotubes, graphene, ferric oxides, manganese oxides, titanium oxides, magnesium oxides, and zinc oxides. They are present in diverse forms, such as particles, tubes, and others. Oxide-based nanomaterials formed by metal or metal oxides are the inorganic nanoparticles, which are broadly used for the removal of the hazardous metal ions like mercury, cadmium, lead, arsenic, and chromium. Nanosized metals or metal oxides include ferric oxides, manganese oxides (Feng et al. 2012), titanium oxides (Gao et al. 2009), magnesium oxides (Gupta et al. 2011), and zinc oxides (Tuzen and Soylak 2007) that possess high surface area and specific affinity for the adsorption of pollutants. Metal oxides own negligible ecological impact and low solubility and even do not lead to any secondary pollution and also have been widely adopted as sorbents to remove heavy metals from wastewater.

Iron nanoparticle is considered to be the first nanoparticle to be used in environmental cleanup (Tratnyek and Johnson 2006). Ever since iron is eco-friendly, nanosized ferric oxides can be pumped directly to contaminated sites with negligible risks of secondary contamination (Li et al. 2003) and are a low-cost adsorbent for noxious metal sorption.

Iron oxide nanoparticle has novel properties such as strong adsorption capacity, chemical inertness, high biocompatibility, and superparamagnetism. In addition, nanoparticle amendment to enhance microbial metabolic activity has gained increased attention in recent years due to the unique surface and quantum size effects of NP (Pan et al. 2010). These interesting features allow their applications as microbial immobilization carrier to enhance biocatalytic efficiency. For example,

Xu et al. (2008) had successfully used iron oxide NP as a cell immobilization carrier with minimal mass transfer resistance. Gadhe et al. (2015) took benefit of NP's high electron transfer rate to boost microbial enzyme activity. The inexpensive iron oxide NP has been employed as effective nanosorbents for the removal of a broad range of environmental contaminants such as metal ions (Liu et al. 2008) or dye.

The adsorption efficiency of Ni^{2+}, Cu^{2+}, Cr^{6+}, and Cd^{2+}ions by Fe_3O_4 nanoparticles is strongly dependent on pH, temperature, adsorbent dose, and incubation time. The Fe_3O_4 nanoparticles have been widely utilized for the elimination of heavy metal ions (Ozmen et al. 2010; Wang et al. 2013) from wastewater by means of carboxyl-, amine-, and thiol-functionalized Fe_3O_4 nanoparticles (succinic acid, ethylenediamine, and 2,3-dimercaptosuccinic acid, respectively). These magnetic nanoadsorbents captured the metal ions depending on the surface functionality (COOH, NH_2, or SH), by either forming chelate complexes by ion exchange process or through electrostatic removal mechanism and the hydrogen bond of adsorbate and adsorbent molecule. It is observed that these modified surface-engineered Fe_3O_4 nanoparticles have a strong affinity for the simultaneous adsorption of Cr^{3+}, Co^{2+}, Ni^{2+}, Cu^{2+}, Cd^{2+}, Pb^{2+}, and As^{3+} from wastewater, and 100% removal rate was observed at pH > 8 (Fig. 13.1).

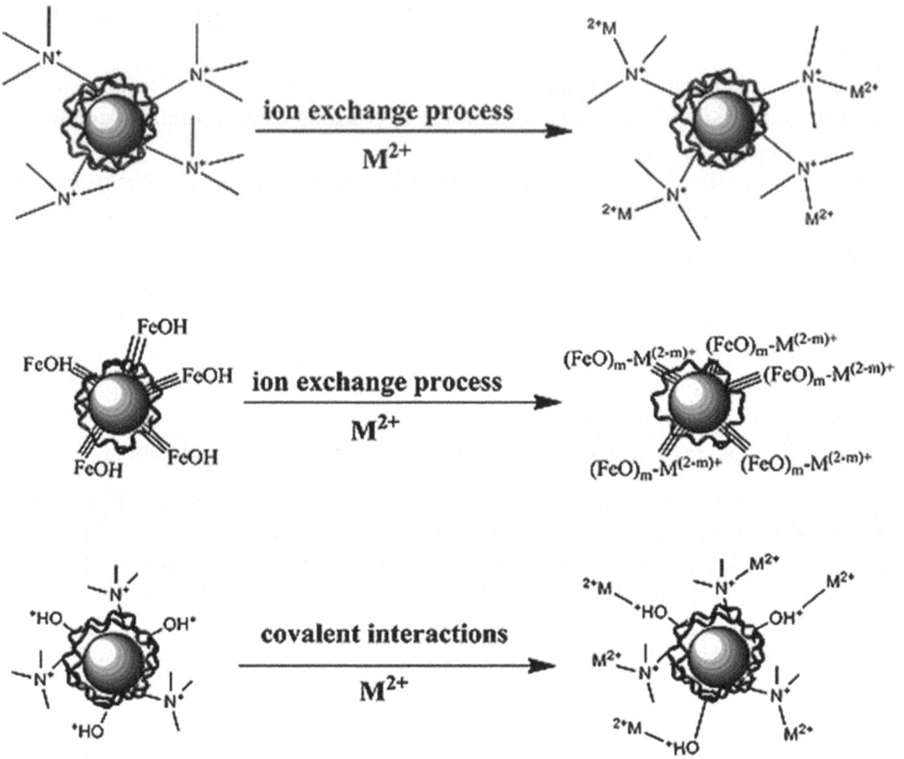

Fig. 13.1 Schematic representations of feasible interactions between metal ions and magnetic nanocomposite

The functional groups on the surface of magnetic nanoparticles provide a higher number of active sites as well as aqueous stability, which is essential for the successful adsorption of metals from wastewater. Furthermore, these particularly modified surface-engineered magnetic nanoparticles are highly effective, efficient, economically viable, and reusable magnetic nanoadsorbents that are used for the removal of heavy metal ions from wastewater.

Rhodamine hydrazide-modified Fe_3O_4, granular ferric hydroxide, zero-valent iron, iron-coated sand, modified iron, and iron oxide-based adsorbents are used for the selective detection and removal of mercury, arsenic, and other toxic heavy metal ions from different environmental samples, such as portable water, lake water, and river water. The synthesized nanocomposites are utilized for the adsorption of Pb (II) (cationic) and Cr (VI) (anionic) metal ions from water.

13.13 Applications of Nanoparticles

Wastewater treatment has been widely investigated with available techniques including precipitation, sedimentation, reverse osmosis, ion exchange, membrane process, electrochemical treatment, and adsorption (Edzwald 2011). Among all the mentioned techniques, the adsorption process has been widely explored because adsorption-based systems are simple to design, easy to operate, and economical and show higher efficiency toward the removal of various toxic pollutants including metals (Faust and Aly 2013).

For high efficient removal of heavy metal ions from wastewater, NPs as adsorbents must satisfy the following criteria:

1. The nanosorbents should be nontoxic.
2. The sorbents should demonstrate high sorption efficiency and selectivity at very low concentration of pollutants.
3. The adsorbed pollutant might be eliminated from the surface of the nanoadsorbent easily.
4. Infinite recycling of the sorbents.
5. The reversible process should be capable of getting back the adsorbent.

The nanoparticles considered in the elimination of heavy metal ions of wastewater should have high adsorption capacity (Savage and Diallo 2005; Cloete 2010). There are a number of techniques for the removal of loaded nanoparticles, i.e., nanoparticle decorated or impregnated on the activated carbon from the water, like filtration, flocculation, coagulation, centrifugation, sedimentation, or magnet deposition, followed by acid treatment, extraction, and combustion, is used to separate the nanoparticle from the water.

Recent advances in nanoscience and nanotechnology have led to the development of a number of eco-friendly nanoparticles for the environmental remediation of various contaminants from wastewater (Zhang 2003). Due to their high specific surface area and reactivity, nanoparticles are considered as a suitable option for fast removal of contaminants from wastewater. Nanostructured metal oxides such as mesoporous alumina (Kim et al. 2004), titanium oxide (Pena et al. 2006), nanocomposites comprised of aluminum oxide nanoparticles (AluNPs), iron oxide (Tang et al. 2011), and hydrous cerium oxide (Li et al. 2012) are mostly used in wastewater management.

The nanoparticles and nanostructure sorbents can be used as an efficient and convenient adsorbent (Sheela and Nayaka 2012; Salmani et al. 2013) due to their unique characteristics, such as a high reaction rate. Environmentally friendly technology such as adsorption of metal ions on iron nanoparticles has been studied as a promising agent for the exclusion of organic pollutants and heavy metal ions from water and wastewater (Mahvi et al. 2011; Xu et al. 2012; Ebrahimi et al. 2013).

Nanoparticles incorporated or deposited on the surface of functioning materials have risk potential since nanoparticles might release and emit to the environment where they can accumulate for long periods of time. Till now, no online monitoring systems exist to provide reliable real-time measurement data on the quality and quantity of nanoparticles present only in trace amounts in water, thus offering a high innovation potential.

13.14 Conclusion

Owing to the complexity involved in the conventional methods for the management of wastewater, the use of microbes has arisen as a time-saver for bioremediation. Furthermore, bioremediation technology has restrictions: a few microorganisms will not be able to disrupt toxic metals into nontoxic metabolites which have inhibitory effects on microbial activity. An advance in nanotechnology is providing new opportunities to develop more low cost-effective and environmentally acceptable water treatment techniques. Nanoparticles have a number of specific physicochemical properties that make them particularly attractive for wastewater purification.

Nanotechnology is improving our everyday lives by enhancing the performance and efficiency by providing a clean environment, safer air and water, and clean renewable energy for a sustainable future. These technologies with bioremediation efficiently remove pollutants by enhancing the activity of microorganisms. Nanotechnology could provide eco-friendly alternatives for environmental management without harming the natural environment and may afford effective solutions for many pollution-related problems like heavy metal contamination, adverse effects of chemical pollutants, oil pollution, and so on.

Nanoparticles obtained from bacteria, fungi, and plants have had actual application in removing some heavy metals from polluted sites through detoxification and bioremediation in a highly polluted environment. Recent researches had indicated that the use of nanomaterials as adsorbents is a very useful and powerful tool for the removal of metal ions, importantly toxic heavy metals, due to their unique structure and specific surface characteristics. These materials are capable of removing toxic heavy metal ions even at a low concentration, i.e., up to parts per billion (ppb) level also, with very high selectivity and adsorption capacity. These unique and specific properties of nanosorbents make them ideal materials for wastewater management. Carbon-based nanomaterials, such as carbon active, carbon nanotubes, graphene, and graphene oxide, and on the other hand oxide-based nanomaterials, such as ferric oxides, manganese oxides, magnesium oxides, titanium oxides, and zinc oxides, are utilized very competently and magnificently in the elimination of toxic metal ions from wastewater, inexpensive iron oxide NP has been employed as effective nanosorbents for the removal of metal ions.

The important technical limitation of nanoengineered water technology is that they are rarely adaptable to mass processes and at present, in many cases, are not competitive with conventional treatment technologies. However, in the coming decades, nanoengineered materials have a great potential for water revolutions, particularly in decentralized treatment systems, point-of-use devices, and deeply degradable pollutants. In the future, the superiority and length of bioremediation will encompass the amendment and adaptation of nanotechnology. The opportunities and potential for innovation, historical track record, and the impact of the possible advantages of nanotechnology lead to the recognition of this area as of increasing importance. With the development of nanotechnology, the exploitation of new efficient adsorption materials is essential and will continue infinitely. The future of nanomaterials in the removal of heavy metal ions in wastewater treatment is comparatively vivid.

References

Abbas SH, Ismail IM, Mostafa TM, Sulaymon AH (2014) Biosorption of heavy metals: a review. J Chem Sci Technol 3:74–102

Abdel Hameed MSA (2006) Continuous removal and recovery of lead by alginate beads, free and alginate-immobilized *Chlorella vulgaris*. Afr J Biotechnol 5(19):1819–1823

Abdel Hameed MSA, Ebrahim OH (2007) Biotechnological potentials uses of immobilized algae. Int J Agri Biol 9(1):183–192

Adler RA, Claassen M, Godfrey L, Turton AR (2007) Water, mining and waste: an historical and economic perspective on conflict management in South Africa. Economic Peace Security J 2:32–41

Afal A, Wiener SW (2014) Metal toxicity. https://www.medscape.org/ (viewed 21 November 2018)

Ahluwalia SS, Goyal D (2007) Microbial and plant derived biomass for removal of heavy metals from wastewater. Bioresour Technol 98:2243–2257

Ahmad N, Sharma S, Singh VN, Shamsi SF, Fatma A, Mehta BR (2011) Biosynthesis of silver nanoparticles from *Desmodium triflorum*: a novel approach towards weed utilization. Biotechnol Res Int 2011:454090. https://doi.org/10.4061/2011/454090

Akcil A, Erust C, Ozdemiroglu S, Fonti V, Beolchini F (2015) A review of approaches and techniques used in aquatic contaminated sediments: metal removal and stabilization by chemical and biotechnological processes. J Clean Prod 86:24–36

Al-Rub FAA, El-Naas, Benyahia MHF, Ashour I (2004) Biosorption of nickel on blank alginate beads free and immobilized algal cells process. Biochemist 39:1767–1773

Astefanei O, Núñez MT (2015) Galceran characterisation and determination of fullerenes: a critical review. Anul Chim Acta 882:1–21

Baird C, Cann M (2012) Environmental chemistry, 5th edn. WH Freeman, New York
Balasubramanian G, Neumann P, Twitchen D, Markham M, Kolesov R, Mizuochi N, Isoya J, Achard J, Beck J, Tissler J, Jacques V, Hemmer PR, Jelezko F, Wrachtrup J (2009) Ultralong spin coherence time in isotopically engineered diamond. Nat Mater 8(5):383–387
Bitton G (2011) Wastewater microbiology, 4th edn. A John Witney and Sons Inc., Hoboken, New Jersey, pp 482–485
Brumfiel G (2003) Nanotechnology: a little knowledge. Nature 424(6946):246–248
Chaturvedi AD, Pal D, Penta S, Kumar A (2015) Ecotoxic heavy metals transformation by bacteria and fungi in aquatic ecosystem. World J Microbiol Biotechnol 31:1595–1603
Chehregani A, Malayeri B, Golmohammadi R (2004) Effect of heavy metals on the developmental stages of ovules and embryonic sac in *Euphorbia macroclada*. Pak J Biol Sci 8:622–625
Chong MN, Jin B, Chow CW, Saint C (2010) Recent developments in photocatalytic water treatment technology: a review. Water Res 44(10):2997–3027
Cloete TE (2010) Nanotechnology in water treatment applications. Horizon Scientific Press, New York
D'Amore J, Al-Abed S, Scheckel K, Ryan J (2005) Methods for speciation of metals in soils. J Environ Qual 34:1707–1745
Dastjerdi R, Montazer M (2010) A review on the application of inorganic nano-structured materials in the modification of textiles: focus on anti-microbial properties. Colloids Surf B Biointerfaces 79(1):5–18
Di Natale F, Erto A, Lancia A, Musmarra D (2008) Experimental and modelling analysis of As(V) ions adsorption on granular activated carbon. Water Res 42:2007–2016
Dixit R, Malaviya D, Pandiyan K, Singh UB, Sahu A, Shukla R, Singh BP, Rai JP, Sharma PK, Lade H (2015) Bioremediation of heavy metals from soil and aquatic environment: an overview of principles and criteria of fundamental processes. Sustainability 7(2):2189–2212. https://doi.org/10.3390/su7022189
Dwivedi S (2012) Bioremediation of heavy metal by algae: current and future perspective. J Adv Lab Res Biol 3(3):195–199
Ebrahimi R, Maleki A, Shahmoradi B, Daraei H, Mahvi AH, Barati AH (2013) Elimination of arsenic contamination from water using chemically modified wheat straw. J Desalin Wat Treat 51(10–12):2306–2316
Edelstein AS, Cammaratra RC (1998) Nanomaterials: synthesis, properties and applications, 2nd edn. CRC Press, New York
Eduardo SB, Ines TA (1988) Heavy metals in rivers and soils of central Chile. Int J Biosci Biochem Bioinfo 2(37):251–255
Edzwald JK (2011) Water quality and treatment: a handbook on drinking water. American Society of Civil Engineers, McGraw-Hill, Las Vegas
Fashola M, Ngole-Jeme V, Babalola O (2016) Heavy metal pollution from goldmines: environmental effects and bacterial strategies for resistance. Int J Environ Res Public Health 13:1047. https://doi.org/10.3390/ijerph13111047
Faust SD, Aly OM (2013) Adsorption processes for water treatment. Elsevier, Amsterdam
Feng M, Cao X, Ma Y, Zhu C (2012) Super paramagnetic high-surface-area Fe_3O_4 nanoparticles as adsorbents for arsenic removal. J Hazard Mater 217:439–446
Ferroudj N, Nzimoto J, Davidson AD, Talbot E, Briot V, Dupuis S (2013) Abramson maghemite nanoparticles and maghemite/silica nanocomposite microspheres as magnetic Fenton catalysts for the removal of water pollutants. App Catal B Environ 136:9–18
Fomina M, Gadd GM (2014) Biosorption: current perspectives on concept, definition and application. Bioresour Technol 160:3–14
Fulekar M, Singh A, Bhaduri AM (2009) Genetic engineering strategies for enhancing phytoremediation of heavy metals. Afr J Biotechnol 8:529–535
Gadhe A, Sonawane SS, Varma MN (2015) Enhanced biohydrogen production from dark fermentation of complex dairy wastewater by sonolysis. Int J Hydrog Energy 40:9942–9951

Gao C, Zhang W, Li H, Lang L, Xu Z (2008) Controllable fabrication of mesoporous MgO with various morphologies and their absorption performance for toxic pollutants in water. Cryst Growth Des 8:3785–3790

Gao Z, Bandosz TJ, Zhao Z, Han M, Qiu J (2009) Investigation of factors affecting adsorption of transition metals on oxidized carbon nanotubes. J Hazard Mater 167(1):357–365

Gaur N, Flora G, Yadav M, Tiwari A (2014) A review with recent advancements on bioremediation-based abolition of heavy metals. Environ Sci Process Impacts 16:180–193

Gavrilescu M (2004) Removal of heavy metals from the environment by biosorption. Eng Life Sci 4:219–232

Gupta A, Singh S, Kundu SS, Jha N (2011) Evaluation of tropical feedstuffs for carbohydrate and protein fractions by CNCP system. Ind J Anim Sci 81(11):1154–1160

Gupta VK, Nayak A, Agarwal S (2015a) Bioadsorbents for remediation of heavy metals: current status and their future prospects. Environ Eng Res 20:1–18

Gupta VK, Tyagi I, Sadegh H, Shahryari-Ghoshekand R, Makhlouf ASH, Maazinejad B (2015b) Nanoparticles as adsorbent; a positive approach for removal of noxious metal ions: a review. Sci Technol Dev 34(3):195–214

Hakim R, Philippe Q (2006) Analytical methods for drinking water: advances in sampling and analysis. John Wiley and Sons Ltd, England

Hristovski K, Baumgardner A, Westerhoff P (2007) Selecting metal oxide nanomaterials for arsenic removal in fixed bed columns: from nanopowders to aggregated nanoparticle media. J Hazard Mater 147:265–274

Hu Z, Wang L (2010) Pan synthesis of monodisperse Fe_3O_4@silica core–shell microspheres and their application for removal of heavy metal ions from water. J Alloys Compd 492:656–661

Hulakoti NI, Taranath TC (2014) Biosynthesis of nanoparticles using microbes: a review. Colloids Surf B Biointerfaces 121:474–483

Iijima S (1991) Helical microtubules of graphitic carbon nature. J Hazard Mater 354:56–58

Iravani S (2011) Green synthesis of metal nanoparticles using plants. Green Chem 13:2638–2650

Kamal S, Prasad R, Varma A (2010) Soil microbial diversity in relation to heavy metals. In: Soil Heavy Metals (eds. Sherameti I, Varma A) Springer-Verlag Berlin Heidelberg, 19:31–64

Kapoor A, Viraraghvan T (1995) Fungal biosorption – an alternative treatment option for heavy metal bearing wastewater: a review. Bioresour Technol 53:195–206

Khayat Z, Sarkar F (2013) Selective removal of lead (II) ion from wastewater using superparamagnetic monodispersed iron oxide (Fe_3O_4) nanoparticles as a effective adsorbent. Int J Nanosci Nanotechnol 9(2):109–114

Kim Y, Kim C, Choi I, Rengaraj S, Yi J (2004) Arsenic removal using mesoporous alumina prepared via a templating method. Environ Sci Technol 38:924–931

Kim BY, Rutka JT, Chan WC (2010) Nanomedicine. N Engl J Med 363(25):2434–2443

Klaus T, Joerger R, Olsson E, Granqvist CG (1999) Silver based crystalline nanoparticles, microbially fabricated. J Proc Natl Acad Sci USA 96:13611–13614

Konishi Y, Uruga T (2007) Bioreductive deposition of platinum nanoparticles on the bacterium *Shewanella* algae. J Biotechnol 128:648–665

Kratochvil D, Volesky B (1998) Advances in the biosorption of heavy metals. Trends Biotechnol 16(7):291–300

Kshirsagar DA (2013) Bioremediation of wastewater by using microalgae: an experimental study. Int J Life Sci Biotech Pharma Res 2(3):339–346

Kyriacou SV, Brownlow WJ, Xu XN (2004) Using nanoparticle optics assay for direct observation of the function of antimicrobial agents in single live bacterial cells. Biochemist 43:140–147

Landis DA, Wratten SD, Gurr GM (2000) Habitat management to conserve natural enemies of arthropod pests in agriculture. Annu Rev Entomol 45:175–201. https://doi.org/10.1146/annurev.ento.45.1.175

Lesmana SO, Febriana N, Soetaredjo FE, Sunarso J, Ismadji S (2009) Studies on potential applications of biomass for the separation of heavy metals from water and wastewater. Biochem Eng J 44:19–41

Letterman R, Mitsch W (1978) Impact of mine drainage on a mountain stream in Pennsylvania. Environ Pollut 17:53–73

Li Y, Li B (2011) Study on fungi-bacteria consortium bioremediation of petroleum contaminated mangrove sediments amended with mixed biosurfactants. Adv Mater Res 183(185):1163–1167

Li YH, Wang S, Wei J, Zhang X, Xu C, Luan Z, Wu D, Wei B (2002) Lead adsorption on carbon nanotubes. Chem Phys Lett 357:263–266

Li YH, Ding J, Luan Z, Di Z, Zhu Y, Xu C, Wu D, Wei B (2003) Competitive adsorption of Pb^{2+}, Cu^{2+} and Cd^{2+} ions from aqueous solutions by multiwalled carbon nanotubes. Carbon 41(14):2787–2792

Li SM, Jia N, Ma MG, Zhang Z, Liu QH, Sun RC (2011) Cellulose-silver nanocomposites: microwave-assisted synthesis, characterization, their thermal stability, and antimicrobial property. Carbohydr Polym 86(2):441–447

Li MH, Oberle DF, Lucas PM (2012) Effects of dietary fiber concentrations supplied by corn bran on feed intake, growth, and feed efficiency of channel catfish. N Am J Aquacult 74(2):148–153

Liu Y, Wang X, Yang F, Yang X (2008) Excellent antimicrobial properties of mesoporous anatase TiO_2 and Ag/TiO_2 composite films. Micropor Mesopor Mater 114(1):431–439

Liu X, Xu W, Pan Y, Du E (2015) Underestimated dissolved organic nitrogen (N) but overestimated total particulate N in wet deposition in China. Sci Total Environ 520(1):300–301. https://doi.org/10.1016/j.scitotenv.2015.03.004

Lubick N, Betts K (2008) Silver socks have cloudy lining | Court bans widely used flame retardant. Environ Sci Technol 42(11):3910–3910

Mahvi AH, Ebrahimi SJA, Mesdaghinia A, Gharibi H, Sowlat MH (2011) Performance evaluation of a continuous bipolar electrocoagulation/electrooxidation–electroflotation (ECEO–EF) reactor designed for simultaneous removal of ammonia and phosphate from wastewater effluent. J Hazard Mater 192(3):1267–1274

Mani D, Kumar C (2014) Biotechnological advances in bioremediation of heavy metals contaminated ecosystems: an overview with special reference to phytoremediation. Int J Environ Sci Technol 11:843–872

Mann S (2001) Biomineralization, principles and concepts in bioinorganic materials chemistry. Oxford University Press, Oxford

Manzer H, Mohamed HS, Whaibi A, Firoz M, Mutahhar Y, Khaishany A (2015) In: Siddiqui MH (ed) Nanotechnology in plant science. Springer Int Publication, Switzerland

McGrath SP (1999) Adverse effects of cadmium on soil microflora and fauna. In: McLaughlin MJ, Singh BR (eds) Cadmium in soils and plants. Kluwer Academic, Dordrecht, pp 85–95

Mishra A, Malik A (2013) Recent advances in microbial metal bioaccumulation. Crit Rev Environ Sci Technol 43:1162–1222

Mohsenzadeh F, Chehregani A, Yousefi N (2011) Effect of the heavy metals on developmental stages of ovule, pollen, and root proteins in *Reseda lutea* L. (Resedaceae). Biol Trace Elem Res 140(3):368–376

Mosa KA, Saadoun I, Kumar K, Helmy M, Dhankher OP (2016) Potential biotechnological strategies for the cleanup of heavy metals and metalloids. Front Plant Sci 7:1–14

Narayanan KB, Sakthivel N (2010) Biological synthesis of metal nanoparticles by microbes. Adv Colloid Interf Sci 156:1–13

Ndeddy Aka RJ, Babalola OO (2016) Effect of bacterial inoculation of strains of *Pseudomonas aeruginosa*, *Alcaligenes feacalis* and *Bacillus subtilis* on germination, growth and heavy metal (Cd, Cr, and Ni) uptake of *Brassica juncea*. Int J Phytoremediation 18:200–209

Olaniran AO, Balgobind A, Pillay B (2013) Bioavailability of heavy metals in soil: impact on microbial biodegradation of organic compounds and possible improvement strategies. Int J Mol Sci 14:10197–10228

Ozmen M, Can K, Arslan G, Tor A, Cengeloglu Y, Ersoz M (2010) Adsorption of Cu (II) from aqueous solution by using modified Fe_3O_4 magnetic nanoparticles. Desalination 254(1):162–169

Pan B, Qiu H, Pan B, Nie G, Xiao L, Lv L (2010) Highly efficient removal of heavy metals by polymer-supported nanosized hydrated Fe (III) oxides: behavior and XPS study. Water Res 44(3):815–824

Pandey S, Saha P, Biswas S, Maiti TK (2011) Characterization of two heavy metal resistant strains isolated from slag disposal site at Burnpur. Ind J Environ Biol 32:773–779

Pang S, Hernandez Y, Feng X, Mullen K (2011) Graphene as transparent electrode material for organic electronics. Adv Mater 23(25):2779–2795

Pena M, Meng V, Korfiatis GP, Jing C (2006) Adsorption mechanism of arsenic on nanocrystalline titanium dioxide. Environ Sci Technol 40(4):1257–1262

Prasad R (2014) Synthesis of silver nanoparticles in photosynthetic plants. Journal of Nanoparticles, Article ID 963961, http://dx.doi.org/10.1155/2014/963961

Prasad R (2016) Advances and Applications through Fungal Nanobiotechnology. Springer, International Publishing Switzerland (ISBN: 978-3-319-42989-2)

Prasad R, Pandey R, Barman I (2016) Engineering tailored nanoparticles with microbes: quo vadis. WIREs Nanomed Nanobiotechnol 8:316–330. doi: https://doi.org/10.1002/wnan.1363

Prasad R (2017) Fungal Nanotechnology: Applications in Agriculture, Industry, and Medicine. Springer Nature Singapore Pte Ltd. (ISBN 978-3-319-68423-9)

Prasad R, Aranda E (2018) Approaches in Bioremediation. Springer International Publishing https://www.springer.com/de/book/9783030002368

Prasad R, Kumar V, Kumar M, Wang S (2018a) Fungal Nanobionics: Principles and Applications. Springer Nature Singapore Pte Ltd. (ISBN 978-981-10-8666-3) https://www.springer.com/gb/book/9789811086656

Prasad R, Jha A, Prasad K (2018b) Exploring the Realms of Nature for Nanosynthesis. Springer International Publishing (ISBN 978-3-319-99570-0) https://www.springer.com/978-3-319-99570-0

Prasad R, Thirugnanasanbandham K (2019) Advances Research on Nanotechnology for Water Technology. Springer International Publishing https://www.springer.com/us/book/9783030023805

Rasmussen LD, Sørensen SJ, Turner RR, Barkay T (2000) Application of a mer-lux biosensor for estimating bioavailable mercury in soil. Soil Biol Biochem 32:639–646

Rawat I, Kumar RR, Mutanda T, Bux F (2011) Dual role of microalgae: phycoremediation of domestic wastewater and biomass production for sustainable biofuels production. Appl Energy 88(10):3411–3424

Sadegh H, Shahryari-Ghoshekandi R, Kazemi M (2014) Study in synthesis and characterization of carbon nanotubes decorated by magnetic iron oxide nanoparticles. Int Nano Lett 4:129–135

Salmani MH, Ehrampoush MH, Aboueian-Jahromi M, Askarishahi M (2013) Comparison between Ag (I) and Ni (II) removal from synthetic nuclear power plant coolant water by iron oxide nanoparticles. J Environ Health Sci Eng 1:11–21. https://doi.org/10.1186/2052-336X-11-21

Sanders T, Liu Y, Buchner V, Tchounwou PB (2009) Neurotoxic effects and biomarkers of lead exposure: a review. Rev Environ Health 24:15–45

Savage N, Diallo MS (2005) Nanomaterials and water purification: opportunities and challenges. J Nanopart Res 7(4–5):331–342

Schwarzenbach RP, Egli T, Hofstetter TB, Gunten UV, Wehrli B (2010) Global water pollution and human health. Annu Rev Environ Resour 35:109–136

Seiler HG, Sigel H, Sigel A (1988) Handbook on toxicity of inorganic compounds. Marcel Dekker Inc., New York

Shankar SS, Ahmed A, Akkamwar B, Sastry M, Rai A, Singh A (2004) Biological synthesis of triangular gold nanoprism. Nat Mater 3(7):482–488

Sheela T, Nayaka YA (2012) Kinetics and thermodynamics of cadmium and lead ions adsorption on NiO nanoparticles. Chem Eng J 91:123–131

Siegert M, Sonawane JM, Ezugwu CI, Prasad R (2019) Economic assessment of nanomaterials in bio-electrical water treatment. In: Advanced Research in Nanosciences for Water Technology (eds. Prasad R, Thirugnanasanbandham K), Springer International Publishing AG 5–23

Singh B, Walker A (2006) Microbial degradation of organophosphorus compounds. FEMS Microbiol Rev 30:428–471

Tak HI, Ahmad F, Babalola OO (2013) Advances in the application of plant growth-promoting rhizobacteria in phytoremediation of heavy metals. Rev Environ Contam Toxicol 223:33–52. https://doi.org/10.1007/978-1-4614-5577-6_2

Tang W, Li Q, Gao S, Shang JK (2011) Arsenic (III, V) removal from aqueous solution by ultrafine α-Fe_2O_3 nanoparticles synthesized from solvent thermal method. J Hazard Mater 192(1):131–138

Theron J, Walker JA, Cloete TE (2008) Nanotechnology and water treatment: applications and emerging opportunities. Crit Rev Microbiol 34:43–69

Tilley E, Ulrich L, Lüthi C, Reymond PO, Zurbrügg C (2016) Compendium of sanitation systems and technologies, 2nd edn. Swiss Federal Institute of Aquatic Science and Technology, Duebendorf, p 175

Tratnyek PG, Johnson RL (2006) Nanotechnologies for environmental cleanup. Nano Today 1(2):44–48

Tuzen M, Soylak M (2007) Multiwalled carbon nanotubes for speciation of chromium in environmental samples. J Hazard Mater 147:219–225

USEPA (2009) Drinking water contaminants. United States Environmental Protection Agency (EPA), Washington, DC

Vigneshwaran N, Ashtaputre NM, Varadarajan PV, Nachane RP, Paralikar KM, Balasubramanya RH (2007) Biological synthesis of silver nanoparticles using the fungus *Aspergillus flavus*. Mater Lett 61(6):1413–1418. https://doi.org/10.1016/j.matlet.2006.07.042

Wang J, Chen C (2009) Biosorbents for heavy metals removal and their future. Biotechnol Adv 27:195–226

Wang S, Wei C, Wang W, Li Q, Zhengping H (2012) Synergistic and competitive adsorption of organic dyes on multiwalled carbon nanotubes. Chem Eng J 197:34–40

Wang S, Sun H, Ang HM, Tadé MO (2013) Adsorptive remediation of environmental pollutants using novel graphene-based nanomaterials. Chem Eng J 226:336–347

Willner I, Baron R, Willner B (2006) Growing metal nanoparticles by enzymes. J Adv Mater 18:1109–1120

Xie FJ, Zhu Amombo E, Lou Y, Chen L, Fu J (2016) Effect of heavy metals pollution on soil microbial diversity and Bermuda grass genetic variation. Front Plant Sci 7:775. https://doi.org/10.3389/fpls.2016.00755

Xu D, Tan X, Chen C, Wang X (2008) Removal of Pb (II) from aqueous solution by oxidized multiwalled carbon nanotubes. J Hazard Mater 154:407–416

Xu P, Zeng GM, Huang DL, Feng CL, Hu S, Zhao MH, Lai C, Wei Z, Huang C, Xie GX, Liu ZF (2012) Use of iron oxide nanomaterials in wastewater treatment: a review. Sci Total Environ 424:1–10

Yadanaparthi SK, Graybill R, von Wandruszka R (2009) Adsorbents for the removal of arsenic, cadmium, and lead from contaminated waters. J Hazard Mater 171:1–15

Yin P, Xu Q, Qu R, Zhao G, Sun Y (2010) Adsorption of transition metal ions from aqueous solutions onto a novel silica gel matrix inorganic–organic composite material. J Hazard Mater 173:710–716

Yousefi N, Chehregani A, Malayeri B, Lorestani B, Cheraghi M (2011) Investigating the effect of heavy metals on developmental stages of anther and pollen in *Chenopodium botrys* L. (Chenopodiaceae). Biol Trace Elem Res 140:368–376

Zare K, Najafi FH, Sadegh (2013) Studies of *ab initio* and *Monte Carlo* simulation on interaction of fluorouracil anticancer drug with carbon nanotube. J Nanostruct Chem 3:1–8

Zhang WX (2003) Nanoscale iron particles for environmental remediation: an overview. J Nanopart Res 5:323–332

Chapter 14
Biogenic Nanoparticles for Degradation of Noxious Dyes

Abhishek Mundaragi, Prashantkumar Chakra, Om Prakash, Ravichandra Hospet, Devarajan Thangadurai, Jeyabalan Sangeetha, and Shivanand Bhat

Contents

A. Mundaragi (✉) · P. Chakra
Department of Microbiology, Davangere University, Davangere, Karnataka, India

O. Prakash
Department of Fruits and Vegetables Technology, CSIR-Central Food Technological Research Institute, Mysore, Karnataka, India

R. Hospet · D. Thangadurai
Department of Botany, Karnatak University, Dharwad, Karnataka, India

J. Sangeetha
Department of Environmental Science, Central University of Kerala, Kasaragod, Kerala, India

S. Bhat
Department of Botany, Government Arts and Science College, Karwar, Uttar Kannada, Karnataka, India

D. Thangadurai et al. (eds.), *Nanotechnology for Food, Agriculture, and Environment*, Nanotechnology in the Life Sciences,
https://doi.org/10.1007/978-3-030-31938-0_14

14.1 Introduction

The industrial dyes are considered to be one of the largest groups of organic compounds used extensively. Despite the significant role played by natural and synthetic dye products in making our world spectacular, its abandoned discharge during the last decade is a substantial cause of the non-aesthetic pollution that leads to the destruction of the ecosystem (Meenakumari and Philip 2015). Dyes can be of many different structural varieties like azo, basic, acidic, anthraquinone and metal complex (Ratna and Padhi 2012). These dyes are the most important class of synthetic organic compounds released by many industries such as plastic, paper, food, tanneries, pharmaceutical, cosmetic and textile industries (Bogireddy et al. 2016). Among the mentioned industries, the textile industry produces most of the wastewater with high concentration of dyes in the range of 10–200 mg L^{-1}. Estimates indicated that 15–20% of consumer paint in this industry enters into the sewage system. The colour content in dye adsorbs at a wavelength of 350–700 nm (visible light region) and reflects sunlight inflowing the polluted water, which prevents penetration of light into the depths, thereby hindering photosynthesis and interfering with the development of aquatic species (Nasrollahzadeh et al. 2018). These organic pollutants may induce skin irritation, blood disorder, liver and kidney damage, mutagenic and carcinogenic effects as well as poisoning of the central nervous system in humans and animals. Hence, the removal of such carcinogenic compounds from water is biologically and environmentally relevant. Degradation of these compounds to non-toxic products is severe because of their high solubility in water and other solvents and high stability (Bogireddy et al. 2016). Conventional water treatment techniques include activated carbon sorption, flocculation, ultrafiltration, chemical, photochemical, electrochemical and biodegradation methods for removing toxic dyes. These dyes are chemically stable, and the dye pollutants are usually resistant to degradation by various physicochemical means. Even though biodegradation methods are cost-effective, they are inherently slow and are not adequate for dye degradation as these are toxic to microorganisms (Joseph and Mathew 2015; Vinothkannan et al. 2015). And also these methods, perhaps effective, could cause the generation of new compounds which require further treatments.

In recent years, nanocatalysis has emerged as an alternative to conventional wastewater treatment methods for proficient degradation and adsorption of toxic dyes. Nanobiotechnology represents the intersection of nanotechnology and biotechnology, which is an emerging field dedicated to the creation, improvement and utility of nanoscale structures for advanced biotechnology (Goodsell 2004). Nanotechnology remediation employs reactive nanomaterials for the transformation and detoxification of chemicals either through chemical reduction or catalytic process. The finite sizes, large surface-area-to-volume ratio and size-dependent reactivity have made metal nanoparticles an efficient catalyst. The degradation of dyes using chemical reduction is thermodynamically favourable, but it is kinetically not. The nanoscale materials like gold, silver and copper provide an alternative path for the reaction to proceed by reducing the activation energy, and hence the kinetic barrier thus makes the response thermodynamically and kinetically feasible (Meenakumari and Philip 2015). In the past, the nanomaterials were synthesized using various physical and chemical methods. The conventional methods for nanoparticle production involve the use of capping agents, toxic solvents, harsh chemicals, and other additives, limiting their use in biomedical and clinical fields. Most of these methods are energy and capital intensive. To overcome these major negatives, most of the researchers focused on developing a greener protocol for chemical synthesis that avoids the use of solvents and toxic reagents, avoids waste by-products and utilizes available renewable, cheap and biodegradable resources which is a need of the chemical industry (Kulkarni and Bhange 2014) (Fig. 14.1).

The biological approaches assist in eliminating harsh processing conditions by allowing the synthesis at physiological pH and temperature and at relatively low cost. These biological methods do not generate hazardous waste, and the products usually do not need purification. The biological methods use the ability of natural reducing agents present in prokaryotes and plant extracts for the reduction of metal ions (Ahluwalia et al. 2016). Green synthesis imparts steric stabilization of nanoparticles against aggregation and helps overcome the concerns related to the use of sodium borohydride as a reducing agent in conventional production reported so far which is corrosive and flammable (Hoag et al. 2009). Several efforts have been devoted towards biosynthesis of metal nanoparticles using bacteria, fungi, actinomycetes, yeast, viruses, exudates from arthropods and plant extracts for the green synthesis of nanoparticles (Kulkarni and Bhange 2014; Prasad et al. 2016, 2018). In

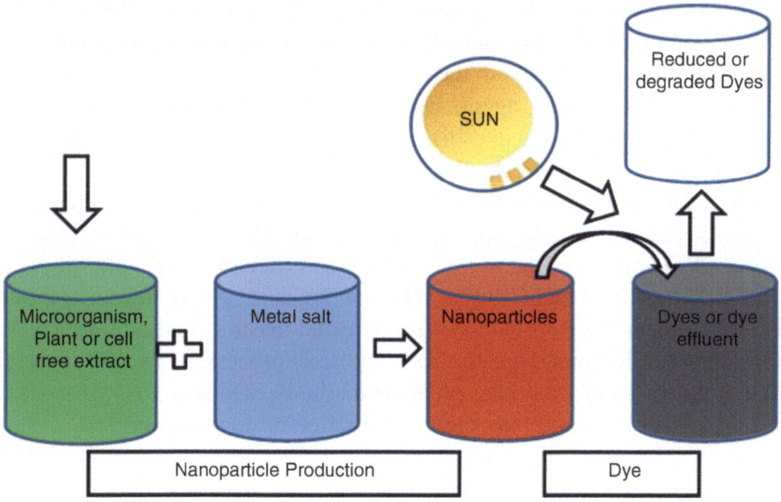

Fig. 14.1 Schematic diagram of nanoparticle production and dye reduction

bacteria, proteins and other biomolecules have the capability of controlling inorganic crystal growth during biomineralization processes. In recent years, some bacterial species such as *Pseudomonas aeruginosa*, *Rhizopus oryzae*, *Zoogloea ramigera* and many more have been explored for the fabrication of metallic nanoparticles (Srivastava and Mukhopadhyay 2014). Recently, eco-friendly synthesis of silver, selenium, titanium dioxide and gold (metal) nanoparticles using various bacterial strains such as *Bacillus* sp., *B. amyloliquefaciens*, *B. clausii* and *Azoarcus* species has been reported (Nadaf and Kanase 2016).

Compared with other microorganisms, fungi are considered as a better resource for industrially producing AuNPs due to their several superiorities, such as high yields, short synthesis period and high ion concentration tolerance (Dhillon et al. 2012; Prasad 2016, 2017, 2018a). Mishra et al. (2014) showed that the AuNPs synthesized by *Trichoderma viride* could serve as an efficient catalyst for complete reduction of 4-nitrophenol in water. Meanwhile, Narayanan et al. (2015) found that the AuNPs intracellularly synthesized by *Flammulina velutipes* could be used as a heterogeneous catalyst in the reduction of organic pollutants, including methylene blue and 4-nitrophenol.

Leaf and fruit extracts of various plants have already been reported to have the impressive ability for reducing silver ions into Ag metallic nanoparticles (Prasad 2014). In a previous study, silver nanoparticles were prepared from silver salt using infusion of potato (*Solanum tuberosum*) tuber that is readily available in common market and cheaper than other vegetables (Roy et al. 2015b). The active chemical constituents such as glucose, fructose, polyphenols, tannin, gallic acid, amino acid, citric acid and alkaloids are responsible for the formation of stable metal nanoparticles (Edison et al. 2016a). The nanoparticles synthesized using plant extracts display greater stability over a prolonged period and do not require the addition of the stabilizing agents, probably due to the presence of integral components which act as both capping and stabilizing agent. Bogireddy et al. (2016) report the reduction/degradation of various organic dyes like phenol red, methyl orange, methylene blue, 4-nitrophenol and DB24 using AgNPs as a catalyst. In another study, green synthesis of AuNPs has been reported using a variety of polysaccharides, including *Acacia nilotica* leaf extract, xanthan gum and gellan gum. The green-synthesized AuNPs using *Salmalia malabarica* gum were proven as efficient catalysts with enhanced rates of reduction of MB and CR dyes (Ganapuram et al. 2015). Thus, the biological synthesis of nanoparticles provides a non-toxic and feasible method for the degradation of aromatic pollutants. Hence, these synthesis processes have full potential to substitute both chemical and other physical methods used in industry for the large-scale production of NPs and thus for dye reduction or degradation.

14.2 Dye Classification

Dyes are the proficient organic compounds of natural or synthetic origin, which are known to absorb visible light radiation from the range 400–700 nm (Hunger 2003). They possess two different groups: the chromophoric group

(–CH=CH–, –N=N–, $-NO_2$), which are compounds that readily absorb the electromagnetic radiations in visible range, and the auxochromic group, which constitutes –OH, –OR and $-NH_2$ groups and is accountable for the interaction of dyes with materials that are to be dyed (Hunger 2003; Wawrzkiewicz 2012).

To meet the demands of textile industries and new technology, new dyes are continuously being developed. Characterization of dyes can be done in terms of structure, colour and application methods, and the most suitable, accepted classification is based on application (Gupta 2009). Table 14.1 highlights the classification of dyes based on various applications. Nevertheless, dyes can also be classified based on their chemical moiety; Table 14.2 lists the major categories of dyes distinguished based on chemical structure.

14.3 Dye Degradation Methods

In recent years, revolution in industries has relentlessly impacted the environment, especially textile industries which use a wide range of dyes to shade their products and as a result generate wastewater containing strong synthetic organic compounds (Mohan et al. 2007). Effluent from textile industries needs the most regular treatment due to partial fixation of synthetic dyes during the dyeing process. It was projected that around 10–15% of the 700,000 tons of synthetic textile dyes used annually were lost to effluents (Khataee et al. 2009). Wastewater from textile industries is very complex and excessive, and the effluent is characterized by high alkalinity and high biological and chemical oxidation demand and by total dissolved solids with dye concentrations of below 1 gm/dm^3 (Kaushik and Malik 2009). If the effluents produced from these industries are not properly treated and if incompletely treated effluent mixes with other pure water sources, even just 1.0 mg/L of dye concentration could affect the colour of drinking water, hence making it unsuitable as potable water (Malik et al. 2007).

The majority of the synthetic organic dyes have serious implication on human health and aquatic life due to their toxic, mutagenic and carcinogenic potentials (Fanchiang and Tseng 2009). If consumed by human beings or animals, they are unsafe and may cause irritation to the skin and eyes and systemic effects including cyanosis, blood changes and gastrointestinal and respiratory tract disturbances (Merouani et al. 2010). Li et al. (2010) reported that functioning of human serum albumin is inhibited by reactive brilliant red by means of causing a conformational change or even precipitation via binding to it. Hence, decolourization of dyes from industry effluents becomes essentially important. Several methods have been proposed until now for efficient dye degradation. The advantages and disadvantages involved in different methods are indicated in Table 14.3.

Table 14.1 Classification of dye based on applications (Forgacs et al. 2004; Le Coz 2005; Hernández-Montoya et al. 2013; Dawood and Sen 2014)

Types of dyes	Chemical characteristics	Examples	Configuration	Substrate
Acid	Water-soluble organic salts of carboxylic and sulphonic acids	Congo red Methyl orange Methyl red Orange I Acid blue Acid black Acid violet	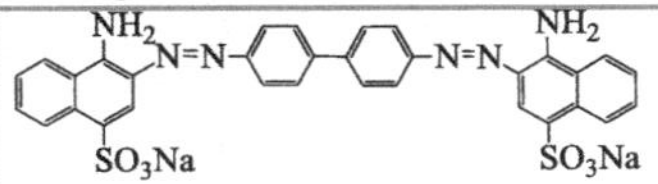 Congo red	Protein fibres, polyamide fibres, paper, leather, cosmetics, food
Basic	Water-dissolvable salts of natural bases	Methylene blue Basic red Basic brown Basic blue 41 Crystal violet Aniline yellow Brilliant green	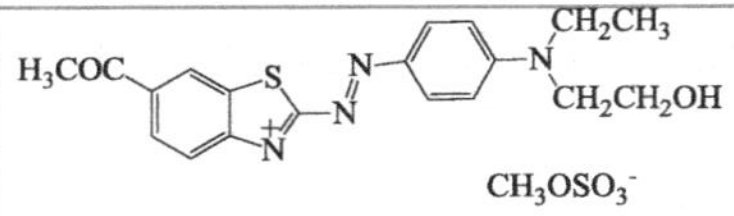 Basic blue 41	Wool, natural silk, polyester and polyacrylonitrile fibres
Direct	Salts of natural sulphonic acids dissolvable in water	Martius yellow Direct black Direct orange 26 Direct blue Direct violet Direct red	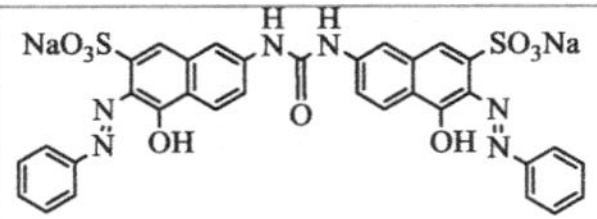 Direct blue	Cellulose fibres like cotton, linen, leather
Disperse	Sparingly dissolved in water or insoluble	Disperse blue Disperse red 60 Disperse orange Disperse yellow Disperse brown	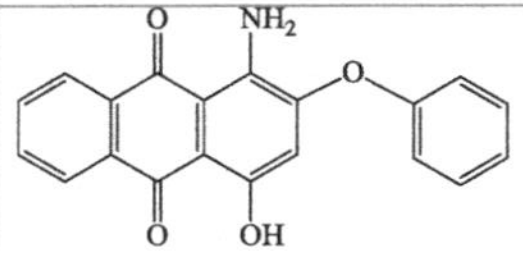 Disperse red 60	Synthetic mainly polyester, polyamide, polyacrylonitrile, acetic

Types of dyes	Chemical characteristics	Examples	Configuration	Substrate
Reactive	Salts of natural bases and acids dissolvable in water	Reactive red Reactive blue Reactive yellow Reactive black 5 Remazol blue Remazol yellow	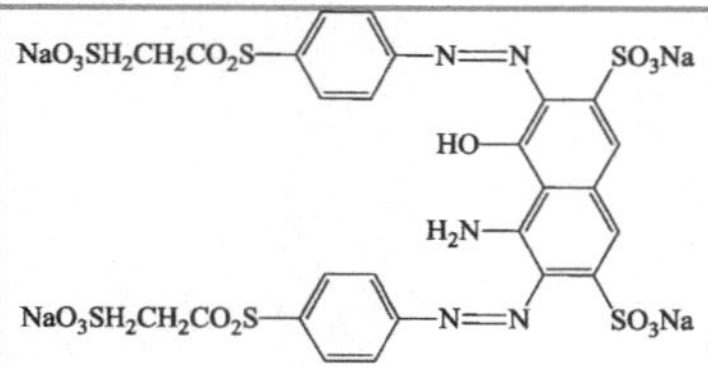 Reactive black 5	Cellulose and protein fibres
Sulphur	Water insoluble	Indophenol, sulphur black, sulphur violet 5	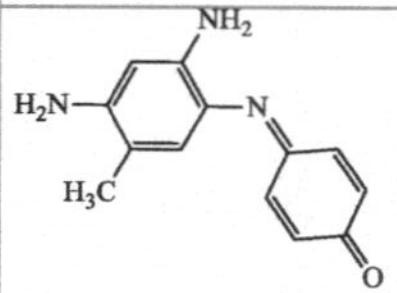 Sulphur violet 5	Cellulose fibres
Vat	Unsolvable in water	Indigo Benzanthrone Vat blue Vat green 1	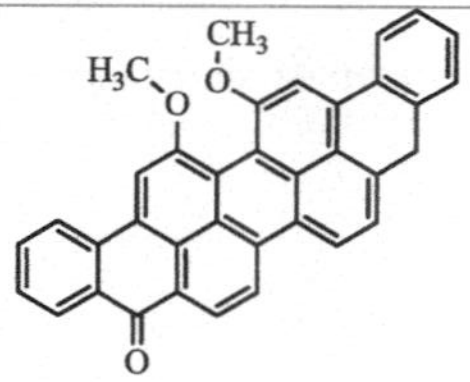 Vat green 1	Cellulose fibres

Table 14.2 Classification of dyes based on chemical structure

Dye classes	Characteristics	Example
Azo dyes	The largest and most important class of dyes. They are categorized by the presence of one or more azo groups (–N=N–), which form connecting link between two or more aromatic rings	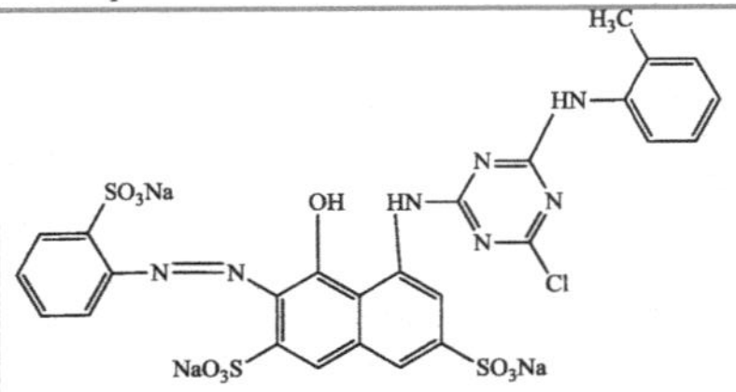 Reactive red 198
Anthraquinone dyes	Two other benzene rings are merged with a *p*-quinoid group	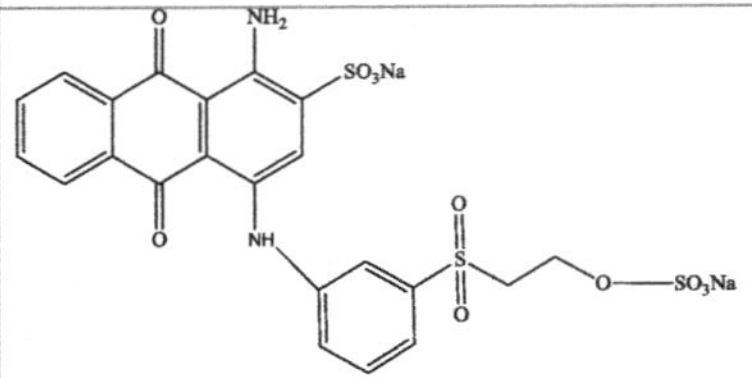 Remazol brilliant blue R
Indigoid dyes	Indigoid is the parent complex of indigoid dyes	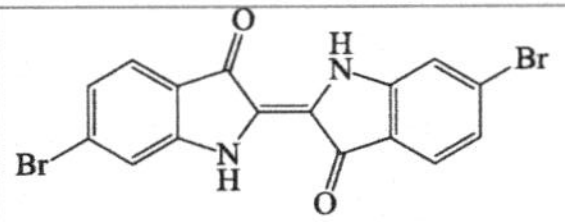 Indigo

Dye classes	Characteristics	Example
Nitro dyes	At least one nitro group, ortho or para to the hydroxyl group of polynitro derivatives of phenols	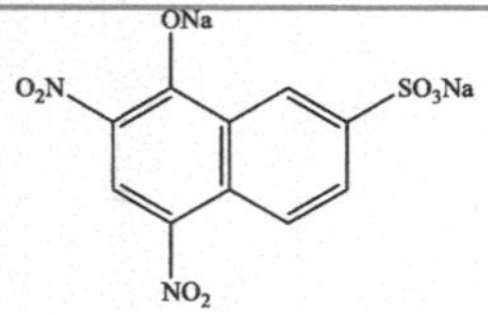 Napthol yellow S
Triarylmethane dyes	Triphenylmethane dyes are not fast to light or washing; central carbon atom is coupled to two benzene rings and to a *p*-quinoid group	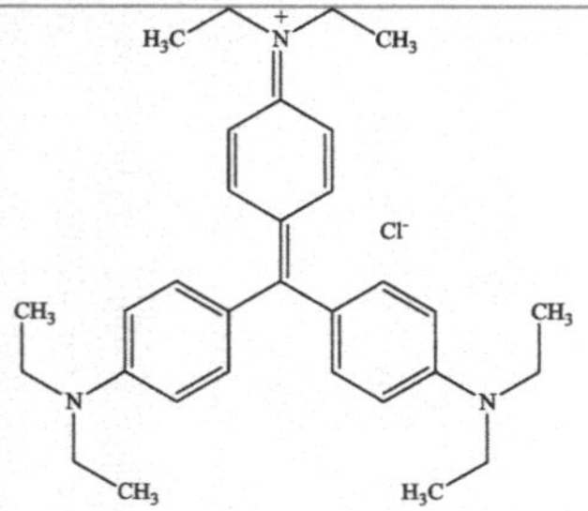 Ethyl violet

Table 14.3 Advantages and disadvantages of methods employed in dye degradation

Different methods of degradation	Advantages	Disadvantages
Adsorption	Adsorption capacity is very high for dyes	High cost of adsorbents, need to dispose of adsorbents, low surface area for some adsorbents
Ion exchange	It also enables particular heavy metals to be reused	Ineffectual for disperse dyes
Membrane filtration	Efficient for all dyes with elevated quality effluent	High purchase price of the membranes and residue has to be collected or further treated
Irradiation	Oxidation is effective at lab scale	Dissolved O_2 is required in high concentration
Electrocoagulation	High efficiency at low capital reduces the need for additional chemicals; in comparison to coagulation, there is drastically decrease in sludge production	High water conductivity is essential to reduce power requirements, development of batch or continuous electrocoagulation reactors is essential
Oxidation	Rapid and efficient process	Toxic products are produced like organochlorine compounds
Photochemical	Sludge is not generated, rapid process and good sorption capacity for dyes	Formation of toxic by-products Formation of by-products
Electrochemical destruction	No utilization of chemicals and there is no sludge buildup	There is direct reduction in dye removal due to high flow rates
Ozonation	Gaseous state of ozone is significant, and volume of wastewater and sludge is reduced	Short half-life (20 min)
Fenton reagent H_2O_2+Fe(II) salts	Effective process and cheap reagent. Ozone can be applied in its gaseous state and does not increase the volume of wastewater and sludge	Short half-life (20 min)
Sodium hypochlorite (NaOCl)	Initiation and acceleration of azo-bond cleavage	Aromatic amines are released
Decolourization by white-rot fungi	Enzymes produced by white-rot fungi are able to degrade dyes	Enzyme production has also been shown to be undependable
Other microbial cultures (mixed bacterial)	Effective decolourization within 24–30 h	Azo dyes are not readily metabolized in aerobic conditions
Adsorption by living/dead microbial biomass	Some dyes having specification in binding affinity with microbial species	Ineffectual for all dyes
Anaerobic textile-dye bioremediation systems	Azo and other water-soluble dyes are able to decolourize	Methane and hydrogen sulphide are produced during anaerobic breakdown

14.4 Importance and Role of Biogenic Nanoparticles in Dye Degradation

Nanoparticles derived from biological source such as plants and microbes and microwave-assisted synthesis have enabled the wide applications across various sectors including environment remediation and reclamation. In addition, reducing/capping/dispersing agents play a significant role in this process; chemical-mediated synthesis involves use of chemicals such as sodium borohydride ($NaHB_4$), poly(γ-glutamic acid), polyethylene glycol (PEG), polyacrylamide, poly-N-vinylpyrrolidone, carboxymethyl cellulose (CMC) sodium and poly(propyleneimine), which act as complexing agents and aid in stabilizing nanoparticles (Iravani et al. 2014; Zhang et al. 2018). However, biogenic nanoparticles comprise the natural derivatives such as the proteins rich in amines, carboxylate ions, enzymes, electrochemically active biofilm (EAB) and antioxidants; polyphenols and flavonols are promising reducing agents and have been significantly effective compared to chemical-mediated synthesis (Ansari et al. 2013; Hebbalalu et al. 2013). A recent study outlined the effect of chemically synthesized and biogenic silver nanoparticles on tomato plant (*Solanum lycopersicum* L.). The study demonstrated that negative effects of chemically derived silver nanoparticles were comparatively higher than that of biogenic silver nanoparticles after evidences were derived assessing various anatomical, physiological and phytochemical tests (Girilal et al. 2018). Furthermore, biogenic nanoparticles exhibit distinctive chemical, biochemical, magnetic and optical properties, due to their relatively small size and high surface-area-to-volume ratio (Prasad et al. 2016). Biogenic nanoparticles have profound applications including bioactivities and catalysis; for instance, metallic nanoparticles such as AgNPs offer antibacterial and antioxidant activities against a wide array of pathogens and oxidative free radicals, respectively (Aziz et al. 2015, 2016, 2019). Moreover, they tend to act as semi-conductors and are excellent redox catalysts that aid in catalytic reactions involved in the degradation of toxic pollutants including dyes. Studies show that metallic nanoparticle-mediated photocatalysis of dyes is efficient, rapid and reliable (Edison and Sethuraman 2013) (Table 14.4).

Dyes and their deleterious effect on the environment have been recently understood. Drastic growth in textile industries and use of multiple dyes for dyeing have apparently increased the effluent treatment, which has become a major challenge. The lack of progressive knowledge on degradation pathways and cost involved in the efficient treatment poses a serious threat to the environment including the human, animal and plant populations. Besides, effluent comprising higher concentration of dyes discharged to water bodies has drastically affected the aquatic fauna. Though bioremediation offers a great platform to these problems, the time involved in degradation is relatively high, and possibilities of formation of novel intermediate compound with multifold increased toxicity are quite more, which limits the effectiveness of this approach under normal conditions (Prasad and Aranda 2018). A remarkable number of studies

Table 14.4 Biogenic nanoparticles for the photocatalysis of various dyes

Nanoparticles	Dye	Reducing agents/ used	Degradation efficiency (%)	Method	Remarks	Reference
Ag	CBB G-250	*Coccinia grandis*	–	Photocatalytic	UV light, 90 min showed maximum degradation	Arunachalam et al. (2012)
	MO	*Hypnea musciformis*	–	Photocatalytic	Visible light at 420 nm showed maximum degradation within 10 h	Selvam and Sivakumar (2015)
	MB	*Chlorella pyrenoidosa*	–	Photocatalytic	Visible light, 150 min showed maximum degradation	Aziz et al. (2015)
	MB MO PR DB24	*Sterculia acuminata*	–	Catalytic	$NaBH_4$ used as reducing agent, rapid degradation within 3 min	Bogireddy et al. (2016)
	MB	*Terminalia chebula*	–	Photocatalytic	Maximum degradation was seen at 60 min	Edison and Sethuraman (2012)
	CR MO	*Anacardium occidentale*	80–85	Catalytic	Degraded within 30 min	Edison et al. (2016b)
	EY	*Camellia japonica*	97	Photocatalytic	Visible light irradiation for 60 min	Karthik et al. (2017)
	CR	*Amaranthus gangeticus*	–	Catalytic	$NaBH_4$ is used as reducing agent; dye was rapidly degraded within 15 min	Kolya et al. (2015)
	MB	*Biophytum sensitivum*	–	Catalytic	$NaBH_4$ is used as reducing agent; dye was rapidly degraded within 10 min	Joseph and Mathew (2015)
	MO MB	Sugarcane juice	100	Catalytic	Within 18 min, complete degradation of MO occurred due to catalytic activity	Kulkarni and Bhanage (2014)

	MO	*Ulva lactuca*	–	Photocatalytic	Visible light was used	Kumar et al. (2013)
	AO	*Erigeron bonariensis*		Catalytic		Kumar et al. (2016)
	MB MO EY	*Punica granatum*	83 99 96	Catalytic	After 12 min, degradation completed	Meenakumari and Philip (2015)
	MB	*Saccharomyces cerevisiae*	Above 80	Photocatalytic	Sunlight was used for irradiation at 27 °C for 6 h	Roy et al. (2015a)
	MO	*Solanum tuberosum*	Above 75	Photocatalytic	Sunlight was used for irradiation at 30 °C for 8 h	Roy et al. (2015b)
AU	MB CR	*Salmalia malabarica* gum	–	Catalytic	$NaBH_4$ acts as reducing agent and degraded dye to the maximum	Ganapuram et al. (2015)
	MB EY 4-NP MR CR EB	*Plumeria alba* flower extract (PAFE)	–	Catalytic	$NaBH_4$ acts as reducing agent and degraded dye to the maximum	Mata et al. (2016)
	MB MO EY	*Punica granatum*	95 94 91	Catalytic	After 12 min, degradation completed	Meenakumari and Philip (2015)
	MB	*Bacillus marisflavi*	88	Catalytic	$NaBH_4$ is used as reducing agent; within 10 min, degradation occurred	Nadaf and Kanase (2016)
	MB	*Pogostemon benghalensis*	–	Photocatalytic	Within 8 min, maximum degradation was seen	Paul et al. (2015)

(continued)

Table 14.4 (continued)

Nanoparticles	Dye	Reducing agents/used	Degradation efficiency (%)	Method	Remarks	Reference
	Ca R X-GRL AO II AS GR Ca R AY 11 RR AR B RR X-3B AO G AB 10B RG KE 4B RB	*Aspergillus* sp.	95.2 94.6 96.4 94.3 92.7 95.3 94.9 94.1 91.0 92.8 94.3 94.9	Catalytic	$NaBH_4$ was used as reducing agent, all dyes degraded within 7 min and fastest was acid Orange G at 20 s	Qu et al. (2017)
Ag-Au alloy	EBT EY MO RB	*Coleus aromaticus*	84.6 96.2 97.4 95.4	Catalytic	$NaBH_4$ is used as reducing agent, and all showed varying time for degradation, but within 15 min, all dyes were degraded	Vilas et al. (2016)
ZnO	MB	*Azadirachta indica*	82	Photocatalytic	UV light irradiation for 180 min showed maximum degradation	Bhuyan et al. (2015)
	MR MB MO RB21	Lemon juice and zinc acetate	50–55 50–55 90–95 75–80	Photocatalytic	UV rays MR, MB and MO showed degradation in less than 40 min, whereas RB21 took 270 min	Davar et al. (2015)
	MB	*Cassia fistula*	98.71	Photocatalytic	UV and sunlight irradiation was done, and pH 4 showed maximum degradation	Suresh et al. (2015a, b)

	MB	*Artocarpus gomezianus*	90–95	Photocatalytic	UV and sunlight irradiation was used, and sunlight irradiation along with pH 10 showed maximum degradation in 120 min	Suresh et al. (2015a, b)
SnO_2	Phenolsulfonphthalein	*Persea americana* seeds	–	Photocatalytic	UV rays of 365 nm took 2 h for maximum degradation	Elango et al. (2015)
	MB	*Cyphomandra betacea* (methanol extract)	75–85	Photocatalytic	UV rays of 365 nm took 70 min for complete degradation	Elango and Roopan (2016)
	MB MO EB-T	*Erwinia herbicola*	93.3 94 97.8	Photocatalytic	UV rays of 254 nm was used, and within 2 h, maximum degradation was observed	Srivastava and Mukhopadhyay (2014)
	RhB	*Plectranthus amboinicus*	95	Photocatalytic	Visible light irradiation for 120 min showed maximum degradation	Fu et al. (2015)
Pd	PR	*Catharanthus roseus*	–	Photocatalytic	Visible light was used, and at pH 8, maximum dye degraded	Kalaiselvi et al. (2015)
	CBB MB MO	*Anogeissus latifolia*	–	Catalytic	$NaBH_4$ was used as reducing agent, and all three showed complete degradation within 2 min	Kora and Rastogi (2015)
CuO	CBB R-250	*Carica papaya*	–	Photocatalytic	Sunlight irradiation was applied, and within 90 min, maximum degradation was observed	Sankar et al. (2014)
	AB 210	*Abutilon indicum*	–	Photocatalytic	Sunlight irradiation was used, and within 1 h, maximum degradation was observed	Ijaz et al. (2017)

(continued)

Table 14.4 (continued)

Nanoparticles	Dye	Reducing agents/used	Degradation efficiency (%)	Method	Remarks	Reference
	MB CR	*Thymus vulgaris*	–	Catalytic	Bentonite was used as a supporting material for CuO, and $NaBH_4$ was used as a reducing agent, and within 40 s and 5 min, maximum degradation was observed for MB and CR, respectively	Issaabadi et al. (2017)
	MB	*Tinospora cordifolia*	90–95	Photocatalytic	Both UV and visible light irradiation were studied; pH 4 showed maximum dye degradation within 2 h of irradiation	Nethravathi et al. (2015)
	MR	*Musa balbisiana*	–	Photocatalytic	Visible light irradiation was used, and within 20 min, maximum degradation was observed	Tamuly et al. (2014)
Fe	MB MO	Green tea extracts	80	Fenton-like catalyst	MB was degraded within 5 min and MO took 1 h	Shahwan et al. (2011)
	MG	Green tea extracts	96	Catalytic	The dye showed maximum degradation	Weng et al. (2013)
Nanocomposite: palladium/sodium borosilicate	Cr(VI) CR MO MB	*Euphorbia milii*	–	Catalytic	$NaBH_4$ used as reducing agent in rapid degradation within 10 min	Nasrollahzadeh et al. (2018)
Se-ZnS	MO	*Bacillus* sp.	95	Photocatalytic	UV rays for 160 min showed maximum degradation	Ahluwalia et al. (2016)
Au/TiO_2	MO	*Cinnamomum tamala*	75	Photocatalytic	Visible light was used, and at pH 9 2 maximum dye degradation	Naik et al. (2013)

RGO/Fe_3O_4	MB	*Solanum trilobatum*	95.18	Catalytic	$NaBH_4$ is used as reducing agent, and the dyes showed maximum degradation in 12 min	Vinothkannan et al. (2015)
Ag decorated on GO nanosheet (GO-Ag)	MB	*Picrasma quassioides*	–	Catalytic	$NaBH_4$ was used, and within 15 min maximum degradation reported	Sreekanth et al. (2016)
Pd/RGO/Fe_3O_4	4-NP	*Withania coagulans*	98	Catalytic	$NaBH_4$ is used as reducing agent, and within 1 min maximum degradation was noted	Atarod et al. (2016)
Natrolite zeolite/Pd	MB MO CR 4-NP RB	*Piper longum*	–	Catalytic	$NaBH_4$ was used as reducing agent, and all showed different timings for maximum degradation 1 s, 2 min, 27 s, 50 s and 8 s, respectively	Hatamifard et al. (2015)
Ag-Au	MG	*Bacillus safensis*	86.1–92.6	Catalytic	Within 48 h, dye was degraded	Ojo et al. (2016)
Ag/RGO/TiO_2	MB CR 4-NP	*Euphorbia helioscopia*	–	Catalytic	$NaBH_4$ is used as reducing agent, and within 195 s, 4-NP was degraded to maximum, whereas within 116 s, CR and MB degraded	Nasrollahzadeh et al. (2016)

4-NP nitrophenol, *AB* acid black, *AO* acid orange, *AR* acid red, *AS* acid scarlet GR, *AY* acid yellow, *CaR* cationic red X-GRL, *CBB* Coomassie brilliant blue G-250, *Cr(VI)* chromium (VI), *CR* Congo red, *CV* crystal violet, *DB* direct blue, *EB* ethidium bromide, *EB-T* Eriochrome black-T, *EMB* eosin methylene blue, *EY* eosin yellow, Fe_2O_3 ferrosoferric oxide/magnetite, *GO* graphene oxide, *MB* methylene blue, *MG* malachite green, *MO* methyl orange, *MR*, methyl red, *MV* methyl violet, *Pd* palladium, *Ph* phenolsulfonphthalein, *PR* phenol red, *RB* reactive blue, *RB* reactive black, *RGO* reduced graphene oxide, *RG* reactive green, *RhB* rhodamine B, *RR*, reactive red, *Se-ZnS* selenium-zinc sulphide, *s* seconds, *min* minutes, *h* hour, *nm* nanometre confirm the potential application of nanoparticles, especially biogenic nanoparticles in degradation of toxic pollutants including noxious dyes (Duan et al. 2012; Huang et al. 2012; Castro et al. 2018; Liu et al. 2018; Teimouri et al. 2018).

Biogenic gold nanoparticles have been extensively studied for various applications. Recent studies indicate that AuNPs have great potential in degradation of dyes of wide chemical nature. Different reducing agents have been used for the synthesis of AuNPs; most studies highlight the use of different parts of plant materials for synthesizing AuNPs such as *kokum* fruit (*Garcinia indica*) (Desai et al. 2018), burdock root (*Arctium lappa*) (Nguyen et al. 2018), *Alpinia nigra* leaves (Baruah et al. 2018), *Dalbergia coromandeliana* roots (Umamaheswari et al. 2018), *Mussaenda glabrata* leaf (Francis et al. 2017), turnip leaf (*Brassica rapa* L.) (Narayanan and Park 2015) and *Pogostemon benghalensis* (B) O. Ktz. leaf extract (Paul et al. 2015). Similarly, other bioresources have also been made use for fabrication of AuNPs, viz., fungi (*Trichoderma harzianum*) (Tripathi et al. 2018), mushroom (*Flammulina velutipes*) (Narayanan et al. 2015), yeast (*Hansenula anomala*) (Amutha et al. 2011), macroalgae (*Padina tetrastromatica*) (Princy and Gopinath 2018) and chemical derivatives of algae such as fucoidans that are also found to be notable reducing agents for AuNP formation (Lirdprapamongkol et al. 2010; Khan et al. 2018). Several bacteria such as *Bacillus subtilis* (Srinath et al. 2018) and halotolerant bacteria *Pseudoalteromonas lipolytica* (Kulkarni et al. 2018) are reported for the synthesis of AuNPs. The bioactive compounds derived from aforementioned organisms including enzymes, amino acids, exopolysaccharides, polyphenols and flavonoids were involved in reducing AuNPs. Besides, these molecules also acted as capping/dispersing agents. The AuNP mechanism of dye degradation has been well documented. AuNPs act as redox catalysts and are known to accelerate the reduction rate of donor (a reducing agent, i.e. $NaBH_4$) and acceptor molecules (a dye) by electron relay effect (Cheval et al. 2012; Kulkarni et al. 2018). Their relatively small size, large surface-to-volume ratio and high Fermi potential enable them to be potential catalysts for dye degradation (Rajan et al. 2015). Further, biocomposites with AuNPs act as efficient heterogeneous catalysts against 4-nitrophenol and have profound influence on dye degradation (dose-dependent degradation). In addition, recovery and recycling of AuNP catalysts have also been demonstrated (Narayanan and Sakthivel 2011). Thus, several studies demonstrate the significance of AuNPs in rendering potential applications in degradation of hazardous organic pollutants such as dyes from industrial wastewater system.

Though several nanoparticles are reported to exhibit excellent catalytic activity, silver nanoparticles receive a great deal of attention as silver shows exclusive properties such as spectral, electrical (surface plasmon resonance) and thermal conductivity, in addition to optical and nonlinear optical properties (Rao et al. 2003; Pandey et al. 2012). Hence, these properties of AgNPs facilitate in important industrial transformations/catalytic reactions such as heterocyclizations, cycloaddition of imines and oxidation of ethylene to ethylene oxide and methanol to formaldehyde (Wiley et al. 2007; Lvarez-Corral et al. 2008; Yamamoto 2008; Nadagouda et al. 2011; Edison and Sethuraman 2013). Moreover, because of simplicity, sensitivity and ease of use, AgNPs are widely used as optical sensors in imaging technology (Pandey et al. 2012). Nevertheless, AgNPs significantly contribute to high surface energy facilitating surface reactivity with increased adsorption and catalysis. A notable study by Veisi et al. (2018) demonstrated synthesis and dye degradation ability of *AgNP/Thymbra* nanocomposite against 4-nitrophenol, rhodamine B and methylene blue at room temperature using plant *Thymbra spicata* leaf extract. Highly dispersed spherical nanoparticles with a size of approximately 7 nm and crystalline in nature were able to catalyse the dyes efficiently. Furthermore, significant degradation in short period was observed in the presence of reducing agent $NaBH_4$ with very minimal concentration of catalyst, which was comparatively more rapid to previous studies reported elsewhere. In addition, their study also indicated that recycled AgNPs were able to produce similar observations without loss in activity. Table 14.3 gives the description of recent studies on dye degradation using various biogenic nanoparticles including nanocomposites and nanomaterials.

14.5 Important Factors Influencing Degradation of Noxious Dyes

Production of metabolites depends vastly on environmental conditions and optimized parameters. Optimization of physical parameters not only enhances the growth of microbes but also improves the product yield. Priyom and Uma (2017) investigated the photocatalytic degradation of methylene blue dye using biosynthesized silver nanoparticles by solar irradiation technique in different time intervals. Optimization parameters of silver nanoparticles such as the effect of $AgNO_3$ concentration, effect of pH, effect of temperature and stability were studied. The different concentration of fungal filtrate and silver nitrate solution was optimized for maximum production of silver nanoparticles. It was found that maximum synthesis was obtained at 1:1 ratio (fungal filtrate/silver nitrate solution). The pH exhibits a vital role in the synthesis of silver nanoparticles; reaction was adjusted with five different levels of pH. Maximum synthesis of silver nanoparticles occurred at neutral pH with the formation of reddish brown colour in the reaction mixture. Temperature is considered to be an essential factor affecting synthesis of silver nanoparticles. The different temperature was maintained for the production of silver nanoparticles. Maximum synthesis was found at 40 °C having a sharp peak at 430 nm. Stability of the synthesized silver nanoparticles is an important factor. In the present study, the synthesized silver nanoparticles were found stable till the 45th day. The UV-visible spectrum of biosynthesized silver nanoparticles showed peak at 432 nm on the 45th day. Proteins might play a role in forming a coat covering the metal nanoparticles. The capping of silver nanoparticles is necessary for preventing agglomeration of nanoparticles, thereby attaining stability. Silver nanoparticles produced by the endophyte *Fusarium oxysporum* was reported to be stable for 60 days.

In the process of photocatalytic degradation of dye, the following operating parameters play a vital role: pH of solution to be degraded and pH of precursor solution, oxidizing agent, dopant content and calcination temperature. These parameters are considered mainly as they influence processes of photocatalytic degradation of dyes (Fig. 14.2).

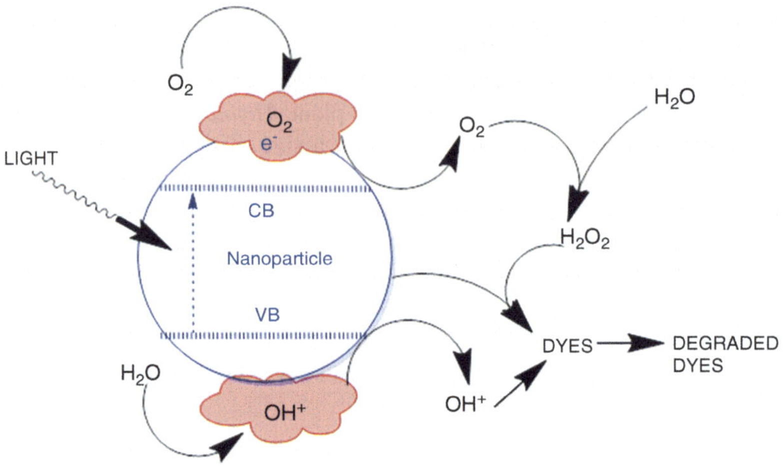

Fig. 14.2 Nanoparticle-mediated photocatalytic reduction of dyes

14.5.1 Effect of pH on the Photodegradation of Dyes

The analysis of pH effects on the process of dye photodegradation is very tedious work because of its complex phenomenon (Konstantinou and Albanis 2004). Firstly, the entire process is related to ionization state of surface as well as to that of reactant dyes and by-products such as acids and amines. The pH changes can thus affect the adsorption of dye molecules to the TiO_2 surfaces (Fox and Dulay 1993). The investigation by Bahnemann et al. (1994) reported that the photocatalytic activity of metal oxide surfaces is induced by its acid-base properties. Baran et al. (2008) also investigated that acidic condition was better than alkaline medium for degradation of bromocresol purple dye. After solution pH was acidified from 8.0 to 4.5, a sixfold increase in the adsorption efficacy was noticed (Wang et al. 2000).

The degradation rate of azo dyes increases with decrease in pH (Konstantinou and Albanis 2004). At pH < 6, efficient adsorption of the dye on TiO_2 particles was noticed as the result of electrostatic attraction of positively charged TiO_2 with dye. At pH > 6.8, as the dye molecules are negatively charged in the alkaline solution, their adsorption also seems to be influenced by enhancement in density of the TiO groups on semi-conductor surface. Hence, due to the coulombic repulsion, scarcely dyes are adsorbed (Lachheb et al. 2002; Stylidi et al. 2003). Sleiman et al. (2007) investigated on the influence of pH on degradation of metanil yellow, an anionic dye with sulphonate group, over TiO_2 photocatalyst under UV illumination. Their results indicated that process efficiency is not majorly affected over a wide range of pH (4–8). Zhiyong et al. (2007) in their investigation showed that $ZnSO_4$-TiO_2-doped catalyst exhibited high efficiency in photocatalytic degradation of dyes and further reported that pH has influential role on photocatalytic degradation of the Orange II, with SO_3 groups. Thus, their study denoted that photocatalytic activity was optimum at a lower pH 3.0 but decreases with inefficient rate at pH 10.0. Henceforth, it is very important to analyse the nature of pollutant to be degraded and standardize the assay for determination of optimum pH level to degrade them.

14.5.2 Influence of Oxidizing Agents on the Degradation of Dyes

Reports show that oxidizing agents have a great deal of influence on photocatalytic degradation of dyes. It was reported by Saquiba et al. (2008) that hydrogen peroxide (H_2O_2), ammonium persulphate $(NH_4)_2S_2O_8$ and potassium bromate ($KBrO_3$) have individual influence on degradation of Fast Green FCF and Patent Blue VF using Hombikat UV 100 and Degussa P25 as photocatalysts. Their investigations noticed that ammonium persulphate and potassium bromate had potential effect on degradation rate for Fast Green FCF dye in the presence of UV 100, whereas in case of Patent Blue VF, electron acceptors were found to increase the rate remarkably in the presence of Degussa P25. Huang et al. (2008) also studied the influence of adding H_2O_2 on decolourization of methyl orange. The decolourization rate was found to increase with increase in H_2O_2 concentration. The experiment was performed at a concentration of 0.4–2 mM/l H_2O_2. They have examined an optimum dose of 1.2 mM/l H_2O_2 for decolourization of methyl orange by Pt-modified TiO_2 on natural zeolite. Actually, the addition of H_2O_2 enhanced the reaction. Zhiyong et al. (2008) also reveal that the addition of H_2O_2 (1 mM) to the methyl orange-mediated TiO_2 Degussa P25 (0.5 g/l) under photocatalyst irradiated by sunlight exhibits degradation of methyl orange in 1 h.

Sun et al. (2006, 2008) have also reported the effect of oxidants in photocatalytic reactions. It was found out as part of their work that using TiO_2 as photocatalysts is undesired electron-hole recombination, in the absence of electron donor or acceptor, which is efficient and hence represents major energy-wasting step, thereby limiting optimum yield. They therefore suggested a unique method to inhibit the electron-hole recombination to initiate irreversible electron acceptors to reaction; for this, they used H_2O_2 to examine its influence on the degradation of Orange G on N-doped TiO_2 under various light sources. The result suggests an optimal dosage of H_2O_2, at which efficient degradation of OG on N-doped TiO_2 attained height. Rengaraj and Li (2007) reported that without the use of sacrificial whole scavenger (formic acid), there was no catalytic activity of either TiO_2 in the nitrate solution. They too noticed that an optimal dosage of the scavenger must be utilized for nitrate photodegradation in specific reactions. Furthermore, there is need to appraise and optimize the influence of the oxidizing agent in photodegradation of the dyes.

14.5.3 Role of Dopant Content on the Photocatalytic Activity of Catalysts

The dopant content's influence on the photocatalytic activity of catalysts has been studied by Wei et al. (2007) and Bouras et al. (2007). The reports of Bouras et al. (2007) revealed that degradation of basic blue 41 dye with UV light was more effective in the presence of the pure TiO_2 than in the presence of the Fe-doped TiO_2. A continuous decrease in percentage degradation of dye from 80 to about 1 as dopant content increased from 0 to 30 at percentage Fe was observed. The same result was noticed for Co-TiO_2 and Cr-TiO_2, at varying percentage reduction, but for Bouras et al. (2007), the optimum dopant, either mol% or wt%, varies from catalyst to catalyst.

14.5.4 Effect of Calcination Temperature on Activity of Photocatalysts

The method of preparation and usage of TiO_2 or doped TiO_2 and the activity of prepared photocatalysts depend mainly on calcination temperatures. Yu et al. (2007) prepared nitrogen-doped TiO_2 nanoparticle catalyst and analysed its catalytic activity under visible light. In their investigation, the influence of temperature of calcination on degradation of the methylene blue under UV irradiation was examined. The N-doped TiO_2 samples were subjected to various calcination temperatures ranging from 300 to 700 °C. The results of their investigations revealed that activity of photocatalysts increased with increase in the temperature level from 300 to 500 °C. At 500 °C, it reached optimum results in complete crystallization of the anatase at this specific temperature range. However, catalytic activity of N-doped TiO_2 shows decreased rate with increase of calcination temperature from 500 to 700 °C. This was also observed to be consistent with TEM analysis. Sun et al. (2006) also reported the influence of calcination temperature on photocatalytic activity of Sn (IV)/TiO_2/AC on degradation of Orange G. Their results showed that photocatalytic efficacy of prepared photocatalyst was significantly influenced by calcination temperature and the optimal calcination temperature recorded was at 550 °C. The XRD results revealed that the sample calcinated at 550 °C contains both rutile and anatase phases of the TiO_2, which further inform higher activity for photocatalytic degradation of Orange G. These results were found contradictory to reports of Yu et al. (2007) and Zhiyong et al. (2008).

14.6 Conclusion and Future Perspectives

A great number of studies indicate that dye degradation using biogenic nanoparticle-assisted photocatalysis and chemical ($NaBH_4$) catalysis offers a myriad of potential applications and could significantly contribute to remediation of pollution due to synthetic dye toxicity. Future studies involving biogenic nanoparticle-mediated photocatalysis/catalysis must be exhibited in an open and larger system to confirm the feasibility and process involving the pathways in detail. Nonetheless, at present scenario, studies confirm the potential benefits of biogenic nanoparticles under laboratory conditions. On-field trials for efficient effluent treatments have to be assessed and validated. Thus, biogenic nanoparticles exhibit excellent stability and catalytic performance for the reduction of various noxious dyes under changing environmental conditions. Nevertheless, novel biotechnological tools may improvise the yield and stability of nanoparticles for repeated applications in dye degradation.

References

Ahluwalia S, Prakash NT, Prakash R, Pal B (2016) Improved degradation of methyl orange dye using bio-co-catalyst Se nanoparticles impregnated ZnS photocatalyst under UV irradiation. Chem Eng J 306:1041–1048

Amutha R, Arumugam P, Berchmans S (2011) Synthesis of gold nanoparticles: an ecofriendly approach using *Hansenula anomala*. ACS Appl Mater Interfaces 5(3):1418–1425

Ansari SA, Khan MM, Ansari MO, Lee J, Cho MH (2013) Biogenic synthesis, photocatalytic, and photoelectrochemical performance of Ag–ZnO nanocomposite. J Phys Chem A 117(51):27023–27030

Arunachalam R, Dhanasingh S, Kalimuthu B, Uthirappan M, Rose C, Mandal AB (2012) Phytosynthesis of silver nanoparticles using *Coccinia grandis* leaf extract and its application in the photocatalytic degradation. Colloids Surf B Biointerfaces 94:226–230

Atarod M, Nasrollahzadeh M, Sajadi SM (2016) Green synthesis of Pd/RGO/Fe_3O_4 nanocomposite using *Withania coagulans* leaf extract and its application as magnetically separable and reusable catalyst for the reduction of 4-nitrophenol. J Colloid Interface Sci 465:249–258

Aziz N, Faraz M, Pandey R, Sakir M, Fatma T, Varma A, Barman I, Prasad R (2015) Facile algae-derived route to biogenic silver nanoparticles: Synthesis, antibacterial and photocatalytic properties. Langmuir 31: 11605–11612 DOI: https://doi.org/10.1021/acs.langmuir.5b03081

Aziz N, Pandey R, Barman I, Prasad R (2016) Leveraging the attributes of Mucor hiemalis-derived silver nanoparticles for a synergistic broad-spectrum antimicrobial platform. Front Microbiol 7:1984. doi: https://doi.org/10.3389/fmicb.2016.01984

Aziz N, Faraz M, Sherwani MA, Fatma T, Prasad R (2019) Illuminating the anticancerous efficacy of a new fungal chassis for silver nanoparticle synthesis. Front Chem 7:65. doi: https://doi.org/10.3389/fchem.2019.00065

Bahnemann D, Cunningham J, Fox MA, Pelizzetti E, Pichat P, Serpone N, Zepp RG, Heltz GR, Crosby DG (1994) Aquatic and surface photochemistry. Lewis, Boca Raton, pp 261–278

Baran W, Makowski A, Wardas W (2008) The effect of UV radiation absorption of cationic and anionic dye solutions on their photocatalytic degradation in the presence of TiO_2. Dyes Pigments 76:226–230

Baruah D, Goswami M, Yadav RNS, Yadav A, Das AM (2018) Biogenic synthesis of gold nanoparticles and their application in photocatalytic degradation of toxic dyes. J Photochem Photobiol B Biol 186:51–58

Bhuyan T, Mishra K, Khanuja M, Prasad R, Varma A (2015) Biosynthesis of zinc oxide nanoparticles from *Azadirachta indica* for antibacterial and photocatalytic applications. Mater Sci Semicon Proc 32:55–61

Bogireddy NKR, Kumar HAK, Mandal BK (2016) Biofabricated silver nanoparticles as green catalyst in the degradation of different textile dyes. J Environ Chem Eng 4(1):56–64

Bouras P, Stathatos E, Lianos P (2007) Pure versus metal-ion-doped nanocrystalline titania for photocatalysis. Appl Catal B Environ 73:51–59

Castro L, Blázquez ML, González F, Muñoz JA, Ballester A (2018) Heavy metal adsorption using biogenic iron compounds. Hydrometallurgy 179:44–51

Cheval N, Gindy N, Flowkes C, Fahmi A (2012) Polyamide 66 microspheres metallised with in situ synthesised gold nanoparticles for a catalytic application. Nanoscale Res Lett 7(1):182. https://doi.org/10.1186/1556-276X-7-182

Davar F, Majedi A, Mirzaei A (2015) Green synthesis of ZnO nanoparticles and its application in the degradation of some dyes. J Am Ceram Soc 98(6):1739–1746

Dawood S, Sen T (2014) Review on dye removal from its aqueous solution into alternative cost effective and non-conventional adsorbents. Chem Process Eng 1(104):1–11

Desai MP, Sangaokar GM, Pawar KD (2018) Kokum fruit mediated biogenic gold nanoparticles with photoluminescent, photocatalytic and antioxidant activities. Process Biochem 70:188–197

Dhillon GS, Brar SK, Kaur S, Verma M (2012) Green approach for nanoparticle biosynthesis by fungi: current trends and applications. Crit Rev Biotechnol 32(1):49–73

Duan Z, Ma G, Zhang W (2012) Preparation of copper nanoparticles and catalytic properties for the reduction of aromatic nitro compounds. Bull Kor Chem Soc 33:4003–4006

Edison TJI, Sethuraman MG (2012) Instant green synthesis of silver nanoparticles using *Terminalia chebula* fruit extract and evaluation of their catalytic activity on reduction of methylene blue. Process Biochem 47(9):1351–1357

Edison TJI, Sethuraman MG (2013) Biogenic robust synthesis of silver nanoparticles using *Punica granatum* peel and its application as a green catalyst for the reduction of an anthropogenic pollutant 4-nitrophenol. Spectrochim Acta A Mol Biomol Spectrosc 104:262–264

Edison TNJI, Atchudan R, Kamal C, Lee YR (2016a) Caulerpa racemosa: a marine green alga for eco-friendly synthesis of silver nanoparticles and its catalytic degradation of methylene blue. Bioprocess Biosyst Eng 39(9):1401–1408

Edison TNJI, Atchudan R, Sethuraman MG, Lee YR (2016b) Reductive-degradation of carcinogenic azo dyes using *Anacardium occidentale* testa derived silver nanoparticles. J Photochem Photobiol B Biol 162:604–610

Elango G, Roopan SM (2016) Efficacy of SnO_2 nanoparticles toward photocatalytic degradation of methylene blue dye. J Photochem Photobiol B Biol 155:34–38

Elango G, Kumaran SM, Kumar SS, Muthuraja S, Roopan SM (2015) Green synthesis of SnO_2 nanoparticles and its photocatalytic activity of phenolsulfonphthalein dye. Spectrochim Acta A Mol Biomol Spectrosc 145:176–180

Fanchiang JM, Tseng DH (2009) Degradation of anthraquinone dye CI Reactive Blue 19 in aqueous solution by ozonation. Chemosphere 77(2):214–221

Forgacs E, Cserhati T, Oros G (2004) Removal of synthetic dyes from wastewaters: a review. Environ Int 30(7):953–971

Fox MA, Dulay MT (1993) Heterogeneous photocatalysis. Chem Rev 93(1):341–356

Francis S, Joseph S, Koshy EP, Mathew B (2017) Green synthesis and characterization of gold and silver nanoparticles using *Mussaenda glabrata* leaf extract and their environmental applications to dye degradation. Environ Sci Pollut Res 24(21):17347–17357

Fu L, Zheng Y, Ren Q, Wang A, Deng B (2015) Green biosynthesis of SnO_2 nanoparticles by *plectranthus amboinicus* leaf extract their photocatalytic activity toward rhodamine B degradation. J Ovonic Res 11(1):21–26

Ganapuram BR, Alle M, Dadigala R, Dasari A, Maragoni V, Guttena V (2015) Catalytic reduction of methylene blue and Congo red dyes using green synthesized gold nanoparticles capped by *Salmalia malabarica* gum. Int Nano Lett 5(4):215–222

Girilal M, Fayaz AM, Elumalai LK, Sathiyaseelan A, Gandhiappan J, Kalaichelvan PT (2018) Comparative stress physiology analysis of biologically and chemically synthesized silver nanoparticles on *Solanum lycopersicum* L. Colloid Interfac Sci Commun 24:1–6

Goodsell DS (2004) Bionanotechnology: lessons from nature. Wiley, New York

Gupta V (2009) Application of low-cost adsorbents for dye removal – a review. J Environ Manag 90(8):2313–2342

Hatamifard A, Nasrollahzadeh M, Lipkowski J (2015) Green synthesis of a natrolite zeolite/palladium nanocomposite and its application as a reusable catalyst for the reduction of organic dyes in a very short time. RSC Adv 5(111):91372–91381

Hebbalalu D, Lalley J, Nadagouda MN, Varma RS (2013) Greener techniques for the synthesis of silver nanoparticles using plant extracts, enzymes, bacteria, biodegradable polymers, and microwaves. ACS Sustain Chem Eng 1(7):703–712

Hernández-Montoya V, Pérez-Cruz MA, Mendoza-Castillo DI, Moreno-Virgen MR, Bonilla-Petriciolet A (2013) Competitive adsorption of dyes and heavy metals on zeolitic structures. J Environ Manag 116:213–221

Hoag GE, Collins JB, Holcomb JL, Hoag JR, Nadagouda MN, Varma RS (2009) Degradation of bromothymol blue by 'greener' nano-scale zero-valent iron synthesized using tea polyphenols. J Mater Chem 19(45):8671–8677

Huang M, Xu C, Wu Z, Huang Y, Lin J, Wu J (2008) Photocatalytic discolorization of methyl orange solution by Pt modified TiO_2 loaded on natural zeolite. Dyes Pigments 77:327–334

Huang C-C, Lo S-L, Lien H-L (2012) Zero-valent copper nanoparticles for effective dechlorination of dichloromethane using sodium borohydride as a reductant. Chem Eng J 203:95–100

Hunger K (2003) Industrial dyes: chemistry, properties, applications. Wiley, New York

Ijaz F, Shahid S, Khan SA, Ahmad W, Zaman S (2017) Green synthesis of copper oxide nanoparticles using *Abutilon indicum* leaf extract: antimicrobial, antioxidant and photocatalytic dye degradation activities. Trop J Pharm Res 16(4):743–753

Iravani S, Korbekandi H, Mirmohammadi SV, Zolfaghari B (2014) Synthesis of silver nanoparticles: chemical, physical and biological methods. Res Pharm Sci 9(6):385–406

Issaabadi Z, Nasrollahzadeh M, Sajadi SM (2017) Green synthesis of the copper nanoparticles supported on bentonite and investigation of its catalytic activity. J Clean Prod 142:3584–3591

Joseph S, Mathew B (2015) Microwave-assisted green synthesis of silver nanoparticles and the study on catalytic activity in the degradation of dyes. J Mol Liq 204:184–191

Kalaiselvi A, Roopan SM, Madhumitha G, Ramalingam C, Elango G (2015) Synthesis and characterization of palladium nanoparticles using *Catharanthus roseus* leaf extract and its application in the photo-catalytic degradation. Spectrochim Acta A Mol Biomol Spectrosc 135:116–119

Karthik R, Govindasamy M, Chen SM, Cheng YH, Muthukrishnan P, Padmavathy S, Elangovan A (2017) Biosynthesis of silver nanoparticles by using *Camellia japonica* leaf extract for the electrocatalytic reduction of nitrobenzene and photocatalytic degradation of Eosin-Y. J Photochem Photobiol B Biol 170:164–172

Kaushik P, Malik A (2009) Fungal dye decolourization: recent advances and future potential. Environ Int 35(1):127–141

Khan AU, Khan M, Malik N, Cho MH, Khan MM (2018) Recent progress of algae and blue–green algae-assisted synthesis of gold nanoparticles for various applications. Bioprocess Biosyst Eng 42(1):1–15

Khataee AR, Pons MN, Zahraa O (2009) Photocatalytic degradation of three azo dyes using immobilized TiO_2 nanoparticles on glass plates activated by UV light irradiation: influence of dye molecular structure. J Hazard Mater 168(1):451–457

Kolya H, Maiti P, Pandey A, Tripathy T (2015) Green synthesis of silver nanoparticles with antimicrobial and azo dye (Congo red) degradation properties using *Amaranthus gangeticus* Linn leaf extract. JAST 6(1):33. https://doi.org/10.1186/s40543-015-0074-1

Konstantinou IK, Albanis TA (2004) TiO_2-assisted photocatalytic degradation of azo dyes in aqueous solution: kinetic and mechanistic investigations – a review. Appl Catal B Environ 49:1–14

Kora AJ, Rastogi L (2015) Green synthesis of palladium nanoparticles using gum ghatti (*Anogeissus latifolia*) and its application as an antioxidant and catalyst. Arab J Chem 11(7):1097–1106

Kulkarni AA, Bhanage BM (2014) Ag@AgCl nanomaterial synthesis using sugar cane juice and its application in degradation of azo dyes. ACS Sustain Chem Eng 2(4):1007–1013

Kulkarni R, Harip S, Kumar AR, Deobagkar D, Zinjarde S (2018) Peptide stabilized gold and silver nanoparticles derived from the mangrove isolate *Pseudoalteromonas lipolytica* mediate dye decolorization. Colloids Surf A Physicochem Eng Asp 555:180–190

Kumar P, Govindaraju M, Senthamilselvi S, Premkumar K (2013) Photocatalytic degradation of methyl orange dye using silver (Ag) nanoparticles synthesized from *Ulva lactuca*. Colloids Surf B: Biointerfaces 103:658–661

Kumar V, Singh DK, Mohan S, Hasan SH (2016) Photo-induced biosynthesis of silver nanoparticles using aqueous extract of *Erigeron bonariensis* and its catalytic activity against Acridine Orange. J Photochem Photobiol B Biol 155:39–50

Lachheb H, Puzenat E, Houas A, Ksibi M, Elaoui E, Guillard G, Hermann JM (2002) Photocatalytic degradation of various types of dyes (alizarin S, crecein Orange G, methyl red, congo red, methylene blue) in water by UV-irradiated titania. Appl Catal B Environ 39:75–90

Le Coz CJ (2005) Dyes. In: Encyclopedia of toxicology, 2nd edn. Elsevier, New York, pp 104–114

Li WY, Chen FF, Wang SL (2010) Binding of reactive brilliant red to human serum albumin: insights into the molecular toxicity of sulfonic azo dyes. Protein Pept Lett 17(5):621–629

Lirdprapamongkol K, Warisnoicharoen W, Soisuwan S, Svasti J (2010) Eco-friendly synthesis of fucoidan-stabilized gold nanoparticles. Am J Appl Sci 7(8):1038

Liu YS, Chang YC, Chen HH (2018) Silver nanoparticle biosynthesis by using phenolic acids in rice husk extract as reducing agents and dispersants. J Food Drug Anal 26(2):649–656

Lvarez-Corral M, Munoz-Dorado M, Rodríguez-García I (2008) Silver-mediated synthesis of heterocycles. Chem Rev 108:3174–3198

Malik R, Ramteke DS, Wate SR (2007) Adsorption of malachite green on groundnut shell waste based powdered activated carbon. Waste Manag 27(9):1129–1138

Mata R, Bhaskaran A, Sadras SR (2016) Green-synthesized gold nanoparticles from *Plumeria alba* flower extract to augment catalytic degradation of organic dyes and inhibit bacterial growth. Particuology 24:78–86

Meenakumari M, Philip D (2015) Degradation of environment pollutant dyes using phytosynthesized metal nanocatalysts. Spectrochim Acta Part A Mol Biomol Spectrosc 135:632–638

Merouani S, Hamdaoui O, Saoudi F, Chiha M (2010) Sonochemical degradation of Rhodamine B in aqueous phase: effects of additives. Chem Eng J 158(3):550–557

Mishra M, Kumari S, Pandey V, Chaudhry KC, Gupta CS, Nautiyal CS (2014) Biocatalytic and antimicrobial activities of gold nanoparticles synthesized by *Trichoderma* sp. Bioresour Technol 166:235–242

Mohan N, Balasubramanian N, Basha CA (2007) Electrochemical oxidation of textile wastewater and its reuse. J Hazard Mater 147(1–2):644–651

Nadaf NY, Kanase SS (2016) Biosynthesis of gold nanoparticles by *Bacillus marisflavi* and its potential in catalytic dye degradation. Arab J Chem. https://doi.org/10.1016/j.arabjc.2016.09.020

Nadagouda MN, Speth TF, Varma RS (2011) Microwave-assisted green synthesis of silver nanostructures. Acc Chem Res 44(7):469–478

Naik GK, Mishra PM, Parida K (2013) Green synthesis of Au/TiO_2 for effective dye degradation in aqueous system. Chem Eng J 229:492–497

Narayanan KB, Park HH (2015) Homogeneous catalytic activity of gold nanoparticles synthesized using turnip (*Brassica rapa* L.) leaf extract in the reductive degradation of cationic azo dye. Korean J Chem Eng 32(7):1273–1277

Narayanan KB, Sakthivel N (2011) Synthesis and characterization of nano-gold composite using *Cylindrocladium floridanum* and its heterogeneous catalysis in the degradation of 4-nitrophenol. J Hazard Mater 189(1–2):519–525

Narayanan KB, Park HH, Han SS (2015) Synthesis and characterization of biomatrixed-gold nanoparticles by the mushroom *Flammulina velutipes* and its heterogeneous catalytic potential. Chemosphere 141:169–175

Nasrollahzadeh M, Atarod M, Jaleh B, Gandomirouzbahani M (2016) *In situ* green synthesis of Ag nanoparticles on graphene oxide/TiO_2 nanocomposite and their catalytic activity for the reduction of 4-nitrophenol, congo red and methylene blue. Ceram Int 42(7):8587–8596

Nasrollahzadeh M, Sajjadi M, Maham M, Sajadi SM, Barzinjy AA (2018) Biosynthesis of the palladium/sodium borosilicate nanocomposite using *Euphorbia milii* extract and evaluation of its catalytic activity in the reduction of chromium (VI), nitro compounds and organic dyes. Mater Res Bull 102:24–35

Nethravathi PC, Kumar MP, Suresh D, Lingaraju K, Rajanaika H, Nagabhushana H, Sharma SC (2015) *Tinospora cordifolia* mediated facile green synthesis of cupric oxide nanoparticles and their photocatalytic, antioxidant and antibacterial properties. Mat Sci Semicon Proc 33:81–88

Nguyen TTN, Vo TT, Nguyen BNH, Nguyen DT, Dang VS, Dang CH, Nguyen TD (2018) Silver and gold nanoparticles biosynthesized by aqueous extract of burdock root, *Arctium lappa* as antimicrobial agent and catalyst for degradation of pollutants. Environ Sci Pollut Res 25(34):34247–34261

Ojo SA, Lateef A, Azeez MA, Oladejo SM, Akinwale AS, Asafa TB, Yekeen TA, Akinboro A, Oladipo IC, Gueguim-Kana EB, Beukes LS (2016) Biomedical and catalytic applications of gold and silver-gold alloy nanoparticles biosynthesized using cell-free extract of *Bacillus safensis* LAU 13: antifungal, dye degradation, anti-coagulant and thrombolytic activities. IEEE Trans Nanobioscience 15(5):433–442

Padhi BS (2012) Pollution due to synthetic dyes toxicity and carcinogenicity studies and remediation. Int J Environ Sci 3(3):940–955

Pandey S, Goswami GK, Nanda KK (2012) Green synthesis of biopolymer–silver nanoparticle nanocomposite: an optical sensor for ammonia detection. Int J Biol Macromol 51(4):583–589

Paul B, Bhuyan B, Purkayastha DD, Dey M, Dhar SS (2015) Green synthesis of gold nanoparticles using *Pogostemon benghalensis* (B) O. Ktz. leaf extract and studies of their photocatalytic activity in degradation of methylene blue. Mater Lett 148:37–40

Prasad R (2014) Synthesis of silver nanoparticles in photosynthetic plants. Journal of Nanoparticles, Article ID 963961, http://dx.doi.org/10.1155/2014/963961

Prasad R (2016) Advances and Applications through Fungal Nanobiotechnology. Springer, International Publishing Switzerland (ISBN: 978-3-319-42989-2)

Prasad R, Pandey R, Barman I (2016) Engineering tailored nanoparticles with microbes: quo vadis. WIREs Nanomed Nanobiotechnol 8:316–330. doi: https://doi.org/10.1002/wnan.1363

Prasad R (2017) Fungal Nanotechnology: Applications in Agriculture, Industry, and Medicine. Springer Nature Singapore Pte Ltd. (ISBN 978-3-319-68423-9)

Prasad R, Aranda E (2018) Approaches in Bioremediation. Springer International Publishing https://www.springer.com/de/book/9783030023683

Prasad R, Jha A, Prasad K (2018) Exploring the Realms of Nature for Nanosynthesis. Springer International Publishing (ISBN 978-3-319-99570-0) https://www.springer.com/978-3-319-99570-0

Prasad R, Kumar V, Kumar M, Wang S (2018a) Fungal Nanobionics: Principles and Applications. Springer Nature Singapore Pte Ltd. (ISBN 978-981-10-8666-3) https://www.springer.com/gb/book/9789811086656

Princy KF, Gopinath A (2018) Optimization of physicochemical parameters in the biofabrication of gold nanoparticles using marine macroalgae *Padina tetrastromatica* and its catalytic efficacy in the degradation of organic dyes. J Nanostructure Chem 8(3):333–342

Priyom B, Uma Gowrie S (2017) Mycosynthesis, optimisation and characterization of silver nanoparticles by endophytic fungus isolated from the root of *Casuarina junghuhniana* Miq. nt. J Pharm Sci Rev Res 43(1):107–115

Qu Y, Pei X, Shen W, Zhang X, Wang J, Zhang Z, Li S, You S, Ma F, Zhou J (2017) Biosynthesis of gold nanoparticles by *Aspergillus* sp. WL-Au for degradation of aromatic pollutants. Physica E Low Dimens Syst Nanostruct 88:133–141

Rajan A, Vilas V, Philip D (2015) Studies on catalytic, antioxidant, antibacterial and anticancer activities of biogenic gold nanoparticles. J Mol Liq 212:331–339

Rao CNR, Kulkarni GU, Thomas PJ, Edwards PP (2003) Size-dependent chemistry: properties of nanocrystals. In: Advances in chemistry: a selection of CNR Rao's Publications (1994–2003), pp 227–233

Rengaraj S, Li XZ (2007) Enhanced photocatalytic reduction reaction over Bi^{3+}–TiO_2 nanoparticles in presence of formic acid as a hole scavenger. Chemosphere 66:930–939

Roy K, Sarkar CK, Ghosh CK (2015a) Photocatalytic activity of biogenic silver nanoparticles synthesized using yeast (*Saccharomyces cerevisiae*) extract. Appl Nanosci 5(8):953–959

Roy K, Sarkar CK, Ghosh CK (2015b) Photocatalytic activity of biogenic silver nanoparticles synthesized using potato (*Solanum tuberosum*) infusion. Spectrochim Acta Part A Mol Biomol Spectrosc 146:286–291

Sankar R, Manikandan P, Malarvizhi V, Fathima T, Shivashangari KS, Ravikumar V (2014) Green synthesis of colloidal copper oxide nanoparticles using *Carica papaya* and its application in photocatalytic dye degradation. Spectrochim Acta Part A Mol Biomol Spectrosc 121:746–750

Saquiba M, Tariqa MA, Faisala M, Muneer M (2008) Photocatalytic degradation of two selected dye derivatives in aqueous suspensions of titanium dioxide. Desalination 219:301–311

Selvam GG, Sivakumar K (2015) Phycosynthesis of silver nanoparticles and photocatalytic degradation of methyl orange dye using silver (Ag) nanoparticles synthesized from *Hypnea musciformis* (Wulfen) JV Lamouroux. Appl Nanosci 5(5):617–622

Shahwan T, Sirriah SA, Nairat M, Boyaci E, Eroğlu AE, Scott TB, Hallam KR (2011) Green synthesis of iron nanoparticles and their application as a Fenton-like catalyst for the degradation of aqueous cationic and anionic dyes. Chem Eng J 172(1):258–266

Sleiman M, Vildozo D, Ferronato C, Chovelon JM (2007) Photocatalytic degradation of azo dye Metanil Yellow: optimization and kinetic modeling using a chemometric approach. Appl Catal B Environ 77:1–11

Sreekanth TVM, Jung MJ, Eom IY (2016) Green synthesis of silver nanoparticles, decorated on graphene oxide nanosheets and their catalytic activity. Appl Surf Sci 361:102–106

Srinath BS, Namratha K, Byrappa K (2018) Eco-friendly synthesis of gold nanoparticles by *Bacillus subtilis* and their environmental applications. Adv Sci Lett 24(8):5942–5946

Srivastava N, Mukhopadhyay M (2014) Biosynthesis of SnO_2 nanoparticles using bacterium *Erwinia herbicola* and their photocatalytic activity for degradation of dyes. Ind Eng Chem Res 53(36):13971–13979

Stylidi M, Kondarides DI, Verykios XE (2003) Pathways of solar light-induced photocatalytic degradation of azo dyes in aqueous TiO_2 suspension. Appl Catal B Environ 40:271–286
Sun J, Wang X, Sun J, Sun R, Sun S, Qiao L (2006) Photocatalytic degradation and kinetics of Orange G using nano-sized Sn(IV)/TiO_2/AC photocatalyst. J Mol Catal A Chem 260:241–246
Sun J, Qiao L, Sun S, Wang G (2008) Photocatalytic degradation of Orange G on nitrogen-doped TiO_2 catalysts under visible light and sunlight irradiation. J Hazard Mater 155:312–319
Suresh D, Nethravathi PC, Rajanaika H, Nagabhushana H, Sharma SC (2015a) Green synthesis of multifunctional zinc oxide (ZnO) nanoparticles using *Cassia fistula* plant extract and their photodegradative, antioxidant and antibacterial activities. Mat Sci Semicon Proc 31:446–454
Suresh D, Shobharani RM, Nethravathi PC, Kumar MP, Nagabhushana H, Sharma SC (2015b) *Artocarpus gomezianus* aided green synthesis of ZnO nanoparticles: luminescence, photocatalytic and antioxidant properties. Spectrochim Acta A Mol Biomol Spectrosc 141:128–134
Tamuly C, Hazarika M, Das J, Bordoloi M, Borah DJ, Das MR (2014) Bio-derived CuO nanoparticles for the photocatalytic treatment of dyes. Mater Lett 123:202–205
Teimouri M, Khosravi-Nejad F, Attar F, Saboury AA, Kostova I, Benelli G, Falahati M (2018) Gold nanoparticles fabrication by plant extracts: synthesis, characterization, degradation of 4-nitrophenol from industrial wastewater, and insecticidal activity – a review. J Clean Prod 184:740–753
Tripathi RM, Shrivastav BR, Shrivastav A (2018) Antibacterial and catalytic activity of biogenic gold nanoparticles synthesised by *Trichoderma harzianum*. IET Nanobiotechnol 12(4):509–513
Umamaheswari C, Lakshmanan A, Nagarajan NS (2018) Green synthesis, characterization and catalytic degradation studies of gold nanoparticles against congo red and methyl orange. J Photochem Photobiol B Biol 178:33–39
Veisi H, Azizi S, Mohammadi P (2018) Green synthesis of the silver nanoparticles mediated by *Thymbra spicata* extract and its application as a heterogeneous and recyclable nanocatalyst for catalytic reduction of a variety of dyes in water. J Clean Prod 170:1536–1543
Vilas V, Philip D, Mathew J (2016) Biosynthesis of Au and Au/Ag alloy nanoparticles using *Coleus aromaticus* essential oil and evaluation of their catalytic, antibacterial and antiradical activities. J Mol Liq 221:179–189
Vinothkannan M, Karthikeyan C, Kim AR, Yoo DJ (2015) One-pot green synthesis of reduced graphene oxide (RGO)/Fe_3O_4 nanocomposites and its catalytic activity toward methylene blue dye degradation. Spectrochim Acta A Mol Biomol Spectrosc 136:256–264
Wang KH, Hsieh YH, Wu CH, Chang CY (2000) The pH and anion effects on the heterogeneous photocatalytic degradation of *O*-methylbenzoic acid in TiO_2 aqueous suspension. Chemosphere 40:389–394
Wawrzkiewicz M (2012) Comparison of the efficiency of Amberlite IRA 478RF for acid, reactive, and direct dyes removal from aqueous media and wastewaters. Ind Eng Chem Res 51(23):8069–8078
Wei CH, Tang XH, Liang JR, Tan S (2007) Preparation, characterization and photocatalytic activity of boron- and cerium-codoped TiO_2. J Environ Sci 19:90–96
Weng X, Huang L, Chen Z, Megharaj M, Naidu R (2013) Synthesis of iron-based nanoparticles by green tea extract and their degradation of malachite. Ind Crop Prod 51:342–347
Wiley B, Sun Y, Xia Y (2007) Synthesis of silver nanostructures with controlled shapes and properties. Acc Chem Res 40(10):1067–1076
Yamamoto Y (2008) Silver-catalyzed Csp– H and Csp– Si bond transformations and related processes. Chem Rev 108(8):3199–3222
Yu H, Zheng X, Yin Z, Tao F, Fang B, Hou K (2007) Preparation of nitrogen-doped TiO_2 nanoparticle catalyst and its catalytic activity under visible light. Chin J Chem Eng 15(6):802–807
Zhang S, Zhang C, Liu M, Huang R, Su R, Qi W, He Z (2018) Poly (γ-glutamic acid) promotes enhanced dechlorination of *p*-chlorophenol by Fe-Pd nanoparticles. Nanoscale Res Lett 13(1):219. https://doi.org/10.1186/s11671-018-2634-y
Zhiyong Y, Bensimon M, Sarria V, Stolitchnov I, Jardim W, Laub D, Mielczarski E, Mielczarski J, Kiwi-Minsker L, Kiwi J (2007) $ZnSO_4$–TiO_2 doped catalyst with higher activity in photocatalytic processes. Appl Catal B Environ 76:185–195
Zhiyong Y, Keppner H, Laub D, Mielczarski E, Mielczarski J, Kiwi-Minsker L, Renken A, Kiwi J (2008) Photocatalytic discoloration of Methyl Orange on innovative parylene–TiO_2 flexible thin films under simulated sunlight. Appl Catal B Environ 79:63–71

Chapter 15
Ecotoxicity of Nanomaterials in Aquatic Environment

Murat Ozmen, Abbas Gungordu, and Hikmet Geckil

Contents

15.1 Toxicological Properties of Nanomaterials

Basically, a water ecosystem contains organisms that act as producers, consumers, and decomposers in a balanced level required for its long-term maintenance. Contaminants, however, cause a shift in this balance toward total collapse through their toxic effects exerted by the death of organisms and depletion in the level of dissolved oxygen. Ecosystems are polluted by a plethora of xenobiotics through human activities year in, year out. A novel source of such pollution in recent years is nano-engineered materials that have become increasingly critical to numerous technologies. Nanomaterials (NMs) have entered our lives through various industries such as cosmetics, food, medicine, pharmacy, chemistry, textile, electronics,

M. Ozmen (✉) · A. Gungordu
Department of Biology, Inonu University, Malatya, Turkey
e-mail: murat.ozmen@inonu.edu.tr

H. Geckil
Department of Molecular Biology and Genetics, Inonu University, Malatya, Turkey

D. Thangadurai et al. (eds.), *Nanotechnology for Food, Agriculture, and Environment*, Nanotechnology in the Life Sciences,
https://doi.org/10.1007/978-3-030-31938-0_15

space studies, water treatment, and more (Cao and Wang 2011). Risks posed by ever-increasing production, transport, and use of NMs are still mostly unknown due to the lack of information on their toxicity.

Organisms are naturally exposed to nanoscale particles (NPs) found in ecosystems through natural phenomena such as erosion and volcanic activities (Strambeanu et al. 2015). However, most of the nano-sized materials can enter and accumulate in the aquatic environment through human activities (Fig. 15.1). As a result, toxicological effects of NMs can readily be observed in aquatic organisms. In this context, most engineered NMs can readily bind to and incorporate into cells and integrate with their constituents, given their physicochemical characteristics and bioavailability. Discharging such materials into the environment may pose detrimental effect on the health of organisms and environment. Today, there is an ever-increasing concern about synthetic NMs given the result of many ecotoxicological studies reporting their adverse effects. While most NPs are considered relevant given that they do not cause an immediate death in organisms, at the molecular level, they

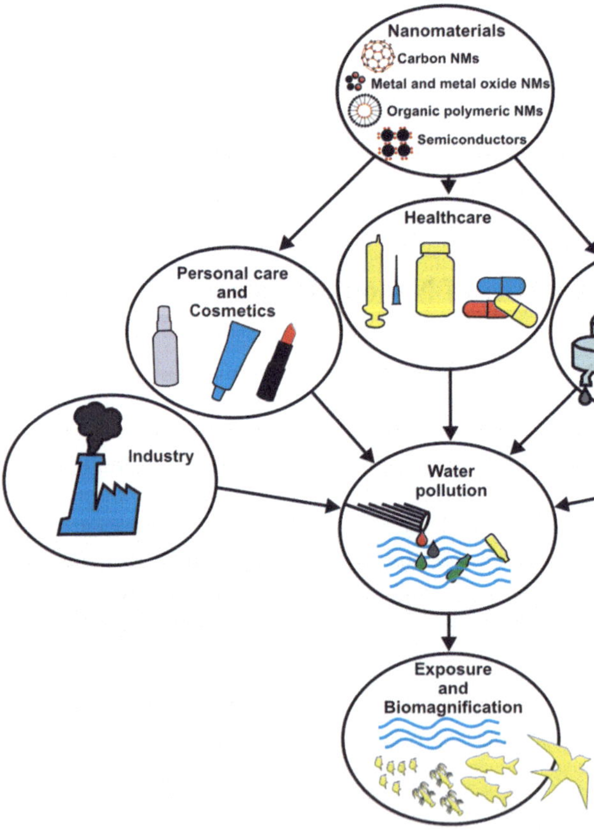

Fig. 15.1 Sources of nano-sized materials given to aquatic environment as a result of human activities

exert a wide range of toxic effects (Prasad 2019). Titanium NPs are the best examples in this regard. Studies have reported that TiO_2 NPs are not toxic or slightly toxic to aquatic organisms, while long-term exposure may cause serious risks (Zhu et al. 2010). NMs including TiO_2 have been shown to have a wide range of adverse effects on the organisms by causing the formation of reactive oxygen species, lipid peroxidation, genotoxicity, mutagenesis, apoptosis, necrosis, changes in mitochondrial dysfunction, and cell morphology (Haynes et al. 2017; Mahaye et al. 2017; Cambier et al. 2018; Freixa et al. 2018). Aquatic organisms are more often exposed to NP toxicity than terrestrial organisms, given their constant contact with these agents through swallowing, gill entry, cellular uptake, and dermal intake. In this regard, the physicochemical characteristics (such as size, aggregation, morphology, surface charge, reactivity, and dissolution) of NMs are key determinants of their toxicity. Accumulated NPs in aquatic organisms move through the food chain and affect all trophic level in any ecosystem (Petersen, 2017). Studies have shown that by translocating within the body and penetrating to various organs and tissues, NMs exert their most toxic effects (Selck et al. 2016). The physicochemical properties of NPs and the properties of intracellular and extracellular environment (e.g., protein or lipid adsorption models) may cause a specific localization of NPs in a target cell and may determine the potential of their toxicity (Yue et al. 2017).

Nanostructures can be classified according to their chemical and structural properties: NPs, dendrimers, mycelia, and drug conjugates are polymeric structures, while carbon nanotubes, metallic NPs, silica, and quantum dots are non-polymeric ones (Bhatia 2016). Another method for classifying NMs may be related to the components that make up their structures such as metallic NPs and metal oxides (such as Ag, Au, TiO_2, CuO_2, ZnO), carbon NMs (such as fullerenes and nanotubes), semiconductors (such as quantum dots), and natural and synthetic polymeric NMs (such as chitosan, poly-lactide-co-glycolic acid) (Gatos and Leong 2017). Metallic NPs are among the most widely used types of engineered NMs; however, relatively little is known about their environmental fate and effects (Hartmann et al. 2014; Dale et al. 2015). Metallic NPs in the environment can be found in a wide variety of forms such as structure, composition, as well as size, surface charge, hydrophobicity, and penetration ability, all important factors affecting their toxicity. Therefore, to understand their toxicological risks, the characteristics of the NMs need to be well analyzed. For example, most of the NPs and nanocomposites may have high photocatalytic activity (Dong et al. 2015; Tissera et al. 2018) and may play a role in photocatalysis of other organic pollutants in the environment (Ozmen et al. 2015; Ozmen et al. 2018). Thus, NMs are likely to react with other components or even may produce more toxic compounds due to their photocatalysis which indirectly leads to ecological risks. In the risk assessment studies of NMs, models have been used as important components of the food chain, representing various trophic levels such as fish, crustacean, and algae.

Ecotoxicology is interested in the harmful effects of chemicals on ecosystems (Walker et al. 1996) where ecological and toxicological effects of chemical pollutants in the environment are integrated (Forbes and Forbes 1994). When an organism is exposed to toxic metallic agents such as NMs in an ecosystem, these substances can also be transmitted to other organisms. In particular, metals and hydrophobic compounds tend to accumulate in organisms. Since various NMs contain metals, they cannot be degraded by physical-chemical or biological mechanisms. Thus, research is focused on the possible toxicological and pathological risks of NMs to human health and to the environment (Yildirimer et al. 2011; Sajid et al. 2015; Almansour et al. 2016; Liu et al. 2017; Indeglia et al. 2018). The bioaccumulation of NMs is common especially in aquatic invertebrates (such as *Daphnia magna, Amphiascus tenuiremis, Chironomus* sp., *Carcinus maenas*), vertebrates (such as *Danio rerio, Carassius auratus, Cyprinus carpio*), and algae and plants (such as *Chlamydomonas reinhardtii, Microcystis aeruginosa, Scenedesmus obliquus, Elodea canadensis*) (Krysanov et al. 2010; Rossbach et al. 2017; Asztemborska et al. 2018; Forouhar Vajargah et al. 2018; Luo et al. 2018; Yu et al. 2018).

Carbon nanotubes are mostly found in digestive tracts of terrestrial organisms such as *Eisenia foetida* (Petersen et al. 2008; Asztemborska et al. 2014; Ates et al. 2015). Due to their specific surface characteristics, carbon nanotubes may bind and transport toxic pollutants and generate reactive radicals (Moore 2006). The particles themselves can be a factor in direct toxicity as most NPs may readily pass through biological membranes and can accumulate in different tissues. However, the tendency of aggregation and poor solubility of NPs in water may relatively limit their uptake by living organisms.

In this chapter, in the light of recent studies, we discussed the toxicological properties and potential risks of NMs in terms of factors contributing to their toxicology, bioavailability, and accumulation in aquatic organisms and in the environment. The possible ecotoxicological risks of NMs in aquatic ecosystems have been reviewed (Table 15.1). The properties and characteristics of NMs are important factors contributing to their toxicity. In addition, the interaction of NMs with environmental components and pollutants may play an important role in their toxicology. A gradual increase of these interactions within the food chain is also possible. Therefore, these complex relationships have been evaluated through the general characteristics of NMs.

15.2 Factors Affecting NM Toxicity

Organisms are continuously being exposed to nanoscale particles such as dust storms, volcanic ash, and other natural sources. Technological advancements further have changed the characteristics of the particulate pollution in the environment. Thus, the proportion of nano-sized particles and the diversity of chemicals are increasing in the ecosystems. Many factors, such as NP size, surface characteristics, shape, chemical composition, and agglomeration in water, may determine their toxicity to organisms and to the environment. Because of their size, most NPs can easily penetrate the biological membranes and readily bind to or mimic certain biological molecules within the cells. This can lead to cellular damage by altering the cellular redox balance, leading to organismal death. In order to better understand

Table 15.1 Some aquatic toxicity studies using different nanoparticles

Type	Size (nm)	Shape	Aim	References
Silica-titania	25–125	Hollow NPs	Cellular uptake, cytotoxicity, and innate immune response	Oh and Park (2014)
TiO_2	20–40	Anatase and rutile	Different toxicity of anatase and rutile form	Yu et al. (2017)
TiO_2	≤20	NP	Comparative toxicity of metal oxide NP aqueous suspensions	Zhu et al. (2008)
Al_2O_3	80			
ZnO	20			
TiO_2	11	NP	The mechanism of toxicity of different NPs based on dissolution and oxidative stress properties	Xia et al. (2008)
CeO_2	8			
ZnO	13			
Au NP	10	Amine-coated	Impacts of gold nanoparticle exposure on two freshwater species	Renault et al. (2008)
Au NP	10–200	Octahedra NP	Impact of gold octahedra NPs on marine clams	Fkiri et al. (2018)
Au NP	30	Nanorod	Understanding size and shape dependent toxicity of gold NMs	Wang et al. (2008)
Ag NMs	10–40	Nanosphere	The hazard potential of silver NP	George et al. (2012)
	45 × 10	Nanoplate		
	20μm × 65	Nanowire		
Silicon carbide	82	Nanowire	Cellular toxicity of silicon carbide NMs as a function of morphology	Chen et al. (2018)
	79	NP		
Polystyrene particle	1000	Spherical and non-spherical	Role of target geometry in phagocytosis	Champion and Mitragotri (2006)
Chitin NMs	8.7–674	Nanocrystal	The effects of the production process on the morphology and properties of chitin NMs produced from the same source of chitin	Larbi et al. (2018)
	9.7–244	Nanofiber		
CuAg bimetallic NM	40/200	Dumbbell shaped	Shape and size-dependent CuAg bimetallic dumbbell structures for organic pollutant hydrogenation	Mallikarjuna and Kim (2018)
Graphene NMs		Nanosheet	In vitro cytotoxicity assessment of graphene nanosheets	Dervin et al. (2018)

(continued)

Table 15.1 (continued)

Type	Size (nm)	Shape	Aim	References
Fe-doped silica nanoshells	500	Nanoshell	In vivo systemic toxicity and biodistribution of iron-doped silica nanoshells	Mendez et al. (2017)
Multiwalled carbon nanotubes	6.6 nm–5 µm	Nanotube	Multiwalled carbon nanotubes toxicity in fish	Cimbaluk et al. (2018)
Carbon nanotubes	26 nm × 1.7 µm	Nanotube	The effects of macrophage exposure to NPs on autophagy	Cohignac et al. (2018)
Nanometric TiO_2	30	NP		
Carbon nanotubes	≥20	Nanotube	Effects of same nanomaterials on marine microalgae	Pikula et al. (2018)
Silicon nanotubes	≥45	Nanotube		
Carbon nanofibers	Length, 5–50 µm	Nanofiber		

the potential ecotoxicological effects of nano-sized metal containing materials released into aqueous environments, it is important to assess their physicochemical characteristics.

15.2.1 Particle Size

Biological membranes are selective semipermeable structures that surround and protect cells (Lodish et al. 2013). Proteins in the membrane structure can serve as channels for the intercellular and extracellular movement of ions and material. The transport of large particles that cannot easily achieved through membrane transporters can be achieved by endocytosis and exocytosis. In this regard, it is important to understand the mechanism of endocytosis and exocytosis to determine the mechanism of toxicity of NPs on organisms in an aquatic ecosystem (Albanese et al. 2012). The cellular uptake pathways can also be utilized for efficient therapeutic applications of NPs (Oh and Park 2014). It is becoming increasingly important to understand the NP transport to determine their possible hazards to health of human and other organisms (Shang et al. 2014). Some membrane transporters play critical roles in infection, parasitism, immunity, and neurodegeneration. Comparative experiments with the bulk counterparts of NMs (i.e., ZnO/bulk, TiO_2/bulk, and Al_2O_3/bulk) were conducted to understand the effect of particle size on toxicity of nanoscale materials on aquatic organisms. However, NPs cannot easily be dispersed in aquatic environment, and several dispersed NPs can re-aggregate rapidly due to their physicochemical properties (Genix and Oberdisse 2018). Individual dispersed NPs may easily pass through the cell membrane, while transport of the aggregates through the membrane barrier may be relatively difficult. Sonication may be an effective method for dispersing the NPs but, they may re-aggregate within a short period (Joo et al. 2013). However, studies have shown that there is no difference between the toxicity level of bulky and nanoscale particles of some metals. In this context, both bulky and nanoparticles of ZnO were found to be equally toxic to embryo and larval development in zebrafish, reducing the survival and hatching rates of embryos and causing tissue damage during development (Zhu et al. 2008). In contrast, neither nano TiO_2 and TiO_2/bulk nor nano Al_2O_3 and Al_2O_3/bulk showed toxicity to zebrafish embryos and larvae under the same experimental conditions (Zhu et al. 2008).

The composition of exposure medium may also affect the toxicity of NPs on organisms. The discrepancies in experimental design, such as life stage of organisms, NP size, and exposure scenarios are known to influence the ecotoxicological effects of NPs (Bondarenko et al. 2016; Minetto et al. 2016). ZnO NPs (13 nm particle size) in RAW-264.7 and BEAS-2B cell lines initiated reactive oxygen species (ROS), oxidative damage, inflammation induction, and cell death-inducing toxicity. Non-dissolved ZnO NPs enter the caveolae in BEAS-2B cells, while dissolved smaller NPs were found in lysosomes in RAW 264.7 cells (Xia et al. 2008). However, CeO_2 NPs (8 nm) suppressed ROS production and induced cellular resistance to an exogenous source of oxidative stress. Studies on various NPs show that certain metal oxide NPs may have cytotoxic effects, while some metals may have cell protective functions.

To determine the toxic effects of gold NPs, different concentrations of 10 nm amine-coated NPs were investigated on phytoplankton algae *Scenedesmus subspicatus* and benthic bivalve *Corbicula fluminea* (Renault et al. 2008). It was shown that amine-coated gold NPs caused 20% mortality in algae even at concentration of 1.6×10^2 NP/cell. Furthermore, cell concentration is decreased, and development is suppressed due to the toxicity of the NPs on algae. However, since no NP residues were found in the algal cells, it was proposed that the lethal effect was due to the weakening of the algae cell walls. It was revealed that the cause of the adsorption of NPs around the cell walls was a consequence of the interaction of the positively charged amine coating with the negatively charged cell wall. The algal cells in the medium are ingested and digested by the mussels. Thus, the exposure of the mussels to the NPs is may be due to trophic levels. It was reported that Au NPs did not enter the stomach cells but only entered into the epithelial cells of the digestive glands and localized in the lysosomal vesicles. Au NPs have also been found in structures called specialized vesicles. Many studies have shown that Au NPs can be cell-mediated by endocytosis (Ng et al. 2015; Behzadi et al. 2017). This is a mechanism by which the NPs' entry can be explained (Gomes et al. 2011; Gomes et al. 2014; Faggio et al. 2018). The increase in the concentration of metallothionein is considered as a defense mechanism against the metallic NPs' toxicity. The significant increase in metallothionein levels was observed in visceral tissues and gills of mussel specimens exposed to metal NPs.

The silver particles are most commonly used in medical and industrial appliances given their physical, chemical, and biological properties. Silver NPs, however, are known to have also toxic effects, and studies relating their size and toxicity are limited. In the literature, toxic effects have been discussed in parallel with increasing concentrations of silver NPs (Kteeba et al. 2017; Liu et al. 2017; Dobrochna et al. 2018; Hu et al. 2018a; Souza et al. 2018). A study described the toxicity of spherical and amine-coated Ag NPs in *Labeo rohita* (Khan et al. 2018). Silver NPs were shown to have toxic effects on organisms by inducing oxidative stress and lipid peroxidation while decreasing the mitochondrial function (Zhang et al. 2014; Zhornik et al. 2014; Khan et al. 2017b).

The toxic effects of smaller NPs are much higher than the larger ones of the same molecule, given their higher ratio of surface area and particle size. This means that the number of bonding sites on an NP surface will increase proportionally. Park et al. (2011) compared the effects of different sizes of silver NPs on cytotoxicity, inflammation, genotoxicity, and developmental toxicity. They found that silver particles were toxic to organisms as they caused substantial change in biomarkers analyzed. The silver NPs in all sizes affected cell membrane integrity and metabolic activity. Most studies have shown that silver NPs show a toxic effect through the oxidative stress. Interestingly, the potential of the 20 nm NPs to generate ROS in a cellular environment as measured by electron spin resonance technique was lower than that of the larger NPs, while their cellular ROS generation was higher. Thus, the interaction of cellular components with NPs in ROS formation may play a critical role (Carlson et al. 2008). This effect has been regarded as a secondary toxic effect. On the other hand, the higher cytotoxicity of smaller particles may relate to the amount of ROS generated at the relatively larger surface area of small NPs (Liu et al. 2010). It was shown that more silver ions can be released from the surface of smaller NPs compared to the larger ones. Toxicological effects may be especially observed more clearly in some organs and tissues. For example, TiO_2 and ZnO NPs affected AChE activity and protein carbonylation in the brain of juvenile *Prochilodus lineatus*. The presence of electron-dense material within the retina is indicative of another uptake route for TiO_2. Gills were important target organs for waterborne toxic agents, and the physical adsorption of NPs on epithelial cells can explain the damages observed in this organ (Miranda et al. 2016).

The studies showed that both Ag NPs and Ag nanowires are toxic to freshwater organisms. However, Ag NPs seem to be more toxic than nanowires, a feature related to their particle size and toxicity. Ag NPs with particle size of 5–25 nm had more than two times toxic effect on *Oryzias latipes* (medaka) than Ag nanowires having a diameter of about 60 nm. The inhibitory rates increased with increasing concentrations of Ag NMs. The median effective concentrations of Ag NPs and Ag nanowires were calculated as 0.012 and 0.139 mg/L for *Daphnia magna*, respectively. Ag nanowires and Ag NP concentration of 0.25 mg/L resulted in flocculation, and 0.5 mg/L and higher concentrations resulted in flocculation, depolarization, rupture, and atrophy. On the other hand, the biomass in the control cultures increased more than 16-fold within the 72-hour test period for *Raphidocelis subcapitata* (Sohn et al. 2015). The amount of silver ions released from the surface of the Ag NPs was greater than the amount of Ag^+ released from the surface of the Ag nanowires; thus, the Ag NPs logically displayed a greater toxicity to the aquatic organisms compared with the nanowires. Thus, it appears that, at least in freshwater environments, the toxicity of Ag NMs may be related to the size and surface area.

Biological barriers are indispensable for the integrity and function of organisms. Blood-brain barrier, small intestine, nose, skin, and mouth mucosa have a protective role against many chemical and biological agents. However, due to their small size, most NPs may easily pass through such barriers (Bennat and Muller-Goymann 2000; Tinkle et al. 2003). Studies have shown that larger Ag NPs (80 nM) have significantly a restricted penetration into rat brain microvessel endothelial cells, while small particles can readily penetrate even at lower concentrations and relatively in short time period. Further, this study suggests that Ag NPs may interact with the cerebral microvasculature producing a pro-inflammatory cascade; these events may further induce brain inflammation and neurotoxicity (Trickler et al. 2010). Therefore, measuring the capacity of NPs to pass biological barriers and show their associated toxic effect is an important aspect in the evaluation of toxicity.

The ecotoxicological effects of particle size and agglomeration of NPs in aqueous media were also investigated using silver NPs in alevin (almost embryonic) and juvenile trout. When NPs were in colloidal form, their toxicity was more than 100-fold lower than in suspended form, 0.25 mg/L vs. 28.2 mg/L, respectively. Thus, the increase in particle size due to agglomeration has been reported to be an important factor in the reduction of toxicity (Johari et al. 2013).

15.2.2 Shape of NPs

Not only the size but also the shape of NPs is an important issue in nano-engineering technology. The shape of NPs depends on the unit cell structure forming the material and the processing conditions. In addition, the shape of NPs may change depending on temperature. Given the ease of their synthesis, most synthetic NPs have a spherical structure (Khan et al. 2017a). However, many NPs with different shapes are synthesized with regard to their chemical composition, procedure of synthesis, and intended use. NPs can be amorphous, as well as found in any other shape such as nanocube, sphere, rod, tube, and wire. Importantly, the biological interactions of NPs are often associated with their shapes which are also an important aspect for toxicity (Chen and Inbaraj 2018). Variations in shapes of NPs may cause a difference in cellular uptake with associated toxicity at different levels. NPs can interact with a particular region in a cell by virtue of their shape. By making use of this association, studies are carried out to understand the acquisition of chemical agents in the treatment of disease such as cancer.

It is reported that the shape of NPs is important in the uptake, distribution, and circulation by the organisms (Caldorera-Moore et al. 2010; Truong et al. 2015). NPs in the spherical structure have been reported to be easier to disperse than irregular or amorphous particles (Champion et al. 2007). In vitro studies showed that nanocylinder and nanosphere particles were taken more promptly than longer filaments. Compared to the cylindrical NPs, the distribution rate of spherical NPs in the body is higher (Geng et al. 2007). Recent studies have shown that the cytotoxic effects of identical NPs with different shapes are different; for example, nanowire particles are more toxic than spherical ones (Truong et al. 2015). Similarly, gold NPs with unstructured surfaces were found to be more toxic to fibroblast cells compared to gold nanostars with multi-segmented surface structures (Favi et al. 2015).

As the production of graphene NPs increases year by year, they also possess an important environmental risk (Schinwald et al. 2012). However, due to changes in their form and other physical-chemical properties, their toxic effects and associated ecotoxicological risks need to be well understood (Handy et al. 2008; Perez et al. 2009; Montagner et al. 2017). However, there are difficulties in revealing the toxic effects of materials such as graphene NPs. Toxicological effects of graphene oxide NPs have been assessed in terms of their cytotoxicity and genotoxicity in adult zebrafish evaluating for various markers (Souza et al. 2017). Graphene oxide (GO) exposure has been reported to cause an increase in the number of gill cells in early apoptotic and necrotic stages. Also, researchers have detected the formation of reactive oxygen species in gill cells exposed to different concentration of GO NPs. Histopathologic analyses performed on many tissues have also revealed various lesions.

15.2.3 Surface Characteristics

The surface charge and intensity are the important determinants of NP interaction with the environment. Surface features can also determine electrostatic interactions with bioactive compounds, particularly in cellular membranes. However, toxicity studies do not adequately reflect the mechanisms of NP uptake by organisms. The intracellular uptake of NPs may be directly related to the particle surface characteristics such as surface charge as well as the cell type. Increasing surface charges have been shown to increase particle uptake in comparison with uncharged NPs. Therefore, ionic forms of many metals have higher toxic effects than those oxidized forms. However, the issue about the higher toxicity of ionic forms of NPs is poorly understood. For example, charged gold NPs are more toxic than their neutral forms (Oh et al. 2010). In general, positively charged NPs are reported to cause a higher cytotoxicity. For example, positively charged ZnO NPs may assert more cytotoxicity on non-phagocytic cells than their negatively charged variants. Previously, it was reported that cationic NPs were more potent than anionic NPs in terms of their destabilizing effect on cell membrane composition, disruption of mitochondrial and lysosomal function, and inducing autophagosome formation (Frohlich 2012). It has been reported that phagocytic cells generally have higher uptake of anionic NPs, while non-phagocytic cells have higher levels of uptake of cationic NPs. Furthermore, surface load density may be an important factor in the uptake of NPs (Ruenraroengsak and Tetley 2015). The surface charge of a specific NP may influence its capacity to produce reactive oxygen species (ROS), determine the binding sites for receptors, and influence its dispersion and aggregation.

Important uptake mechanisms for eukaryotic cells are pinocytosis, receptor-mediated endocytosis, and phagocytosis (Kettler et al. 2014). But, it is not clear yet how NP characteristics in the nano-bio interface interact with biological receptor properties, a potential mediator of toxicity (Silva et al. 2014).

Silver NPs are the most widely used and well-known NPs, and their toxic effect on organisms is induced by various mechanisms (Guigas et al. 2017; Koser et al. 2017; Bouallegui et al. 2018; Hu et al. 2018b; Aziz et al. 2015, 2016, 2019); (i) they accumulate in the membrane surface and change the cellular membrane properties; (ii) they can penetrate the cell membrane, enter cell, and cause mutations in DNA; and (iii) they may interact with both proteins and phosphorus moieties in the DNA through released Ag ions and eventually cause cell death. In this context, there is a lack of consensus regarding the form(s) of Ag that lead to the cell toxicity (Hu et al. 2018b). It was shown that negatively charged Ag NPs have the ability to interact with positively charged choline found in the cellular membrane. It has been reported that this affects and induces the flip-flop of phospholipids in the membrane. Hence, NPs with more negative charge will induce a higher level of flip-flop of membrane phospholipids. In this respect, Au NPs were shown to cause more flip-flop than Ag NPs. It has also been shown that metal oxide NPs with different chemical compositions may cause different toxic effects in the developing zebrafish. Exposure to nano TiO_2, ZnO, and Al_2O_3 in zebrafish embryos has been reported to lead to significantly different toxic and severe effects (Zhu et al. 2008).

15.2.4 Shape

The fullerene (nC60) NMs are pure carbon-based molecules with at least 60 atoms of carbons, which have hollow sphere, ellipsoid, tubular, and many other shapes. They are widely used in a variety of fields such as semiconductors and transport vehicles for a variety of drugs. It was shown that when exposed to water-soluble fullerene NMs, juvenile *Micropterus salmoides* had a significant cellular lipid peroxidation and total protein oxidation, while the same was not observed in adult fish (Oberdorster 2004). Researchers have reported that fish exposed to fullerene NPs have a significant increase in lipid peroxidation in brain tissue but a decrease in lipid peroxidation due to the induction of repair enzymes in gill and liver tissues. In subsequent studies, toxicity differences of nC60 dissolved in tetrahydrofuran (THF) and mixed with water were evaluated by acute toxicity tests on *Daphnia magna* for further assays in adult male fathead minnow (*Pimephales promelas*) (Zhu et al. 2006). Water-stirred-nC60 did not show any significant physical effect in fish after 48 hours, whereas with a tetrahydrofuran-fullerene (THF-nC60) mixture, 100% mortality was observed between 6 and 18 hours. However, it has been reported that nC60 mixed with water increases lipid peroxidation (LPO) level in the brain, significantly increases LPO in the gill, and significantly increases expression of CYP2 family isozymes in the liver compared to control fish. This is important because it shows that the same NPs found in different environments may lead to different toxic effects in the same organism.

Nickel NPs with different particle sizes and dendritic structure were also compared. Soluble nickel salts showed almost equal toxicity in zebrafish embryos compared to nano-sized particles. The dendritic structure (branching) of Ni NPs was found to be very toxic. Intestinal defects were not observed in embryos exposed to soluble nickel. However, the separation of skeletal muscle was at concentrations even higher than the LD50 value detected (Ispas et al. 2009).

The toxicity of NPs shows significant differences with regard to their solubility in an aqueous environment. ZnO and CuO NPs are highly soluble materials and readily release ions in an aqueous solution. This property makes them highly toxic to aquatic organisms (Cong et al. 2017; Thit et al. 2017). In this context, toxicity and bioavailability of TiO_2 and CeO_2 NPs are limited, given their insolubility (Keller et al. 2010). However, TiO_2 is one of most widely used NPs that may have toxic effects on aquatic organisms (Sharma 2009). But, most studies have shown that TiO_2 NPs may be more relevant than other NPs and that their toxicity to aquatic organisms is relatively low (Mukherjee and Acharya 2018). The properties of nano TiO_2 differ significantly from bulk-TiO_2 of the same composition because of an increase in surface area (Sendra et al. 2017). However, TiO_2 NPs found in sediments have been also reported to be present in plants. This shows that aquatic plants, a major trophic level of the food chain, can accumulate NPs and become NP sources for higher organisms (Asztemborska et al. 2018).

15.3 Biomarkers of Ecotoxicological Effects

The contamination of the environment comes from almost every aspect of human activities, being agricultural or industrial (Barrick et al. 2017). As with many other environmental chemicals, the negative effects of NMs on the health of organisms and environment are of great concern. Thus, there is an urgent need to determine the toxic effects of NMs sensitively, rapidly, and cost-effectively. However, assessing the effects of these substances on the ecosystems is challenging and time-consuming, as the complexity of ecosystems and a plethora of factors affecting the chemical contamination cannot easily be simulated in a laboratory setting. In addition, some conditions caused by environmental factors cannot be easily determined in the laboratory environment because the gradual accumulation of trophic pathways and toxic substances in water ecosystems and chemical interactions can lead to fast and effective toxicity (Bundschuh et al. 2016).

Most NMs have properties as conductivity, high tensile strength, high heat tolerance, high stability, hydrophobic or hydrophilic properties, electric current transport, superconductivity, ultraviolet light blocking or absorbing capacity, and antimicrobial activity (Kashiwada 2006). Depending on the differences between their composition and properties, NMs can affect organisms at different levels and in different ways, such as cellular, subcellular, and molecular. Historically, toxicology is regarded as the concentration of substance and the response of the exposed organism. Today, concentration and exposure time are two important parameters in toxicology. However, these factors are not easily measurable for most chemicals. A threshold concentration at which the chemical can be regarded as "safe" or "harmful" is often not readily detectable (Elsaesser and Howard 2012). In the case of NMs, the situation is even more complicated. NMs can be very different from one another in terms of their effect even if they are classified under the same name. For example, toxicity mechanism of three different NPs on freshwater alga *Pseudokirchneriella subcapitata* can be quite different; ZnO toxicity is mostly due to the Zn^{2+} ions, while CeO_2 toxicity is mainly the result of direct interaction with algal cells, whereas Ag toxicity is caused by the oxidative dissolution of Ag to toxic Ag^+ (Batley et al. 2013). In this regard, whether the properties of NMs require a new toxicological science is a good question posed by Nel et al. (2006). Today, studies evaluating the toxicity of NMs have begun to be considered as a separate subfield of toxicology under the name of nanotoxicology.

In order to better identify the toxicological risks of NMs, one has to determine not only their environmental impacts but their genotoxic effects, as well as endocrine disruption capabilities, and their effect on immune and reproductive systems (Kashiwada 2006). However, despite the rapid development in nanotoxicology, there is still a lack of standardization of experimental conditions, as well as the exposure routes and times, all contributing to difficulties in interpretation of the ecotoxicity of NMs (Rocha et al. 2015b). As with other chemical substances, toxic effects of NMs are not only determined by mortality. In particular, to determine the health effects of sublethal concentrations of chemicals, physiological and bio-

Table 15.2 Useful biomarkers for determining toxicity of NPs to aquatic organisms

Organism	Species	Nanomaterial	Concentration	Exposure time	Biomarker												References
					SOD	GST	GPx	GR	GSH	CAT	MDA	AChE	LDH	DNA damage	Na/K ATPase	LPO	
Amphibian	*Xenopus laevis*	TiO_2	1500 mg/L	96 h		–		=	–				=			=	Ozmen et al. (2015)
	Xenopus laevis	TiO_2	320 mg/L	96 h		=		=				=	=				Birhanli et al. (2014)
	Rhinella arenarum	SiO_2	0.001 mg/L	48 h		+											Lajmanovich et al. (2018)
Fish	*Danio rerio*	Graphene oxide	10 mg/L	96 h	+				–	+	+						Chen et al. (2016)
	Danio rerio	Carbone nanotubes	50 mg/L	96 h	=	–				=				+			Cimbaluk et al. (2018)
	Astyanax altiparanae	Carbone nanotubes	50 mg/L	96 h	=	=				=				+			
	Chapalichthys pardalis	Ag NP	4.08 mg/L	21 d	–		–			–	=						Valerio-Garcia et al. (2017)
	Catla catla	Ag NP	64 mg/L	24 h	–				–	–				+		+	Taju et al. (2014)
	Labeo rohita	Ag NP	64 mg/L	24 h	–				–	–				+		+	
	Danio rerio	Ag NP	20 ppt	–						+							Yeo and Kang (2008)
	Hoplias intermedius	Ag nanospheres	0.2 mg/kg	96 h								+		–			Klingelfus et al. (2017)
	Micropterus salmoides	C60	1 mg/L	48 h					–		+						Oberdorster (2004)
	Cyprinus carpio	C60	1 mg/L	4 h				–			+						Ferreira et al. (2012)

	Cyprinus carpio	CuO NP	0.1 mg/L	7 d	+	+				+							Gupta et al. (2016)
	Apistogramma agassizii	CuO NP	0.058 mg/L	96 h	+	=				+					=	+	Braz-Mota et al. (2018)
	Paracheirodon axelrodi	CuO NP	0.070 mg/L	96 h	=	=				=					=	+	
Invertebrates	*Daphnia pulex*	C60	100 mg/L	24 h		+				+							Klaper et al. (2009)
	Daphnia magna	C60	20 mg/L	24 h	–											+	Lv et al. (2017)
	Mytilus galloprovincialis	Cd-based quantum dots	0.01 mg Cd/L	7 d	+	+	+			=							Rocha et al. (2015a)
	Mytilus coruscus	TiO_2 NP	10 mg/L	24 h	+	+	+		+	+	+						Huang et al. (2018)
	Mytilus edulis	CuO NP	0.01 mg/L	24 h	+	+				+							Chatel et al. (2018)
	Scrobicularia plana	CuO NP	0.01 mg/L	16 d	+	+				+	=	=	=				Buffet et al. (2011)
	Nereis diversicolor	CuO NP	0.01 mg/L	7 d	=	+				+	=	=	=				

Symbols and abbreviations: – inhibition, = no (significant) response, + strong induction, *h* hour, *d* day chemical responses may be considered as useful early warning systems (Ozmen et al. 2015). The biomarkers may change according to biologic responses (van der Oost et al. 2003). Today, there are a variety of biomarkers for assessing the effects of NMs as well as other environmental pollutants on aquatic organisms (Table 15.2).

15.3.1 Molecular and Biochemical Effects

Many studies have shown that the toxicity of some NMs is dependent on the generation of reactive oxygen species (ROS) including the superoxide anion, hydroperoxyl radical, hydrogen peroxide, and the hydroxyl radical (Klaunig and Kamendulis 2008). The production of ROS has been shown to result in oxidative stress, lipid peroxidation, protein, and DNA damage (Yu et al. 2017; Chatel et al. 2018). Some NPs can promote the formation of pro-oxidants that degrade the delicate balance between the ability of the biological system to produce and detoxify ROS (Arora et al. 2012). Due to oxidative damage caused by NMs, the disruption of the antioxidant defense system [glutathione peroxidase (GPx), catalase (CAT), and superoxide dismutase (SOD)], the induction or inhibition of detoxification enzymes [cytochrome p4501A (CYP1A), glutathione *S*-transferase (GST)], LPO level, protein oxidation, mitochondrial degradation, cytoskeletal disorganization, and DNA damage may occur (Rocha et al. 2015b; Chatel et al. 2018). In this sense, Ag NPs may directly interact with DNA, inducing DNA-DNA and DNA-protein crosslinks (Klingelfus et al. 2017). The short-term exposure to CuO NPs was shown to cause oxidative stress in fish. It was reported that Cu NPs caused an increase in the activity of enzymes such as GST, SOD, and CAT in the kidney, liver, and gill, due to overproduction of free radicals and impaired internal homeostasis. Also, it was determined that a protein expression associated with oxidative stress and steroid biosynthesis may result (Gupta et al. 2016). Ag NP also was shown to cause important changes in antioxidant biomarkers such as increased lipid peroxidation and significantly reduced GSH, SOD, and CAT activities (Taju et al. 2014). The fullerenes (nC60), the widely used carbon-containing NMs, can also cause oxidative damage and lipid peroxidation in fish brain and deplete the GSH (Oberdorster 2004). Another study reported that the graphene oxide exposure caused an increased malondialdehyde (MDA) level, SOD and CAT activities, and decreased glutathione level, all indicating the oxidative stress (Chen et al. 2016).

The toxicological effects of NMs and non-nanoscale bulk materials may not only be different in different organisms but different in different tissues of the same organism. A study reported nano and non-nano Cd-induced tissue-specific responses. The SOD, GPx, and GST activities have been identified as the most sensitive biomarkers of oxidative stress caused by Cd-based quantum dots (QDs) in mussels. Exposure to Cd-QDs and dissolved Cd produced different results in GPx activities in different tissues. The Cd-QDs increased the GPx activities in the gill and digestive glands of fish, while the dissolved Cd decreased the activity of the enzyme (Rocha et al. 2015a).

Carbon nanotubes (CNTs) are also new NMs used in the production of new technologies given their advantageous physical and chemical features (Martinez et al. 2014). However, as with other NMs, CNTs also induce oxidative stress biomarkers and cause DNA damage. In addition, CNTs can bind to macromolecules like proteins and cause the formation of residual metabolites. The impossibility of GSH to establish a conjugation with these metabolites is thought to cause of decreased levels of GST (Cimbaluk et al. 2018).

15.3.2 Cellular Effects

The type, size, and shape of NMs affect their entry into cells where they induce ROS production through interactions with intracellular components. Such interactions may damage not only the plasma membrane but also organelle membranes. This has been confirmed by observing severe necrosis closely associated with direct damage to the plasma and lysosomal membranes (Yu et al. 2017). The toxicity mechanism of NMs can be generalized in three respects, taking into account the TiO_2 toxicity for (1) ROS production, (2) cell membrane damage and lipid peroxidation, and (3) ensuing cell damage (Hou et al. 2018).

The graphene oxide NPs were shown to cause a series of cellular changes such as vacuolation, disintegration of cell boundaries, loose arrangement of cells, and histolysis (Chen et al. 2016). Cells exposed to the Ag NPs showed mitochondrial damage and lysosomal dysfunction and caused a rounded, fusiform, and irregular shape in cells (Taju et al. 2014). Cerium dioxide (CeO_2) NPs were shown to induce changes in lysosome membrane stability (LMS), phagocytic capacity, and extracellular ROS in hemocytes of *Mytillus galloprovincialis*. These NMs also lead to downregulation of the immune system as evidenced by loss of LMS and a reduction in phagocytosis capacity. The smaller agglomerates of these NMs did not cause significant changes in biomarkers of stress and immunological parameters in mussel hemocytes (Sendra et al. 2018). The short-term exposure to copper NPs (CuO NPs) caused altered morphology and cellular damage in fish (Gupta et al. 2016). CuO NPs were shown to accumulate in the gills of mussels and cause serious effects on biochemical, organelle, cell, and tissue levels. At the organelle level, there is clear evidence that CuO NPs affect lysosome membrane stability in mussel, *M. edulis*, whereas at the tissue level, they cause discoloration and pigmentation in cells. Thus, the NPs encountered in the natural environment have the potential to affect filter feeders at multiple biological organizational levels (Hu et al. 2014).

15.3.3 *Physiological Effects*

When the organism is exposed to NMs in the aquatic system, they may be taken up, transported, and bioaccumulated in various tissues. Several sub-effects such as respiratory toxicity, impairment of trace elements in the tissue, inhibition of enzymes required for physiological functions (such as Na^+/K^+ ATPase), and oxidative stress have been demonstrated in NM-exposed organisms. Pathologies originating from NP exposure have been observed in a number of organs including the gill, liver, and brain and hematologic, ocular, and visual systems (Liu et al. 2016).

Exposure of fish to xenobiotics such as aqueous metal ions may alter its physiological functions including alterations in the osmoregulation, acid-base balance, and respiratory functions. NM exposure increased mucus secretions and epithelial edema. Histological changes in the gill epithelium result in loss of blood flow with subsequent cardiovascular collapse and ultimately organismal death. The evidence is now also emerging that some nanometals can also affect the gill in similar ways to dissolved metals. Exposure to TiO_2 NPs in rainbow trout caused edema in the gills. Exposure to nano Cu caused damage to the lamellae characterized by proliferation of epithelial cells and edema of primary and secondary filaments due to inhibition of Na^+/K^+-ATPase (Shaw and Handy 2011). Furthermore, exposure to CuO NPs affected both mitochondrial physiology and gill osmoregulatory mechanisms. The Cu accumulation in the body and lipid peroxidation in gills significantly increased, and the activity of antioxidant enzymes (SOD and CAT) increased (Braz-Mota et al. 2018). Excessive iron exposure may result in accumulation of iron masses that can cause blockage in the gills (Shaw and Handy 2011).

The graphene oxide NMs in a species of water flea *Ceriodaphnia dubia* after both acute and chronic exposures may cause ROS generation, reduction on feeding rates, reproduction inhibition, accumulation on gut tract, and death of the organism. Accumulation of NPs in the gut tract may possibly cause a reduced nutrient uptake and interfere with reproduction. In addition to its effect on energy availability, graphene oxide accumulation on the gut tract of *C. dubia* may cause more serious consequences, such as trophic transfer, since predators may be indirectly exposed to NMs. For this reason, graphene oxide NM presence in water may cause strong adverse effects on the aquatic organisms not only at the organismal level but also at population level (Souza et al. 2018).

15.4 Accumulation of Nanomaterials

Bioaccumulation occurs when the rate of uptake of the chemical substance by the organism is faster than the excretion rate of the same substance (Freixa et al. 2018). However, there are some differences in bioavailability and in particular the uptake of metal NMs compared to other environmental pollutants. For example, the chemistry and behavior of NPs include the dynamic aspects of the collection theory, far from the equilibrium models traditionally used for free metal ions. Biological uptake of NPs is not possible through ion transport. Endocytosis and receptor-mediated transport are two possible mechanism for the uptake of NPs (Shaw and Handy 2011; Yue et al. 2017).

Although many NMs used today are present at low nondetectable concentrations in the environment, the increase in concentration at a trophic level is a concern, given their accumulation in aquatic organisms. For example, since there have been no major releases reported for many NPs, such as nano Fe_2O_3, little is known about their true distribution levels in the aquatic environment. However, some NMs have been found to accumulate in organisms and also in the sediment environment.

Dispersed NPs such as ZnO have been detected in the intestine, epithelial cells, and even tissues beyond the epithelial barrier. Several ZnO NPs were identified in the ovaries of *D. magna*. ZnO NPs ingested by *D. magna* in the soluble form were sufficient to directly damage the mitochondria, causing low energy availability and consequently reduction in growth and reproduction (Bacchetta et al. 2017).

Nano-sized silver is a commonly used substituent in NPs used for antimicrobial activity. Ag NPs reaches water resources through human activities, causing ecotoxicological effects on living organisms (Yeo and Kang 2008; Asztemborska et al. 2014; Khan et al. 2015; Afifi et al. 2016). The accumulation of Ag NPs in the water ecosystem and thus in the organism is an important aspect of their bioaccumulation via the food chain (Petersen 2017).

As was the case for toxicity, particle size and particulate type are also important factors for bioaccumulation of NPs. Au NPs are generally considered to be inert and biocompatible nontoxic metals such as bulk gold. However, in the aquatic ecosystem, Au NPs were found to be bioaccumulated in both mussels and zebrafish (Joubert et al. 2013; Dedeh et al. 2015; Dayal et al. 2017).

15.5 Conclusion

In conclusion, while they are becoming more involved in our daily lives, nanotechnology products must be given a higher consideration as important environmental pollutants and toxicity agents in our ecosystems. Hence, there is a need for proper test methods and equipment to assess the risks of NPs to the aquatic environment taking into account their physical, chemical, and biological aspects and mode of interaction with organisms, cells, and environment.

References

Afifi M, Saddick S, Abu Zinada OA (2016) Toxicity of silver nanoparticles on the brain of *Oreochromis niloticus* and *Tilapia zillii*. Saudi J Biol Sci 23:754–760

Albanese A, Tang PS, Chan WC (2012) The effect of nanoparticle size, shape, and surface chemistry on biological systems. Annu Rev Biomed Eng 14:1–16

Almansour M, Sajti L, Melhim W, Jarrar BM (2016) Ultrastructural hepatocytic alterations induced by silver nanoparticle toxicity. Ultrastruct Pathol 40:92–100

Arora S, Rajwade JM, Paknikar KM (2012) Nanotoxicology and *in vitro* studies: the need of the hour. Toxicol Appl Pharmacol 258:151–165

Asztemborska M, Jakubiak M, Ksiazyk M, Steborowski R, Polkowska-Motrenko H, Bystrzejewska-Piotrowska G (2014) Silver nanoparticle accumulation by aquatic organisms - neutron activation as a tool for the environmental fate of nanoparticles tracing. Nukleonika 59:169–173

Asztemborska M, Jakubiak M, Steborowski R, Chajduk E, Bystrzejewska-Piotrowska G (2018) Titanium dioxide nanoparticle circulation in an aquatic ecosystem. Water Air Soil Poll 229:208

Ates M, Arslan Z, Demir V, Daniels J, Farah IO (2015) Accumulation and toxicity of CuO and ZnO nanoparticles through waterborne and dietary exposure of goldfish (*Carassius auratus*). Environ Toxicol 30:119–128

Aziz N, Faraz M, Pandey R, Sakir M, Fatma T, Varma A, Barman I, Prasad R (2015) Facile algae-derived route to biogenic silver nanoparticles: Synthesis, antibacterial and photocatalytic properties. Langmuir 31:11605–11612. DOI: https://doi.org/10.1021/acs.langmuir.5b03081

Aziz N, Pandey R, Barman I, Prasad R (2016) Leveraging the attributes of Mucor hiemalis-derived silver nanoparticles for a synergistic broad-spectrum antimicrobial platform. Front Microbiol 7:1984. doi: https://doi.org/10.3389/fmicb.2016.01984

Aziz N, Faraz M, Sherwani MA, Fatma T, Prasad R (2019) Illuminating the anticancerous efficacy of a new fungal chassis for silver nanoparticle synthesis. Front Chem 7:65. doi: https://doi.org/10.3389/fchem.2019.00065

Bacchetta R, Santo N, Marelli M, Nosengo G, Tremolada P (2017) Chronic toxicity effects of $ZnSO_4$ and ZnO nanoparticles in *Daphnia magna*. Environ Res 152:128–140

Barrick A, Chatel A, Bruneau M, Mouneyrac C (2017) The role of high-throughput screening in ecotoxicology and engineered nanomaterials. Environ Toxicol Chem 36:1704–1714

Batley GE, Kirby JK, McLaughlin MJ (2013) Fate and risks of nanomaterials in aquatic and terrestrial environments. Acc Chem Res 46:854–862

Behzadi S, Serpooshan V, Tao W, Hamaly MA, Alkawareek MY, Dreaden EC, Brown D, Alkilany AM, Farokhzad OC, Mahmoudi M (2017) Cellular uptake of nanoparticles: journey inside the cell. Chem Soc Rev 46:4218–4244

Bennat C, Muller-Goymann CC (2000) Skin penetration and stabilization of formulations containing microfine titanium dioxide as physical UV filter. Int J Cosmet Sci 22:271–283

Bhatia S (2016) Nanoparticles types, classification, characterization, fabrication methods and drug delivery applications. In: Bhatia S (ed) Natural polymer drug delivery systems: nanoparticles, plants, and algae. Springer, Cham, pp 33–93

Birhanli A, Emre FB, Sayilkan F, Gungordu A (2014) Effect of nanosized TiO_2 particles on the development of *Xenopus laevis* embryos. Turk J Biol 38:283–288

Bondarenko OM, Heinlaan M, Sihtmäe M, Ivask A, Kurvet I, Joonas E, Jemec A, Mannerström M, Heinonen T, Rekulapelly R, Singh S, Zou J, Pyykkö I, Drobne D, Kahru A (2016) Multilaboratory evaluation of 15 bioassays for (eco)toxicity screening and hazard ranking of engineered nanomaterials: FP7 project NANOVALID. Nanotoxicology 10:1229–1242

Bouallegui Y, Ben Younes R, Oueslati R, Sheehan D (2018) Role of endocytotic uptake routes in impacting the ROS-related toxicity of silver nanoparticles to *Mytilus galloprovincialis*: a redox proteomic investigation. Aquat Toxicol 200:21–27

Braz-Mota S, Campos DF, MacCormack TJ, Duarte RM, Val AL, Almeida-Val VMF (2018) Mechanisms of toxic action of copper and copper nanoparticles in two Amazon fish species: Dwarf cichlid (*Apistogramma agassizii*) and cardinal tetra (*Paracheirodon axeirodi*). Sci Total Environ 630:1168–1180

Buffet PE, Tankoua OF, Pan JF, Berhanu D, Herrenknecht C, Poirier L, Amiard-Triquet C, Amiard JC, Bérard JB, Risso C, Guibbolini M, Roméo M, Reip P, Valsami-Jones E, Mouneyrac C (2011) Behavioural and biochemical responses of two marine invertebrates *Scrobicularia plana* and *Hediste diversicolor* to copper oxide nanoparticles. Chemosphere 84:166–174

Bundschuh M, Seitz F, Rosenfeldt RR, Schulz R (2016) Effects of nanoparticles in fresh waters: risks, mechanisms and interactions. Freshw Biol 61:2185–2196

Caldorera-Moore M, Guimard N, Shi L, Roy K (2010) Designer nanoparticles: incorporating size, shape and triggered release into nanoscale drug carriers. Expert Opin Drug Del 7:479–495
Cambier S, Røgeberg M, Georgantzopoulou A, Serchi T, Karlsson C, Verhaegen S, Iversen TG, Guignard C, Kruszewski M, Hoffmann L, Audinot JN, Ropstad E, Gutleb AC (2018) Fate and effects of silver nanoparticles on early life-stage development of zebrafish (*Danio rerio*) in comparison to silver nitrate. Sci Total Environ 610-611:972–982
Cao G, Wang Y (2011) Nanostructures and Nanomaterials: synthesis, properties, and applications. World scientific series in nanoscience and nanotechnology: Vol. 2. World Scientific, Singapore, p 596
Carlson C, Hussain SM, Schrand AM, Braydich-Stolle LK, Hess KL, Jones RL, Schlager JJ (2008) Unique cellular interaction of silver nanoparticles: size-dependent generation of reactive oxygen species. J Phys Chem B 112:13608–13619
Champion JA, Mitragotri S (2006) Role of target geometry in phagocytosis. Proc Natl Acad Sci U S A 103:4930–4934
Champion JA, Katare YK, Mitragotri S (2007) Particle shape: a new design parameter for micro- and nanoscale drug delivery carriers. J Control Release 121:3–9
Chatel A, Lievre C, Barrick A, Bruneau M, Mouneyrac C (2018) Transcriptomic approach: a promising tool for rapid screening nanomaterial-mediated toxicity in the marine bivalve *Mytilus edulis* – application to copper oxide nanoparticles. Comp Biochem Physiol C 205:26–33
Chen BH, Inbaraj BS (2018) Various physicochemical and surface properties controlling the bioactivity of cerium oxide nanoparticles. Crit Rev Biotechnol 38:1003–1024
Chen M, Yin J, Liang Y, Yuan S, Wang F, Song M, Wang H (2016) Oxidative stress and immunotoxicity induced by graphene oxide in zebrafish. Aquat Toxicol 174:54–60
Chen F, Li G, Zhao ER, Li J, Hableel G, Lemaster JE, Bai Y, Sen GL, Jokerst JV (2018) Cellular toxicity of silicon carbide nanomaterials as a function of morphology. Biomaterials 179:60–70
Cimbaluk GV, Ramsdorf WA, Perussolo MC, Santos HKF, Da Silva De Assis HC, Schnitzler MC, Schnitzler DC, Carneiro PG, Cestari MM (2018) Evaluation of multiwalled carbon nanotubes toxicity in two fish species. Ecotoxicol Environ Saf 150:215–223
Cohignac V, Landry MJ, Ridoux A, Pinault M, Annangi B, Gerdil A, Herlin-Boime N, Mayne M, Haruta M, Codogno P, Boczkowski J, Pairon JC, Lanone S (2018) Carbon nanotubes, but not spherical nanoparticles, block autophagy by a shape-related targeting of lysosomes in murine macrophages. Autophagy 14:1323–1334
Cong Y, Jin F, Wang J, Mu J (2017) The embryotoxicity of ZnO nanoparticles to marine medaka, *Oryzias melastigma*. Aquat Toxicol 185:11–18
Dale AL, Casman EA, Lowry GV, Lead JR, Viparelli E, Baalousha M (2015) Modeling nanomaterial environmental fate in aquatic systems. Environ Sci Technol 49:2587–2593
Dayal N, Singh D, Patil P, Thakur M, Vanage G, Joshi DS (2017) Effect of bioaccumulation of gold nanoparticles on ovarian morphology of female zebrafish (*Danio rerio*). World J Pathol 6:1–12
Dedeh A, Ciutat A, Treguer-Delapierre M, Bourdineaud JP (2015) Impact of gold nanoparticles on zebrafish exposed to a spiked sediment. Nanotoxicology 9:71–80
Dervin S, Murphy J, Aviles R, Pillai SC, Garvey M (2018) An *in vitro* cytotoxicity assessment of graphene nanosheets on alveolar cells. Appl Surf Sci 434:1274–1284
Dobrochna A, Jerzy S, Teresa O, Magda F, Malgorzata R, Yuichiro M, Kacper M (2018) Effect of copper and cilver nanoparticles on trunk muscles in rainbow trout (*Oncorhynchus mykiss*, Walbaum, 1792). Turk J Fish Aquat Sci 18:781–788
Dong ZH, Zhang F, Wang D, Liu X, Jin J (2015) Polydopamine-mediated surface-functionalization of graphene oxide for heavy metal ions removal. J Solid State Chem 224:88–93
Elsaesser A, Howard CV (2012) Toxicology of nanoparticles. Adv Drug Deliver Rev 64:129–137
Faggio C, Tsarpali V, Dailianis S (2018) Mussel digestive gland as a model tissue for assessing xenobiotics: an overview. Sci Total Environ 636:220–229
Favi PM, Gao M, Johana Sepulveda Arango L, Ospina SP, Morales M, Pavon JJ, Webster TJ (2015) Shape and surface effects on the cytotoxicity of nanoparticles: gold nanospheres versus gold nanostars. J Biomed Mater Res A 103:3449–3462

Ferreira JRF, Barros DM, Geracitano LA, Fillmann G, Fossa CE, De Almeida EA, Prado MC, Neves BRA, Pinheiro MVB, Monserrat JM (2012) Influence of in vitro exposure to fullerene C60 in redox state and lipid peroxidation of brain and gills of carp *Cyprinus carpio* (Cyprinidae). Environ Toxicol Chem 31:961–967

Fkiri A, Sellami B, Selmi A, Khazri A, Saidani W, Imen B, Sheehan D, Hamouda B, Smiri LS (2018) Gold octahedra nanoparticles ($Au_{\_0.03}$ and $Au_{\_0.045}$): synthesis and impact on marine clams *Ruditapes decussatus*. Aquat Toxicol 202:97–104

Forbes VE, Forbes TL (1994) Ecotoxicology in theory and practice. Chapman and Hall, London

Forouhar Vajargah M, Mohamadi Yalsuyi A, Hedayati A, Faggio C (2018) Histopathological lesions and toxicity in common carp (*Cyprinus carpio* L. 1758) induced by copper nanoparticles. Microsc Res Tech 81:724–729

Freixa A, Acuna V, Sanchis J, Farre M, Barcelo D, Sabater S (2018) Ecotoxicological effects of carbon based nanomaterials in aquatic organisms. Sci Total Environ 619:328–337

Frohlich E (2012) The role of surface charge in cellular uptake and cytotoxicity of medical nanoparticles. Int J Nanomedicine 7:5577–5591

Gatos KG, Leong YW (2017) Classification of nanomaterialsand nanocomposites. In: Parameswaranpillai J, Hameed N, Kurian T (eds) Nanocomposite materials: synthesis, properties and applications. CRC Press, Boca Raton

Geng Y, Dalhaimer P, Cai S, Tsai R, Tewari M, Minko T, Discher DE (2007) Shape effects of filaments versus spherical particles in flow and drug delivery. Nat Nanotechnol 2:249–255

Genix AC, Oberdisse J (2018) Nanoparticle self-assembly: from interactions in suspension to polymer nanocomposites. Soft Matter 14:5161–5179

George S, Lin S, Ji Z, Thomas CR, Li L, Mecklenburg M, Meng H, Wang X, Zhang H, Xia T, Hohman JN, Lin S, Zink JI, Weiss PS, Nel AE (2012) Surface defects on plate-shaped silver nanoparticles contribute to its hazard potential in a fish gill cell line and zebrafish embryos. ACS Nano 6:3745–3759

Gomes T, Pinheiro JP, Cancio I, Pereira CG, Cardoso C, Bebianno MJ (2011) Effects of copper nanoparticles exposure in the mussel *Mytilus galloprovincialis*. Environ Sci Technol 45:9356–9362

Gomes T, Pereira CG, Cardoso C, Sousa VS, Teixeira MR, Pinheiro JP, Bebianno MJ (2014) Effects of silver nanoparticles exposure in the mussel *Mytilus galloprovincialis*. Mar Environ Res 101:208–214

Guigas C, Walz E, Graf V, Heller KJ, Greiner R (2017) Mutagenicity of silver nanoparticles in CHO cells dependent on particle surface functionalization and metabolic activation. J Nanopart Res 19(207):1–14

Gupta YR, Sellegounder D, Kannan M, Deepa S, Senthilkumaran B, Basavaraju Y (2016) Effect of copper nanoparticles exposure in the physiology of the common carp (*Cyprinus carpio*): biochemical, histological and proteomic approaches. Aquac Fish 1:15–23

Handy RD, Henry TB, Scown TM, Johnston BD, Tyler CR (2008) Manufactured nanoparticles: their uptake and effects on fish--a mechanistic analysis. Ecotoxicology 17:396–409

Hartmann N, Skjolding L, Hansen S, Baun A, Kjølholt J, Gottschalk F (2014) Environmental fate and behaviour of nanomaterials: new knowledge on important transformation processes. Environmental Project No. Danish Environmental Protection Agency, Copenhagen, p 1594. https://doi.org/10.13140/2.1.1943.4240

Haynes VN, Ward JE, Russell BJ, Agrios AG (2017) Photocatalytic effects of titanium dioxide nanoparticles on aquatic organisms - current knowledge and suggestions for future research. Aquat Toxicol 185:138–148

Hou J, Wang L, Wang C, Zhang S, Liu H, Li S, Wang X (2018) Toxicity and mechanisms of action of titanium dioxide nanoparticles in living organisms. J Environ Sci (China) 75:40–53

Hu W, Culloty S, Darmody G, Lynch S, Davenport J, Ramirez-Garcia S, Dawson KA, Lynch I, Blasco J, Sheehan D (2014) Toxicity of copper oxide nanoparticles in the blue mussel, *Mytilus edulis*: a redox proteomic investigation. Chemosphere 108:289–299

Hu J, Zhang Z, Zhang C, Liu S, Zhang H, Li D, Zhao J, Han Z, Liu X, Pan J, Huang W, Zheng M (2018a) Al_2O_3 nanoparticle impact on the toxic effect of Pb on the marine microalga *Isochrysis galbana*. Ecotoxicol Environ Saf 161:92–98

Hu PP, Zhang XX, Li YX, Pichan C, Chen Z (2018b) Molecular interactions between silver nanoparticles and model cell membranes. Top Catal 61:1148–1162

Huang X, Liu Z, Xie Z, Dupont S, Huang W, Wu F, Kong H, Liu L, Sui Y, Lin D, Lu W, Hu M, Wang Y (2018) Oxidative stress induced by titanium dioxide nanoparticles increases under seawater acidification in the thick shell mussel *Mytilus coruscus*. Mar Environ Res 137:49–59

Indeglia PA, Georgieva AT, Krishna VB, Martyniuk CJ, Bonzongo JCJ (2018) Toxicity of functionalized fullerene and fullerene synthesis chemicals. Chemosphere 207:1–9

Ispas C, Andreescu D, Patel A, Goia DV, Andreescu S, Wallace KN (2009) Toxicity and developmental defects of different sizes and shape nickel nanoparticles in zebrafish. Environ Sci Technol 43:6349–6356

Johari SA, Kalbassi MR, Soltani M, Yu IJ (2013) Toxicity comparison of colloidal silver nanoparticles in various life stages of rainbow trout (*Oncorhynchus mykiss*). Iran J Fish Sci 12:76–95

Joo NY, Lee J, Kim SJ, Hong S, Park HM, Yun WS, Yoon M, Song NW (2013) Preparation of an aqueous suspension of stabilized TiO_2 nanoparticles in primary particle form. J Nanosci Nanotechnol 13:6153–6159

Joubert Y, Pan JF, Buffet PE, Pilet P, Gilliland D, Valsami-Jones E, Mouneyrac C, Amiard-Triquet C (2013) Subcellular localization of gold nanoparticles in the estuarine bivalve *Scrobicularia plana* after exposure through the water. Gold Bull 46:47–56

Kashiwada S (2006) Distribution of nanoparticles in the see-through medaka (*Oryzias latipes*). Environ Health Persp 114:1697–1702

Keller AA, Wang H, Zhou D, Lenihan HS, Cherr G, Cardinale BJ, Miller R, Ji Z (2010) Stability and aggregation of metal oxide nanoparticles in natural aqueous matrices. Environ Sci Technol 44:1962–1967

Kettler K, Veltman K, van de Meent D, van Wezel A, Hendriks AJ (2014) Cellular uptake of nanoparticles as determined by particle properties, experimental conditions, and cell type. Environ Toxicol Chem 33:481–492

Khan MS, Jabeen F, Qureshi NA, Asghar MS, Shakeel M, Noureen A (2015) Toxicity of silver nanoparticles in fish: a critical review. J Bio Environ Sci 6:211–227

Khan I, Saeed K, Khan I (2017a) Nanoparticles: properties, applications and toxicities. Arab J Chem. https://doi.org/10.1016/j.arabjc.2017.05.011

Khan MS, Qureshi NA, Jabeen F, Asghar MS, Shakeel M, Fakhar-e-Alam M (2017b) Eco-friendly synthesis of silver nanoparticles through economical methods and assessment of toxicity through oxidative stress analysis in the *Labeo rohita*. Biol Trace Elem Res 176:416–428

Khan MS, Qureshi NA, Jabeen F, Shakeel M, Asghar MS (2018) Assessment of waterborne amine-coated silver nanoparticle (Ag-NP)-induced toxicity in *Labeo rohita* by histological and hematological hrofiles. Biol Trace Elem Res 182:130–139

Klaper R, Crago J, Barr J, Arndt D, Setyowati K, Chen J (2009) Toxicity biomarker expression in daphnids exposed to manufactured nanoparticles: changes in toxicity with functionalization. Environ Pollut 157:1152–1156

Klaunig JE, Kamendulis LM (2008) Chemical Carcinogenesis. In: Klaassen CD (ed) Casarett and Doull's toxicology: the basic science of poisons, 7th edn. McGraw Hill, New York, pp 329–379

Klingelfus T, Lirola JR, Oya Silva LF, Disner GR, Vicentini M, Nadaline MJB, Robles JCZ, Trein LM, Voigt CL, Silva de Assis HC, Mela M, Leme DM, Cestari MM (2017) Acute and long-term effects of trophic exposure to silver nanospheres in the central nervous system of a neotropical fish *Hoplias intermedius*. Neurotoxicology 63:146–154

Koser J, Engelke M, Hoppe M, Nogowski A, Filser J, Thoming J (2017) Predictability of silver nanoparticle speciation and toxicity in ecotoxicological media. Environ Sci Nano 4:1470–1483

Krysanov EY, Pavlov DS, Demidova TB, Dgebuadze YY (2010) Effect of nanoparticles on aquatic organisms. Biol Bull 37:406–412

Kteeba SM, El-Adawi HI, El-Rayis OA, El-Ghobashy AE, Schuld JL, Svoboda KR, Guo L (2017) Zinc oxide nanoparticle toxicity in embryonic zebrafish: mitigation with different natural organic matter. Environ Pollut 230:1125–1140

Lajmanovich RC, Peltzer PM, Martinuzzi CS, Attademo AM, Colussi CL, Basso A (2018) Acute toxicity of colloidal silicon dioxide nanoparticles on amphibian larvae: emerging environmental concern. Int J Environ Res 12:269–278

Larbi F, Garcia A, del Valle LJ, Hamou A, Puiggali J, Belgacem N, Bras J (2018) Comparison of nanocrystals and nanofibers produced from shrimp shell alpha-chitin: from energy production to material cytotoxicity and Pickering emulsion properties. Carbohyd Polym 196:385–397

Liu W, Wu Y, Wang C, Li HC, Wang T, Liao CY, Cui L, Zhou QF, Yan B, Jiang GB (2010) Impact of silver nanoparticles on human cells: effect of particle size. Nanotoxicology 4:319–330

Liu W, Long Y, Yin N, Zhao X, Sun C, Zhou Q, Jiang G (2016) Toxicity of engineered nanoparticles to fish. In: Xing B, Vecitis CD, Senesi N (eds) Engineered nanoparticles and the environment: biophysicochemical processes and toxicity, 1st edn. John Wiley and Sons, Hoboken, pp 347–366

Liu YX, Yan ZH, Xia J, Wang K, Ling XC, Yan B (2017) Potential toxicity in crucian carp following exposure to metallic nanoparticles of copper, chromium, and their mixtures: a comparative study. Pol J Environ Stud 26:2085–2094

Lodish H, Berk A, Kaiser CA, Krieger M, Bretscher A, Ploegh H, Amon A, Scott MP (2013) Molecular Cell Biology, 7th edn. W. H. Freeman, New York

Luo Z, Wang Z, Yan Y, Li J, Yan C, Xing B (2018) Titanium dioxide nanoparticles enhance inorganic arsenic bioavailability and methylation in two freshwater algae species. Environ Pollut 238:631–637

Lv XH, Huang B, Zhu X, Jiang Y, Chen B, Tao Y, Zhou J, Cai Z (2017) Mechanisms underlying the acute toxicity of fullerene to *Daphnia magna*: energy acquisition restriction and oxidative stress. Water Res 123:696–703

Mahaye N, Thwala M, Cowan DA, Musee N (2017) Genotoxicity of metal based engineered nanoparticles in aquatic organisms: a review. Mutat Res 773:134–160

Mallikarjuna K, Kim H (2018) Synthesis of shape and size-dependent CuAg bimetallic dumbbell structures for organic pollutant hydrogenation. Phys E 102:44–49

Martinez DST, Franchi LP, Ferreira CM, Filho AGS, Alves OL, Takahashi CS (2014) Carbon nanotubes: from synthesis to genotoxicity. In: Durán N, Guterres S, Alves OL (eds) Nanotoxicology, materials, methodologies, and assessments. Springer, New York, pp 125–152

Mendez N, Liberman A, Corbeil J, Barback C, Viveros R, Wang J, Wang-Rodriguez J, Blair SL, Mattrey R, Vera D, Trogler W, Kummel AC (2017) Assessment of *in vivo* systemic toxicity and biodistribution of iron-doped silica nanoshells. Nanomed-Nanotechnol 13:933–942

Minetto D, Volpi Ghirardini A, Libralato G (2016) Saltwater ecotoxicology of Ag, Au, CuO, TiO_2, ZnO and C60 engineered nanoparticles: an overview. Environ Int 92-93:189–201

Miranda RR, Damaso da Silveira AL, de Jesus IP, Grötzner SR, Voigt CL, Campos SX, Garcia JR, Randi MA, Ribeiro CA, Filipak Neto F (2016) Effects of realistic concentrations of TiO_2 and ZnO nanoparticles in *Prochilodus lineatus* juvenile fish. Environ Sci Pollut R 23:5179–5188

Montagner A, Bosi S, Tenori E, Bidussi M, Alshatwi AA, Tretiach M, Prato M, Syrgiannis Z (2017) Ecotoxicological effects of graphene-based materials. 2D Mater 4:012001

Moore MN (2006) Do nanoparticles present ecotoxicological risks for the health of the aquatic environment? Environ Int 32:967–976

Mukherjee K, Acharya K (2018) Toxicological effect of metal oxide nanoparticles on soil and aquatic habitats. Arch Environ Contam Toxicol 75:175–186

Nel A, Xia T, Madler L, Li N (2006) Toxic potential of materials at the nanolevel. Science 311:622–627

Ng CT, Tang FM, Li JJ, Ong C, Yung LL, Bay BH (2015) Clathrin-mediated endocytosis of gold nanoparticles *in vitro*. Anat Rec (Hoboken) 298:418–427

Oberdorster E (2004) Manufactured nanomaterials (fullerenes, C60) induce oxidative stress in the brain of juvenile largemouth bass. Environ Health Perspect 112:1058–1062

Oh N, Park JH (2014) Endocytosis and exocytosis of nanoparticles in mammalian cells. Int J Nanomedicine 9(S1):51–63

Oh WK, Kim S, Choi M, Kim C, Jeong YS, Cho BR, Hahn JS, Jang J (2010) Cellular uptake, cytotoxicity, and innate immune response of silica - titania hollow nanoparticles based on size and surface functionality. ACS Nano 4:5301–5313

Ozmen M, Gungordu A, Erdemoglu S, Ozmen N, Asilturk M (2015) Toxicological aspects of photocatalytic degradation of selected xenobiotics with nano-sized Mn-doped TiO_2. Aquat Toxicol 165:144–153

Ozmen N, Erdemoglu S, Gungordu A, Asilturk M, Turhan DO, Akgeyik E, Harper SL, Ozmen M (2018) Photocatalytic degradation of azo dye using core@shell nano-TiO_2 particles to reduce toxicity. Environ Sci Pollut Res 25:29493–29504

Park MV, Neigh AM, Vermeulen JP, de la Fonteyne LJ, Verharen HW, Briedé JJ, van Loveren H, de Jong WH (2011) The effect of particle size on the cytotoxicity, inflammation, developmental toxicity and genotoxicity of silver nanoparticles. Biomaterials 32:9810–9817

Perez S, Farre M, Barcelo D (2009) Analysis, behavior and ecotoxicity of carbon-based nanomaterials in the aquatic environment. Trac-Trend Anal Chem 28:820–832

Petersen SJ (2017) Silver nanoparticle fate and accumulation in the aquatic food web of stream microcosms. Master Thesis. Georgia Southern University, USA

Petersen EJ, Huang Q, Weber WJ (2008) Ecological uptake and depuration of carbon nanotubes by *Lumbriculus variegatus*. Environ Health Perspect 116:496–500

Pikula KS, Zakharenko AM, Chaika VV, Vedyagin AA, Orlova TY, Mishakov IV, Kuznetsov VL, Park S, Renieri EA, Kahru A, Tsatsakis AM, Golokhvast KS (2018) Effects of carbon and silicon nanotubes and carbon nanofibers on marine microalgae *Heterosigma akashiwo*. Environ Res 166:473–480

Prasad R (2019) Plant Nanobionics: Approaches in Nanoparticles Biosynthesis and Toxicity. Springer International Publishing (ISBN 978-3-030-16379-2) https://www.springer.com/gp/book/9783030163785

Renault S, Baudrimont M, Mesmer-Dudons N, Gonzalez P, Mornet S, Brisson A (2008) Impacts of gold nanoparticle exposure on two freshwater species: a phytoplanktonic alga (*Scenedesmus subspicatus*) and a benthic bivalve (*Corbicula fluminea*). Gold Bull 41:116–126

Rocha TL, Gomes T, Mestre NC, Cardoso C, Bebianno MJ (2015a) Tissue specific responses to cadmium-based quantum dots in the marine mussel *Mytilus galloprovincialis*. Aquat Toxicol 169:10–18

Rocha TL, Gomes T, Sousa VS, Mestre NC, Bebianno MJ (2015b) Ecotoxicological impact of engineered nanomaterials in bivalve molluscs: an overview. Mar Environ Res 111:74–88

Rossbach LM, Shaw BJ, Piegza D, Vevers WF, Atfield AJ, Handy RD (2017) Sub-lethal effects of waterborne exposure to copper nanoparticles compared to copper sulphate on the shore crab (*Carcinus maenas*). Aquat Toxicol 191:245–255

Ruenraroengsak P, Tetley TD (2015) Differential bioreactivity of neutral, cationic and anionic polystyrene nanoparticles with cells from the human alveolar compartment: robust response of alveolar type 1 epithelial cells. Part Fibre Toxicol 12(19):1–20

Sajid M, Ilyas M, Basheer C, Tariq M, Daud M, Baig N, Shehzad F (2015) Impact of nanoparticles on human and environment: review of toxicity factors, exposures, control strategies, and future prospects. Environ Sci Pollut Res Int 22:4122–4143

Schinwald A, Murphy FA, Jones A, MacNee W, Donaldson K (2012) Graphene-based nanoplatelets: a new risk to the respiratory system as a consequence of their unusual aerodynamic properties. ACS Nano 6:736–746

Selck H, Handy RD, Fernandes TF, Klaine SJ, Petersen EJ (2016) Nanomaterials in the aquatic environment: a European Union-United States perspective on the status of ecotoxicity testing, research priorities, and challenges ahead. Environ Toxicol Chem 35:1055–1067

Sendra M, Yeste MP, Gatica JM, Moreno-Garrido I, Blasco J (2017) Homoagglomeration and heteroagglomeration of TiO_2, in nanoparticle and bulk form, onto freshwater and marine microalgae. Sci Total Environ 592:403–411

Sendra M, Volland M, Balbi T, Fabbri R, Yeste MP, Gatica JM, Canesi L, Blasco J (2018) Cytotoxicity of CeO_2 nanoparticles using *in vitro* assay with *Mytilus galloprovincialis* hemocytes: relevance of zeta potential, shape and biocorona formation. Aquat Toxicol 200:13–20

Shang L, Nienhaus K, Nienhaus GU (2014) Engineered nanoparticles interacting with cells: size matters. J Nanobiotechnol 12:5

Sharma VK (2009) Aggregation and toxicity of titanium dioxide nanoparticles in aquatic environment – a review. J Environ Sci Health A Tox Hazard Subst Environ Eng 44:1485–1495

Shaw BJ, Handy RD (2011) Physiological effects of nanoparticles on fish: a comparison of nanometals versus metal ions. Environ Int 37:1083–1097

Silva T, Pokhrel LR, Dubey B, Tolaymat TM, Maier KJ, Liu XF (2014) Particle size, surface charge and concentration dependent ecotoxicity of three organo-coated silver nanoparticles: comparison between general linear model-predicted and observed toxicity. Sci Total Environ 468:968–976

Sohn EK, Johari SA, Kim TG, Kim JK, Kim E, Lee JH, Chung YS, Yu IJ (2015) Aquatic toxicity comparison of silver nanoparticles and silver nanowires. Biomed Res Int 2015:893049

Souza JP, Baretta JF, Santos F, Paino IMM, Zucolotto V (2017) Toxicological effects of graphene oxide on adult zebrafish (*Danio rerio*). Aquat Toxicol 186:11–18

Souza JP, Venturini FP, Santos F, Zucolotto V (2018) Chronic toxicity in *Ceriodaphnia dubia* induced by graphene oxide. Chemosphere 190:218–224

Strambeanu N, Demetrovici L, Dragos D (2015) Natural sources of nanoparticles. In: Lungu M, Neculae A, Bunoiu M, Biris C (eds) Nanoparticles' promises and risks. Springer International Publishing, Cham, Switzerland, pp 9–19

Taju G, Majeed SA, Nambi KSN, Hameed ASS (2014) *In vitro* assay for the toxicity of silver nanoparticles using heart and gill cell lines of *Catla catla* and gill cell line of *Labeo rohita*. Comp Biochem Phys C 161:41–52

Thit A, Skjolding LM, Selck H, Sturve J (2017) Effects of copper oxide nanoparticles and copper ions to zebrafish (*Danio rerio*) cells, embryos and fry. Toxicol In Vitro 45:89–100

Tinkle SS, Antonini JM, Rich BA, Roberts JR, Salmen R, DePree K, Adkins EJ (2003) Skin as a route of exposure and sensitization in chronic beryllium disease. Environ Health Persp 111:1202–1208

Tissera ND, Wijesena RN, Sandaruwan CS, de Silva RM, de Alwis A, de Silva KMN (2018) Photocatalytic activity of ZnO nanoparticle encapsulated poly(acrylonitrile) nanofibers. Material Chem Phys 204:195–206

Trickler WJ, Lantz SM, Murdock RC, Schrand AM, Robinson BL, Newport GD, Schlager JJ, Oldenburg SJ, Paule MG, Slikker W Jr, Hussain SM, Ali SF (2010) Silver nanoparticle induced blood-brain barrier inflammation and increased permeability in primary rat brain microvessel endothelial cells. Toxicol Sci 118:160–170

Truong NP, Whittaker MR, Mak CW, Davis TP (2015) The importance of nanoparticle shape in cancer drug delivery. Expert Opin Drug Deliv 12:129–142

Valerio-Garcia RC, Carbajal-Hernandez AL, Martinez-Ruiz EB, Jarquin-Diaz VH, Haro-Perez C, Martinez-Jeronimo F (2017) Exposure to silver nanoparticles produces oxidative stress and affects macromolecular and metabolic biomarkers in the goodeid fish *Chapalichthys pardalis*. Sci Total Environ 583:308–318

van der Oost R, Beyer J, Vermeulen NP (2003) Fish bioaccumulation and biomarkers in environmental risk assessment: a review. Environ Toxicol Pharmacol 13:57–149

Walker CH, Hopkin SP, Sibly RM, Peakall DB (1996) Principles of ecotoxicology. Taylor and Francis, London

Wang SG, Lu WT, Tovmachenko O, Rai US, Yu HT, Ray PC (2008) Challenge in understanding size and shape dependent toxicity of gold nanomaterials in human skin keratinocytes. Chem Phys Lett 463:145–149

Xia T, Kovochich M, Liong M, Mädler L, Gilbert B, Shi H, Yeh JI, Zink JI, Nel AE (2008) Comparison of the mechanism of toxicity of zinc oxide and cerium oxide nanoparticles based on dissolution and oxidative stress properties. ACS Nano 2:2121–2134

Yeo MK, Kang M (2008) Effects of nanometer sized silver materials on biological toxicity during zebrafish embryogenesis. Bull Kor Chem Soc 29:1179–1184

Yildirimer L, Thanh NTK, Loizidou M, Seifalian AM (2011) Toxicological considerations of clinically applicable nanoparticles. Nano Today 6:585–607

Yu Q, Wang H, Peng Q, Li Y, Liu Z, Li M (2017) Different toxicity of anatase and rutile TiO_2 nanoparticles on macrophages: involvement of difference in affinity to proteins and phospholipids. J Hazard Mater 335:125–134

Yu Z, Hao R, Zhang L, Zhu Y (2018) Effects of TiO_2, SiO_2, Ag and CdTe/CdS quantum dots nanoparticles on toxicity of cadmium towards *Chlamydomonas reinhardtii*. Ecotoxicol Environ Saf 156:75–86

Yue Y, Li X, Sigg L, Suter MJ, Pillai S, Behra R, Schirmer K (2017) Interaction of silver nanoparticles with algae and fish cells: a side by side comparison. J Nanobiotechnol 15:16

Zhang T, Wang L, Chen Q, Chen C (2014) Cytotoxic potential of silver nanoparticles. Yonsei Med J 55:283–291

Zhornik EV, Baranova LA, Drozd ES, Sudas MS, Chau NH, Buu NQ, Dung TT, Chizhik SA, Volotovskiĭ ID (2014) Silver nanoparticles induce lipid peroxidation and morphological changes in human lymphocytes surface. Biofizika 59:466–473

Zhu S, Oberdorster E, Haasch ML (2006) Toxicity of an engineered nanoparticle (fullerene, C60) in two aquatic species, *Daphnia* and fathead minnow. Mar Environ Res 62S:S5–S9

Zhu XS, Zhu L, Duan ZH, Qi RQ, Li Y, Lang YP (2008) Comparative toxicity of several metal oxide nanoparticle aqueous suspensions to Zebrafish (*Danio rerio*) early developmental stage. J Environ Sci Heal A 43:278–284

Zhu XS, Chang Y, Chen YS (2010) Toxicity and bioaccumulation of TiO_2 nanoparticle aggregates in *Daphnia magna*. Chemosphere 78:209–215

Chapter 16
Impact of Nanomaterials on Beneficial Insects in Agricultural Ecosystems

Malaichamy Kannan, Kolanthasamy Elango, Thangavel Tamilnayagan, Sundharam Preetha, and Govindaraju Kasivelu

Contents

16.1 Introduction

Nanotechnology has created a global impact by the remarkable achievement in the area of medicine, engineering and biological and agricultural science. The application of nanoscience and technology in plant pest and disease management is in infant stage. Nano is defined as one billionth part of a metre, and size of the material in any

M. Kannan (✉)
Department of Plant Protection, Horticultural College and Research Institute, Periyakulam, Tamil Nadu Agricultural University, Tamil Nadu, India

S. Preetha
Department of Nano Science and Technology, Tamil Nadu Agricultural University, Coimbatore, Tamil Nadu, India

K. Elango · T. Tamilnayagan
Department of Entomology, Tamil Nadu Agricultural University, Coimbatore, Tamil Nadu, India

G. Kasivelu
Centre for Ocean Research, Sathyabama Institute of Science and Technology, Chennai, Tamil Nadu, India

D. Thangadurai et al. (eds.), *Nanotechnology for Food, Agriculture, and Environment*, Nanotechnology in the Life Sciences,
https://doi.org/10.1007/978-3-030-31938-0_16

one dimension could be less than 100 nm across. Nanomaterials are produced naturally, incidentally or through engineering; their forms could be amorphous, crystalline, polymeric or composites with different shapes, viz. spheres, tubes, rods, cones and fibres. It can be non-metal (e.g. carbon), metallic (e.g. Au, Ag), semiconductor (e.g. Cd, Se) or a combination (Huang et al. 2007). Nanomaterials are formed atom by atom through building process or from macro, micro to nano size material via breaking process. Using the above methods, scientific communities developed inorganic, organic and composite nanomaterials and effectively utilized them in almost all the science sector to social sector. Even then, the applicability of nanomaterials in agriculture is still under explored, especially in crop protection and production. Nanomaterial has been widely used as conductor, semiconductor, medical device, sensors, coating, catalytic agents and also as pesticides in the recent (Salata 2004).

Nanotechnological principles and concepts can be exploited for early detection, monitoring and management of insect pests. It set to offer a platform to transform agriculture sector from production, protection, processing and storage. The key contributions due to application of nanotechnology in agriculture are delivery systems that aid in slow release with more use efficiency in agro-input usage. In India, the post-Green Revolution demands to reduce the use of higher quantities of fertilizers and pesticides to safeguard the future agriculture and environment. A challenge to balance between crop production and environmental protection can be achieved by adopting nanotechnology.

16.2 Nanoparticles as Novel Insecticides

Nanoparticles are effective against leaf feeding and sap sucking insect pests of crops. Different types of nanomaterials, viz. nano silica, silver, zinc oxide, titanium dioxide and aluminium oxide nanoemulsion, can be developed as novel insecticide against insect pests in field and storage. Chandrashekharaiah et al. (2015) have more recently reviewed the modes of action and application of nanomaterials in pest management. The nanoparticles applied on the plant get adsorbed in insect body when they crawl over the surface and absorbed into the cuticular lipids. Further, the peristaltic movement of the insect facilitates the nanoparticle enter in to the body tissue and cause physical death to cell organelle, followed by oozing out of body fluid leading to desiccational death. Surface charge-modified hydrophobic nanoparticles can be successfully advocated for management of ectoparasites in animals and insect pests of agriculture (Ulrichs et al. 2005).

Bhattacharyya et al. (2010) compiled the progress and scope in nanotechnology as potent tool in changing agriculture. Many scientific workers have reported that several nanoparticles like aluminium oxide, silver, gold, silicon dioxide, zinc oxide, *titanium dioxide* (TiO_2) and silica nanoparticles biologically derived from plant have proved its insecticidal activity; larval mortality; larval, pupal and adult developmental periods; and fecundity against rice weevil, *Sitophilus oryzae* L.; red flour beetle, *Tribolium castaneum* H., larvae of *Spodoptera litura* L.; oleander aphid, *Aphis nerii* B.; bruchid beetle, *Callosobruchus maculatus* F.; and diamondback moth, *Plutella xylostella* L. (Wang et al. 2007; Barik et al. 2008; Yang et al. 2009;

Goswami et al. 2010; Chakravarthy et al. 2012; Rouhani et al. 2012; Sahayaraj et al. 2016). These studies can expand the cutting edge for nanoparticle-based technologies in pest management. The nanomaterial-associated insecticides can change the activity through improved formulations, targeting specific easy applications, improved efficiency, reduced dose and less environmental threats over conventional insecticides. The main objective of this chapter is accounting a comprehensive knowledge on the impacts of nanomaterials on the growth, development, parasitism or predatory efficiency and adult emergence of effective beneficial insects and also identifying the way to conserve parasitoids, predators and pollinators from nanotoxicity in the cropping system.

16.3 Beneficial Insects and Their Role in Agroecosystem

Beneficial insects are important component in the food chain, and they afford amenity for maintaining the ecological imbalance in the agroecosystem. The members in the group of beneficial insects are pollinators, natural enemies, weed killers, productive insects (honey bees, silkworms and lac insects), scavengers and decomposers. Pollinators like honey bees, solitary bees, bumblebees, pollen wasps, ants, hoverflies, mosquitoes, butterflies, moths and flower beetles play a vital role in pollinating the agricultural crop species for seed and fruit set for enhancing agricultural productivity. Eradication of weeds by weed-feeding insects also increase the productivity of agriculture (Cochineal insect, Aristolochia butterfly, Calotropis butterfly, AK grasshopper, Water hyacinth weevil and Parthenium weed killer feeds prickly pear, aristalochia, calotropis, eichhornia and parthenium, respectively (Capinera 2005)).

Furthermore, insects like scavenger beetles, soil ants, larvae of cutworms, flies, crickets, termites and wasps live in soil and make tunnels. Insect activities in soil, improves soil aeration, beneficial microorganisms and earthworm populations. They indeed improves the soil properties such as texture, structure and pore space. The adult and larvae insects like rove beetles, chafer beetles, darkling beetles, nitidulids, water scavenger beetle, daddy long legs, muscid flies, termites and ants feed on decaying matter and convert the complex material into simple substances through biochemical cycling which prevent environment hazards.

The predators and parasitoids of agricultural important insect pests have played significant role to decrease pest outbreaks. These natural enemies devour the insect pests at egg, larval/nymphal, pupal and adult stages and bring down the pest load below economic threshold level. In the agroecosystem, entomophagous insects have the ability of reducing the pest load below 30% in field condition even without insecticide application. The role of natural enemies in pest management was very well achieved by classical biological control through natural enemies of invasive insect pests like pink hibiscus mealybug, *Maconellicoccus hirsutus* G.; cassava mealybug, *Phenacoccus manihoti* M.; sugarcane woolly aphid, *Ceratovacuna lanigera* Z.; and papaya mealybugs, *Paracoccus marginatus* W. were managed by *Cryptolaemus montrouzieri* M., *Apoanagyrus* (*Epidinocarsis*) *lopezi* D., *Dipha aphidivora* M., *Acerophagus papayae* N., respectively (Ballal and Verghese 2015) (Tables 16.1 and 16.2).

Table 16.1 Important insect predators in agricultural ecosystem (Ballal and Verghese 2015)

Order	Predator	Prey insect
Odonata Dragonflies	*Sympetrum frequens* *S. infuscatum*	Mosquito
Coleoptera Ladybird beetles	*Coccinella septempunctata*	Aphids
	Cryptolaemus montrouzieri	Papaya mealybug, cottony cushion scales
	C. transversalis	Aphids
	Cheilomenes sexmaculatus	Aphids
	Scymnus coccivora	Aphids and mealy bugs
	Chilochorus nigritus	Scales
	Rodolia cardinalis	Cottony cushion scale
Neuroptera	*Chrysoperla* spp.	All soft body insects, eggs of many lepidopterans
	Micromus igrotus	Sugarcane woolly aphid
Lepidoptera	*Dipha aphidivora*	Sugarcane woolly aphid
Hemiptera	*Cyrtorhinus lividipennis*	Rice hoppers
	Platymeris laevicollis	Coconut rhinoceros beetle
	Eucanthecona furcellata	Red hairy caterpillar
Predatory spider	*Lycosa pseudoannulata*	Rice BPH
Predatory mite	*Phytoseiulus persimilis*	Red spider mite

Table 16.2 Important insect parasitoids in agricultural ecosystem (Ballal and Verghese 2015)

Family	Example	Target	Category
Hymenoptera			
Trichogrammatidae	*Trichogramma chilonis*	Many lepidopterous pests	Egg parasitoid
Scelionidae	*Telenomus remus*	*Spodoptera litura*	Egg parasitoid
Braconidae	*Bracon brevicornis*	*Opisina arenosella*	Larval parasitoid
	Chelonus blackburni	Cotton spotted bollworms, *Earias* spp.	Egg larval parasitoid
Encyrtidae	*Copidosoma koehleri*	Potato tuber moth	Egg larval parasitoid
Platygasteridae	*Platygaster oryzae*	Rice gall midge	Larval parasitoid
Chalcididae	*Brachymeria nephantidis*	*Opisina arenosella*	Larval parasitoid
Bethylidae	*Goniozus nephantidis*	*Opisina arenosella*	Larval parasitoid
Ichneumonidae	*Isotima javensis*	Sugarcane top borer	Larval pupal parasitoid
Encyrtidae	*Acerophagus papayae*	Papaya mealybug	Nymphal adult parasitoid
Aphelinidae	*Encarsia formosa*	Cotton whitefly	Nymphal adult parasitoid
Eulophidae	*Trichospilus pupivora*	*Opisina arenosella*	Pupal parasitoids
	Tetrastichus israeli	*Opisina arenosella*	Pupal parasitoids

(continued)

Table 16.3 (continued)

Family	Example	Target	Category
Diptera			
Tachinidae	*Sturmiopsis inferens*	Sugarcane shoot borer *Chilo infuscatellus*	Larval parasitoid
	Spoggosia bezziana	*Opisina arenosella*	Larval parasitoid
	Eucelatoria bryani	*H. armigera*	Larval pupal parasitoid
Lepidoptera			
Epiricanidae	*Epiricania melanoleuca*	Sugarcane leafhopper, *Pyrilla perpusilla*	Nymphal adult parasitoid

16.4 Impact of Nanomaterials and Chemical Pesticides on Natural Enemies

The green revolution in Indian agriculture paved a way for the application of broad-spectrum insecticides, and fertilizer-responsive varieties have created an impact on the reduction of natural enemy populations which increased pest resurgence, resistance and residues. Further, the incorrect use of agrochemicals is also threatening the sustainability of agriculture around the globe (Bueno and Freitas 2004). Several broad-spectrum insecticides and their metabolites are used unsafe to natural enemies when exposed directly. Further, studies conducted by Lord et al. (1968) and Stapel et al. (2000) evidenced that systemic novel group of insecticides caused ill effects to pollinators, predators and parasitoids (Gill and Garg 2014).

In recent years, the possible usages and benefits of nanotechnological tools in pest management are huge. Nanotechnology facilitate techniques for slow release of fertilizer through nanoporous zeolites; nanoencapsulates nanomaterials for delivery of plant protection inputs in pest management; nanosensors for pest detection and monitoring (Prasad et al. 2014, 2017a, b; Bhattacharyya et al. 2016; Abd-Elsalam Prasad 2018, 2019). The possible impacts of different nano inputs on the beneficial insects in agricultural ecosystem are to be studied in detail to reduce the adverse effects and conserve them for maintaining the ecological balance.

16.5 Impact of Nanomaterials on General Predator

The dragonflies, damsel flies, praying mantis, giant water bug, robber flies, hover flies and wasps are the general predators of insect pests. The exposure of low-concentration silver nanoparticles altered the behaviour, survival and reproduction in common natural predatory dragonfly (Green Darner, *Anax junius* Drury: Odonata) naiads (Pokhrel and Dubey 2012). It clearly indicates that Ag nanoparticle (NP) exposures could influence the life history and behavioural traits in Odonata (Lovern et al. 2007; Brausch et al. 2011). Murugan et al. (2015) also documented that AgNPs synthesized from *Datura metel* increased the predator potential of young instars of dragonfly (blue Darner, *Anax immaculifrons* Rambur) against malarial vector *Anopheles stephensi* Liston from 75.5 to 95.5 and 53.5 to 78% in second and third instars of dragonfly naiads. At lower doses, the thorn apple *Datura metel* L.-synthesized AgNPs reduce the larval population of malaria vectors and have no adverse effect on the behaviour of dragonfly naiads.

Uses of nanomaterials as soil amendment affect not only the soil environment but also the entire soil microbial community, and only little effects are known and many on the interaction of nanomaterials with soil microbial biota are unknown. McKee and Filser (2016) concluded that use of metal-based engineered nanomaterials caused adverse effects on plant- fungi, plant-bacteria interactions, bioaccumulation, biomagnification decreased the nitrogen turnover and increased carbon emissions in soil which indeed reduced microbial population. Further, nanomaterial affecting the food web through biomagnification is also possible. Similarly, the short-term exposure of applied chemical insecticides, namely, imidacloprid, fipronil and its metabolite fipronil sulfone, in rice ecosystem caused significant harm such as acute toxicity, feeding inhibition and delayed toxicity to dragonfly naiads (Autumn darter, *Sympetrum frequens* S. and *Sympetrum infuscatum* S.) (Jinguji et al. 2018). Accordingly, paralysis due to fipronil and fipronil sulfone sublethal effects may cause feeding inhibition which leads to death in dragonfly species, and realistic application of fipronil may eliminate dragonflies in paddy ecosystem. The findings of the above studies suggested that the chemical insecticides imidacloprid, fipronil and fipronil sulfone caused mortality to dragonfly naiads. On the contrary, the AgNPs synthesized from *Datura metel* increased the predator potential of young instars of dragonfly *A. immaculifrons*.

16.6 Impact of Nanomaterials on Lacewings (*Chrysoperla* spp.)

The green lacewings are efficient and successful predators in the management of insect pests. Karthika et al. (2015) tested the safety of hexanal nanoemulsion to the *Chrysoperla carnea* Stephens grubs and stated that nanoemulsion was not toxic to the predator grub, and 100 per cent pupation and adult emergence was also noticed. Similarly, Kannan and Elango (2019) studied the effect of silica nanoparticles at different doses (20000, 15000, 10000, 5000, 4500, 1000, 500, 100 and 50 ppm) on the emergence potential of eggs of green lacewing, *Chrysoperla zastrowi sillemi* (Esben-Peterson), and revealed that the emergence was only 29.41 per cent at 20000 and 15,000 ppm comparing to the untreated check (96.51%) (Table 16.3 and Fig. 16.1).

Quite the reverse, compared to nanomaterials, even at low doses, chemical insecticides caused negative impact on the growth and development of the common green

Table 16.3 Effect of silica nanoparticles to egg parasitoids, *Trichogramma chilonis* Ishii, and green lacewing, *Chrysoperla zastrowi sillemi* (Esben-Peterson) (Kannan and Elango 2019)

	Egg parasitoids, *Trichogramma chilonis*			Green lacewing, *Chrysoperla zastrowi sillemi*		
Dose (ppm)	% emergence of eggs	% reduction over control	% emergence of eggs	% reduction over control	% emergence of eggs	Category[a]
20000	16.25	19.78	83.20	29.41	69.53	Moderately harmful
15000	30.41	43.72	68.55	49.07	49.15	Slightly harmful
10000	52.56	53.92	45.64	52.54	45.56	Harmless
5000	63.92	76.38	33.89	58.75	39.13	Harmless
4500	67.97	79.44	29.70	71.95	25.45	Harmless
1000	76.43	82.65	20.96	76.32	20.93	Harmless
500	78.86	87.53	18.44	76.81	20.41	Harmless
100	84.95	91.34	12.14	77.36	19.85	Harmless
50	93.10	90.38	3.72	93.15	3.48	Harmless
Control	96.69	93.48	–	96.51	–	Harmless

[a]<50 mortality, harmless; 50–79, slightly harmful; 80–89, moderately harmful; >90, harmful

lacewing, *Chrysoperla carnea* (Stephens) (Neuroptera: Chrysopidae). The fenoxycarb has been reported to cause the prolonged development time in all the stages of the predator *Chrysoperla rufilabris* Burmeister (Liu and Chen 2001), and imidacloprid, propargite and pymetrozine also caused significant reductions in fecundity (Rezaei et al. 2007). Similarly, the direct contact of spinosad caused reduced progeny production and malformation of biological stages in *C. carnea* (Medina et al. (2001, 2003; Desneux et al. 2007). Mandour (2009) and Maia and Moore (2011) also stated that direct spray of spinosad on immature stages of *C. carnea* affected the reproduction and survival of adult.

16.7 Impact of Nanomaterials on Egg Parasitoids

Trichogrammatidae are effective egg parasitoids and parasitize the eggs of several insects belonging to more than eight orders in aquatic and terrestrial habitats. Karthika et al. (2015) observed lesser toxicity by nanoemulsion of hexanal at 0.02% to the immature stages of *Trichogramma japonicum* Ashmead. The mean adult emergence of two experiments ranged from 97.15 to 93.05 per cent in different doses. The hexanal treatments had little impact on the parasitization of *Trichogramma chilonis* Ishii. Further, Mohan et al. (2017a, b) revealed that nanoemulsion of hexanal was safer to *Trichogramma pretiosum* Riley and *T. chilonis* with 98.53 and 97.88% parasitization and 97.57 and 96.60 adult emergence, respectively. In addi-

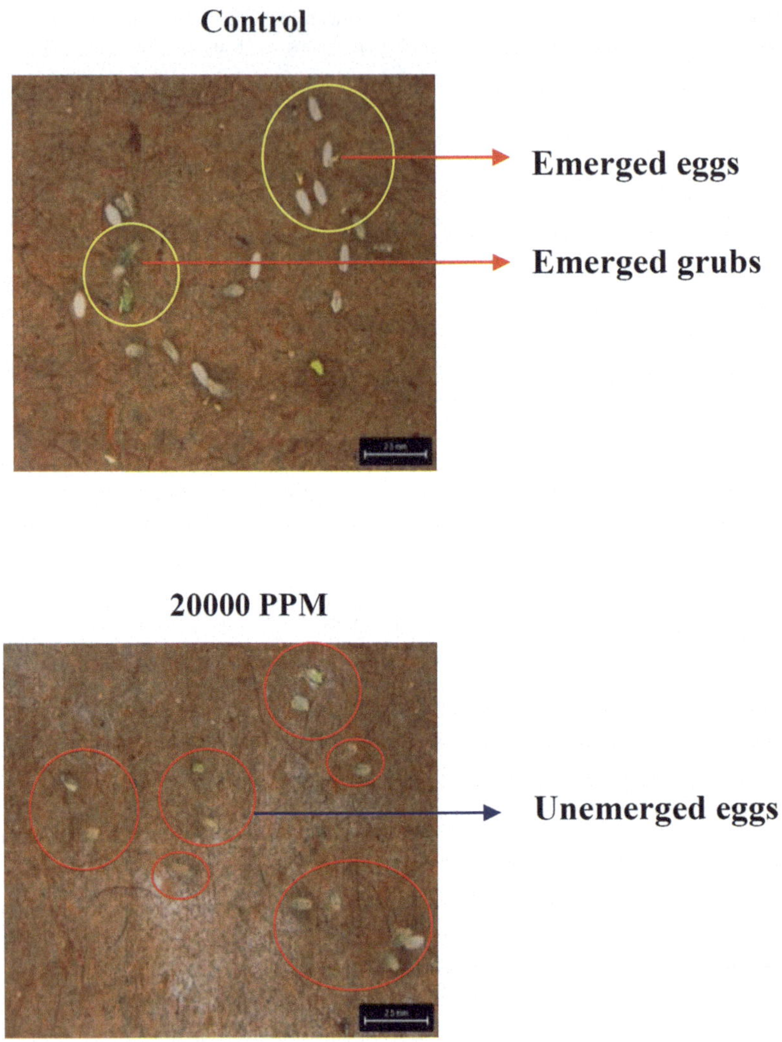

Fig. 16.1 Effect of silica nanoparticles on the emergence potential of green lacewing, *Chrysoperla zastrowi sillemi* (Esben-Peterson)

tion, after the treatment, the second-generation *T. pretiosum* and *T. chilonis* adults also showed 96.12 and 97.65 per cent parasitization with 94.29 and 96.40% adult emergence, respectively. Finally, they decided that hexanal being a naturally occurring alkyl aldehyde compound in plants, the nanoformulation was safer to egg parasitoids (Table 16.4).

Likewise, Preetha et al. (2018) reported that the neem oil-based nanoemulsion was safer to egg parasitoid, *T. chilonis*, than macroemulsion. The parasitism (86.00%) and adult emergence (79.98%) were significantly higher in neem oil

Table 16.4 Effect of hexanal nanoemulsion to egg parasitoids, *Trichogramma* spp. (Mohan et al. 2017a, b)

Treatment	*Trichogramma pretiosum*		*Trichogramma chilonis*	
	% parasitization	% adult emergence	% parasitization	% adult emergence
T1 – Nanoemulsion of hexanal @ 0.02%	97.57	98.53	96.60	97.88
T2 – Nanoemulsion of hexanal @ 0.04%	96.47	96.92	94.92	97.38
T3 – Nanoemulsion of hexanal @ 0.06%	95.72	96.01	93.58	95.85
T4 – Pure hexanal @ 0.02%	94.90	95.87	93.39	93.72
T5 – Pure hexanal @ 0.04%	93.78	94.74	91.74	93.15
T6 – Pure hexanal @ 0.06%	92.88	94.00	90.41	92.51
T7 – Tween 20 @ 0.02%	92.35	93.83	88.87	90.38
T8 – Ethanol @ 0.02%	92.86	94.62	89.28	91.21
T9 – Control	98.97	99.31	99.05	98.93

nanoemulsion; the lowest per cent of adult emergence (48.45%) and parasitism (66.78%) was recorded in the highest concentration of neem oil macroemulsion. The adult emergence was found on increasing rate by decreasing the concentrations of neem nano- and macroemulsions. Equally, Kannan and Elango (2019) also studied the effect of silica nanoparticles at different doses (20000, 15000, 10000, 5000, 4500, 1000, 500, 100 and 50 ppm) on the parasitization and adult emergence of egg parasitoids of *T. chilonis* and revealed maximum parasitization in untreated check (96.69 per cent). At 20000 and 15000 ppm concentration, only silica nanoparticles showed maximum effect on parasitization of *T. chilonis* (16.25 and 30.41%, respectively) adult emergence (19.78 and 93.48%) compared to untreated check (Table 16.3; Figs. 16.2 and 16.3).

In contrast, except methoxyfenozide and tebufenozide, lambda cyhalothrin, cypermethrin, thiodicarb, profenofos and spinosad adversely affected the emergence, adult survival, life span and fitness parameters of *Trichogramma exiguum* (Pinto and Platner) in *Helicoverpa zea* (Boddie) host eggs (Suh et al. 2000). Similarly, Shoeb (2010) also observed the effects when they treated the immature stages of *Trichogramma evanescens* (Westwood) with Lambda-cyhalothrin, spinosad and fenitrothion. Subsequently, they concluded that negative effects on parasitism and emergence of parasitoids were observed. But the neem product-based bioinsecticides achieved nil serious side effect on parasitism and emergence rates of *T. pretiosum* (Riley) and *T. minutum* (Riley) and have given good control against

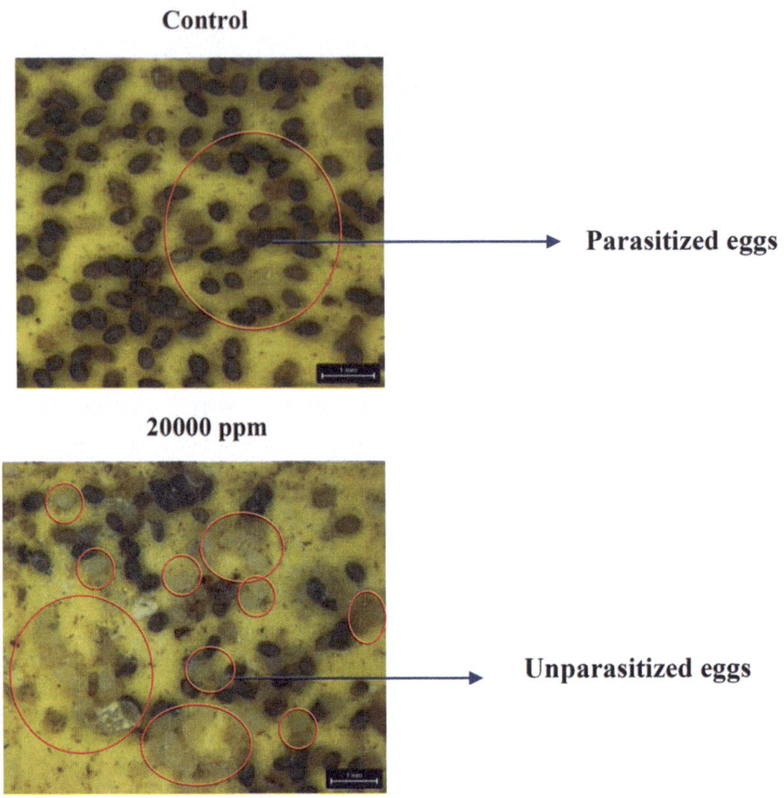

Fig. 16.2 Effect of silica nanoparticles on the parasitization potential of *Trichogramma chilonis* Ishii

Helicoverpa armigera (Hübner) under greenhouse condition (El-Wakeil et al. 2006).

16.8 Impact of Nanomaterials on Pollinator (Honey Bees)

Plant pollination through insects is the natural event that happens in the reproduction of the common flowering plants of domesticated species. Many plant species insects are the only source of pollinating agent, and their contribution in pollination was around 36% which enhances the fruit set and yield of the cultivated crops. In the near future, nanomaterial-based agricultural input application may yield effect on the beneficial aspects of crop cultivation but also permits the possibility of adverse effects on the beneficials in the agroecosystem. Some of the laboratory studies summarized below enlist the effects of nanomaterials on the pollinators. Karthika et al. (2015) concluded that no mortality was found

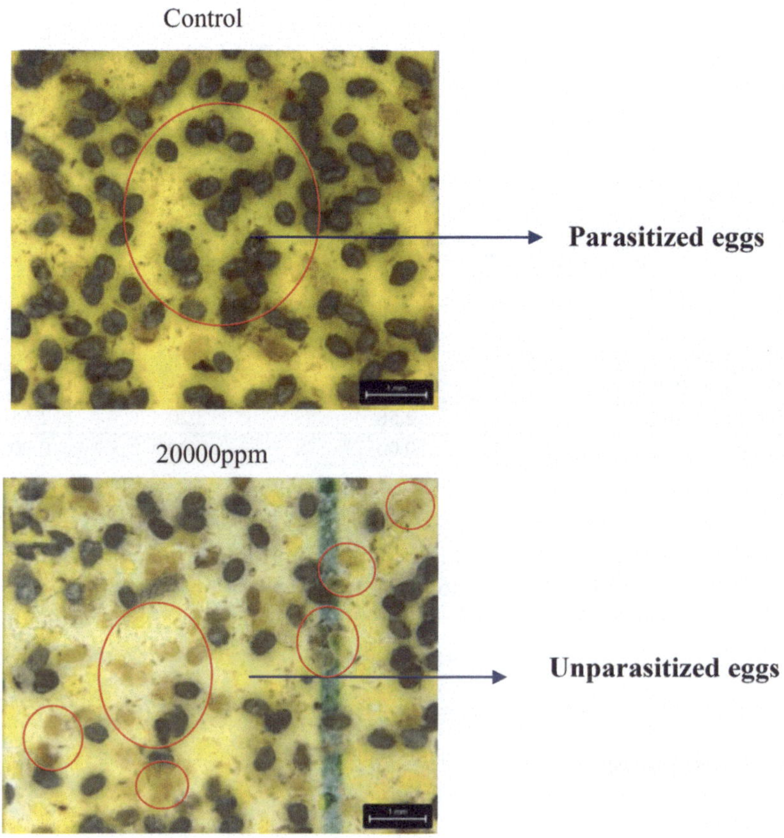

Fig. 16.3 Effect of silica nanoparticles on the emergence potential of *Trichogramma chilonis* Ishii

in any of the nano hexanal formulation treatments even after 48 hours after treatment. The results are in comparable with Flesar et al. (2010) who reported that acute oral toxicity of the most active natural products was determined on adult honey bees, showing them as non-toxic at concentrations as high as 100 μg/bee. Likewise Mohan et al. (2017a, b) reported that all the concentration of nanoformulation of hexanal was harmless to worker honey bees (maximum mortality of 7.65% in *Apis cerana indica* (Fabricius), 5.80% in *Apis mellifera* (Linnaeus) and 5.20% in *Apis florea* (Fabricius) at 24 hours after treatment) which recorded a mortality of <30 per cent as per the threshold prescribed by IOBC (Table 16.5).

On the contrary, compared to nanoformulations, chemical insecticides caused considerable adverse effect on the pollinators. Bees exposed to imidacloprid insecticides altered the foraging behaviour in *Apis mellifera* and bees are not returned to the hive then control bees (Yang et al. 2008). In the same way, bees intoxicated with the sublethal doses of neonicotinoid insecticides affected the navigation, foraging efficiency of bees and adult of natural enemies. The persis-

Table 16.5 Effect of hexanal nanoemulsion to honey bees (Mohan et al. 2017a, b)

Treatments	(%) Mortality of worker bees (24 hours after treatment)		
	Italian bees *Apis mellifera* L.	Indian bees *Apis cerana indica* F.	Little bees *Apis florea* F.
T1 – Nanoemulsion of hexanal @ 0.02%	4.75	4.33	
T2 – Nanoemulsion of hexanal @ 0.04%	5.20	4.75	4.20
T3 – Nanoemulsion of hexanal @ 0.06%	5.80	7.65	5.20
T4 – Pure hexanal @ 0.02%	6.67	8.20	6.67
T5 – Pure hexanal @ 0.04%	8.90	10.00	9.60
T6 – Pure hexanal @ 0.06%	9.10	11.10	12.80
T7 – Tween 20 @ 0.2%	2.58	3.33	2.58
T8 – Ethanol @ 0.2%	2.58	2.20	2.65
T9 – Control	0.00	0.00	0.00

tence toxicity of insecticides, viz. imidacloprid, fipronil, indoxacarb, dimethoate, cypermethrin and azadirachtin, to honey bees assert that fipronil followed by imidacloprid recorded highest residual toxicity than other insecticides (Morandin and Winston (2003).

16.9 Conclusion

Nanotechnology is an emerging field of science capable of resolving issues and problems that are impossible to tackle in engineering, agriculture and other biological sciences. Nanotechnological principles and concepts can be exploited for early detection, monitoring and management of insect pests. The growing population around the world demands more use of agricultural inputs to meet the profuse food stock with low cost of production. The increasing use of pesticides had a negative impact on farmers, consumers, nontarget organism and environment.

Different types of available nanomaterials, viz. nano SiO_2, Ag, Al_2O_3, ZnO, TiO_2 and nanoemulsion, can be effectively used as a novel insecticide in the management of tobacco caterpillar, *S. litura*; oleander aphid, *A. nerii*; bruchid beetle, *C. maculatus*; diamondback moth, *P. xylostella*; lesser grain borer, *R. dominica*; rice weevil, *S. oryzae*; and red flour beetle, *T. castaneum*. Development of novel nanoformualtions pave a way for the advent of new plant protection products in insect pest management, which will be an alternative to chemical insecticides to conserve, enhance the toxicity, persistence to UV radiation, safer to environment and extended shelf-life of biomolecules and bio-agents in field for sustainable agriculture. The knowledge on the impacts of nanomaterials on growth, development, parasitism or predatory efficiency and emergence capacity as well as to safeguard beneficial insects are essential to conserve parasitoids, predators and pollinators from nanotoxicity and at the same time to suppress the insect pests in the cropping system.

Based on the recent available literature on the impacts of nanomaterials against beneficial insects especially predators, parasitoids and pollinators, it is very clear that nanomaterials are not causing deleterious effects to the beneficial insects. There was a little harmfulness observed in 500-fold (20000 ppm) increased LC_{50} dose of host insect pests. Compared to nanomaterials, the synthetic chemical insecticides, even at very low doses, caused toxicity and impaired the development of natural enemies from host insects. Hence, the knowledge on the effects of nanomaterials on biological parameters, parasitism or predatory efficiency and emergence capacity of beneficial insects are essential to conserve parasitoids, predators and pollinators from nanotoxicity to suppress the insect pests in the cropping system. This chapter clearly briefed that nanomaterials are not causing harmful effects to the beneficial insects but synthetic chemical insecticides, even at very low doses, caused acute toxicity, feeding inhibition, growth impairment and delayed toxicity to beneficial insects.

References

Abd-Elsalam KA, Prasad R (2018) Nanobiotechnology Applications in Plant Protection. Springer International Publishing (ISBN 978-3-319-91161-8) https://www.springer.com/us/book/9783319911601

Abd-Elsalam KA, Prasad R (2019) Nanobiotechnology Applications in Plant Protection. Springer International Publishing

Ballal CR, Verghese A (2015) Role of parasitoids and predators in the management of insect pests. In: Chakravarthy AK (ed) New horizons in insect science: towards sustainable pest management. Springer, New Delhi, pp 307–326

Barik TK, Sahu B, Swain V (2008) Nanosilica – from medicine to pest control. Parasitol Res 103(2):253

Bhattacharyya A, Bhaumik A, Rani PU, Mandal S, Epidi TT (2010) Nano-particles – a recent approach to insect pest control. Afr J Biotechnol 9(24):3489–3493

Bhattacharyya A, Duraisamy P, Govindarajan M, Buhroo AA, Prasad R (2016) Nano-biofungicides: Emerging trend in insect pest control. In: Advances and Applications through Fungal Nanobiotechnology (ed. Prasad R), Springer International Publishing Switzerland 307–319

Brausch KA, Anderson TA, Smith PN, Maul JD (2011) The effect of fullerenes and functionalized fullerenes on *Daphnia magna* phototaxis and swimming behavior. Environ Toxicol Chem 30(4):878–884

Bueno AF, Freitas S (2004) Effect of the insecticides abamectin and lufenuron on eggs and larvae of *Chrysoperla externa* under laboratory conditions. BioControl 49(3):277–283

Capinera JL (2005) Relationships between insect pests and weeds: an evolutionary perspective. Weed Sci 53(6):892–901

Chakravarthy AK, Kandakoor SB, Atanu B, Dhanabala K, Gurunatha K, Ramesh P (2012) Bio efficacy of inorganic nanoparticles CdS, Nano-Ag and Nano-TiO_2 against *Spodoptera litura* (Fabricius) (Lepidoptera: Noctuidae). Curr Biotica 6(3):271–281

Chandrashekharaiah M, Kandakoor SB, Gowda GB, Kammar V, Chakravarthy AK (2015) Nanomaterials: a review of their action and application in pest management and evaluation of DNA-tagged particles. In: Chakravarthy AK (ed) New horizons in insect science: towards sustainable pest management. Springer, New Delhi, pp 113–126

Desneux N, Decourtye A, Delpuech JM (2007) The sublethal effects of pesticides on beneficial arthropods. Annu Rev Entomol 52:81–106

El-Wakeil NE, Gaafar NM, Vidal S (2006) Side effect of some neem products on natural enemies of *Helicoverpa* (*Trichogramma* spp.) and *Chrysoperla carnea*. Arch Phytopathol Plant Protect 39(6):445–455

Flesar J, Havlik J, Kloucek P, Rada V, Titera D, Bednar M, Stropnicky M, Kokoska L (2010) *In vitro* growth-inhibitory effect of plant-derived extracts and compounds against *Paenibacillus* larvae and their acute oral toxicity to adult honey bees. Vet Microbiol 145(1–2):129–133

Goswami A, Roy I, Sengupta S, Debnath N (2010) Novel applications of solid and liquid formulations of nanoparticles against insect pests and pathogens. Thin Solid Films 519(3):1252–1257

Gill HK, Garg H (2014) Pesticides: environmental impacts and management strategies. In: Larramendy ML, Soloneski S (eds) Pesticides – toxic aspects. IntechOpen, Croatia. https://doi.org/10.5772/57399

Huang J, Li Q, Sun D, Lu Y, Su Y, Yang X, Wang H, Wang Y, Shao W, He N, Hong J (2007) Biosynthesis of silver and gold nanoparticles by novel sundried *Cinnamomum camphora* leaf. Nanotechnology 18(10):105104

Jinguji H, Ohtsu K, Ueda T, Goka K (2018) Effects of short-term, sublethal fipronil and its metabolite on dragonfly feeding activity. PLoS One 13(7):1–23

Kannan M, Elango K (2019) Effect of Silica nano particles to egg parasitoids, *Trichogramma chilonis* Ishii and green lacewing, *Chrysoperla zastrowi sillemi* (Esben-Peterson) (unpublished data)

Karthika S, Nandakumar NB, Gunasekaran K, Subramanian KS (2015) Biosafety of nanoemulsion of hexanal to honey bees and natural enemies. J Sci Technol 8(30):1–7

Liu TX, Chen TY (2001) Effects of the insect growth regulator fenoxycarb on immature *Chrysoperla rufilabris* (Neuroptera: Chrysopidae). Fla Entomol 84(4):628–633

Lord KA, May MA, Stevenson JH (1968) The secretion of the systemic insecticides dimethoate and phorate into nectar. Ann Appl Biol 61(1):19–27

Lovern SB, Strickler JR, Klaper R (2007) Behavioral and physiological changes in *Daphnia magna* when exposed to nanoparticle suspensions (titanium dioxide, nano-C60, and $C_{60}H_xC_{70}H_x$). Environ Sci Technol 41(12):4465–4470

Maia MF, Moore SJ (2011) Plant-based insect repellents: a review of their efficacy, development and testing. Malar J 10(1):11

Mandour NS (2009) Influence of spinosad on immature and adult stages of *Chrysoperla carnea* (Stephens) (Neuroptera: Chrysopidae). BioControl 54(1):93

McKee MS, Filser J (2016) Impacts of metal-based engineered nanomaterials on soil communities. Environ Sci Nano 3(3):506–533

Medina P, Budia F, Tirry L, Smagghe G, Vinuela E (2001) Compatibility of spinosad, tebufenozide and azadirachtin with eggs and pupae of the predator *Chrysoperla carnea* (Stephens) under laboratory conditions. Biocontrol Sci Tech 11(5):597–610

Medina P, Budia F, Del Estal PE, Viñuela E (2003) Effects of three modern insecticides, pyriproxyfen, spinosad and tebufenozide, on survival and reproduction of *Chrysoperla carnea* adults. Ann Appl Biol 142(1):55–61

Mohan C, Sridharan S, Gunasekaran K, Subramanian KS, Natarajan N (2017a) Biosafety of hexanal as nanoemulsion on egg parasitoid *Trichogramma* spp. J Entomol Zool Stud 5(2):1541–1544

Mohan C, Sridharan S, Subramanian KS, Natarajan N, Nakkeeran S (2017b) Effect of nanoemulsion of hexanal on honey bees (Hymenoptera; Apidae). J Entomol Zool Stud 5(3):1415–1418

Morandin LA, Winston ML (2003) Effects of novel pesticides on bumble bee (Hymenoptera: Apidae) colony health and foraging ability. Environ Entomol 32(3):555–563

Murugan K, Dinesh D, Kumar PJ, Panneerselvam C, Subramaniam J, Madhiyazhagan P, Suresh U, Nicoletti M, Alarfaj AA, Munusamy MA, Higuchi A (2015) *Datura metal*-synthesized silver nanoparticles magnify predation of dragonfly nymphs against the malaria vector *Anopheles stephensi*. Parasitol Res 114(12):4645–4654

Pokhrel LR, Dubey B (2012) Potential impact of low-concentration silver nanoparticles on predator–prey interactions between predatory dragonfly nymphs and *Daphnia magna* as a prey. Environ Sci Technol 46(14):7755–7762

Prasad R, Kumar V, Prasad KS (2014) Nanotechnology in sustainable agriculture: present concerns and future aspects. Afr J Biotechnol 13(6):705–713

Prasad R, Bhattacharyya A, Nguyen QD (2017a) Nanotechnology in sustainable agriculture: Recent developments, challenges, and perspectives. Front Microbiol 8:1014. doi: https://doi.org/10.3389/fmicb.2017.01014

Prasad R, Kumar M, Kumar V (2017b) Nanotechnology: An Agriculture paradigm. Springer Nature Singapore Pte Ltd. (ISBN: 978-981-10-4573-8)

Preetha S, Kannan M, Lokesh S, Gowtham V (2018) Effect of neem oil based nanoemulsion on egg parasitoid, *Trichogramma chilonis* (Ishii) (Hymenoptera: Trichogrammatidae). J Biol Control 2(2):103–107

Rezaei M, Talebi K, Naveh VH, Kavousi A (2007) Impacts of the pesticides imidacloprid, propargite, and pymetrozine on *Chrysoperla carnea* (Stephens) (Neuroptera: Chrysopidae): IOBC and life table assays. BioControl 52(3):385–398

Rouhani M, Samih MA, Kalantari S (2012) Insecticide effect of silver and zinc nanoparticles against *Aphis nerii* Boyer De Fonscolombe (Hemiptera: Aphididae). Chilean J Agric Res 72(4):590–594

Sahayaraj K, Madasamy M, Radhika SA (2016) Insecticidal activity of bio-silver and gold nanoparticles against *Pericallia ricini* Fab. (Lepidaptera: Archidae). J Biopest 9(1):63

Salata OV (2004) Applications of nanoparticles in biology and medicine. J Nanobiotechnol 2(1):3

Shoeb MA (2010) Effect of some insecticides on the immature stages of the egg parasitoid *Trichogramma evanescens* West. (Hym.; Trichogrammatidae). Egypt Acad J Biol Sci 3(1):31–38

Stapel JO, Cortesero AM, Lewis WJ (2000) Disruptive sublethal effects of insecticides on biological control: altered foraging ability and life span of a parasitoid after feeding on extra floral nectar of cotton treated with systemic insecticides. Biol Control 17(3):243–249

Suh CP, Orr DB, VanDuyn JW (2000) Effect of insecticides on *Trichogramma exiguum* (Trichogrammatidae: Hymenoptera) preimaginal development and adult survival. J Econ Entomol 93(3):577–583

Ulrichs C, Mewis I, Goswami A (2005) Crop diversification aiming nutritional security in West Bengal: biotechnology of stinging capsules in nature's water-blooms. Ann Tech Issue of State Agri Technologists Service Assoc: 1–18

Wang L, Li X, Zhang G, Dong J, Eastoe J (2007) Oil-in-water nanoemulsions for pesticide formulations. J Colloid Interface Sci 314(1):230–235

Yang EC, Chuang YC, Chen YL, Chang LH (2008) Abnormal foraging behavior induced by sublethal dosage of imidacloprid in the honey bee (Hymenoptera: Apidae). J Econ Entomol 101(6):1743–1748

Yang FL, Li XG, Zhu F, Lei CL (2009) Structural characterization of nanoparticles loaded with garlic essential oil and their insecticidal activity against *Tribolium castaneum* (Herbst) (Coleoptera: Tenebrionidae). J Agric Food Chem 7(21):10156–10162

Index

D. Thangadurai et al. (eds.), *Nanotechnology for Food, Agriculture, and Environment*, Nanotechnology in the Life Sciences,
https://doi.org/10.1007/978-3-030-31938-0

O

The manufacturer's authorised representative in the EU is Springer
Nature Customer Service Centre GmbH, Europaplatz 3, 69115 Heidelberg,
Germany. If you have any concerns regarding our products, please
contact ProductSafety@springernature.com

Printed and bound by CPI Group (UK) Ltd, Croydon, CR0 4YY
05/10/2026
02228130-0015